ELECTRO-OPTICAL DEVICES AND SYSTEMS

Mohammad A. Karim, Ph.D.
The University of Dayton

Department of Electrical Engineering and Center for Electro-Optics

PWS-KENT Publishing Company
Boston, Massachusetts

PWS-KENT
Publishing Company

20 Park Plaza
Boston, Massachusetts 02116

His **light** may be compared to that of a concave mirror behind a lamp; the lamp is inside a glass; the glass is just as if it were a radiating star and the fuel thereof comes from a blessed olive tree, that is neither Eastern nor Western, whose oil is self-radiating without flame. **Light** upon **light**.

Al-Qur'an

To the memory of my grandfather, Khadim Rahmani Nuri, whose whole life at the foot of the Himalayas was spent in the quest of **light** and in the service of **An-Nur**.

PWS-KENT Publishing Company is a division of Wadsworth, Inc.

Library of Congress Cataloging-in-Publication Data

Karim, Mohammad A.
Electro-optical devices and systems/Mohammad A. Karim.
p. cm.
Includes bibliographical references.
ISBN 0-534-91630-9
1. Electrooptical devices. I. Title.
TA1750.K38 1990 89-27286
621.381'52—dc20 CIP

Sponsoring Editor: Jonathan Plant
Assistant Editor: Mary Thomas
Production Coordinator: Helen Walden
Production: Carol Dondrea/Bookman Productions
Manufacturing Coordinator: Margaret Sullivan Higgins
Interior Design: Nancy Benedict/Graphic Design
Cover Design: Helen Walden
Interior Illustrations: John Foster
Typesetting: H. Charlesworth & Co. Ltd.
Printing and Binding: Maple-Vail Book Mfr. Group

Printed in the United States of America

90 91 92 93 94 — 10 9 8 7 6 5 4 3 2 1

PREFACE

Optics is perhaps the most sought-after interdisciplinary area in both sciences and engineering. All-optical as well as optoelectronic devices are being used to improve the performance of many systems to a degree that was unthinkable in the past. The surge of activity has initiated new and exciting technological possibilities. This evolutionary process actually was begun in the 10th century by Abu 'Ali Al-Hasan Ibn Al-Haytham, the first person to give an accurate account of vision and a not-so-accurate characterization of reflection principles and of spherical and parabolic mirrors. His extensive treatise on geometrical optics laid the foundation for all subsequent milestones in optics. Electro-optics, the prized end-product of this thousand-year-long search, now attracts students and researchers from almost all backgrounds. Numerous educational programs under various names, both graduate and undergraduate, are being defined and refined to channel this interest. One of the elements that is essential in these programs is a variety of updated texts with pragmatic approaches and emphases. While there are several texts on classical optics, there are but a few on the more modern aspects of optics, and the majority of these are good only as reference books.

The goal of this book is to give the readers an introduction to the principles of electro-optical devices and systems, and its main emphasis is on applying the characteristics of devices and systems to the solution of optical engineering problems. This book resulted from several years of lectures that were prepared for a first course on this subject, and is intended for beginning graduate and advanced undergraduate students. Only an adequate course in electro-optical devices can prepare the readers to understand and appreciate the workings of systems related to Fourier optics, fiber optics, integrated optics, optical computing, and so on. To facilitate the understanding of these devices, we have developed and used the pertinent concepts of atomic physics, quantum mechanics, and solid-state physics in this book.

As a minimum prerequisite, readers are expected to be equipped with an adequate background in differential equations and basic physics. Familiarity with basic optics is considered helpful but not essential for this fairly self-contained text. Each chapter is equipped with a collection of sample exercise problems, some of which are quite involved. A list of references and materials for further reading is included at the end of each chapter.

Chapter 1 reviews the essentials of electromagnetic field theory, reflection and refraction, atomic physics, and quantum mechanics. Chapter 2 outlines and introduces the concepts of radiation in terms of both radiometric and photometric units. Based on the understanding of the first two chapters, the physics of semiconductor junctions is developed in Chapter 3. Chapter 4 describes and highlights the characteristics of a number of incoherent photoemitters, display devices, and photodetectors—both solid-state-based and tube-based. The physics and engineering of coherent lasers are covered next in Chapter 5, and Chapter 6 introduces the concept of noise and how it affects the characteristics of detection systems. Chapter 7 develops the concepts of modulation and polarization and then describes various recording devices, as well as devices based on electro-optic, magneto-optic, and acousto-optic effects. Chapter 8 presents the concept of Fourier optics and then discusses a few of the Fourier-optics-based systems, including those suitable for optical computing. Finally, Chapter 9 discusses the physics and engineering of devices and systems that are based on optical fibers and integrated optics.

As with most authors, I am particularly indebted to a number of people who have contributed to bringing this text to fruition. Most importantly, I am indebted to my teacher and guide, Hua-Kuang Liu of Jet Propulsion Laboratory, California Institute of Technology, who introduced me to the basics of electro-optics. Numerous authors have influenced my understanding of the subject and have added lustre to the materials of this book. It would take pages to list them all. These authors and their publications are listed at the end of every chapter. I am also grateful to my students at the University of Dayton who have helped to hone this book. The contributions of Abdul Awwal and Abdallah Cherri in preparing some of the figures deserve special mention. I would like to express my appreciation to Professor Donald L. Moon for his continuing encouragement in this endeavor and to the following reviewers for their input: Robert M. Bunch, Rose-Hulman Institute of Technology; Allen Drake, University of New Hampshire; Brian J. Thompson, University of Rochester; and Alok K. Berry, George Mason University. In addition, I would like to thank Carol Dondrea and the people at Bookman Productions for all their care and help in getting the book through the production process. Last but not least, a particular debt goes to my wife, Setara, and our children, Lutfi and Lamya, for their unfailing support, patience, and tolerance of all those late hours.

CONTENTS

CHAPTER ONE

Fundamentals of Quantum Electronics 1

CHAPTER TWO

Optical Radiation 43

CHAPTER FIVE

Lasers 175

CHAPTER SIX

Noise in Optical Detection 223

CHAPTER SEVEN

Modulation of Light 263

CHAPTER ONE

Fundamentals of Quantum Electronics

1.1 Introduction

This chapter reviews the basic principles pertaining to both electromagnetic wave theory and quantum mechanics. The concepts are essential to an understanding of several electro-optic topics of interest: reflection and refraction at plane and spherical boundaries; radiation associated with electromagnetic fields; optoelectronic devices, many of which are solid state in nature; and waveguiding principles that govern the design and performance of fiber optic and integrated optic devices.

The wave theory of light is described by a set of equations, referred to as Maxwell's equations, that describes the space and time behavior of the electromagnetic field vectors and that supports the existence of transverse electromagnetic waves. Electromagnetic wave theory provides an effective explanation for both interference and diffraction. However, it fails to account for the emission and absorption of light. This apparent dilemma is overcome by reviving Newton's corpuscular theory of light in the form of quantum theory. This latter form basically asserts that the energy of a light beam is quantized. It is interesting to note that the wave and quantum theories of light are not contradictory, but rather completely complementary.

For the sake of simplicity, in this chapter we explore the behavior of electromagnetic waves that occur only in homogeneous isotropic media; in a later chapter we discuss the response of electromagnetic waves to anisotropic media. Immediately following the review of electromagnetic theory, we explore the essentials of quantum mechanics. One point about the scope of our discussion on quantum mechanics needs to be stressed: while many of the electro-optical devices are quantum mechanical in nature, we need not dwell on the philosophical and physical aspects of quantum mechanics. In this chapter we treat only those topics that are vital to the study of electro-optical devices. We first focus on the behavior of the hydrogen atom, which is essential to understanding the nature of more complex atoms. Then we consider the behavior of electrons in the presence of potential fields in order to understand the charge-carrying capacity of a semiconductor—whether electro-optic or not—when it is properly organized and stimulated. The significance of this review will become obvious in the subsequent chapters.

1.2 Maxwell's Equations

An important feature of Maxwell's equations for a time-varying electromagnetic field is a traveling wave solution that is responsible for the transfer of energy from one point to another. The electric field **E**, electric flux density **D**, magnetic field **H**, magnetic flux density **B**, and electric current density **J**, are all functions of both space (x, y, z) and time (t). They are related by the following Maxwell's equations:

$$\nabla \times \mathbf{E} = -\frac{\partial \mathbf{B}}{\partial t} \tag{1.1}$$

$$\nabla \times \mathbf{H} = \frac{\partial \mathbf{D}}{\partial t} + \mathbf{J} \tag{1.2}$$

$$\nabla \cdot \mathbf{D} = \rho \tag{1.3}$$

$$\nabla \cdot \mathbf{B} = 0 \tag{1.4}$$

such that

$$\mathbf{D} = \varepsilon_0 \mathbf{E} + \mathbf{P} \qquad \mathbf{B} = \mu_0 \mathbf{H} \qquad \mathbf{J} = \sigma \mathbf{E} \tag{1.5}$$

where ε_0 is the electric permittivity (dielectric constant), μ_0 is the magnetic permeability in vacuum, σ is the electric conductivity of the medium, ρ is the charge density, and **P** is the electric polarization of the medium. The constants ε_0 and μ_0 are, respectively, 8.85×10^{-12} F/m (farads per meter) and $4\pi \times 10^{-7}$ H/m (henries per meter). These equations, it must be stressed, are *mathematical abstractions of experimental results*, but as we show later, the description of a wide range of phenomena can be deduced by means of these equations without any contradiction.

Equations (1.1) through (1.4), respectively, reduce to what are referred to as Faraday's law, Ampère's law, Gauss's law for the electric field, and Gauss's law for the magnetic field. In particular, Eqs. (1.1) and (1.2) show that the variation in time of the magnetic field generates an electric field and that a time-varying electric field gives rise to a part of the magnetic field. In addition, the vector cross product relationship suggests that the two field vectors, **E** and **H**, are not only orthogonal to each other but also orthogonal to the direction of propagation.

By taking the curl of Eq. (1.1) and making use of Eqs. (1.2) and (1.5), we obtain

$$\nabla \times (\nabla \times \mathbf{E}) = -\frac{\partial}{\partial t}(\nabla \times \mathbf{B}) = -\mu_0 \frac{\partial}{\partial t}(\nabla \times \mathbf{H})$$

$$= -\mu_0 \frac{\partial}{\partial t}[\mathbf{J} + \frac{\partial}{\partial t}(\mathbf{D})] \tag{1.6}$$

Equation (1.6) can now be made a function of only **E** by incorporating Eq. (1.5) as follows:

$$\nabla \times (\nabla \times \mathbf{E}) = -\mu_0 \frac{\partial}{\partial t}[\sigma \mathbf{E} + \varepsilon_0 \frac{\partial}{\partial t}(\mathbf{E})] \tag{1.7}$$

where, in a source-free medium, both **P** and ρ are taken to be zero for simplicity. However, using the vector operation, we find that

$$\nabla \times (\nabla \times \mathbf{E}) = \nabla(\nabla \cdot \mathbf{E}) - (\nabla \cdot \nabla)\mathbf{E}$$

$$= \nabla(\nabla \cdot \mathbf{E}) - \nabla^2 \mathbf{E} = -\nabla^2 \mathbf{E} \tag{1.8}$$

where the first term on the right-hand side vanishes according to Eq. (1.3). Thus, Eq. (1.7) reduces to

$$\nabla^2 \mathbf{E} = \mu_0 \sigma \frac{\partial}{\partial t}(\mathbf{E}) + \mu_0 \varepsilon_0 \frac{\partial^2}{\partial t^2}(\mathbf{E}) \qquad \text{1.9)}$$

which, in lossless free space ($\sigma = 0$), further simplifies to

$$\nabla^2 \mathbf{E} = \mu_0 \varepsilon_0 \frac{\partial^2}{\partial t^2}(\mathbf{E}) \tag{1.10}$$

namely,

$$\left[\frac{\partial^2}{\partial x^2} + \frac{\partial^2}{\partial y^2} + \frac{\partial^2}{\partial z^2} - \mu_0 \varepsilon_0 \frac{\partial^2}{\partial t^2}\right]\mathbf{E} = 0 \tag{1.11}$$

Equation (1.11) is characteristic of waves in a vacuum propagating with a velocity

$$c = \left(\frac{1}{\mu_0 \varepsilon_0}\right)^{1/2} = 2.99792458 \times 10^8 \text{ m/s} \tag{1.12}$$

which is identical to the speed of light in a vacuum. In a material medium, however, the speed is reduced to

$$v = \frac{c}{n} \tag{1.13}$$

where n is the refractive index of the medium. In a vacuum, n is unity. Since $n > 1$ in material media, the propagation velocity in a material medium is always less than c.

Any function of the form $f(t - \mathbf{d} \cdot \mathbf{r}/c)$ can be considered as a solution of Eq. (1.10), where $\mathbf{d}$ is a unit vector and $\mathbf{r}$ is the radial vector in space (x, y, z). Such a solution implies, justifiably, that the electromagnetic wave propagates with a velocity of c along $\mathbf{d}$. Consequently, the electric field vector is expressed as

$$\begin{aligned}\mathbf{E}(\mathbf{r}, t) &= \mathrm{Re}(\mathbf{E}_0)\, e^{j\omega[t - (\mathbf{d} \cdot \mathbf{r}/c)]} \\ &= \mathrm{Re}(\mathbf{E}_0)\, e^{j(\omega t - \mathbf{k} \cdot \mathbf{r})} \end{aligned} \tag{1.14}$$

where

$$\mathbf{k} = \left(\frac{\omega}{c}\right)\mathbf{d} \tag{1.15}$$

is the propagation vector, or wave vector, $\mathbf{E}_0$ is a constant vector associated with the maximum amplitude of oscillation and Re operation is equivalent to extracting the real part of a complex function. Similarly the magnetic field vector is expressed as

$$\mathbf{H}(\mathbf{r}, t) = \mathrm{Re}(\mathbf{H}_0)\, e^{j(\omega t - \mathbf{k} \cdot \mathbf{r})} \tag{1.16}$$

The wave functions as represented by Eqs. (1.14) and (1.16) are monochromatic, since the associated oscillation takes place with a single angular frequency ω. There is absolutely no reason to limit the frequency to a single value, and so, in general, we must integrate over the range of allowed frequencies to produce a general solution. We shall, however, limit our current discussion to the monochromatic case only.

The constant value, $\omega t - \mathbf{k} \cdot \mathbf{r}$, of the argument of the two field equations defines a set of planes in space referred to as the surfaces of constant phase. The direction cosines of such a plane are proportional to the components of the propagation vector $\mathbf{k}$. The rate at which this wave surface moves along $\mathbf{k}$ is usually called the phase velocity and is equal to $\omega/|\mathbf{k}|$. We may rewrite Eq. (1.1) using Eqs. (1.14) and (1.16) to produce

$$\nabla \times \mathbf{E} = -j\mathbf{k} \times \mathbf{E} = -j\mu_0\omega\mathbf{H} \tag{1.17}$$

which results in

$$\mathbf{k} \times \mathbf{E} = \mu_0\omega\mathbf{H} \tag{1.18}$$

Similarly, for the case of free space, Eq. (1.2) can be reorganized to yield

$$\mathbf{k} \times \mathbf{H} = -\omega\varepsilon_0\mathbf{E} \tag{1.19}$$

Equations (1.18) and (1.19) reveal that **E**, **H**, and **k** are mutually orthogonal vectors and that light (electromagnetic waves) propagates in a direction perpendicular to the plane of **E** and **H**, as shown in Fig. 1.1. The lengths of the field vectors, **E** and **H**, are generally proportional to their respective magnitude in space. The electromagnetic field pattern moves along **k** with speed c as time progresses. A traveling electromagnetic wave is thus characterized by **E** and **H** vectors when they oscillate in phase. However, for a standing wave, produced as

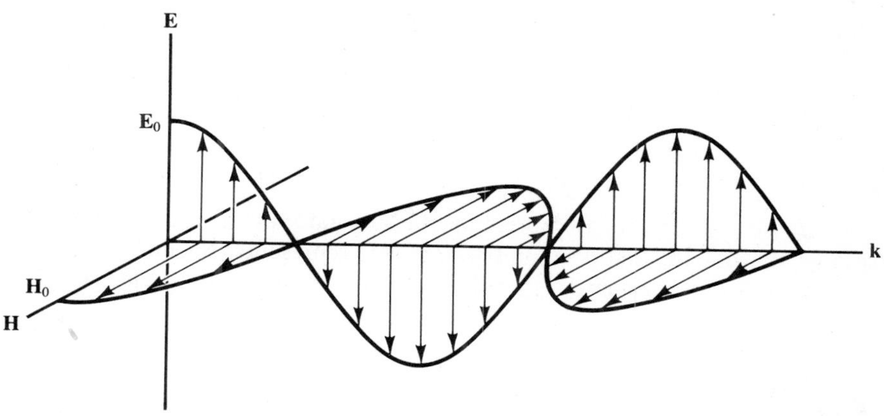

FIGURE 1.1

E and H field vectors frozen in time

a result of superimposing two waves with the same frequency and amplitude but oppositely directed **k** vectors, the phase relations are such that **E** and **H** reach their peak values out of phase (by $\pi/2$) in both space and time.

One of the important aspects of electromagnetic fields is that the flow of energy is directly associated with the flow of electromagnetic fields. The vector dot product of the left-hand side of Eq. (1.1) and **H** results in

$$\mathbf{H}\cdot(\nabla\times\mathbf{E}) = -\mu_0\mathbf{H}\cdot\frac{\partial\mathbf{H}}{\partial t} = -\frac{1}{2}\mu_0\frac{\partial|\mathbf{H}|^2}{\partial t} = -\frac{1}{2}\mu_0\frac{\partial\mathbf{H}^2}{\partial t} \tag{1.20}$$

and similarly from Eq. (1.2) one may obtain

$$\begin{aligned}\mathbf{E}\cdot(\nabla\times\mathbf{H}) &= \frac{1}{2}\varepsilon_0\frac{\partial|\mathbf{E}|^2}{\partial t} + \mathbf{E}\cdot\frac{\partial\mathbf{P}}{\partial t}\\ &= \frac{1}{2}\varepsilon_0\frac{\partial\mathbf{E}^2}{\partial t} + \mathbf{E}\cdot\frac{\partial\mathbf{P}}{\partial t}\end{aligned} \tag{1.21}$$

for a nonconducting medium ($\sigma = 0$), where both **E** and **H** are instantaneous values of the fields.

Equations (1.20) and (1.21) may now be combined to give

$$-\nabla\cdot(\mathbf{E}\times\mathbf{H}) = \frac{1}{2}\frac{\partial}{\partial t}(\varepsilon_0\mathbf{E}^2 + \mu_0\mathbf{H}^2) + \mathbf{E}\cdot\frac{\partial\mathbf{P}}{\partial t} \tag{1.22}$$

where we have used the vector identity

$$\nabla\cdot(\mathbf{A}\times\mathbf{B}) = \mathbf{B}\cdot(\nabla\times\mathbf{A}) - \mathbf{A}\cdot(\nabla\times\mathbf{B}) \tag{1.23}$$

With Gauss's theorem, Eq. (1.22) can be integrated over an arbitrary volume V, which is bounded by the surface S, to yield

$$-\int_S (\mathbf{E} \times \mathbf{H}) \cdot d\mathbf{S} = -\int_V \nabla \cdot (\mathbf{E} \times \mathbf{H})\, dV$$

$$= \int \left[\frac{1}{2} \frac{\partial}{\partial t} (\varepsilon_0 \mathbf{E}^2 + \mu_0 \mathbf{H}^2) + \mathbf{E} \cdot \frac{\partial \mathbf{P}}{\partial t} \right] dV \tag{1.24}$$

The left side of Eq. (1.24) represents the net amount of power flowing into volume V, while the first term on the right side represents the sum of the power, or rate of change of energy, stored within the volume due to electric energy density u_e, and magnetic energy density u_m, where

$$u_e = \frac{1}{2} \varepsilon_0 \mathbf{E}^2 \tag{1.25}$$

$$u_m = \frac{1}{2} \mu_0 \mathbf{H}^2 \tag{1.26}$$

The last term of Eq. (1.24) corresponds to the power given out by the field to the electric dipoles. This power, as can be shown, goes on to change the electric polarization, which is ultimately responsible for many of the important electro-optic device characteristics. The associated integrand refers to the time rate, or flow of electromagnetic energy.

A vector **S**, having the dimensions of energy per unit area per unit time, is known as the **Poynting vector** and is defined as

$$\mathbf{S} = \mathbf{E} \times \mathbf{H} \tag{1.27}$$

It is used to represent the energy flow. For real fields, the Poynting vector is given by

$$\mathbf{S} = \mathbf{E}_0 \times \frac{\mathbf{k} \times \mathrm{E}_0}{\mu_0 \omega} \cos^2(\omega t - \mathbf{k} \cdot \mathbf{r})$$

$$= \frac{|\mathbf{E}_0|^2 \mathbf{k}}{\mu_0 \omega} \cos^2(\omega t - \mathbf{k} \cdot \mathbf{r}) \tag{1.28}$$

using Eqs. (1.14), (1.16), and (1.18). The rate of flow of energy is thus proportional to the square of the amplitude of the electric field. Of more importance is the time-averaged Poynting vector $\langle \mathbf{S} \rangle$ that corresponds to the detectable electromagnetic energy. It is evaluated to give

$$\langle \mathbf{S} \rangle = \frac{1}{2} \sqrt{\frac{\varepsilon_0}{\mu_0}} |\mathbf{E}_0|^2 \mathbf{s} \tag{1.29}$$

such that the time-averaged energy flows along $\mathbf{s}$ $(\mathbf{k}/|\mathbf{k}|)$— that is, in the direction of propagation. The magnitude of the time-averaged Poynting vector is referred to as the **irradiance**. In an isotropic medium, the direction of the energy flow is specified by **s**, which is the same as that of **k**. However, in nonisotropic media, **s** and **k** may not always be in the same direction (see Problem 1.2).

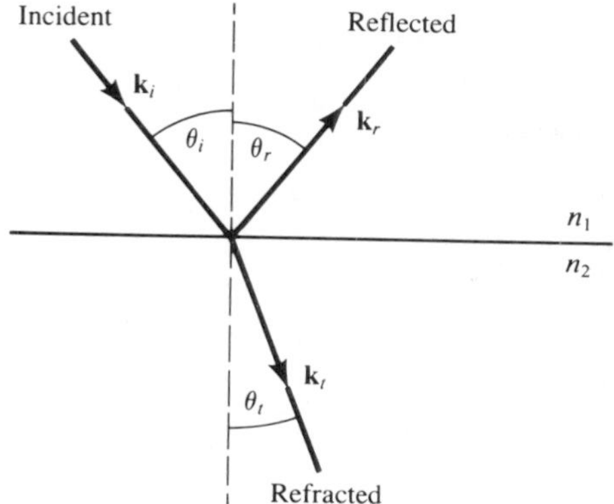

FIGURE 1.2

Propagation vectors at a plane interface

1.3 Reflection and Refraction

In this short section, we investigate the aspects of reflection and refraction of electromagnetic waves incident upon a plane of discontinuity between two different optical media. It may be noted that the laws of reflection and refraction are also deducible from Huygens principle, according to which each point on a propagating wavefront itself radiates a small spherical wavelet, but we avoid taking that route. For the simplicity of presentation, we consider plane interfaces first.

Consider a plane electromagnetic wave incident upon a plane interface that separates two linear optical mediums of refractive indices, n_1 and n_2, respectively, as shown in Fig. 1.2. Experience reveals that when an incident wave reaches the interface, a wave is reflected back to the first medium, while another wave is refracted (or transmitted) into the second medium. It is also an experimental fact that the frequency of wave motion does not change on reflection or refraction. In order for the sum of the incident and reflected waves to be equivalent to the refracted wave at the interface, it is necessary that the respective phases be identical; that is,

$$\mathbf{k}_i \cdot \mathbf{r} = \mathbf{k}_r \cdot \mathbf{r} = \mathbf{k}_t \cdot \mathbf{r} \tag{1.30}$$

where $\mathbf{k}_i$, $\mathbf{k}_r$, and $\mathbf{k}_t$ are the propagation vectors associated with the incident, reflected, and transmitted waves, respectively. Equation (1.30) requires that $\mathbf{k}_i$, $\mathbf{k}_r$, and $\mathbf{k}_t$ be coplanar and that their projections onto the planar interface be equal. Thus

$$k_i \sin\theta_i = k_r \sin\theta_r = k_t \sin\theta_t \tag{1.31}$$

where θ_i, θ_r, and θ_t are the angles made by the incident, reflected, and refracted propagation vectors, respectively, and the normal to the plane of separation. The two vectors $\mathbf{k}_i$ and $\mathbf{k}_r$ are both in the first medium and, therefore, according to Eqs. (1.15) and (1.13), Eq. (1.31) reduces to

$$\theta_i = \theta_r \tag{1.32}$$

and

$$\frac{k_t}{k_i} = \frac{n_2}{n_1} \equiv n_{21} = \frac{\sin\theta_i}{\sin\theta_t} \tag{1.33}$$

where n_{21} is the refractive index of the second medium with respect to the first. Equation (1.32) is **Alhazen's law of reflection**, which states that the angles of incidence and reflection are equal, whereas Eq. (1.33) is referred to as **Snell's law of refraction.**

Electromagnetic waves require special attention because they involve two fields. Both the electric and the magnetic fields are normal to the direction of propagation of each wave. Consequently, it is more convenient to think of each field as having a component parallel to the plane of incidence and a component perpendicular to the plane of incidence. These two cases are referred to as the transverse electric (TE) oscillation mode and transverse magnetic (TM) oscillation mode and are depicted in Fig. 1.3 where the x-y plane is the plane of incidence.

The boundary conditions require that the tangential components of the electric and magnetic fields be continuous as the plane interface is crossed. Consequently, for the TE case

$$E_i + E_r = E_t \tag{1.34a}$$

$$-H_i\cos\theta_i + H_r\cos\theta_r = -H_t\cos\theta_t \tag{1.34b}$$

and for the TM case

$$H_i - H_r = H_t \tag{1.35a}$$

$$E_i\cos\theta_i + E_r\cos\theta_r = E_t\cos\theta_t \tag{1.35b}$$

Note that according to Eq. (1.18) H_i, H_r, and H_t are respectively proportional to k_iE_i, k_rE_r, and k_tE_t. Accordingly, it is possible to evaluate the coefficients of reflection, r_σ and r_π, as follows:

$$r_\sigma = \left.\frac{E_r}{E_i}\right|_{\mathrm{TE}} = \frac{\cos\theta_i - \sqrt{n_{21}^2 - \sin^2\theta_i}}{\cos\theta_i + \sqrt{n_{21}^2 - \sin^2\theta_i}} \tag{1.36a}$$

and

$$r_\pi = \left.\frac{E_r}{E_i}\right|_{\mathrm{TM}} = \frac{\sqrt{n_{21}^2 - \sin^2\theta_i} - n_{21}^2\cos\theta_i}{\sqrt{n_{21}^2 - \sin^2\theta_i} + n_{21}^2\cos\theta_i} \tag{1.36b}$$

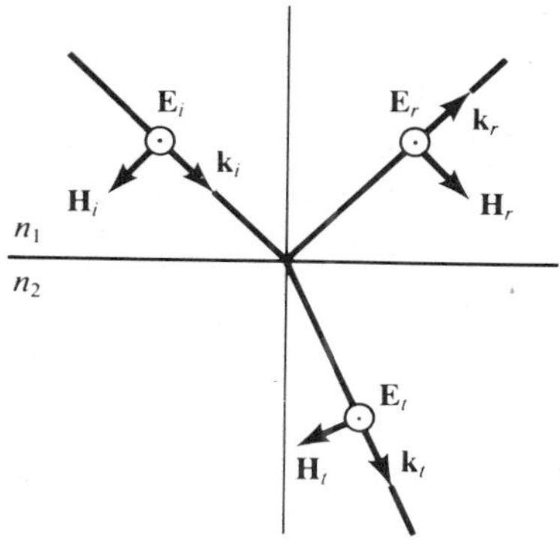

(a) **E** ⊥ Plane of incidence

(b) **E** ∥ Plane of incidence

FIGURE 1.3

Vector geometries for two orientations of E when it is (a) perpendicular to the plane of incidence and (b) parallel to the plane of incidence

using Eqs. (1.32) and (1.33). These equations are also referred to as **Fresnel's equations**. Both r_σ and r_π are identical for the case of normal incidence, with a positive or negative sign that depends on whether $n_2 > n_1$ or $n_2 < n_1$, respectively. A negative value of the amplitude ratio indicates that the phase of the incident wave is 180° out of phase to that of the reflected wave.

The ratio of reflected and incident energy, which is equivalent to the square of the coefficient of reflection, is known as the **reflectance**. As a physical parameter, reflectance is often preferred over the coefficient of reflection since it is more easily measurable. Figure 1.4 shows the ratios of the amplitudes as well as their squares as functions of the incident angle for the two cases: $n_1 < n_2$ and $n_1 > n_2$. The angle at which the TM reflectance becomes zero is referred to as **Brewster's angle**, whereas the angle at which both TE and TM reflectances approach unity—when $n_1 > n_2$—is referred to as the **critical angle**. The significance of both of these angles will become obvious in the later chapters (for example, in Chapters 4, 7, and 9).

Equations (1.32) and (1.33) can be conveniently used to trace the path of light even when it is incident upon a surface that is not plane. The great majority of optical focusing operations are realized by surfaces or combinations of surfaces that are sections of spheres. Figure 1.5 shows the path of a ray originating at the point O in front of (a) a spherical mirror and (b) a spherical refracting surface such that P is a point located at the interface on the axis of the optical element. In the former, all of the incident light returns to the originating medium, while in the latter the refracted ray travels through the second medium before converging at point I on the axis. In addition to Eqs. (1.32) and (1.33), the

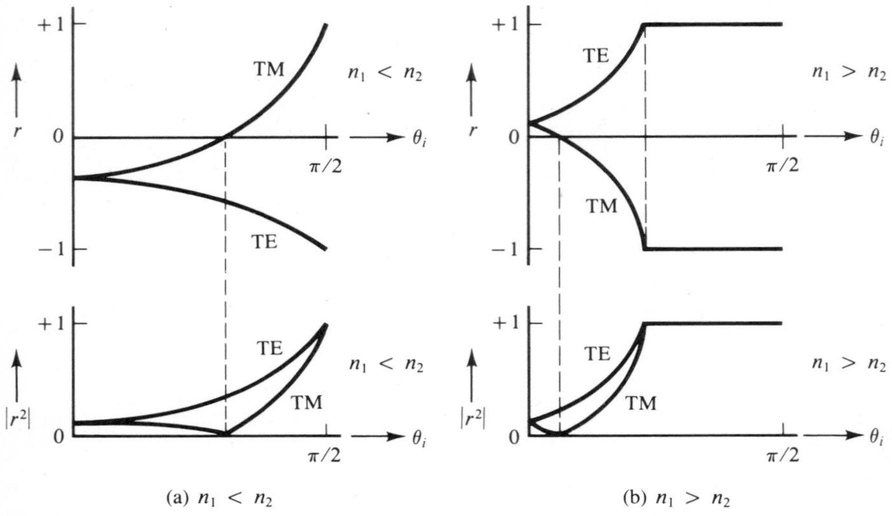

FIGURE 1.4

Plots of r and $|r|^2$ versus the incident angle when (a) $n_1 < n_2$ and (b) $n_1 > n_2$

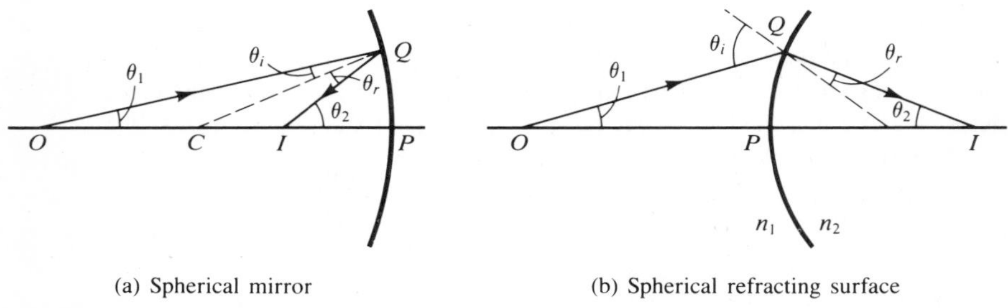

FIGURE 1.5

Ray geometry for spherical (a) reflecting surface and (b) refracting surface

trigonometric law of sines can be used to find that

$$\frac{\sin\theta_1}{\sin\theta_2} = \begin{cases} \dfrac{i-R}{R-o} & \text{spherical mirror} \\[2ex] \dfrac{i-n_2R}{o+n_1R} & \text{spherical refracting surface} \end{cases} \tag{1.37}$$

where o, i, and R are, respectively, the object distance (OP), image distance (PI), and the radius of curvature (CQ).

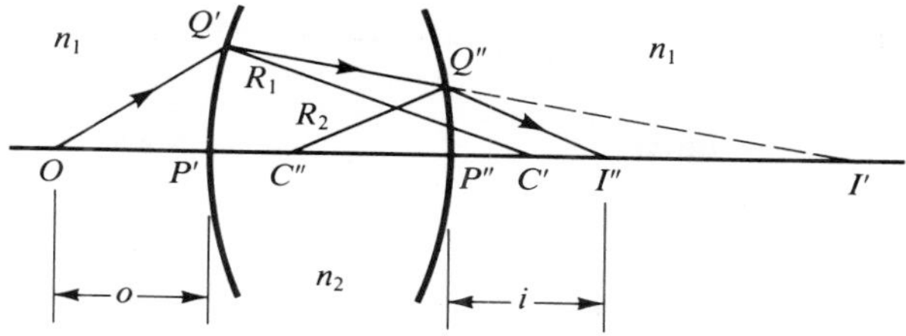

FIGURE 1.6

Ray geometry of a biconvex thick lens

It must be noted that even though the ray tracing characterizations are accurate, they are limited in providing insights into the process of image formation. When a bundle of rays originates from an object point, the corresponding image points may not all coincide. This phenomenon is referred to as **spherical aberration**. To yield a simplified result, a simplification known as **paraxial approximation** is often made to Eq. (1.37). In this approximation, all of the incident rays are assumed to make small angles with the optic axis such that the sines of the angles can simply be replaced by the angles. Consequently, Eq. (1.37) reduces to

$$\frac{1}{o} + \frac{1}{i} = \frac{2}{R} \tag{1.38}$$

for the spherical mirror and

$$\frac{n_1}{o} + \frac{n_2}{i} = \frac{n_2 - n_1}{R} \tag{1.39}$$

for the spherical refracting surface. Equations (1.38) and (1.39) are both applicable to cases other than those depicted in Fig. 1.5. To make these equations applicable to all cases, it is appropriate to introduce a sign convention for depicting i and R. Equation (1.38) is universal as long as R is treated as positive for the concave mirror, shown in Fig. 1.5(a), and negative for the convex mirror. Similarly, Eq. (1.39) is universal as long as R is treated as positive for the convex refracting surface, shown in Fig. 1.5(b), and negative for the concave refracting surface. If, for any combination of o and R, i turns out to be negative, then the rays do not converge at an image point. Instead, they diverge and appear to have originated from a point (location of the virtual image) behind the optical surface in question.

Consider the case of a thick lens of refractive index n_2 (Fig. 1.6) that has two refracting surfaces. For simplicity, we assume a case where both of the surfaces are convex to the external medium of refractive index n_1. For the first spherical interface, we can write

$$\frac{n_1}{OP'} + \frac{n_2}{P'I'} = \frac{n_2 - n_1}{R_1} \tag{1.40}$$

and for the second interface,

$$\frac{n_2}{-(P'I' - P'P'')} + \frac{n_1}{P''I''} = \frac{n_1 - n_2}{-R_2} \tag{1.41}$$

where R_1 and R_2 are the two radii of curvature of the two surfaces, respectively. In the case of a thin lens, $P'P''$ approaches zero and accordingly Eqs. (1.40) and (1.41) can be combined to give

$$\frac{1}{o} + \frac{1}{i} = (n_{21} - 1)\left(\frac{1}{R_1} + \frac{1}{R_2}\right) \tag{1.42}$$

where OP' and $P''I''$ are identified, respectively, with object and image distance.

The distance of the point of convergence (or divergence) from the interface when the object is at infinity (for example, for a parallel input beam of light) is referred to as the **focal length**. Thus the focal length f for the cases of a spherical mirror and a thin lens is given by

$$\frac{1}{f} = \begin{cases} \dfrac{2}{R} & \text{spherical mirror} \\ (n_{21} - 1)\left(\dfrac{1}{R_1} + \dfrac{1}{R_2}\right) & \text{thin lens} \end{cases} \tag{1.43}$$

The paraxial approximation does not provide the most accurate answers for characterizing all imaging systems, but it is often enough for many systems, and in the worst cases it serves as a starting point for more exact calculations. In the chapters to come, the conclusions of this section will be recalled or used when appropriate. But first, let us review the basics of quantum mechanics.

1.4 Basics of Quantum Mechanics

The fundamentals of quantum mechanics are customarily introduced by a description of the simplest of all atoms: the hydrogen atom. This atom consists of only one peripheral negatively charged electron and only one positively charged proton located in its nucleus. Classical physicists envisioned the electron as a highly localized particle that whirls around the nucleus to keep itself from being dragged into the nucleus. But to subject itself to the centripetal motion, the electron must continually radiate electromagnetic energy. According to classical theory, the continual loss of energy is bound to cause the collapse of the electron into the nucleus, but in practice, the electron never collapses. Thus classical electromechanical concepts are incompatible with the

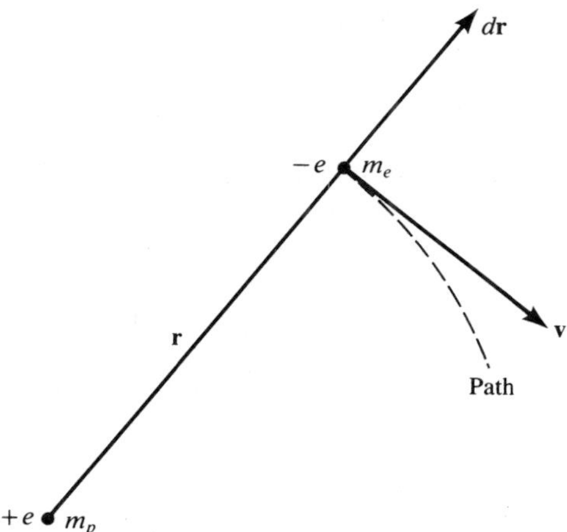

FIGURE 1.7

Classical configuration of a hydrogen atom

experimentally observed stability of the hydrogen atom. This phenomenon and several other paradoxes forced Niels Bohr to introduce several semiclassical assumptions to account for the discrepancies between theory and observations. However, Bohr's semiclassical treatment was also discarded in favor of a quantum mechanical viewpoint that provided accurate insights into the atomic microcosm. Nevertheless, Bohr's theory provided a satisfactory interpretation of the discreteness of the spectrum of radiation of the hydrogen atom. In addition, it served as an intermediate step in the final transition to quantum mechanics. Consequently, three approaches to the physics of an electron in a hydrogen atom—classical, semiclassical, and quantum-mechanical—are discussed in this section.

1.4.1 The Classical Hydrogen Atom

In classical Newtonian physics, the hydrogen atom has an electron that moves around the proton (nucleus) in a perfectly circular orbit, as shown in Fig. 1.7. The electron, which has a negative charge e with a value of 1.6021×10^{-19} C and a finite mass m_e with a value of 9.1091×10^{-31} kg, is located at a radial distance r from the proton. The proton has an equal but positive charge and a comparatively larger mass m_p, the value of which is 1.6725×10^{-27} kg. In this model, the electron would not be stationary in space because of the presence of electrostatic attractive force. If the electron were subjected to either circular or elliptical motion around the proton, however, a stable system would be possible. A circular orbit is thus assumed for the electron for the sake of convenience.

According to Coulomb's electrostatic law, the force $\mathbf{F}_e$ between the peripheral electron and the nuclear proton is expressed as

$$\mathbf{F}_e = -\frac{e^2}{4\pi\varepsilon_0 r^2}\mathbf{r} \tag{1.44}$$

where $\mathbf{r}$ is a unit vector directed outward from the proton along a radial line joining the two charged bodies. The mass of a proton, for all practical purposes, is much much larger (approximately 1836 times) than that of the electron. Therefore, the nucleus may be assumed to be rigidly fixed in space, while the electron is considered to move with a linear velocity v tangential to the circular orbit (of radius r). To maintain equilibrium, it is necessary that the centrifugal force (directed outward), given by $m_e v^2/r$, and the Coulomb force (directed inward as given in Eq. 1.44) on the electron are equal in magnitude. Therefore,

$$\frac{m_e v^2}{r} = \frac{e^2}{4\pi\varepsilon_0 r^2} \tag{1.45}$$

which yields

$$v = \frac{e}{\sqrt{4\pi\varepsilon_0 m_e r}} \tag{1.46}$$

The linear velocity of the peripheral electron in the classical model is thus determined in terms of its mass, charge, and radial position.

Because of the electron's position, the potential energy PE is determined by calculating the total work expended as the electron moves from infinity to a distance r from the proton as follows:

$$\text{PE} = -\int_{\infty}^{r} \mathbf{F}\cdot d\mathbf{r} = \int_{\infty}^{r} \frac{e^2}{4\pi\varepsilon_0 r^2}\, dr = -\frac{e^2}{4\pi\varepsilon_0 r} \tag{1.47}$$

where $d\mathbf{r} = \mathbf{r}\, dr$.The negative sign indicates that the force on the electron is in the $-\mathbf{r}$ direction due to the presence of an attractive force. It can be reasoned that the potential energy becomes zero only when $r = \infty$. It is obvious that work would have to be done on the electron (instead of the electron doing the work) if it has to be taken away from the proton.

The total energy of the electron E can now be calculated as the sum of its kinetic energy

$$\text{KE} = \frac{1}{2} m_e v^2 \tag{1.48}$$

and its potential energy as given by Eq. (1.47). Hence, using Eq. (1.46) we obtain for the total energy of the atomic electron

$$E = \frac{1}{2} m_e v^2 - \frac{e^2}{4\pi\varepsilon_0 r} = -\frac{e^2}{8\pi\varepsilon_0 r} \tag{1.49}$$

The total energy of the electron is negative, which indicates that the electron is bound to the proton. As the radius of the electron orbit becomes larger, the total energy of the electron becomes less negative. This conclusion is generally correct and agrees with the experimental fact that the transition of an electron from an orbit with a large r to one with a small r involves the emission of energy.

However, classical theory allows the electron to have any positive value for the orbit radius r, which is in direct contradiction to realistic atomic systems. If the total energy of the system were greater than zero, it could move freely on its own and would wander off from the proton. Experiments indicate that approximately 13.6 eV are necessary to free an electron from the central proton.

Classical observation is indeed in crisis as soon as one tries to incorporate the findings of electromagnetic theory. An electron orbiting in a circular path is continually changing its direction and is thus being accelerated. If the electron continuously radiates energy, it must gradually collapse into the proton; yet this collapse does not take place. Consequently, it is argued that classical physics begins to lose its validity when it is applied to atomic systems. As was shown later, this failure is due to the fact that classical physics is applicable to either only pure particles or only pure waves but not to a combination of both.

1.4.2 The Semiclassical Hydrogen Atom

While trying to explain the phenomenon of radiation emitted by bodies hot enough to be luminous, Planck found that the radiation is emitted only discretely as little bursts of energy known as **quantas**. It was noted that each quanta of energy E is associated with a photon particle and is directly proportional to its frequency ν. Thus we can express

$$E = h\nu \tag{1.50}$$

where h, known as Planck's constant, has a value of 6.626176×10^{-34} J-s. This finding was shortly followed by de Broglie's famous conclusion about the duality of matter, which postulated that waves and particles are merely two manifestations of the same physical reality. Consequently, it was reasoned that all matter has particlelike as well as wavelike characteristics.

A photon of frequency ν has the momentum p given by

$$p = E/c \tag{1.51}$$

which can be rewritten using Eq. (1.50) to yield the wavelength of photon λ as

$$\lambda = h/p \tag{1.52}$$

where $c = \nu\lambda$. de Broglie argued that nature is symmetric and, therefore, he theorized that the general formula of Eq. (1.52) is applicable to both material particles and photons. For a particle of mass m and velocity v, the de Broglie wavelength is generally given by

$$\lambda = h/mv \tag{1.53}$$

As expected, for a particle having a large momentum, the wavelength is small. Consequently for larger bodies, λ becomes relatively insignificant. In particular, when Eq. (1.53) is applied to the case of an electron, its de Broglie wavelength is found to be

$$\lambda = \frac{h}{e}\sqrt{\frac{4\pi\varepsilon_0 r}{m_e}} \tag{1.54}$$

using Eq. (1.46). For electrons accelerating through a potential of v volts, the wavelength is found to be $12.13 \times 10^{-10}\,\text{v}^{-0.5}$ m. Thus for voltages on the order of 10 V, λ is comparable to the interatomic distances in semiconductors. A remarkable conclusion, however, can be made by comparing the calculated values of λ and the circumference of the classical electron orbit. It turns out that the value of the circumference of the electron orbit in a hydrogen atom is identical to the value of the electron's wavelength (see Problem 1.5). This observation leads to an important extrapolation that an electron can indefinitely circle around a nucleus without irradiation as long as the length of its orbit is equivalent to an integral multiple of its de Broglie wavelength. Instead, if the orbit circumference were equal to a nonintegral multiple of its de Broglie wavelength, destructive interference would set in. This extrapolation, therefore, combines both the particlelike and wavelike characteristics of an electron and limits the atomic radius to certain discrete values only. The circumference of a circular orbit of radius r_n is consequently given by

$$2\pi r_n = n\lambda \qquad n = 1, 2, 3, \ldots \tag{1.55}$$

The integer n is often referred to as the total quantum number. Using Eqs. (1.54) and (1.55), we may obtain the corresponding stable orbit radii as

$$r_n = \frac{n^2 h^2 \varepsilon_0}{\pi m_e e^2} \qquad n = 1, 2, 3, \ldots \tag{1.56}$$

The minimum allowable orbit radius can be calculated to be 5.3×10^{-11} m by setting $n = 1$ in Eq. (1.56). It may also be concluded that the spacing between the permissible adjacent electron orbits increases with the increase of n.

Equations (1.49) and (1.56) may now be used to provide the electron energy value E_n for different values of n as

$$E_n = -\frac{m_e e^4}{8\varepsilon_0^2 h^2 n^2} \tag{1.57}$$

These are the energy levels of a hydrogen atom where E_1, in particular, is generally referred to as the **ground state**, and the remaining energy levels are called **excited states**. These energy levels are shown in Fig. 1.8. As the orbital radius increases with the increase of n, the corresponding excited energy value approaches zero. When ultimately n becomes infinity, the electron becomes completely free.

Bohr's semiclassical characterization of the hydrogen atom not only involved the concept of discretely distributed energy but also touched upon

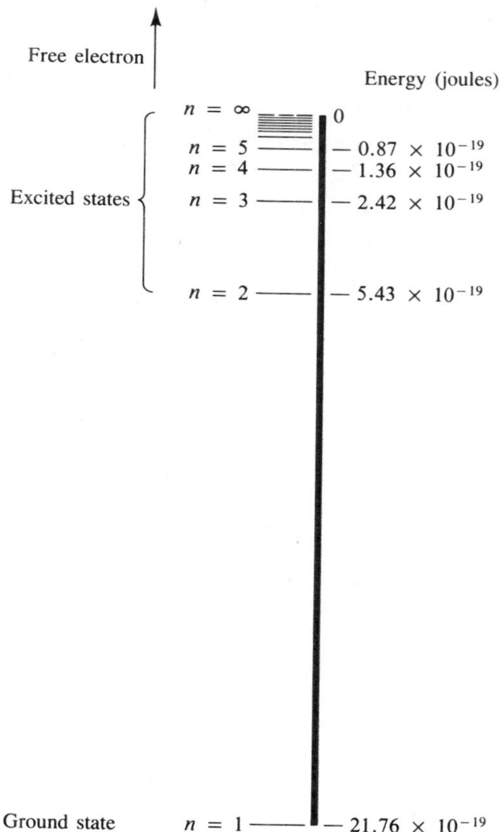

FIGURE 1.8

Hydrogen energy levels

electromagnetic radiation. First, it was assumed that an electron moving continually in any one of the allowed orbits would not irradiate, thus maintaining a constant total energy. According to this assumption, electrons cannot survive in an atomic system except when they are subjected to certain specific energy states. Second, whenever an electron moving initially in an excited orbit of total energy E_i jumps abruptly to another permissible but less excited orbit of total energy E_f, it results in an irradiation the amount of which is equal to the difference of the initial and final energies. The frequency v of the emitted photon is given by

$$v = \frac{E_i - E_f}{h} = \frac{m_e e^4}{8\varepsilon_0^2 h^3}\left(\frac{1}{n_f^2} - \frac{1}{n_i^2}\right) \tag{1.58}$$

and correspondingly the photon wavelength λ is given by

$$\frac{1}{\lambda} = \frac{m_e e^4}{8\varepsilon_0^2 c h^3}\left(\frac{1}{n_f^2} - \frac{1}{n_i^2}\right) \tag{1.59}$$

The wavelengths present in an atomic spectra are classified into several definite sets called **spectral series**. When excited, every chemical element displays a unique line spectrum. The spectral lines are thus often used for analyzing the composition of unknown substances. Whereas the unexcited gases and vapors do not radiate their characteristic spectral lines, they do absorb light from certain of those wavelengths when samples are exposed to white light. Consequently, the absorption spectrum of an element is similar to its emission spectrum.

The prediction of Eq. (1.59) agrees exceedingly well with the experimentally observed seven series of spectral lines (Lyman, Balmer, Paschen, Brackett, Pfund, Humphreys, and Hansen and Strong series) in the hydrogen spectrum. These series are listed in Table 1.1 and are shown in Fig. 1.9.

When an electron receives activating energy from an external source, such as one emitting either light or heat or sound, the electron may make a transition to an excited state of higher energy. However, an electron is not necessarily stable in an excited state. The excitation energy is often reradiated almost at once and, as a result, the electron can make a transition directly to the ground state *or* go through one or more intermediate states, if available, before reaching the ground state. The transition of an electron from a state of higher energy level to a state of lower energy level is always accompanied by the release of an equivalent amount of radiation. But the photons associated with the released energy come off in random directions, not all in the same direction as is often the case in the original activating energy. Often it is a blessing in disguise. A neon sign is an example of how a strong electric field applied between the electrodes in a gas-filled tube results in the emission of a particular wavelength (and thus of a particular color).

TABLE 1.1

Hydrogen Spectral Series and Their Characteristics

Series	*Nature*	*Conditions*
Lyman	UV	$n_f = 1$; $n_i = 2, 3, 4, \ldots$
Balmer	near UV and visible	$n_f = 2$; $n_i = 3, 4, 5, \ldots$
Paschen	IR	$n_f = 3$; $n_i = 4, 5, 6, \ldots$
Brackett	IR	$n_f = 4$; $n_i = 5, 6, 7, \ldots$
Pfund	IR	$n_f = 5$; $n_i = 6, 7, 8, \ldots$
Humphreys	IR	$n_f = 6$; $n_i = 7, 8, 9, \ldots$
Hansen and Strong	IR	$n_f = 7$; $n_i = 8, 9, 10, \ldots$

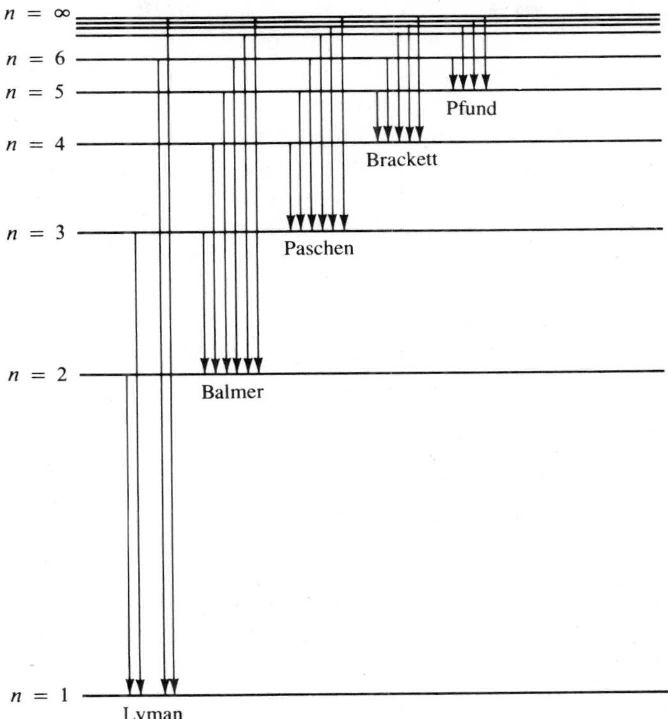

FIGURE 1.9

Hydrogen spectral lines

1.4.3 The Schrödinger Equation and Quantum Model of Atoms

While Bohr's semiclassical model correctly predicts the spectral series of hydrogen, hydrogen isotopes, and hydrogenic atoms (He^+ and Li^{++}), it is generally incapable of explaining the characteristics of complex atoms having two or more electrons. Moreover, the Bohr model can provide no reasonable explanation as to why certain spectral lines are more intense than others. The answers to this shortcoming and several more paradoxes are best provided by solving the quantum counterpart of the classical conservation of energy. This much-celebrated quantum expression is commonly referred to as the **Schrödinger equation**. The quantum characteristics of the Schrödinger equation are derived almost exclusively from de Broglie's postulate of wave–particle duality.

The total energy E of a particle of mass m traveling along the x axis with a momentum p and potential V is obtained by summing its kinetic and potential energies. Thus

$$E = \frac{p^2}{2m} + V(x, t) \tag{1.60}$$

where V, in general, is a function of both position x and time t. This equation is very much like Eq. (1.49) and serves as the basis for formulating the Schrödinger equation. Schrödinger explored the characteristics of a wave function associated with a particle as just described and found the most suitable one to be in the form of a plane wave:

$$\Psi(x, t) = \Psi_0 e^{j\omega(x/\nu\lambda - t)} \tag{1.61}$$

where ω is the angular frequency. The amplitude and the phase angle of the wave function are respectively given by

$$|\Psi(x, t)| = (\Psi_0 \Psi_0^*)^{1/2} \tag{1.62}$$

and

$$\alpha(x, t) = \omega\left(\frac{x}{\nu\lambda} - t\right) \tag{1.63}$$

Such a wave function indicates that the associated constant phase plane moves with a constant velocity of $\nu\lambda$. Consequently, Eq. (1.60) may be rewritten using Eqs. (1.50) and (1.52) to give

$$h\nu = \frac{1}{2m}\left(\frac{h}{\lambda}\right)^2 + V(x, t) \tag{1.64}$$

But from Eq. (1.61), we find that

$$\frac{\partial^2}{\partial x^2}[\Psi(x, t)] = -\frac{4\pi^2}{\lambda^2}\Psi(x, t) \tag{1.65}$$

and

$$\frac{\partial}{\partial t}[\Psi(x, t)] = -j2\pi\nu\Psi(x, t) \tag{1.66}$$

since $\omega = 2\pi\nu$. Thus Eq. (1.64) can be rewritten as

$$h\left[-\frac{1}{j2\pi\Psi(x, t)}\frac{\partial\Psi(x, t)}{\partial t}\right] = \frac{h^2}{2m}\left[-\frac{1}{4\pi^2\Psi(x, t)}\frac{\partial^2\Psi(x, t)}{\partial x^2}\right] + V(x, t) \tag{1.67}$$

or

$$-\frac{\hbar^2}{2m}\frac{\partial^2\Psi}{\partial x^2} + V(x, t)\Psi = j\hbar\frac{\partial\Psi}{\partial t} \tag{1.68}$$

where $\hbar = h/2\pi$. Equation (1.68) is the well-known Schrödinger equation. Schrödinger postulated that a particle, whether free or not, having a potential energy $V(x, t)$, whether constant or not, is always associated with a wave function $\Psi(x, t)$. This postulate can be easily extended to accommodate a three-dimensional Cartesian coordinate system. In the case of a hydrogen atom, one may extend this, for simplicity, to a spherical coordinate system by a simple coordinate transformation.

The wave function $\Psi(x, y, z, t)$ has no physical interpretation because of its inherent complex nature. However, it was shown by Max Born (the proof is not treated here) that $\Psi(x, y, z, t)\Psi^*(x, y, z, t)$ evaluated at a particular position at a particular time is proportional to the probability of experimentally locating the corresponding particle there at that time. In quantum-mechanical treatment, one often determines $\Psi(x, y, z, t)$ for a body when it is subjected to an action of external stimuli. At first, the external stimulus, often in the form of a potential distribution, is characterized and is incorporated into the Schrödinger equation for the body in question. Then the equation is solved with the aid of suitable boundary conditions. The wave function is thereafter employed to determine the probability, $p(x, y, z, t)$, of locating the body in the differential volume $dx\, dy\, dz$ surrounding the point (x, y, z). The probability of finding the particle must always be both real and nonnegative. The coefficient of $\Psi(x, y, z, t)$ is often normalized to unity to meet the requirement that the probability of finding a particle present somewhere in the space is unity; i.e.,

$$\int_{-\infty}^{\infty}\int_{-\infty}^{\infty}\int_{-\infty}^{\infty} \Psi^*(x, y, z, t)\Psi(x, y, z, t)\, dx\, dy\, dz = 1 \tag{1.69}$$

The Schrödinger equation of Eq. (1.68) is a partial differential equation in both time and space. Whenever the potential energy in the equation is a function of both time and space, it becomes very difficult to solve analytically for $\Psi(x, y, z, t)$. However, in a great many problems of practical interest, as we shall see later, the potential energy does not depend on time explicitly. When this is true, we note that the wave function of Eq. (1.61) can be written as

$$\begin{aligned}\Psi(x, t) &= \Psi_0 e^{j\omega(x/v\lambda - t)} \\ &= \Psi_0 e^{-j\omega t} e^{j2\pi x/\lambda} \\ &\equiv \Psi(x) e^{-j\omega t} \\ &= \Psi(x) e^{-jEt/\hbar}\end{aligned} \tag{1.70}$$

where $\Psi(x)$ is a position-dependent function. Consequently, Eq. (1.68) may be written as

$$-\frac{\hbar^2}{2m}\frac{d^2\Psi(x)}{dx^2} + V\Psi(x) = j\hbar\left(-\frac{jE}{\hbar}\right)\Psi(x) \tag{1.71}$$

or

$$\frac{d^2\Psi(x)}{dx^2} + \frac{2m}{\hbar^2}(E - V)\Psi(x) = 0 \tag{1.72}$$

Equation (1.72) is the time-independent, one-dimensional Schrödinger equation where all references to time t have been totally eliminated.

For the case of an electron moving around a fixed nucleus, the potential energy of the electron is generally given by

$$V(r) = -\frac{Ze^2}{4\pi\varepsilon_0 r} \tag{1.73}$$

For the particular case of the hydrogen atom, $Z = 1$. Since the electron is not necessarily traveling in only one direction, Eq. (1.72) is often modified so that it can account for all three dimensions. Thus

$$\nabla^2\Psi(x, y, z) + \frac{2m}{\hbar^2}[E - V(x, y, z)]\Psi(x, y, z) = 0 \tag{1.74}$$

where

$$\nabla^2 = \frac{\partial^2}{\partial x^2} + \frac{\partial^2}{\partial y^2} + \frac{\partial^2}{\partial z^2} \tag{1.75}$$

In terms of the spherical coordinate system (r, θ, ϕ), as shown in Fig. 1.10, the Cartesian coordinates can be expressed as

$$x = r \sin\theta \cos\phi \tag{1.76}$$

$$y = r \sin\theta \sin\phi \tag{1.77}$$

$$z = r \cos\theta \tag{1.78}$$

By making use of Eqs. (1.76) through (1.78), Eq. (1.74) can be written as

$$\frac{1}{r^2}\frac{\partial}{\partial r}\left[r^2\frac{\partial\Psi}{\partial r}\right] + \frac{1}{r^2\sin\theta}\frac{\partial}{\partial\theta}\left[\sin\theta\frac{\partial\Psi}{\partial\theta}\right] + \frac{1}{r^2\sin^2\theta}\frac{\partial^2\Psi}{\partial\phi^2} + \frac{2m_e}{\hbar^2}(E - V)\Psi = 0 \tag{1.79}$$

where $\Psi = \Psi(r, \theta, \phi)$.

The Schrödinger equation for the electron in a hydrogen atom is therefore obtained using Eq. (1.73) as

$$\frac{1}{r^2}\frac{\partial}{\partial r}\left(r^2\frac{\partial\Psi}{\partial r}\right) + \frac{1}{r^2\sin^2\theta}\frac{\partial^2\Psi}{\partial\phi^2} + \frac{1}{r^2\sin\theta}\frac{\partial}{\partial\theta}\left[\sin\theta\frac{\partial\Psi}{\partial\theta}\right] + \frac{2m_e}{\hbar^2}\left[E + \frac{Ze^2}{4\pi\varepsilon_0 r}\right]\Psi = 0 \tag{1.80}$$

When the potential is a function of the radial coordinate r only, $\Psi(r, \theta, \phi)$ can be considered to be a product of three different functions:

$$\Psi(r, \theta, \phi) = R(r)\Theta(\theta)\Phi(\phi) \tag{1.81}$$

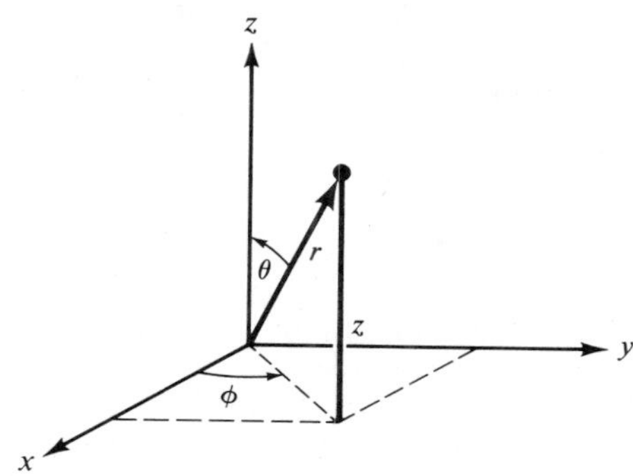

FIGURE 1.10

Spherical coordinate system

where $R(r)$ is a function of r alone, Θ is a function of θ alone, and Φ is a function of ϕ alone. The formidable procedure of solving Eq. (1.80) for Ψ thus reduces to solving for R, Θ, and Φ separately. Using Eq. (1.81), Eq. (1.80) can be rewritten as

$$\frac{\sin^2\theta}{R}\frac{\partial}{\partial r}\left(r^2\frac{\partial R}{\partial r}\right)+\frac{\sin\theta}{\Theta}\frac{\partial}{\partial\theta}\left(\sin\theta\frac{\partial\Theta}{\partial\theta}\right)+\frac{2m_e r^2\sin^2\theta}{\hbar^2}\left(E+\frac{Ze^2}{4\pi\varepsilon_0 r}\right)$$
$$=-\frac{1}{\Phi}\frac{\partial^2\Phi}{\partial\phi^2} \quad \textbf{(1.82)}$$

Note that the left-hand side of Eq. (1.82) is a function of only r and θ, while the right-hand side is a function of ϕ alone. A multivariable differential equation such as this, with one side being a function of certain variables while the other side is a function of the remaining variables, is reasonably easier to solve. Such an equation is said to have valid solutions only when both sides of the equation are equal to the same arbitrary constant. Conveniently, we choose this constant to be m_l^2. Thus

$$\frac{1}{\Phi}\frac{d^2\Phi}{d\phi^2}+m_l^2=0 \quad \textbf{(1.83)}$$

and

$$\frac{1}{R}\frac{\partial}{\partial r}\left(r^2\frac{\partial R}{\partial r}\right)+\frac{2m_e r^2}{\hbar^2}\left(\frac{Ze^2}{4\pi\varepsilon_0 r}+E\right)=\frac{m_l^2}{\sin^2\theta}-\frac{1}{\Theta}\frac{1}{\sin\theta}\frac{\partial}{\partial\theta}\left[\sin\theta\frac{\partial\Theta}{\partial\theta}\right]$$
$$\textbf{(1.84)}$$

Note again that Eq. (1.84) also has the desirable characteristic of separable variables. Its left-hand side is a function of only r while its right-hand side is a function of only θ. The two sides are, therefore, set equal to an arbitrary constant

$l(l+1)$, for convenience, where l is a positive integer. Consequently, Eq. (1.84) reduces to

$$\frac{1}{\sin\theta}\frac{d}{d\theta}\left[\sin\theta\frac{d\Theta}{d\theta}\right]+\left[l(l+1)-\frac{m_l^2}{\sin^2\theta}\right]\Theta=0 \tag{1.85}$$

and

$$\frac{1}{r^2}\frac{d}{dr}\left(r^2\frac{dR}{dr}\right)+\left[\frac{2m_e}{\hbar^2}\left(E+\frac{Ze^2}{4\pi\varepsilon_0 r}\right)-\frac{l(l+1)}{r^2}\right]R=0 \tag{1.86}$$

Each of these equations, Eqs. (1.83), (1.85) and (1.86), is an ordinary differential equation of only one variable. The complexity of Eq. (1.80) is thus avoided at the cost of solving three separate but simpler equations. The solution of the first equation is quite straightforward and is given by

$$\Phi(\phi)=\frac{1}{\sqrt{2\pi}}e^{\pm jm_l\phi} \tag{1.87}$$

The coefficient of the solution is deliberately chosen to be $(2\pi)^{-1/2}$. Such a choice guarantees the fact that Φ is normalized, as per Eq. (1.69), over the complete range of ϕ. The normalized solutions of the remaining two equations, however, are not so simple. Without going through the complexities of various intermediate steps, the solutions (i.e., wavefunctions) and the physical meanings of the associated probability densities are discussed in this section. The normalized solution of Eq. (1.85) is found to be a polynomial in $\cos\theta$ as given by

$$\Theta_{lm_l}(\theta)=N_{lm_l}P_l^{m_l}(\cos\theta) \tag{1.88}$$

where

$$N_{lm_l}=\varepsilon\left[\frac{(2l+1)(l-|m_l|)!}{2(l+|m_l|)!}\right]^{1/2} \tag{1.89}$$

ε is equal to $(-1)^{m_l}$ for positive nonzero m_l and unity otherwise, and $P_l^{m_l}$, referred to as the **associated Legendre function**, is given by

$$P_l^{m_l}(\xi)=(1-\xi)^{m_l/2}\frac{d^{m_l}}{d\xi^{m_l}}P_l(\xi) \tag{1.90}$$

where $P_l(\xi)$ is the **Legendre polynomial function**, which in turn is a solution of the standard differential equation

$$\frac{d}{d\xi}\left[(1-\xi^2)\frac{dP_l}{d\xi}\right]+\left[l(l+1)-\frac{m_l^2}{1-\xi^2}\right]P_l=0 \tag{1.91}$$

where $m_l=0$. For nonzero m_l, however, Eq. (1.91) provides acceptable solutions as long as $|m_l|\leqslant l$. In like manner, the normalized solution of Eq. (1.86) is given by

$$R_{nl}(r)=-\left\{\left(\frac{2Z}{nr_1}\right)^3\frac{(n-l-1)!}{2n[(n+l)!]^3}\right\}^{1/2}e^{-\rho/2}\rho^l L_{n+l}^{2l+1}(\rho) \tag{1.92}$$

where

$$\rho = \frac{2Z}{nr_1}r = \frac{2Zr}{n}\left(\frac{m_e e^2}{4\pi\hbar^2\varepsilon_0}\right) = \frac{Zm_e e^2 r}{2\pi\varepsilon_0\hbar^2}\left(\frac{1}{n}\right) \tag{1.93}$$

r_1 is the radius of the first Bohr orbit of the hydrogen atom, as described by the old quantum theory of Eq. (1.41), n is any integer, and $L_{n+l}^{2l+1}(\rho)$, commonly known as the **associated Laguerre polynomial**, is

$$L_q^p(\rho) = \frac{d^p}{d\rho^p}L_q(\rho) \tag{1.94}$$

where the **Laguerre polynomial** $L_q(\rho)$ is expressed in terms of the Rodrigues formula,

$$L_q(\rho) = e^p\frac{d^q}{d\rho^q}(e^{-\rho}\rho^q) \tag{1.95}$$

A rigorous analysis will show that for the solution of Eq. (1.92) to be valid, the condition

$$\frac{\rho}{r} = \frac{\sqrt{8m_e|E|}}{\hbar} \tag{1.96}$$

must be met. An additional requirement for the validity of Eq. (1.92) is that the total quantum number n must be equal to or greater than $l+1$. This latter constraint allows for an acceptable radial solution whose dominating radial dependence basically comes from the exponential factor. Consequently, for the case of hydrogen at the ground state (i.e., when $Z=1$ and $n=1$), the electron is confined within the classically allowed region ($r<r_1$) even though the tail of exponential distribution appreciably penetrates the potential barrier. With a larger value of Z, the electron is generally confined to a narrower region, and with increasing n the electron is likely to be located over a wider range of r. Incorporation of Eq. (1.93) into Eq. (1.96) reveals that the total energy E of the electron can have only discrete values as given by

$$E_n = -\frac{Zm_e e^4}{32\hbar^2\pi^2\varepsilon_0^2}\left(\frac{1}{n^2}\right) \tag{1.97}$$

where $n = 1, 2, 3, \ldots$. It is interesting to note that for $Z=1$, Eq. (1.97) and the semiclassical discrete energy value as given by Eq. (1.57) are identical. Equation (1.87) is correct as long as m_l is an integer, positive or negative, since as understood from Fig. 1.10, $\Phi(\phi) = \Phi(\phi + 2p\pi)$ where p is any integer. The constant m_l is often referred to as the magnetic quantum number. On the other hand, Eq. (1.88) has a valid solution as long as the absolute value of m_l does not exceed the orbital quantum number l.

The energy values as per Eq. (1.97) depend only on n and so are degenerate with respect to both l and m_l. The three quantum numbers n, l, and m_l are interrelated according to the following selection rules:

$$n = 1, 2, 3, 4, \ldots$$
$$l = 0, 1, 2, 3, \ldots, n-1 \tag{1.98}$$
$$m_l = 0, \pm 1, \pm 2, \ldots, \pm l$$

The set of these quantum numbers is used to specify the energy levels of an atom. For any given energy, as specified by n, there exists a range of allowable discrete values for the total angular momentum l as determined by Eq. (1.98). Given an l, there is then a range of discrete values of the z component of angular momentum as specified by m_l. Note that a rigorous treatment of angular momentum would further reveal that, in addition to the aforementioned three quantum numbers, there exists the need for an additional quantum number, referred to as the spin angular momentum number m_s. This fourth quantum number is considered essential in explaining the existence of fine structures in atomic spectral lines. The electron possesses an intrinsic angular momentum independent of any orbital angular momentum it might have had and, associated with this angular momentum, a certain magnetic moment. The allowed value for $m_s = \pm 1/2$ for an electron. Consequently, the total degeneracy of the energy level E_n is

$$2\sum_{l=0}^{n-1}(2l+1) = 2\left[2\frac{n(n-1)}{2} + n\right] = 2n^2 \tag{1.99}$$

The well-known **Pauli's exclusion principle** governs the configuration of multielectron atoms. It states that no two electrons in an atom can exist in the same quantum state. Each electron in an atom, therefore, must have a unique set of quantum numbers, n, l, m_l, and m_s. A system of electrons in an atom is most stable when its total energy is at a minimum. However, the exclusion principle limits all of the electrons from taking the same quantum state. While the electrons of a nonhydrogen atom are expected to interact directly with one another, it is quite reasonable to model each of the electrons as though it were experiencing a constant field. However, this constant field would have to account for the fraction of the actual field that is being shielded by those other electron(s) that are closer to the heavy nucleus.

All electrons having the same total quantum number are approximately the same distance away from the nucleus and, therefore, have nearly the same energies and interact with almost the same field. Thus electrons having the same n are said to occupy the same atomic shell. Shells corresponding to $n = 1, 2, 3, 4, 5, \ldots$ are generally designated as K, L, M, N, O, ..., respectively. The energy of an electron also depends a little on l as well since l often determines the characteristics of electron probability distribution function. Electrons having smaller l are found comparatively closer to the nucleus and thus have a comparatively higher binding energy. Electrons that share the same value of l in a shell are said to be the members of the same subshell. The subshells corresponding to $l = 0, 1, 2, 3, 4, \ldots$ are conventionally represented by

TABLE 1.2

Symbolic Representation of Atomic States

	$l=0$	$l=1$	$l=2$	$l=3$	$l=4$	$l=5$
$n=1$	1s					
$n=2$	2s	2p				
$n=3$	3s	3p	3d			
$n=4$	4s	4p	4d	4f		
$n=5$	5s	5p	5d	5f	5g	
$n=6$	6s	6p	6d	6f	6g	6h

s, p, d, f, ..., respectively. The subshell electrons have approximately identical energies since the effects of m_l and m_s on electron energy are somewhat insignificant.

The electron configuration is often designated by a series of numerals (corresponding to n), each followed by a letter (corresponding to l) as illustrated in Table 1.2. A numeral superscript is sometimes placed next to l to indicate the total number of subshell electrons. For example, the electronic configuration of Chlorine (Cl), having the atomic number 17, is expressed as $1s^2 2s^2 2p^6 3s^2 3p^5$. We discuss the significance of these characterizations in Chapter 3 when we classify the elements into conductors, semiconductors, and insulators.

1.5 Electrons in Potential Barriers

In the previous section, we reviewed the quantum characteristics of electrons in an atom. In the current section, however, we intend to study the behavior of electrons in the absence *and* in the presence of potential distributions of various kinds. Such an exploration is essential to understanding the intricacies of both free and bound electrons. The relevant concept will then be used to understand the behaviors of semiconductors, which in turn will lead to an understanding of different electro-optic devices.

The formulation and solution of the Schrödinger equation is often the first vital step to understanding electronic behaviors. As we demonstrated earlier, the potential characteristic in question has to be introduced in the Schrödinger equation so that the resulting solutions can identify the behaviors of the involved electron(s). In the process, however, we must make sure to satisfy the normalization condition so that the resulting wave function is single-valued and its real part is finite for all values of position. Furthermore, for finite potential energies, the derivative of the wave function must be continuous so as to be applicable to regions of potential discontinuities.

1.5.1 Free Electrons

An electron of mass m_e is referred to as free when it has been subjected to a potential field $V = 0$. A free electron is generally unconfined and is allowed to move freely. For simplicity, let us assume that it moves along the positive x direction only. The time-independent Schrödinger equation of Eq. (1.72) may be used to describe such a motion as

$$\frac{d^2\Psi(x)}{dx^2} + \frac{2m_e}{\hbar^2} E\Psi(x) = 0 \tag{1.100}$$

The solution of Eq. (1.100) is reasonably straightforward and is given by

$$\Psi(x) = Ae^{jkx} \tag{1.101}$$

where k is defined as

$$k = \frac{\sqrt{2m_e E}}{\hbar} \tag{1.102}$$

But the total energy E of the free electron is $p^2/2m_e$ where p is its momentum. Consequently,

$$k = \frac{p}{\hbar} = \frac{2\pi}{\lambda} \tag{1.103}$$

using Eq. (1.52). We find that k, often called the **wave vector**, is inversely proportional to the wavelength λ.

The wave function $\Psi(x, t)$ of the free particle can be obtained now using Eq. (1.70):

$$\Psi(x, t) = Ae^{jkx}e^{-jEt/\hbar} = Ae^{j(kx-\omega t)} \tag{1.104}$$

where ω is the angular frequency. It follows from Eq. (1.102) that

$$E = \frac{\hbar^2 k^2}{2m_e} \tag{1.105}$$

Because we did not take boundary conditions into consideration when we obtained the plane wave solution of Eq. (1.104), the energy, as described by Eq. (1.105), can take any positive value. Free electrons are, therefore, not constrained in any way as to what energy values they may possess. Such an energy distribution is often referred to as **continuum energy**. Note, however, that each value of E is associated with two possible values for the momentum.

1.5.2 Electrons in Finite Potential Wells

We now consider the case of a bound electron traveling along a path (the x axis, for example). For the sake of illustration, we assume that the electron encounters

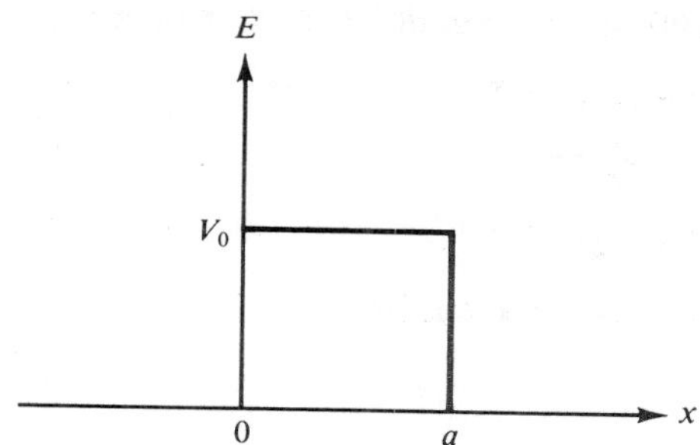

FIGURE 1.11

A finite potential well

a finite potential barrier that is described as follows:

$$V(x) = \begin{cases} V_0 & 0 \leqslant x \leqslant a \\ 0 & x < 0 \text{ and } x > a \end{cases} \tag{1.106}$$

where a is the width of the potential well as shown in Fig. 1.11. This potential distribution is different from that of the last subsection since the electron may now be reflected, partially or totally, by the potential well and, thus, the electron may move along the positive as well as along the negative x direction. The Schrödinger equation describing the given potential distribution may now be written as

$$\frac{d^2\Psi(x)}{dx^2} + \frac{2m_e}{\hbar^2} E\Psi(x) = 0 \qquad \text{for } x \leqslant 0 \text{ and } x \geqslant a \tag{1.107a}$$

and

$$\frac{d^2\Psi(x)}{dx^2} + \frac{2m_e}{\hbar^2}(E - V_0)\Psi(x) = 0 \qquad \text{for } 0 \leqslant x \leqslant a \tag{1.107b}$$

Classical physics is capable of characterizing only those electrons whose total energy is greater than their potential energy. Accordingly, the classical treatment of such an electron results in a plane wave solution for its wave function. However, in the context of quantum mechanics, a particle is also allowed to have a total energy that is less than its potential energy. Consequently, there are two distinct cases for the potential distribution under consideration: (a) $E < V_0$ and (b) $E > V_0$. These two cases are sequentially treated next.

CASE 1. $E < V_0$:

Equation (1.107) can be solved in the same manner as that employed in solving Eq. (1.100). The general solutions corresponding to the three regions (left, central, and right) of the potential well are, respectively, obtained as follows:

$$\Psi_l(x) = A_1 e^{jk_1x} + A_2 e^{-jk_1x} \tag{1.108a}$$

$$\Psi_c(x) = B_1 e^{k_2x} + B_2 e^{-k_2x} \tag{1.108b}$$

$$\Psi_r(x) = C_1 e^{jk_1x} + C_2 e^{-jk_1x} \tag{1.108c}$$

where the wave numbers are given by

$$k_1 = \frac{\sqrt{2m_e E}}{\hbar} \tag{1.109a}$$

for $x \leqslant 0$ and $x \geqslant a$, and

$$k_2 = \frac{\sqrt{2m_e(V_0 - E)}}{\hbar} \tag{1.109b}$$

for the central region ($0 \leqslant x \leqslant a$). For simplicity, we consider the electron to be incident from the left and, therefore, no wave will be traveling to the left in the region defined by $x \leqslant 0$. Accordingly, C_2 is set to 0. This leaves five coefficients, A_1, A_2, B_1, B_2, and C_1, yet to be determined in Eq. (1.108). In this evaluation process, we can make use of both the boundary conditions and the normalization condition. They are, respectively,

$$\Psi_l(0) = \Psi_c(0) \tag{1.110a}$$

$$\Psi_c(a) = \Psi_r(a) \tag{1.110b}$$

$$\left.\frac{d\Psi_l}{dx}\right|_{x=0} = \left.\frac{d\Psi_c}{dx}\right|_{x=0} \tag{1.110c}$$

$$\left.\frac{d\Psi_c}{dx}\right|_{x=a} = \left.\frac{d\Psi_r}{dx}\right|_{x=a} \tag{1.110d}$$

and

$$\int_{-\infty}^{\infty} |\Psi(x)|^2 \, dx = 1 \tag{1.110e}$$

The first four conditions result in

$$A_1 + A_2 = B_1 + B_2 \tag{1.111a}$$

$$B_1 e^{k_2a} + B_2 e^{-k_2a} = C_1 e^{jk_1a} \tag{1.111b}$$

$$jk_1A_1 - jk_1A_2 = k_2B_1 - k_2B_2 \tag{1.111c}$$

and

$$k_2B_1e^{k_2a} - k_2B_2e^{-k_2a} = jk_1C_1e^{jk_1a} \tag{1.111d}$$

An algebraic manipulation of the preceding equations yields

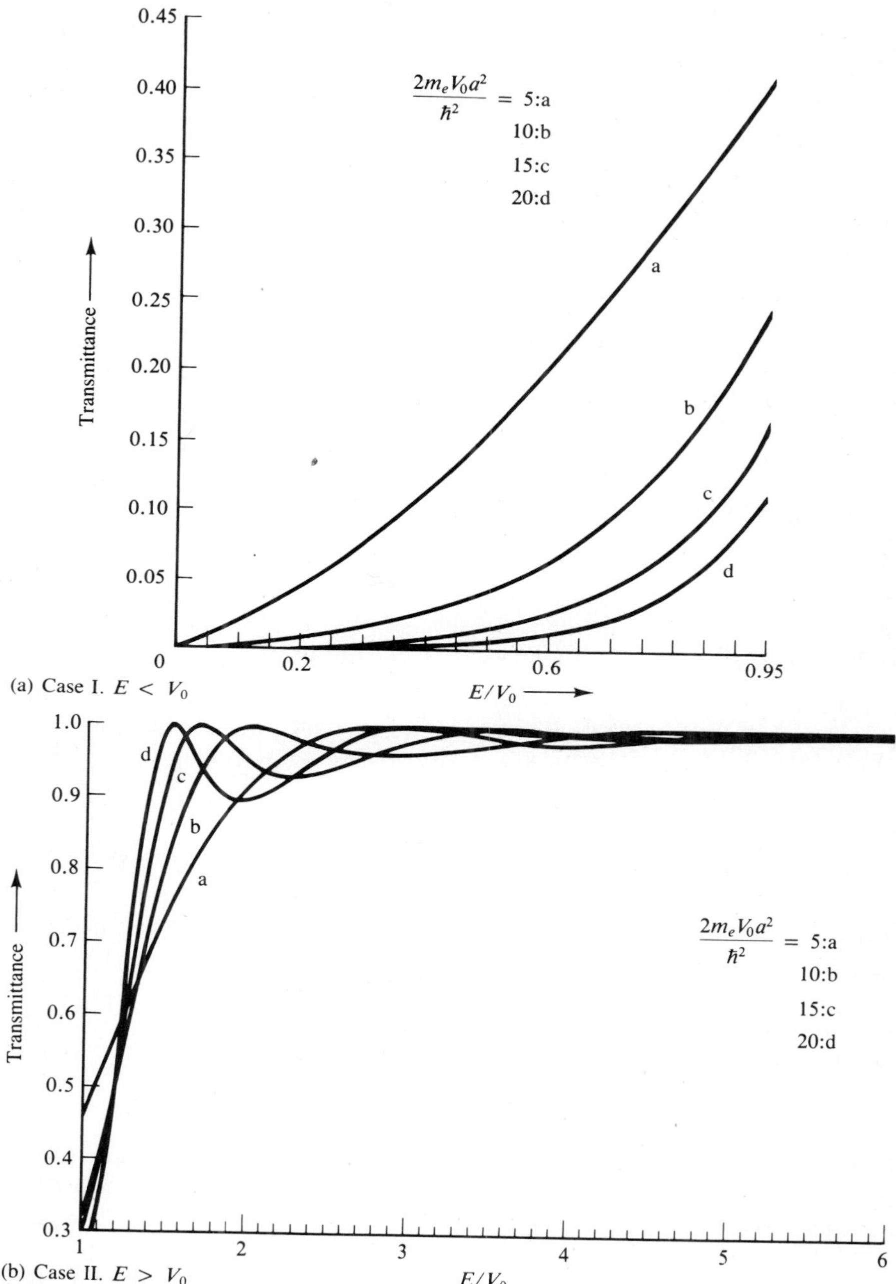

(a) Case I. $E < V_0$

(b) Case II. $E > V_0$

FIGURE 1.12

Transmission coefficient characteristics of a finite potential well for Cases I and II

$$\frac{C_1}{A_1} = \frac{e^{-jk_1 a}}{\cosh(k_2 a) - j\dfrac{k_1^2 - k_2^2}{2k_1 k_2}\sinh(k_2 a)} \tag{1.112}$$

The transmission coefficient can now be defined as the ratio of the intensity of the transmitted probability flux (into the region $x > a$) and the intensity of the incident probability flux (from the region $x < 0$). The ratio C_1/A_1 is very important because it can be used to derive the transmission coefficient T. For the case of $E < V_0$, T is obtained as

$$T = \frac{C_1 C_1^*}{A_1 A_1^*} = \frac{4k_1^2 k_2^2}{4k_1^2 k_2^2 + (k_1^2 + k_2^2)^2 \sinh^2(k_2 a)} \tag{1.113}$$

It can be rewritten using Eqs. (1.109a) and (1.109b) as

$$T = \frac{4\left(\dfrac{E}{V_0}\right)\left[1 - \left(\dfrac{E}{V_0}\right)\right]}{4\left(\dfrac{E}{V_0}\right)\left[1 - \left(\dfrac{E}{V_0}\right)\right] + \sinh^2\left[\dfrac{1}{\hbar}\left(2m_e a^2 V_0 \left\{1 - \left(\dfrac{E}{V_0}\right)\right\}\right)^{1/2}\right]} \tag{1.114}$$

CASE II. $E > V_0$:

For this classical case, Eqs. (1.108a) and (1.108c) are still valid solutions for the left and the right of the potential well, respectively. But for the central region ($0 \leqslant x \leqslant a$), Eq. (1.108b) has to be corrected by replacing k_2 with jk_3, where

$$k_3 = \frac{\sqrt{2m_e(E - V_0)}}{\hbar} \tag{1.115}$$

The transmission coefficient for the potential distribution under consideration ($E > V_0$) is evaluated to give

$$T = \frac{4\left(\dfrac{E}{V_0}\right)\left[\left(\dfrac{E}{V_0}\right) - 1\right]}{4\left(\dfrac{E}{V_0}\right)\left[\left(\dfrac{E}{V_0}\right) - 1\right] + \sin^2\left[\dfrac{1}{\hbar}\left\{2m_e a^2 V_0 \left\{\left(\dfrac{E}{V_0}\right) - 1\right\}\right\}^{1/2}\right]} \tag{1.116}$$

It will be useful to define the reflection coefficient as a ratio of the intensity of the reflected probability flux and the intensity of the incident probability flux. The reflection coefficient R is given by

$$R = \frac{A_2 A_2^*}{A_1 A_1^*} \tag{1.117}$$

The transmission coefficient is plotted versus E/V_0 for a variety of values of $2m_e V_0 a^2/\hbar^2$, as shown in Fig. 1.12. The transmission coefficient is nonzero

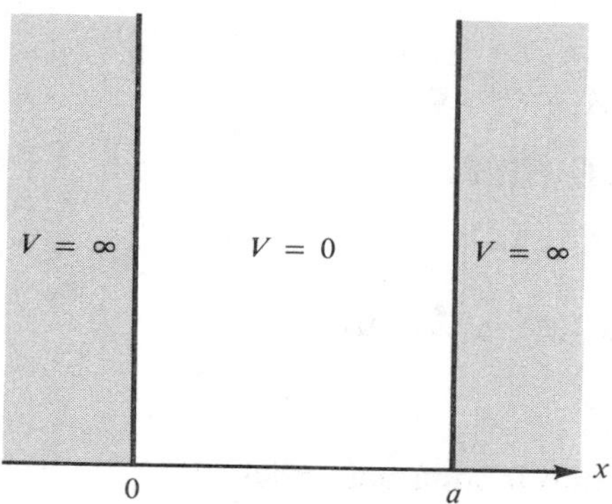

FIGURE 1.13

An infinite potential well

when E is less than V_0 and is nonunity when E is larger than V_0. Our findings thus contradict the expectations of classical physics. The quantum-mechanical conclusion pertaining to the finite potential barrier suggests that an electron can occasionally be found outside the potential well. This characteristic of electrons is referred to as the **tunneling effect**. As we show later, the tunneling effect explains the characteristics of many important devices of significance, such as that of the tunnel diode. The probability of locating the electron outside the potential well increases as E approaches V_0. The probability of locating the electron, however, reaches a maximum when $E = V_0$. Furthermore, the electrons are reflected when $E > V_0$.

1.5.3 Electrons in Infinite Potential Wells

An interesting case is when the electron is subjected to an infinite square potential well. The potential energy is zero in the inside of the said well (of width a) while it is infinite outside the well. In mathematical terms,

$$V(x) = \begin{cases} 0 & 0 \leqslant x \leqslant a \\ \infty & \text{otherwise} \end{cases} \tag{1.118}$$

The corresponding potential well is shown in Fig. 1.13. The potential barriers are expected to forbid the electron from escaping the well. This fact implies that the wave function $\Psi(x)$ should be zero in the regions described by $x < 0$ and $x > a$. The Schödinger equation for the well region thus has the solution

$$\Psi(x) = A_1 e^{jkx} + A_2 e^{-jkx} \tag{1.119}$$

where k has already been described by Eq. (1.102). At the potential discontinuities, Eq. (1.119) gives

$$\Psi(0) = A_1 + A_2 = 0 \tag{1.120}$$

and

$$\Psi(a) = A_1 e^{jka} + A_2 e^{-jka} = 0 \tag{1.121}$$

Equation (1.120) readily yields

$$A_1 = -A_2 \tag{1.122}$$

and, therefore, Eq. (1.121) reduces to

$$\sin(ka) = 0 \tag{1.123}$$

Equation (1.123) in turn is valid only when

$$ka = n\pi \qquad n = 0, 1, 2, 3, \ldots \tag{1.124}$$

Consequently, Eq. (1.102) can be used to obtain the allowable energy values for the infinite potential well as given by

$$E_n = \frac{\hbar^2 \pi^2 n^2}{2m_e a^2} \tag{1.125}$$

The cases corresponding to both $n = 0$ and $A_1 = 0$ are excluded from our treatment as they both lead to $\Psi = 0$. The corresponding probability for finding the electron in the well is zero. Studies involving the infinite potential well problem reveal an important fact: that the energy for the bound electrons can only be discrete, as dictated by Eq. (1.125). It therefore follows that the energy which is emitted or absorbed during the transition of states is also discrete. In comparison, a free electron can assume any energy value.

1.5.4 Electrons in Periodic Potential Wells

In this section, we intend to characterize electrons that are present within a crystalline structure. It is well known that crystalline atoms are arranged in a regular geometric pattern. An electron passing by an atom in the crystal is first subjected to an acceleration and then to a deacceleration, and this acceleration-deacceleration cycle is repeated as the electron gets into and out of the field of the next atom. Thus, to approximately describe a crystal field, we consider a periodic arrangement of potential wells and potential barriers, as shown in Fig. 1.14. Such a periodic potential distribution is often characterized by its lattice constant, l, which for the case of Fig. 1.14 is equal to $a + b$. This picture is an oversimplification of the true nature of a crystal field, but it is enough for our purpose. This model is commonly known as the one-dimensional **Krönig-Penney model**. It does not consider the fact that the electrons are strongly bound to the core or that the individual potential distributions tend to overlap. Yet, this simplistic model is able to give some reasonable insights into the behavior of electrons in crystals.

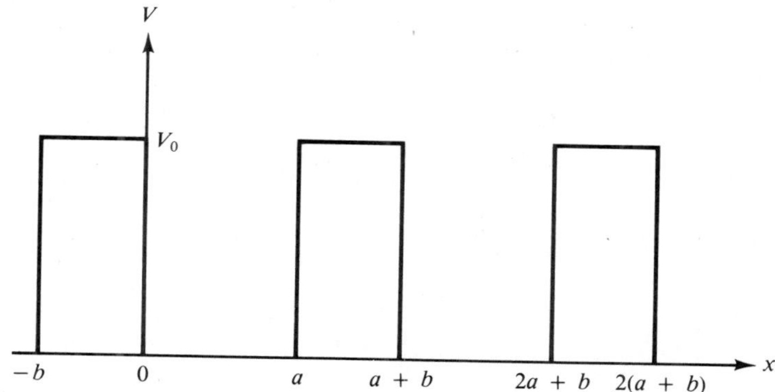

FIGURE 1.14

A periodic potential well

The one-dimensional Schrödinger equations for such a potential distribution are given by

$$\frac{d^2\Psi}{dx^2} + \frac{2m_e}{\hbar^2} E\Psi = 0 \tag{1.126}$$

for regions where $V = 0$, and

$$\frac{d^2\Psi}{dx^2} + \frac{2m_e}{\hbar^2}(E - V_0)\Psi = 0 \tag{1.127}$$

for regions where $V = V_0$. It was shown by Bloch that the nature of a solution of this type of differential equation is given by

$$\Psi(x) = g(x)e^{jkx} \tag{1.128}$$

where $g(x)$ is a periodic function whose period is equal to the crystal lattice so that

$$g(x) = g(x + l) = g(x + 2l) = \dots \tag{1.129}$$

and k is the wave vector having a real value. It can be shown that two distinct values of k exist for each of the allowed values of energy in a one-dimensional potential well. Using Eq. (1.128), Eqs. (1.126) and (1.127) may be rewritten as

$$\frac{d^2g(x)}{dx^2} + 2jk\frac{dg(x)}{dx} + (k_1^2 - k^2)g(x) = 0 \tag{1.130}$$

and

$$\frac{d^2g(x)}{dx^2} + 2jk\frac{dg(x)}{dx} - (k_2^2 + k^2)g(x) = 0 \tag{1.131}$$

The parameters k_1 and k_2 were defined earlier in Eqs. (1.109a) and (1.109b), respectively. The differential equations given in Eqs. (1.130) and (1.131) are both comparable to the well-known differential equation for damped oscillation:

$$\frac{d^2f}{dx^2} + 2a\frac{df}{dx} + bf = 0 \tag{1.132}$$

whose solution is

$$f = [D_1 e^{jsx} + D_2 e^{-jsx}]e^{-ax} \tag{1.133}$$

where

$$s = (b - a^2)^{1/2} \tag{1.134}$$

and D_1 and D_2 are two arbitrary constants. Consequently, both Eqs. (1.130) and (1.131) can be solved by analogy to give

$$g(x) = \Lambda_1 e^{j(k_1 - k)x} + \Lambda_2 e^{-j(k_1 + k)x} \tag{1.135a}$$

and

$$g(x) = \Lambda_3 e^{-(k_2 + jk)x} + \Lambda_4 e^{(k_2 - jk)x} \tag{1.135b}$$

where Λ_1, Λ_2, Λ_3, and Λ_4 are four arbitrary constants. However, since the wave function Ψ and its derivative $d\Psi/dx$ are continuous at $x = 0$, we find that

$$\Lambda_1 + \Lambda_2 = \Lambda_3 + \Lambda_4 \tag{1.136}$$

and

$$j\Lambda_1(k_1 - k) - j\Lambda_2(k_1 + k) = -\Lambda_3(k_2 + jk) + \Lambda_4(k_2 - jk) \tag{1.137}$$

respectively, from the continuity of g and dg/dx at $x = 0$. Again both $g(x)$ and dg/dx are also expected to be periodically continuous. This requirement of the periodic continuity of $g(x)$, for example, implies that Eq. (1.135a), when evaluated at $x = a$, should be identical to Eq. (1.135b), when evaluated at $x = -b$. Thus

$$\Lambda_1 e^{j(k_1 - k)a} + \Lambda_2 e^{-j(k_1 + k)a} = \Lambda_3 e^{(jk + k_2)b} + \Lambda_4 e^{(jk - k_2)b} \tag{1.138}$$

Similarly, from the periodic continuity of the derivative dg/dx, we get

$$j\Lambda_1(k_1 - k)e^{j(k_1 - k)a} - j\Lambda_2(k_1 + k)e^{-j(k_1 + k)a} = -\Lambda_3(k_2 + jk)e^{(jk + k_2)b} + \Lambda_4(k_2 - jk)e^{(jk - k_2)b} \tag{1.139}$$

Equations (1.136) through (1.139) are used to determine the constants Λ_1, Λ_2, Λ_3, and Λ_4. By eliminating the constants, we obtain the transcendental equation as given by

$$\frac{k_2^2 - k_1^2}{2k_1k_2}\sinh(k_2 b)\sin(k_1 a) + \cosh(k_2 b)\cos(k_1 a) = \cos[k(a + b)] \tag{1.140}$$

which can give information about energy since the quantity k_1 is defined in terms of energy.

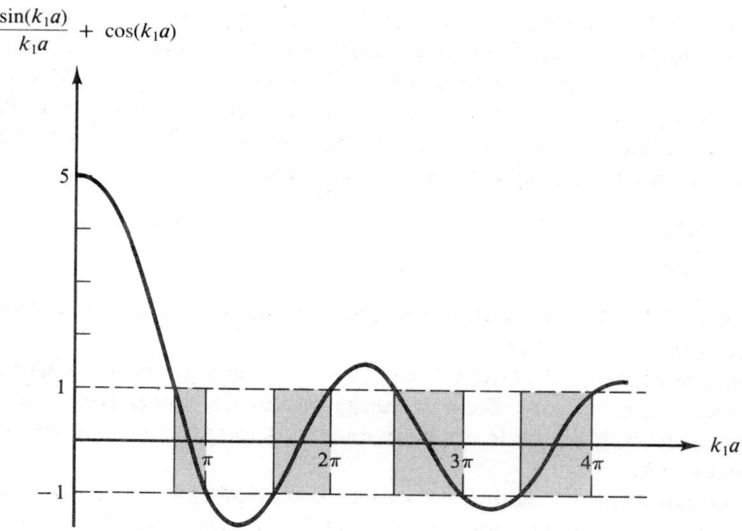

FIGURE 1.15

The allowed energy band characteristics in a periodic potential distribution

To realize a meaningful graphic solution from Eq. (1.140), the potential barrier width b is reduced while at the same time the potential height V_0 is increased such that the product of the two $V_0 b$ is constant. Thus, as k_2 approaches infinity, Eq. (1.140) reduces to

$$\Omega \frac{\sin(k_1 a)}{k_1 a} + \cos(k_1 a) = \cos(ka) \tag{1.141}$$

where

$$\Omega = \frac{m_e a V_0 b}{\hbar^2} \tag{1.142}$$

The quantity Ω is often referred to as the **scattering power of the barrier** because it identifies the affinity with which electrons are tied to the crystal ions. It is obvious that k_1 is constrained by Eq. (1.141), and consequently the electrons of a crystalline solid are allowed to have only specific values of energy.

Figure 1.15 shows the plot of the left-hand side of Eq. (1.141) against $k_1 a$ when, for example, $\Omega = 4$. The right-hand side of the equation, however, limits the extent of the function defined by the left-hand side since $\cos(ka)$ is bounded between 1 and -1, except when k is imaginary. The allowed regions are indicated in the figure as shaded areas. An electron in a periodically varying potential well occupies only certain allowed ranges of energy levels. This observation gives rise to the concept of **bands of energy**, whereby electrons are forbidden to have energies outside of these allowed zones. With increasing $k_1 a$ or Ω, the forbidden bands become narrower. On the other hand, the allowed bands become narrower with larger $V_0 b$. In fact, as Ω (or $V_0 b$) approaches infinity, the permitted energy bands reduce to discrete energy states similar to those encountered in Section 1.5.3. As $V_0 b$ (or when Ω) becomes smaller and smaller, k_1 approaches k and consequently it leads to

$$E = \frac{\hbar^2 k^2}{2m_e} \tag{1.143}$$

Note that Eq. (1.143) is no different from Eq. (1.105) which was derived for the familiar case of free electrons.

In summary, free electrons can take any energy value whereas bound electrons can take only discrete energy values. In comparison, the electron present in crystalline solids can have energy as long as its value lies within the allowed bands.

For clarity of discussion, the energy is plotted against the wave vector, as shown in Fig. 1.16. The allowed energy bands are shaded in the figure. In particular, these ranges are referred to as the **Brillouin zones**. The characteristics of a free electron, for comparison, are indicated in the figure by the dotted parabolic curve. Note that this particular curve is symmetric in k such that there are two and only two k values associated with each of the allowable energies. Furthermore, the energy discontinuities are found to occur at values of $k = n\pi/l$ where n is a nonzero integer. The solid E-k curve represents the allowed energy bands. The discontinuities present in the diagram correspond to certain energy regions, already referred to as the forbidden regions, where the probability of finding an electron is zero. In the diagram, the first Brillouin zone corresponds to the range $-\pi/l < k < \pi/l$, the second Brillouin zone corresponds to the range $\pi/l < |k| < 2\pi/l$, and so on. A feature to be pointed out is that dE/dk is zero at all of the Brillouin zone boundaries.

The motion of a crystalline electron is often determined using the equation of motion of a free electron. But to provide correct results, the equation of motion needs to be corrected. This is commonly accomplished when the rest mass of the electron is replaced with a quantity referred to as its **effective mass**. It can be shown that the slope of the energy band is directly usable in calculating the effective mass.

An electron moving with a velocity v is often represented by a moving group of waves, known also as a wave packet. Each of the waves of this group is characterized by a slightly different frequency. The velocity with which the center of the wave packet moves is referred to as the **group velocity**, and it is given by

$$v_g = \frac{d\omega}{dk} \tag{1.144}$$

where ω is the center frequency and k is the center wave number. While the

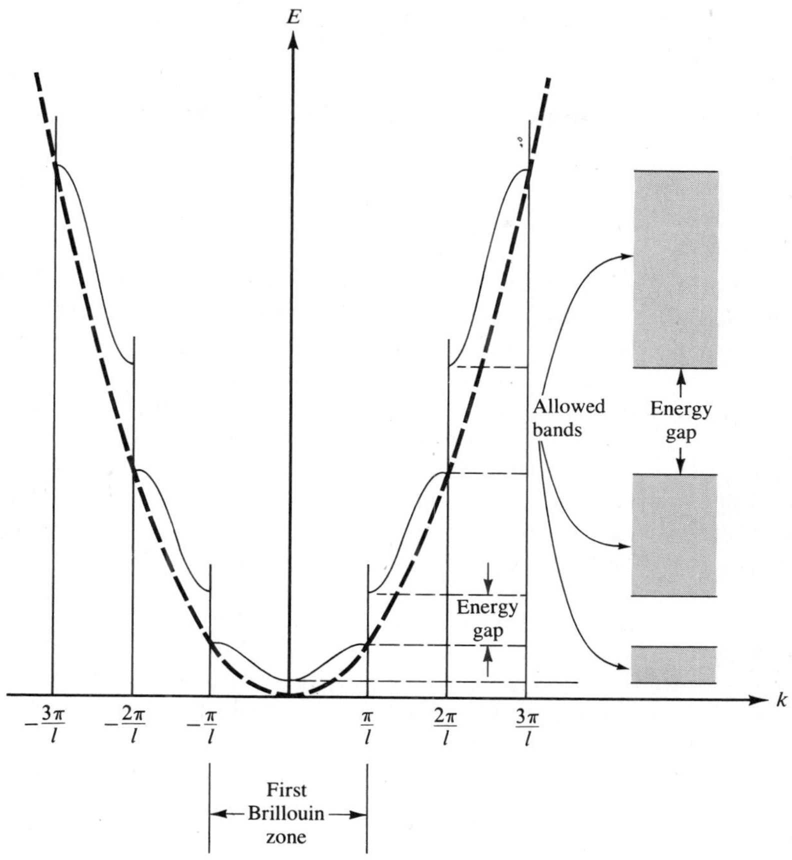

FIGURE 1.16

Brillouin zones in a periodic potential distribution

group velocity describes the velocity of the wave packet, the phase velocity v_p describes the velocity of the wave and is defined as $v_p = \omega/k$.

We can relate the increase in energy dE present in the wave packet with an externally applied force F by

$$F = \frac{1}{v_g}\frac{dE}{dt} = \frac{1}{v_g}\frac{dE}{dk}\frac{dk}{dt} = \left(\frac{1}{\hbar}\frac{dE}{dk}\right)^{-1}\frac{dE}{dk}\frac{dk}{dt} = \hbar\frac{dk}{dt} \tag{1.145}$$

since $E = \hbar\omega$. But

$$\frac{dv_g}{dt} = \frac{1}{\hbar}\frac{d}{dt}\left(\frac{dE}{dk}\right) = \frac{1}{\hbar}\frac{d^2E}{dk^2}\frac{dk}{dt} \tag{1.146}$$

Consequently, Eq. (1.145) can be rewritten as

$$F = m^* \frac{dv_g}{dt} \tag{1.147}$$

where

$$m^* = \frac{\hbar^2}{\dfrac{d^2E}{dk^2}} \tag{1.148}$$

Equation (1.147) is essentially the same as the equation that characterizes Newton's second law of motion. Therefore, the significance of Eq. (1.147) in our study of electro-optic devices is remarkable. It implies that a minimal knowledge of quantum mechanics will be sufficient for the study of basic electro-optic phenomena because an electron can be treated simply as a semiclassical particle as long as we have accounted for its mass by the quantity of Eq. (1.148). This quantity is referred to as the effective mass of an electron. It is positive near the band-energy minimum and negative near the band-energy maximum. It is also noteworthy that an electron whose energy corresponds to either the top or the bottom of a band has an effective mass which is energy independent.

Problems

1.1 Obtain $\nabla^2\mathbf{H} - (1/c^2)\,\partial^2\mathbf{H}/\partial t^2 = 0$ from Maxwell's equation.

1.2 For many dielectric materials, the polarization vector **P** is not collinear with **E**. Show that the wave vector **k** is perpendicular to the plane of **D** and **B**, and also show that its magnitude is given by $k^2 = [\omega^2\mu_0(\mathbf{D}\cdot\mathbf{D})/(\mathbf{E}\cdot\mathbf{D})]$. Show that the Poynting vector could have a direction different from that of the wave vector.

1.3 Show that the relative phase change, $\Delta \equiv \delta_\pi - \delta_\sigma$, for the case of total internal reflection is given by

$$\tan\left(\frac{\Delta}{2}\right) = \cos\theta_i \frac{\sqrt{\sin^2\theta_i - n_{21}^2}}{\sin^2\theta_i}$$

where θ_i is the angle of incidence, n_{21} is the relative refractive index, and δ_σ and δ_π are the phase changes for the TE and TM cases, respectively.

1.4 Determine the coefficient of transmission (analogous to Eqs. 1.36a and 1.36b) and its square, i.e., transmittance, for both TE and TM cases, in terms of the incident angle and the relative refractive index, at a planar interface. Show further that the sum of reflectance and transmittance is unity.

1.5 What is the atomic radius according to the classical model of the hydrogen atom? Verify that the corresponding atomic circumference is equal to the de Broglie wavelength of the electron.

1.6 Calculate the de Broglie wavelength for (a) a 1 eV and a 1 MeV electron; (b) a 1 eV and a 1 MeV proton; and (c) a 5 kg metal ball with a velocity of 10 m/sec.

1.7 Derive Eq. (1.79) from Eq. (1.74).

1.8 Derive Eq. (1.82) from Eq. (1.80).

1.9 Obtain Eq. (1.112) from Eqs. (1.111a) through (1.111d).

1.10 Evaluate the reflection coefficient of Eq. (1.117).

1.11 Determine both the wave function and the corresponding probability function for an electron that is in an infinite square well for the smallest three values of n. Also plot both of these functions.

1.12 Determine the constants of Eqs. (1.135a) and (1.135b).

1.13 Obtain Eq. (1.140).

1.14 Plot $\Omega(\sin k_1 a)/k_1 a + \cos(k_1 a)$ versus $k_1 a$ for different values (large and small) of Ω. *Note:* Use Eq. (1.141).

1.15 Show that the phase velocity can exceed the speed of light. Justify its apparent contradiction with the theory of relativity.

1.16 Calculate the expected values of (a) x, $\langle x \rangle$; (b) x^2, $\langle x^2 \rangle$; (c) p, $\langle p \rangle$; and (d) p^2, $\langle p^2 \rangle$ for an electron in an infinite potential well. Compare $\langle p^2 \rangle / 2m_e$ to the discrete energy level E_n. Note that

$$\langle \sigma \rangle \equiv \int_{-\infty}^{\infty} \Psi^* \sigma \Psi \, dx$$

where σ is defined only in x.

1.17 Calculate $\langle x \rangle$ and $\langle x^2 \rangle$ for the electron in the finite potential well as described by Eq. (1.106).

References and Suggested Readings

Colclaser, R. A., and S. Diehl-Nagle. *Materials and Devices for Electrical Engineers and Physicists.* New York: McGraw-Hill, 1985.

Fowles, G. R. *Introduction to Modern Optics* (2nd ed.). New York: Holt, Rinehart & Winston, 1975.

Fromhold, Jr., A. T. *Quantum Mechanics for Applied Physics and Engineering.* New York: Academic Press, 1981.

Hecht, E., and A. Zajac. *Optics.* Reading, MA: Addison-Wesley, 1979.

Hummel, R. E. *Electronic Properties of Materials: An Introduction for Engineers.* Berlin: Springer-Verlag, 1985.

Klein, M. V., and T. E. Furtak. *Optics* (2nd ed.). New York: John Wiley & Sons, 1986.

Paul, C. R., and S. A. Nasar. *Introduction to Electromagnetic Fields* (2nd ed.). New York: McGraw-Hill, 1987.

Pierret, R. F. *Advanced Semiconductor Fundamentals* (Vol. VI of Modular Series on Solid State Devices). Reading, MA: Addison-Wesley, 1987.

Seeger, K. *Semiconductor Physics: An Introduction* (3rd ed.). Berlin: Springer-Verlag, 1985.

Shen, L. C., and J. A. Kong. *Applied Electromagnetism* (2nd ed.). Boston: PWS-Kent Publishing Co., 1987.

Streetman, B. G. *Solid State Electronic Devices* (2nd ed.). Engelwood Cliffs, NJ: Prentice-Hall, 1980.

CHAPTER TWO

Optical Radiation

2.1 Introduction

Evaluation of the characteristics of an electro-optic system requires an effective means for quantitative measurement of optical radiation. Measurement must include a description of the light sources and optical systems that are instrumental in transferring light from one spatial location to another. There are inherent problems in such measurements, however, because there are two basic types of measurement units, physical and psychophysical, for quantifying radiation-related quantities.

Physical units deal with measurement of electromagnetic radiation, whether or not it is in the visible spectrum. On the other hand, the domain of psychophysical units does not extend beyond the visible spectrum. In spite of its limitation, the latter unit system continues to be used. Since light is basically a manifestation of an electromagnetic field to which an observer's eye responds through direct stimulation of the retina, the sensation of light can also be considered a psychophysical phenomenon.

Radiometry essentially deals with the measurement of quantities associated with radiant energy. **Photometry** is the study of the way in which light is transferred between spatial locations, and it involves the measurement of visually evaluated radiant energy. In the notations used in this text, the subscript e (for energy) is used to denote physical or radiometric units, while the subscript v (for visual) is used to denote psychophysical or photometric units. In practice, most electro-optic detectors of radiation operate in the realm of physical units only.

The use of photometric units of measure causes severe problems because measurement is based primarily on the spectral response of an "average" young

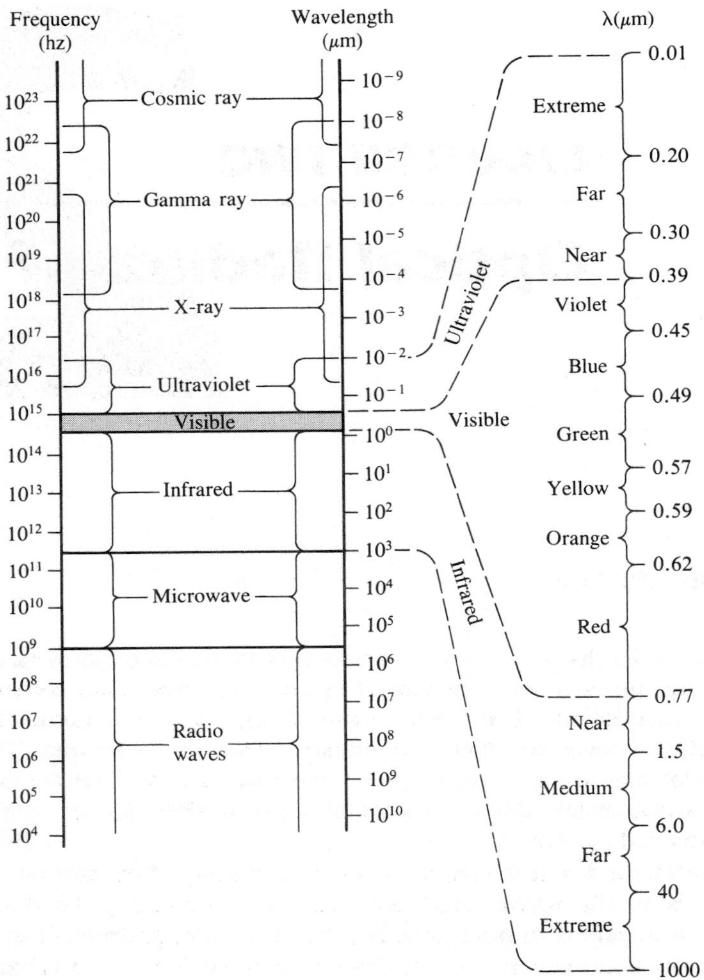

FIGURE 2.1

The electromagnetic spectrum

human observer. All photometric measurements are weighted by the spectral response of the "average" eye, which itself is a vague statistical quantity. In addition, many of the electro-optic devices operate at wavelengths well beyond the range of response of the human eye. Thus it would be meaningless indeed, for example, to describe an infrared device with photometric units. The total visible domain is insignificant in comparison to the whole electromagnetic spectrum, as shown in Fig. 2.1, which clearly illustrates the limitations of using photometric units. Further, photometric measurement is complicated by a profusion of redundant terms. This overabundance of terms has crippled the subject of photometry and made it much more difficult than it should be. In comparison, radiometric units provide unambiguous characterization of all detectors, including that of the human eye.

In spite of the clear advantage of using radiometric units over photometric units, both systems are used, often simultaneously, disregarding international efforts to standardize radiation measurement. The notation in this text follows the standard international (SI) symbols and units.

This chapter is directed toward assimilating both radiometric and photometric measurement systems and providing an essential understanding of all aspects of radiation, including its measurement. For simplicity, our treatment of topics is limited mostly to unpolarized, incoherent radiation.

2.2 Radiometric and Photometric Definitions

2.2.1 Energy and Flux

We have already seen in Chapter 1 that the time-averaged energy density due to an electromagnetic field, as given by Eqs. (1.25) and (1.26), is

$$U = \tfrac{1}{2}(\varepsilon_0 |\mathbf{E}|^2 + \mu_0 |\mathbf{H}|^2) \tag{2.1}$$

Most detectors are functionally effective only in a time-averaged sense because they are unable to follow the rapid fluctuation of the electromagnetic field in most instances.

Radiant energy Q_e is energy in joules (J) that is emitted, transferred or collected in a radiation process. It is the most fundamental radiometric quantity encountered in the measurement of electromagnetic radiation. It is possible to derive most of the other radiometric quantities from radiant energy. Note, however, that radiant energy has an understandable dependence on wavelength, time, and spatial coordinates.

The photometric equivalent of radiant energy is called **luminous energy**, expressed in lumen-seconds (lm-s), and is defined as visually evaluated radiant energy.

The radiant energy contained within a volume of space is generally specified by the **radiant energy density** U_e in joules per cubic meter (J/m^3) and defined as the energy density per unit volume by

$$U_e = \frac{\partial Q_e}{\partial V} \tag{2.2}$$

where dQ is the radiant energy contained in a volume element dV of the radiation field. The photometric equivalent of the radiant energy density is known as the **luminous energy** in lumen-seconds per cubic meter (lm-s/m^3), which in turn is defined as the luminous energy per unit volume.

Radiant flux or **radiant power** Φ_e, expressed in watts (W), is the time rate of change of radiant energy:

$$\Phi_e = \frac{\partial Q_e}{\partial t} \tag{2.3}$$

at which the radiated energy is exported from one region to another by the field. It is also equivalent to the closed surface integral of the Poynting vector given by

$$\Phi_e = \iint \mathbf{S} \cdot d\mathbf{A} \tag{2.4}$$

where $d\mathbf{A}$ is the differential surface area vector. Note that the Poynting vector $\mathbf{S}$ has already been defined in Eq. (1.27). **Luminous flux** Φ_v in lumens (lm) is the corresponding photometric unit and is defined as the time rate of change of luminous energy. Luminous energy and luminous flux are similarly related by

$$\Phi_v = \frac{\partial Q_v}{\partial t} \tag{2.5}$$

The interrelationship between the radiometric and the corresponding photometric quantities involves time, wavelength, and response of the "average" eye. For a given electromagnetic field, the radiometric and the corresponding photometric quantities have similar angular and areal characteristics.

All of the radiation quantities introduced so far are known as **nonspectral quantities** since they do not involve wavelength. On the other hand, there are many quantities that are not necessarily independent of wavelength. To correctly identify such characteristics, we need to make use of **spectral quantities**. When the nonspectral quantity is divided by the value of the wavelength (at which the measurement was made), we obtain the corresponding spectral quantity. For distinction, all of the spectral quantities are customarily written with subscript λ (in addition to either e or v).

Table 2.1 lists the energy and flux quantities. Spectral quantities are often used to evaluate the nonspectral radiation quantities. For example, if the

TABLE 2.1

Energy and Flux Symbols and Units

	Radiometric		*Photometric*	
	Nonspectral	*Spectral*	*Nonspectral*	*Spectral*
Energy	Q_e (J)	$Q_{e\lambda}$ (J/λ)	Q_v (lm-s)	$Q_{v\lambda}$ (lm-s/λ)
Flux	Φ_e (W)	$\Phi_{e\lambda}$ (W/λ)	Φ_v (lm)	$\Phi_{v\lambda}$ (lm/λ)

spectral radiant flux $\Phi_{e\lambda}$ is known (possibly by using a narrow-band spectroradiometer) over a range of λ, λ_1 through λ_2, the nonspectral radiant flux $\Phi_e(\lambda_1, \lambda_2)$ within that range can be evaluated using

$$\Phi_e(\lambda_1, \lambda_2) = \int_{\lambda_1}^{\lambda_2} \Phi_{e\lambda}(\lambda)\, d\lambda \tag{2.6}$$

Such nonspectral quantities are referred to as **broadband quantities**. However, the total radiant flux Φ_e is obtained only by evaluating the integral over the complete wavelength range, from zero through infinity.

Often the photon flux in an optical beam needs to be specified. Since energy per proton is hc/λ [Eq. (1.50)], all of the spectral quantities involving energy could be converted to equivalent photon flux quantities simply by dividing the corresponding radiant expression by hc/λ. Thus the photon flux in a range of wavelengths from λ to $\lambda + d\lambda$ is obtained as follows:

$$N(\lambda)\, d\lambda = \frac{\lambda}{hc} \Phi_{e\lambda}(\lambda)\, d\lambda \tag{2.7}$$

Note that the spectral density quantities can also be expressed in terms of frequency instead of wavelength. This alternate route follows according to the following relationship:

$$c = v\lambda \tag{2.8}$$

where v is the frequency of radiation.

2.2.2 Spatial Densities

Radiant intensity expressed in watts per steradian (W/sr) is defined as the radiant flux per unit solid angle emitted by an entire source in a given direction. The solid angle $d\Omega$ subtended by an area element dA_d at the source is dA_d/r^2 where r represents the distance between the source and the area element. The radiating geometry involving the solid angle is illustrated in Fig. 2.2. The intensity often is of interest, particularly when the source is located far away and when the detector aperture is too small. For a cone of rays making a solid angle $d\Omega$ about the axis of radiation propagation $\mathbf{r}$, the radiant intensity is given by

$$I_e = \frac{\partial \Phi_e}{\partial \Omega} = \frac{\partial^2 Q_e}{\partial t\, \partial \Omega} \tag{2.9}$$

The corresponding photometric term is known as **luminous intensity** I_v and is expressed in either lumens per steradian (lm/sr) or candela (cd). One candela is defined as equivalent to an intensity that is caused by a flux of 1 lm as it spreads through 1 sr.

Radiant intensity of a planar surface generally varies as the cosine of the angle that exists between the viewing direction and the surface normal. Surfaces for which such cosine dependence is true are generally called **Lambertian surfaces**. White paper is a perfect example of a Lambertian reflector when it is

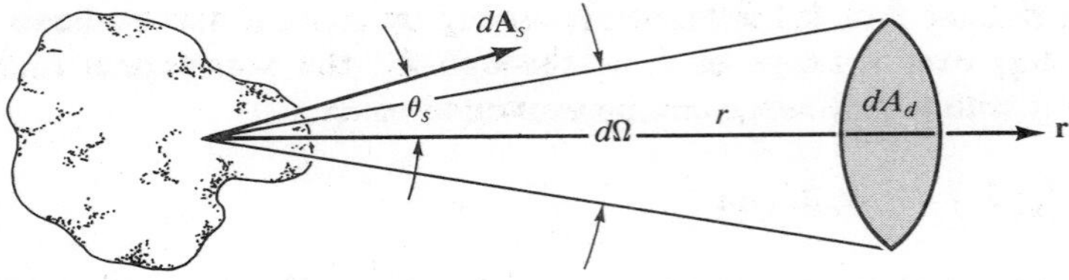

FIGURE 2.2
Radiating geometry

held at an angle of incidence other than 90°. Even though the solid angle changes as one looks at such a reflector from different angles, the brightness is invariant as long as the observer maintains a constant viewing distance.

Another quantity of interest is **radiance**, which describes the differential angular dependence of the radiant flux density. Radiance L_e, expressed in watts per square meter steradian (W/m^2 sr) at a given point and a given direction, is defined as the radiant flux per unit solid angle per unit projected area perpendicular to the propagation direction. If $d^2\Phi_e$ is the radiant flux due to a source that is propagating through a solid angle $d\Omega$, the radiance is defined as

$$L_e = \frac{\partial^2\Phi_e}{\cos\theta_s \partial\Omega \partial A_s} \tag{2.10}$$

where $d\mathbf{A}_s$ is the elemental area around the point and θ_s is the angle between $d\mathbf{A}_s$ and the direction of propagation, as shown in Fig. 2.2. Note that $dA_s \cos\theta_s$ is equivalent to the area projected by the source. Radiance can also be defined as radiant intensity per unit projected area:

$$L_e = \frac{1}{\cos\theta_s}\frac{\partial I_e}{\partial A_s} \tag{2.11}$$

using Eqs. (2.9) and (2.10). It turns out that for the Lambertian surfaces, radiance is independent of both position and viewing angle; i.e., $L_e(\theta) =$ constant. The photometric equivalent of radiance is known as **luminance**, and it is expressed in candela per square meter (cd/m^2) or nit (nt).

Consider the geometry of a detector of area dA_d that is facing a source in a manner so that the normal to the detector surface makes an angle of θ_d with the direction of propagation. In that case, the solid angle subtended by the detector is given by

$$d\Omega = \frac{dA_d \cos\theta_d}{r^2} \tag{2.12}$$

The radiance of the source is then obtained as

$$L_e = \frac{r^2\, d^2\Phi_e}{dA_s \cos\theta_s\, dA_d \cos\theta_d} \tag{2.13}$$

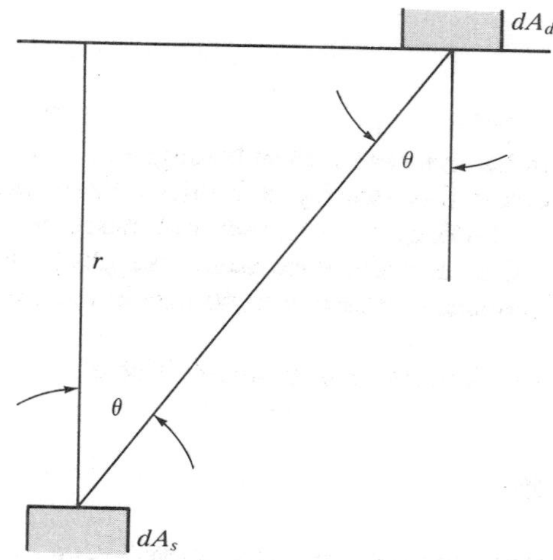

FIGURE 2.3
Radiating geometry of a laterally displaced detector

using Eq. (2.10). Equation (2.13) shows a unique reciprocity relationship between the source and the detector. It implies that the locations of the source and the detector can be interchanged to give an identical value of radiance. When both the source and the detector are facing each other, the source radiance reduces to

$$L_e = \frac{r^2\, d^2\Phi_e}{dA_s\, dA_d} = \frac{d^2\Phi_e}{dA_s d\Omega_d} = \frac{d^2\Phi_e}{dA_d\, d\Omega_s} \tag{2.14}$$

where $d\Omega_d$ is the solid angle subtended by the detector at the source and $d\Omega_s$ is the solid angle subtended by the source at the detector. Next consider the case where the detector and the source are not directly facing each other but are relatively displaced, as shown in Fig. 2.3. In this geometry, the direction of propagation of radiation makes identical angles (say, θ) with the normals to dA_d and dA_s. The radiance for this geometry is given by

$$L_e = \left(\frac{r}{\cos\theta}\right)^2 \frac{d^2\Phi_e}{\cos^2\theta\, dA_s\, dA_d} = \frac{d^2\Phi_e}{dA_s\, dA_d}\left(\frac{r^2}{\cos^4\theta}\right) \tag{2.15}$$

Equation (2.15) is often referred to as the "cosine-to-the-fourth" law of radiometry.

An important physical quantity in radiometry is **irradiance**, which is expressed in watts per square meter (W/m^2) and is defined as the total flux of radiant energy incident per unit area of a real or imaginary surface and is

given by

$$M_e, E_e = \frac{\partial \Phi_e}{\partial A} \tag{2.16}$$

where M_e is used when the involved surface belongs to a radiating source and E_e is used when the radiant flux density is incident either on to a real or an imaginary surface. In particular, the irradiance corresponding to a source surface is often referred to as **radiant exitance**. The photometric equivalent of irradiance is called **illuminance** which is expressed in lumens per square meter (lm/m^2) or lux (lx).

The radiance can be integrated over one-half of the full solid angle to give irradiance as

$$M_e = \int_{\Omega} L_e \cos\theta \, d\Omega \tag{2.17}$$

where $L_e \cos\theta \, d\Omega$ represents the elemental contribution of flux incident from only a single direction. For Lambertian sources, Eq. (2.17) reduces to a very simple relationship (see Problem 2.2).

Table 2.2 summarizes the definitions of various areal and angular densities, whereas Table 2.3 lists expressions for each of these definitions in terms of the spectral radiance. It is obvious from Table 2.3 that the spectral radiance is a very useful quantity.

A source is considered a **point source** when its dimensions are small compared to its distance from the observer. The intensity of a uniform point source is identical in all directions. Thus the total power radiated by a point source is obtained simply by integrating I_e over the total solid angle. Making use of Eq. (2.9), we find that the total radiated power of the point source is $4\pi I_e$. However, if a differential area element dA_d is irradiated with this point source

TABLE 2.2

Spatial Densities

	Areal Density	*Intensity*	*Specific Intensity*
Radiant energy	$\frac{\partial Q_e}{\partial A}$	$\frac{\partial Q_e}{\partial \Omega}$	$\frac{\partial^2 Q_e}{\partial \Omega (\cos\theta \, \partial A)}$
Radiant flux	$\frac{\partial \Phi_e}{\partial A}$	$\frac{\partial \Phi_e}{\partial \Omega}$	$\frac{\partial^2 \Phi_e}{\partial \Omega (\cos\theta \, \partial A)}$
Luminous energy	$\frac{\partial Q_v}{\partial A}$	$\frac{\partial Q_v}{\partial \Omega}$	$\frac{\partial^2 Q_v}{\partial \Omega (\cos\theta \, \partial A)}$
Luminous flux	$\frac{\partial \Phi_v}{\partial A}$	$\frac{\partial \Phi_v}{\partial \Omega}$	$\frac{\partial^2 \Phi_v}{\partial \Omega (\cos\theta \, \partial A)}$

TABLE 2.3

Spatial Densities in Terms of Spectral Radiance

	Areal Density	*Intensity*	*Specific Intensity*
Radiant energy	$\iiint L_{e\lambda} \cos\theta \, d\Omega \, d\lambda \, dt$	$\iiint L_{e\lambda} \cos\theta \, dA \, d\lambda \, dt$	$\iint L_{e\lambda} \, d\lambda \, dt$
Radiant flux	$\iint L_{e\lambda} \cos\theta \, d\Omega \, d\lambda$	$\iint L_{e\lambda} \cos\theta \, dA \, d\lambda$	$\int L_{e\lambda} \, d\lambda$
Luminous energy*	$\iiint V_{d\lambda} L_{e\lambda} \cos\theta \, d\Omega \, d\lambda \, dt$	$\iiint V_{d\lambda} L_{e\lambda} \cos\theta \, dA \, d\lambda \, dt$	$\iint V_{d\lambda} L_{e\lambda} \, d\lambda \, dt$
Luminous flux*	$\iint V_{d\lambda} L_{e\lambda} \cos\theta \, d\Omega \, d\lambda$	$\iint V_{d\lambda} L_{e\lambda} \cos\theta \, dA \, d\lambda$	$\int V_{d\lambda} L_{e\lambda} \, d\lambda$

*For nighttime operation, $V_{d\lambda}$ is replaced by $V_{n\lambda}$.

from a distance r, the differential flux incident on dA_d is found to be

$$d\Phi_e = \frac{I_e \, dA_d \cos\theta_d}{r^2} \tag{2.18}$$

where $dA_d \cos\theta_d / r^2$ is the solid angle and θ_d is the angle between the normal to $d\mathbf{A}_d$ and the radial vector $\mathbf{r}$.

Using Eqs. (2.18) and (2.16), we find that

$$E_e = \frac{I_e \cos\theta_d}{r^2} \tag{2.19}$$

This result illustrates both Lambert's cosine law (which states that the irradiance decreases with the angle of incidence as $\cos\theta_d$) and the inverse square law (which states that the irradiance decreases with distance as $1/r^2$) of irradiance. Any source irradiating in accordance with Eq. (2.19) is generally referred to as a point source. Most sources, when they are either small or are located at great distances from the detector, are also classed as point sources.

2.2.3 Image Radiance and Irradiance

We now concentrate on methods of computing radiance and irradiance of images in an optical system. As we show later, radiance is always conserved in an elementary beam. If no losses are incurred, both the image and the object have identical radiant fluxes. Consider the aplanatic optical system of Fig. 2.4 where $d\phi$ is the azimuthal differential angle. For simplicity, the source object can be considered as rectangular, having an area of dA_o. The corresponding rectangular image is assumed to have an area of dA_i.

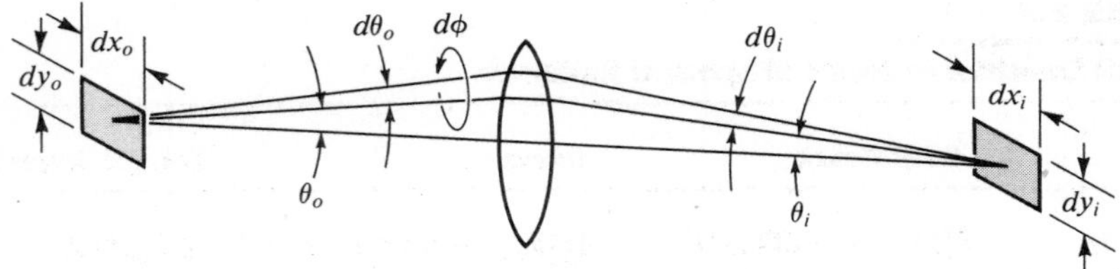

FIGURE 2.4

An imaging system

The source radiance flowing in to the solid angle $d\Omega_o = \sin\theta_o\, d\theta_o\, d\phi$ is calculated in terms of the object flux as

$$L_{eo} = \frac{d^2\Phi_e}{\cos\theta_o\, dx_o\, dy_o\, d\Omega_o}$$

$$= \frac{d^2\Phi_e}{dx_o\, dy_o \sin\theta_o \cos\theta_o\, d\theta_o\, d\phi} \tag{2.20}$$

The image radiance is found likewise to be

$$L_{ei} = \frac{d^2\Phi_e}{\cos\theta_i\, dx_i\, dy_i\, d\Omega_i}$$

$$= \frac{d^2\Phi_e}{dx_i\, dy_i \sin\theta_i \cos\theta_i\, d\theta_i\, d\phi} \tag{2.21}$$

But in aplanatic optical systems,

$$dx_o \sin\theta_o = dx_i \sin\theta_i \tag{2.22a}$$

and

$$dy_o \sin\theta_o = dy_i \sin\theta_i \tag{2.22b}$$

Consequently, we find by comparing Eqs. (2.20) and (2.21) that indeed

$$L_{eo} = L_{ei} \tag{2.23}$$

Thus no optical system generates an image whose radiance exceeds that of the input object. In a generalized system where n_o and n_i are the refractive indices of object and image spaces, respectively, the radiances are related by

$$Tn_i^2 L_{eo} = n_o^2 L_{ei} \tag{2.24}$$

where T is the transmission of the optical system. This mathematical relationship (referred to as the **Brightness theorem**) is true regardless of the amount of aberration or aperture of the optical system that is lying between the object and the image.

For the irradiance calculation, let us limit ourselves to a Lambertian source

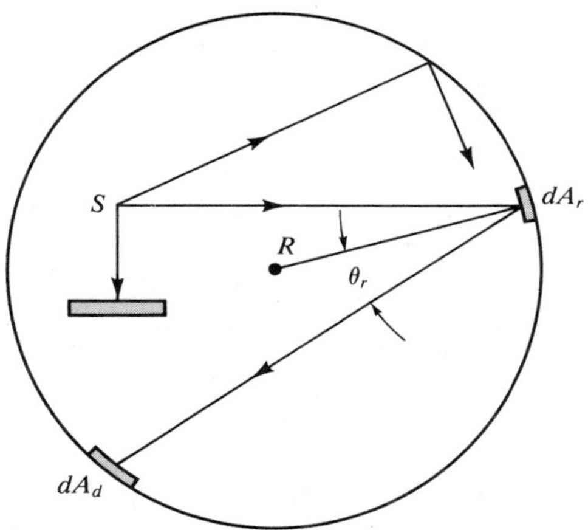

FIGURE 2.5

An integrating sphere enclosing a source *S*

of radiance L_{eo} and area dA_o. Assume further that the maximum angle subtended by the lens at the object is $2\theta_o$ and that subtended by the image is $2\theta_i$. The total radiant flux arriving at the bi-convex lens is obtained as

$$\begin{aligned} d\Phi_e &= \int_0^{\theta_o}\int_0^{2\pi} L_{eo}\, dA_o \sin\theta \cos\theta \, d\phi \, d\theta \\ &= \pi L_{eo}\, dA_o \sin^2\theta_o \end{aligned} \tag{2.25}$$

Thus the image irradiance is found to be given by

$$E_{ei} = \frac{d\Phi_e}{dA_i} = \pi L_{eo}\frac{dA_o}{dA_i}\sin^2\theta_o = \pi L_{eo}\sin^2\theta_i \tag{2.26}$$

using Eqs. (2.22a) and (2.22b).

2.2.4 Integrating Sphere

An **integrating sphere** is a hollow sphere whose inner surface is coated with a material of highly diffuse reflectivity to give it the characteristic of a Lambertian reflector. An integrating sphere is often employed to measure the total power of a source of arbitrary radiance distribution. The source is placed anywhere within the integrating sphere. The manner in which the flux is distributed within the sphere is of no real consequence. A detector of area dA_d is placed anywhere on the inner surface of the sphere. As shown in Fig. 2.5, a stop is placed between the source and the detector to prevent direct irradiation of the detector. The to-be-measured flux can be introduced through a small aperture in the side of the sphere. The radiation introduced thereby will be reflected many times before being finally absorbed by the detector. The detector detects reflected irradiations from all over the Lambertian inner surface. An integrating sphere consequently results in a uniform irradiance that is proportional to the total flux of light energy. When the differential area element of the reflecting surface is represented by dA_r, the radiant power incident on the detector is given by

$$d\Phi_{e1} = \cos\theta_r \, d\Omega \int L_e \, dA_r \tag{2.27}$$

using Eq. (2.10), where L_e is the apparent radiance of dA_r, $d\Omega$ is the solid angle subtended by the detector area at dA_r, and θ_r is the angle of reflection as shown in Fig. 2.5. Since the distance between the two area elements is $2R\cos\theta_r$,

$$d\Phi_{e1} = \frac{dA_d}{4R^2} \int L_e \, dA_r \tag{2.28}$$

Consequently, the radiant power is independent of both the location of dA_r and the value of θ_r. If the inner surface reflects r fraction of the total incident flux, the net radiance L_e is given by (see Problem 2.2)

$$L_e = \frac{rE_e}{\pi} \tag{2.29}$$

Thus Eq. (2.28) can be rewritten as

$$d\Phi_{e1} = \frac{r\,dA_d}{4\pi R^2} \int E_e \, dA_r \equiv \frac{r\,dA_d}{4\pi R^2} \Phi_{es} \tag{2.30}$$

using Eq. (2.16), where Φ_{es} is the total radiant power of the source S. A part of the radiant power is, however, reflected by the inner surface of the integrating sphere repeatedly before it actually reaches the detector. The total power received by the detector can now be caculated as

$$\begin{aligned} d\Phi_e &= \sum d\Phi_{ei} \\ &= \frac{dA_d}{4\pi R^2} \Phi_{es}(r + r^2 + r^3 + \cdots) \\ &= \frac{dA_d}{4\pi R^2} \Phi_{es} r(1 + r + r^2 + \cdots) \\ &= \frac{dA_d}{4\pi R^2} \Phi_{es} \left[\frac{r}{(1-r)} \right] \end{aligned} \tag{2.31}$$

where $d\Phi_{ei}$ is the power detected after the incident beam has gone through the ith reflection off the inner surface. Accordingly, an integrating sphere can be used for measuring flux in terms of the reflectivity of its inner surface and its radius. Conversely, the reflectivity can be measured by introducing a known amount of flux in the integrating sphere.

The efficiency of the integrating sphere, defined as $d\Phi_e/\Phi_{es}$, is relatively small even for a relatively large value of r. A relatively small change of reflectivity, say from 92% to 96%, could change the illumination by a factor of 2.

It must be noted that the detector occupies a finite area, however small. Thus the measurement made with an integrating sphere is not exactly correct, but it is remarkably close to the actual value. For dependable measurement, it is desirable that the dimensions of both the source and the stop should be small compared to that of the integrating sphere.

2.2.5 Conversion between Radiometric and Photometric Units

A photometric quantity is commonly defined in terms of the visual stimulus of an average human eye. Sources of equal spectral radiance may appear to have different photometric spectral luminance to an observer as the wavelength is varied. Accordingly, the wavelength of the radiating light plays an important role in the conversion of a radiometric quantity into its equivalent photometric quantity and vice versa. However, the central problem plaguing the conversion mechanism is linked to identifying the average human eye. Results of field experimentations involving numerous observers have been statistically averaged to obtain standardized spectral luminous efficiency curves like those shown in Fig. 2.6. The photopic curve corresponds to the daylight-adapted cases and the scotopic curve corresponds to the night-adapted cases. Both of these curves ($V_{d\lambda}$ and $V_{n\lambda}$, respectively) are functions of wavelength λ having peak values of unity at 555 nm and 510 nm, respectively. This implies, for example, that a daylight of wavelength λ gives a factor of $V_{d\lambda}$ as much photometric sensation as the light corresponding to the peak wavelength. The maximum value of the conversion factor is 673 lm/W and 1725 lm/W, respectively, for the photopic and scotopic cases. Consequently, for example, 1 W of flux at 510 nm during the night gives the same sensation as 1725 lm. For wavelengths other than the peak wavelength, the conversion factor is 673 $V_{d\lambda}$ lm/W and 1725 $V_{n\lambda}$ lm/W, respectively, for the daylight and nightlight conditions. Note that both $V_{d\lambda}$ and $V_{n\lambda}$ are zero beyond the visible region, 390 nm $\leqslant \lambda \leqslant$ 770 nm.

To obtain the equivalent photometric quantity, the corresponding spectral physical quantity is multiplied by the respective conversion factor. For instance, the luminous flux Φ_v is expressed in terms of daylight spectral flux $\Phi_{e\lambda}$ by

$$\Phi_v = 673 \int_{390\,\text{nm}}^{770\,\text{nm}} V_{d\lambda} \Phi_{e\lambda}(\lambda)\, d\lambda \tag{2.32}$$

The average daylight conversion factor $\langle V_{d\lambda} \rangle$ is related to the luminous flux by

$$\Phi_v = \langle V_{d\lambda} \rangle \int_{390\,\text{nm}}^{770\,\text{nm}} \Phi_{e\lambda}(\lambda)\, d\lambda \tag{2.33}$$

thus resulting in

$$\langle V_{d\lambda} \rangle = \frac{673 \int_{390\,\text{nm}}^{770\,\text{nm}} V_{d\lambda} \Phi_{e\lambda}(\lambda)\, d\lambda}{\int_0^{\infty} \Phi_{e\lambda}(\lambda)\, d\lambda} \;(\text{lm/W}) \tag{2.34}$$

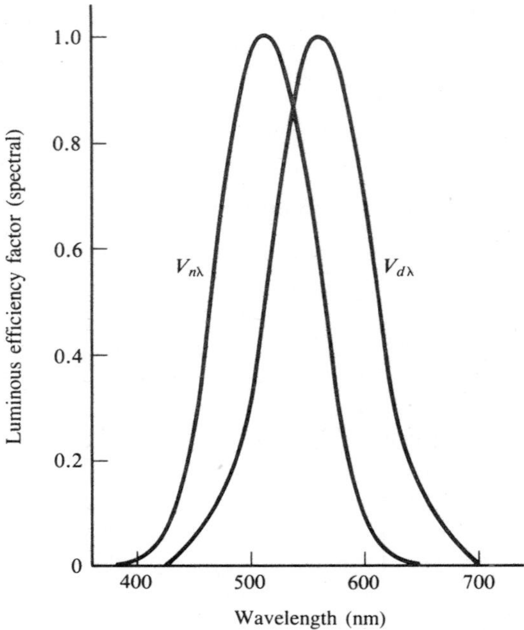

FIGURE 2.6

Spectral luminous efficiency curves. The photopic curve $V_{d\lambda}$ corresponds to the daylight-adapted case while the scotopic curve $V_{n\lambda}$ corresponds to the night-adapted case.[Adapted from R. W. Boyd, *Radiometry and the Detection of Optical Radiation* (New York: Wiley, 1983), p. 100. © 1983 by John Wiley & Sons, Inc.]

Note that it may not always be possible to convert the luminous flux to an equivalent radiant flux. Such conversion is possible only when the spectral distribution of the photometric quantity is known for all of the wavelengths and only if the wavelength range in question is within the visible region of the spectrum.

2.3 Blackbody Radiation

Optical radiating sources are classified according to their wavelength contents into three groups: (a) discrete; (b) narrow-band; and (c) broadband. The discrete source radiates energy at one or only a few neighboring frequencies. The best example for discrete sources is a laser (acronym for Light Amplification by Stimulated Emission of Radiation). The light energy emitted by a laser is usually associated with a single wavelength. Thus the laser light is also often referred to as monochromatic.

The narrow-band sources on the other hand have most of their radiation confined within a narrow, but fairly nondiscrete, range of frequencies. A light-emitting diode (LED) happens to be an example of such a source. Most of the remaining light sources are classified as broadband because they are associated with frequencies that are spread across a good portion of the electromagnetic spectrum. All objects radiate when they are at a temperature above absolute zero. The blackbody radiator that emits radiation over all ranges of frequencies is a perfect example of a broadband source. A blackbody completely absorbs all radiation incident upon it. However, due to its emission characteristics, the blackbody appears black only at very low temperatures. The blackbody, for example, will appear yellow at about 4000 K. All practical sources fall short of the blackbody requirement and are, therefore, referred to as "graybody" radiators.

When the radiation trapped within a cavity reaches an equilibrium with the atoms of the cavity walls, the energy emission rate of the atoms is equal to the corresponding energy absorption rate of the atoms. Therefore, at equilibrium, the energy density of the electromagnetic field remains a constant. Experiments have revealed that at equilibrium the radiation trapped within the cavity has well-defined distribution characteristics. The radiation distribution is such that, corresponding to each of the frequencies, there exists an energy density that strictly depends only on the temperature of the cavity walls and not on the chemical composition of the cavity wall. Note that the monochromatic energy density $E(\nu)$, when multiplied by the frequency spread $d\nu$, provides the energy density of the radiation in the frequency range between ν and $\nu + d\nu$.

The exact physical phenomenon responsible for atomic radiation is best explained by the quantum-mechanical model. This model had its historical roots in the efforts of Max Planck. As described in Chapter 1, individual atoms behave like oscillators and, therefore, participate actively in energy exchanges. The associated energy E of the electromagnetic radiation, absorbed or emitted in any single process of interaction, is given by $h\nu$ according to Eq. (1.50).

In accordance with the quantum-mechanical model, the blackbody possesses numerous quantized energy states, spread over all ranges of energy, so that any photon, whatever its frequency, is readily absorbed by it. A hollow container, completely enclosed except for a small hole through which radiation can enter or leave, is the best approximation to a blackbody. The blackbody is black in the normal sense of the word in that it absorbs all incident radiation. But yet its temperature remains unchanged even though it absorbs energy because the blackbody is not only a perfect absorber but also a perfect emitter. The general nature of the monochromatic energy density of a blackbody radiator as a function of wavelength at different temperatures is shown in Fig. 2.7. The blackbody radiation spectrum is a universal function that is independent of the material of which the radiator is made. As the temperature increases, a larger fraction of the emitted radiation is carried by the higher

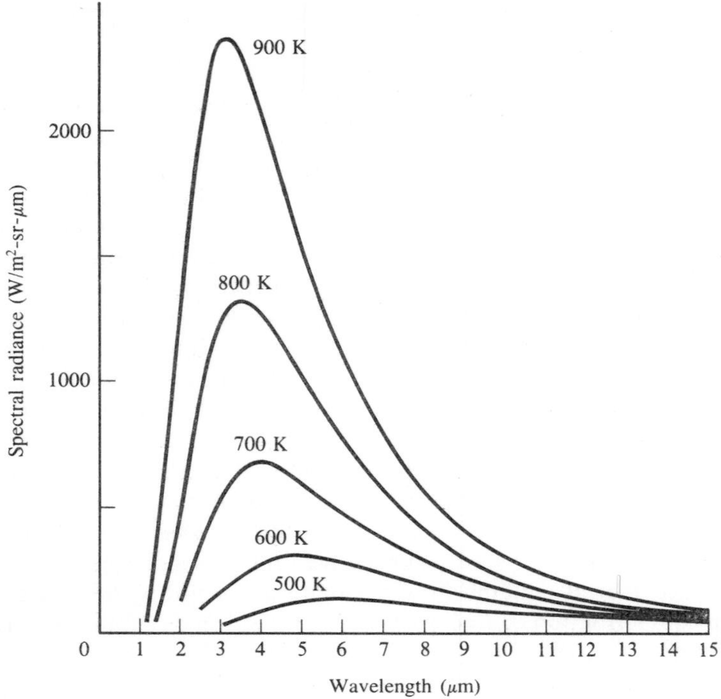

FIGURE 2.7

Monochromatic energy density of a blackbody radiator. [From R. D. Hudson, *Infrared Systems Engineering* (New York: Wiley, 1969). © 1969 by John Wiley & Sons, Inc. Also printed in R. W. Boyd, *Radiometry and the Detection of Optical Radiation* (New York: Wiley, 1983). © 1983 by John Wiley & Sons, Inc.]

frequency components. The frequency at which the energy density is a maximum also changes as the temperature changes. The energy density curve, therefore, is often used to explain the change of color in a radiator when its temperature is varied.

Consider a configuration like that of Fig. 2.3 so that the radiance L_e of a blackbody could be determined in terms of the energy density u_e. The energy passing through the area dA_s in time dt is given by

$$d^3Q_e = d^2\Phi_e\,dt = L_e\,dA_s \cos\theta\,d\Omega\,dt \tag{2.35}$$

using Eqs. (2.3) and (2.10). This energy can be considered to be packed inside a cylindrical volume whose base area is dA_s and whose slant height is $c\,dt$, where c is the speed of propagation. Consequently, the differential energy density is

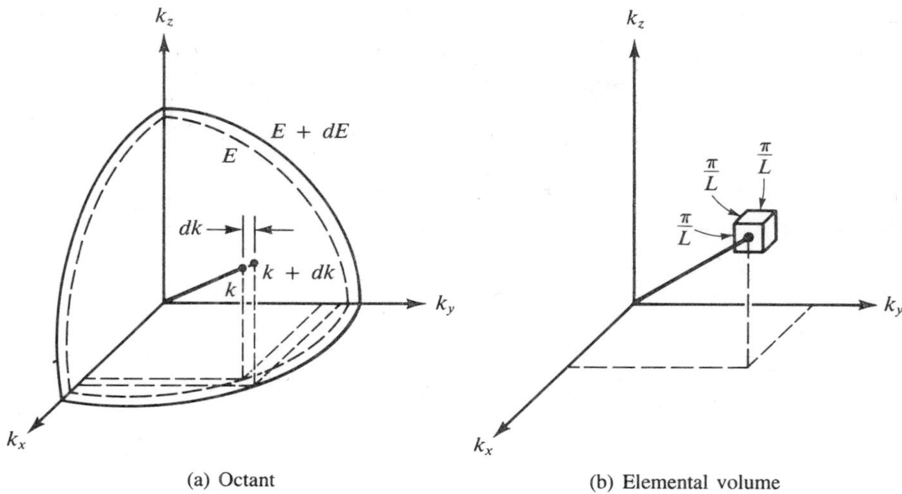

FIGURE 2.8

The three-dimensional *k* space

found to be

$$du_e = \frac{d^3Q_e}{(dA_s \cos\theta)(c\,dt)} = \frac{L_e\,d\Omega}{c} \tag{2.36}$$

Equation (2.36) could be used next to calculate the total energy density as

$$u_e = \int du_e = \frac{4\pi L_e}{c} \tag{2.37}$$

Similarly we can evaluate the energy density in terms of radiant exitance M_e and find, using the solution of Problem 2.2, that

$$u_e = \frac{4M_e}{c} \tag{2.38}$$

To calculate the energy density value, we need to determine the number of energy states present within the energy range E and $E + dE$, where $E = \hbar kc$ and the wave number $k = (k_x^2 + k_y^2 + k_z^2)^{1/2}$. This value can be evaluated using Fig. 2.8, which shows an octant in the three-dimensional k space and a surface of the sphere that corresponds to the constant energy of the free electrons. The density of energy states ρ_e can be determined by accounting for the total number of energy states available within a spherical shell of radius k and thickness dk:

$$\rho_{ek}\,dk = \frac{2(\frac{1}{8})4\pi k^2\,dk}{\left(\frac{\pi}{L}\right)^3 L^3} = \frac{k^2\,dk}{\pi^2} \tag{2.39}$$

where L^3 is the volume of a cube (having electrically conducting walls) within which Maxwell's equations have valid solutions as long as k_x, k_y, and k_z are positive integral multiples of π/L. In Eq. (2.39), the factor 2 accounts for the two available polarization directions, while the factor 1/8 accounts for the fact that only one-eighth of the shell has positive dimensions and $(\pi/L)^3$ is the volume surrounding each point in the k space. Often $\rho_{ek}\,dk$ is referred to as the number of modes per unit volume. Since the propagation constant $k = 2\pi/\lambda$, the mode density can also be expressed as

$$\rho_{e\lambda}\,d\lambda = \frac{8\pi}{\lambda^4}\,d\lambda \tag{2.40}$$

The energy density in the wavelength interval $d\lambda$ can now be found by multiplying the three terms—the mode density, the average number of photons per state, f_{BE}, and the energy per photon. Thus

$$\begin{aligned} u_{e\lambda}\,d\lambda &= \rho_{e\lambda}\,d\lambda f_{\mathrm{BE}}\frac{hc}{\lambda} \\ &= \frac{8\pi hc\,d\lambda}{\lambda^5}\frac{1}{e^{hc/k\lambda T}-1} \end{aligned} \tag{2.41}$$

where

$$f_{\mathrm{BE}} = \frac{1}{e^{hc/k\lambda T}-1} \tag{2.42}$$

is the well-known **Bose-Einstein's occupation number** and $k = 1.380662 \times 10^{-23}$ J/K is the **Boltzmann constant**. The temperature-dependent spectral radiance in Watts per square meter steradian per micrometer (W/m^2-sr-μm) and spectral radiant exitance in Watts per square meter per micrometer (W/m^2-μm) can now be obtained from Eq. (2.41) using Eqs. (2.37) and (2.38), respectively, as

$$L_{e\lambda}(T) = \frac{cu_{e\lambda}}{4\pi} = \frac{2hc^2}{\lambda^5}\frac{1}{e^{hc/k\lambda T}-1} \tag{2.43}$$

and

$$M_{e\lambda}(T) = \frac{cu_{e\lambda}}{4} = \frac{2\pi hc^2}{\lambda^5}\frac{1}{e^{hc/k\lambda T}-1} \tag{2.44}$$

The two equations are two versions of what is known as **Planck's radiation law**. The spectral radiance for a blackbody radiator is shown as a function of wavelength for different values of temperature in Fig. 2.9.

When $hc/k\lambda T \ll 1$, Eq. (2.44) can be reduced as follows:

$$\begin{aligned} M_{e\lambda}(T) &= \frac{2\pi hc^2}{\lambda^5}\left[\frac{hc}{k\lambda T} + \frac{1}{2!}\left(\frac{hc}{k\lambda T}\right)^2 + \cdots\right]^{-1} \\ &= \frac{2\pi ckT}{\lambda^4} \end{aligned} \tag{2.45}$$

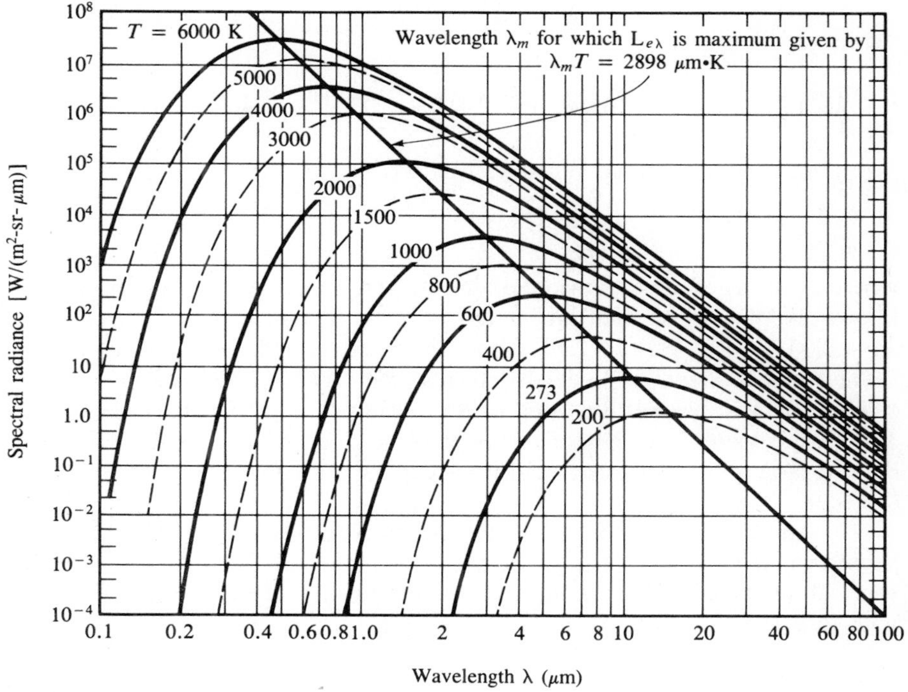

FIGURE 2.9

Spectral radiance of a blackbody radiator. [Adapted from S. L. Valley, *Handbook of Geophysics and Space Environments* (New York: McGraw-Hill, 1965). © 1965 by McGraw-Hill Book Co.]

This simplified result, known commonly as the **Rayleigh-Jeans' law**, can also be derived from the classical electromagnetic field theory. This approximated result is correct within about 1% of the actual exitance characteristics as long as $\lambda T > 0.72$ m-K. Equation (2.45) becomes extremely unreliable, however, as λT approaches zero. This failure is generally referred to as the **ultraviolet catastrophe**. The relative error encountered in the Rayleigh-Jeans' approximation is obtained as follows:

$$\begin{aligned} \text{error} &= \frac{M_{e\lambda}(T)|_{\text{RJ}} - M_{e\lambda}(T)|_{\text{P}}}{M_{e\lambda}(T)|_{\text{P}}} \\ &= \frac{kT\lambda}{hc}[e^{hc/k\lambda T} - 1] - 1 \\ &= \frac{hc}{2k\lambda T} + \frac{2}{3}\left(\frac{hc}{2k\lambda T}\right)^2 + \cdots \end{aligned} \qquad (2.46)$$

Planck's radiation law of Eq. (2.44) can also be simplified for the smaller values of λT since

$$M_{e\lambda}(T) = \frac{2\pi hc^2}{\lambda^5} e^{-hc/k\lambda T}[1 + e^{-hc/k\lambda T} + e^{-2hc/k\lambda T} + \cdots] \tag{2.47}$$

using binomial expansion. When $hc/k\lambda T \gg 1$,

$$M_{e\lambda}(T) \simeq \frac{2\pi hc^2}{\lambda^5} e^{-hc/k\lambda T} \tag{2.48}$$

where all terms of the expansion except the first are neglected. Equation (2.48), known as **Wien's radiation law**, is found to be correct within 1% as long as $\lambda T < 0.0031$ m-K. Figure 2.10 shows the spectral radiance of a blackbody radiator due to Planck's, Rayleigh-Jeans', and Wien's laws. Figure 2.11 shows the relative error in percent for using either of the two approximations, Rayleigh-Jeans' and Wien's laws. These curves are used often to determine whether the desired ranges of temperature and wavelengths fall within the acceptable error limits.

It can be seen from Fig. 2.9 that the blackbody radiation characteristics show a single maximum. The particular value of the wavelength λ_m at which the spectral radiance is a maximum can be determined by differentiating Eq. (2.44) with respect to λ and then setting the derivative equal to zero. Thus

$$\left.\frac{dM_{e\lambda}}{d\lambda}\right|_{\lambda=\lambda_m} = 0 \tag{2.49}$$

or

$$\frac{2\pi hc^2}{\lambda_m^6} \frac{1}{e^{hc/k\lambda_m T} - 1} \left[\frac{\left(\dfrac{hc}{k\lambda_m T}\right) e^{hc/k\lambda_m T}}{e^{hc/k\lambda_m T} - 1} - 5 \right] = 0 \tag{2.50}$$

Therefore, either

$$\frac{2\pi hc^2}{\lambda_m^6 (e^{hc/k\lambda_m T} - 1)} = 0 \tag{2.51}$$

or

$$\frac{\left(\dfrac{hc}{k\lambda_m T}\right) e^{hc/k\lambda_m T}}{e^{hc/k\lambda_m T} - 1} - 5 = 0 \tag{2.52}$$

Equation (2.51) implies that $\lambda_m T = \infty$, but such a solution is noninteresting. On the other hand, Eq. (2.52) yields a transcendental equation

$$\frac{\beta}{5} + e^{-\beta} - 1 = 0 \tag{2.53}$$

where $\beta = hc/k\lambda_m T$. This equation can be solved only by the successive

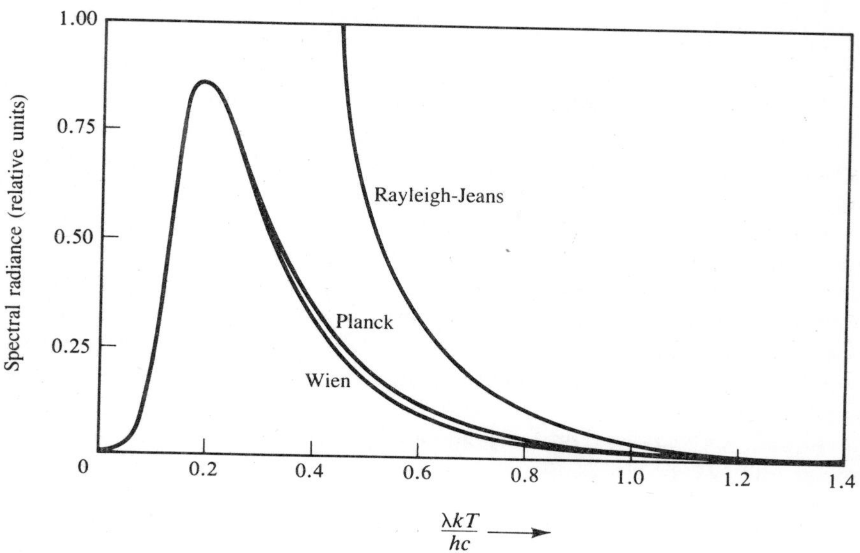

FIGURE 2.10

Comparative spectral radiance characteristics

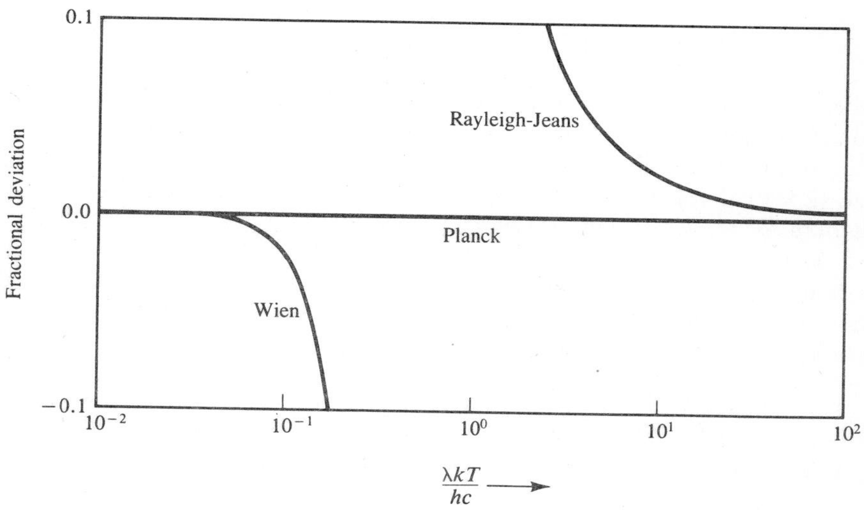

FIGURE 2.11

Relative errors in Rayleigh-Jeans' and Wien's approximations

approximation method, resulting in $\beta = 4.9651142317\ldots$ which in turn will provide

$$\lambda_m T = 2897.8 \ \mu\text{m-K} \tag{2.54}$$

This constant is often called **Wien's displacement constant**. The physical significance of Eq. (2.54) is that the wavelength at which the maximum spectral radiation response occurs decreases as the temperature increases. A blackbody (or graybody, too) thus becomes progressively red, white, and blue with increasing temperature. Theoretically, the spectral distribution of radiant exitance extends over all ranges of wavelength, but in practice $M_{e\lambda}$ becomes very small on either side of λ_m. One percent of its maximum value occurs when the wavelength is either $0.33\lambda_m$ or $6.5\lambda_m$ for all T. Note also that the radiant exitance from $\lambda = 0$ to $\lambda = \lambda_m$ is approximately equal to one-fourth of the total radiant exitance.

Equation (2.44) can be integrated over λ to yield time-dependent total radiance $M_e(T)$ as

$$\begin{aligned} M_e(T) &= \int_0^\infty M_{e\lambda}(T)\, d\lambda \\ &= 2\pi hc^2 \int_0^\infty \frac{1}{\lambda^5[e^{hc/k\lambda T} - 1]}\, d\lambda \end{aligned} \tag{2.55}$$

By substituting $x = hc/k\lambda T$, Eq. (2.55) can be evaluated as

$$\begin{aligned} M_e(T) &= -\frac{2\pi k^4 T^4}{h^3 c^2} \int_0^\infty \frac{x^3}{e^x - 1}\, dx \\ &= 6.4938 \frac{k^4}{h^3 c^2} T^4 \equiv \sigma T^4 \end{aligned} \tag{2.56}$$

where $\sigma = 5.67032 \times 10^{-8}$ W/m^2-K^4. It is interesting to note that Eq. (2.56), commonly known as the **Stefan-Boltzmann law**, was experimentally determined long before the publication of Planck's theoretical work. For nongray bodies, Eq. (2.56) cannot be used for M_e since the integration of Eq. (2.55) did not account for the wavelength-dependent emissivity.

The ratio of spectral radiant exitance $M_{e\lambda}$ and the maximum spectral radiant exitance $M_{e\lambda_m}$ results in a normalized spectral distribution function, known commonly as the universal blackbody ratio. This ratio, $f(\lambda, T)$, is a function of only the product λT, and is given by

$$f(\lambda, T) = \frac{M_{e\lambda}(\lambda, T)}{M_{e\lambda_m}(\lambda_m, T)} \tag{2.57}$$

In like manner, the normalized cumulative spectral distribution function $F(\lambda, T)$ is given by

$$F(\lambda, T) = \frac{\int_0^\lambda M_{e\lambda}(\lambda, T)\, d\lambda}{\int_0^\infty M_{e\lambda}(\lambda, T)\, d\lambda} \tag{2.58}$$

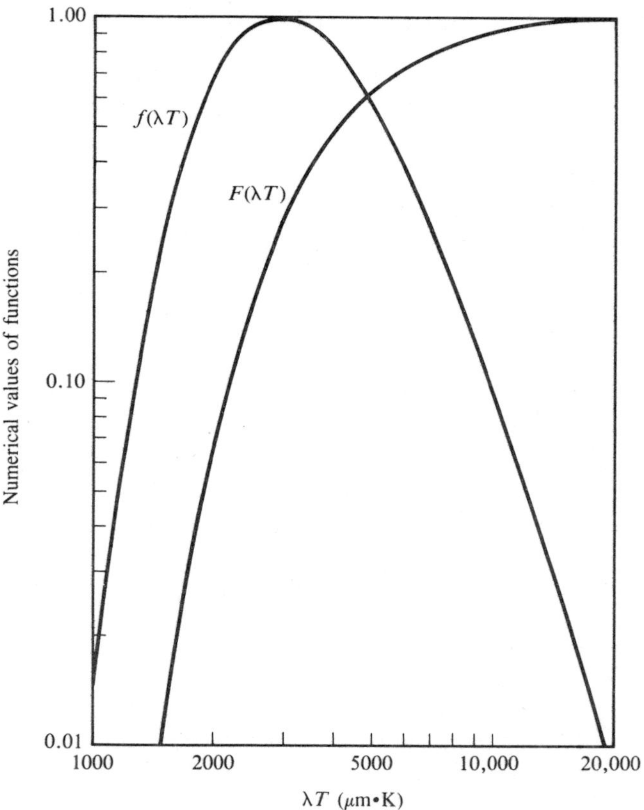

FIGURE 2.12

Normalized and normalized cumulative spectral distribution functions of blackbody radiation

$F(\lambda, T)$ is also a universal function, and it is equivalent to the fraction of total blackbody radiation that has wavelengths smaller than λ and a temperature T.

The wavelength (in micrometers (μm)) having maximum radiant exitance is given by $\lambda_m = 2897.8/T$. Consequently, Eq. (2.57) can be evaluated as

$$f(\lambda, T) = \left(\frac{\lambda_m}{\lambda}\right)^5 \frac{e^{hc/k\lambda_m T} - 1}{e^{hc/k\lambda T} - 1}$$

$$= \left(\frac{2897.8}{\lambda}\right)^5 \frac{e^{hc/(2897.8k)} - 1}{e^{hc/k\lambda T} - 1} \tag{2.59}$$

which provides that

$$M_{e\lambda}(\lambda, T) = M_{e\lambda_m}(\lambda_m, T)\left(\frac{2897.8}{\lambda T}\right)^5 \left[\frac{e^{hc/(2897.8k)} - 1}{e^{hc/k\lambda T} - 1}\right] \tag{2.60}$$

Consequently, Eq. (2.58) can be written as

$$F(\lambda, T) = \frac{M_{e\lambda_m}(\lambda_m, T) \int_0^{\lambda T} \left(\frac{2897.8}{\zeta}\right)^5 \frac{e^{hc/(2897.8k)} - 1}{e^{hc/k\lambda T} - 1} \frac{d\zeta}{T}}{M_{e\lambda_m}(\lambda_m, T) \int_0^{\infty} \left(\frac{2897.8}{\zeta}\right)^5 \frac{e^{hc/(2897.8k)} - 1}{e^{hc/k\lambda T} - 1} \frac{d\zeta}{T}} \quad \textbf{(2.61)}$$

where $\zeta = \lambda T$. $F(\lambda, T)$ is found to increase monotonically with λT finally approaching a value of unity. Figure 2.12 shows the plot of both $f(\lambda T)$ and $F(\lambda T)$ as functions of the product λT.

2.4 Material Interactions

Optical radiation from nonblackbody sources is associated with reflection, absorption, transmission, and emission of energy, as shown in Fig. 2.13. Accordingly, the concepts of reflectance, absorptance, transmittance, and emissivity are discussed in this section. These concepts are basically tied together by means of the principle of conservation of energy, which requires that the energy incident on a body must be equal to the sum of absorbed, reflected, and transmitted energies. The material radiation characteristics are usually defined in terms of radiometric quantities.

A radiator is referred to as a temperature radiator when its radiant exitance is a function of only its temperature and material content. A radiator that does not fall in this category is usually characterized by its emissivity ratio which indicates how closely the radiant exitance of the radiator in question matches the radiant exitance of a blackbody radiator. The emissivity ε_e of a radiator is defined by

$$\varepsilon_e = \frac{M_e}{M_{e,b}} \quad \textbf{(2.62)}$$

where M_e is the radiant exitance of the radiator and $M_{e,b}$ is the radiant exitance of the blackbody radiator, both measured at the same temperature. Spectral emissivity is similarly obtained by taking the ratio of the spectral radiant exitances of the radiator in question and the spectral radiant exitance of a blackbody radiator. Thus

$$\varepsilon_{e\lambda} = \frac{M_{e\lambda}}{M_{e\lambda,b}} \quad \textbf{(2.63)}$$

Over an interval of wavelength $\Delta\lambda$, the mean emissivity is

$$\langle \varepsilon_e \rangle = \frac{\int_{\Delta\lambda} L_{e\lambda}\, d\lambda}{\int_{\Delta\lambda} L_{e\lambda,b}\, d\lambda} = \frac{\int_{\Delta\lambda} \varepsilon_{e\lambda} L_{e\lambda,b}\, d\lambda}{\int_{\Delta\lambda} l_{e\lambda,b}\, d\lambda} \quad \textbf{(2.64)}$$

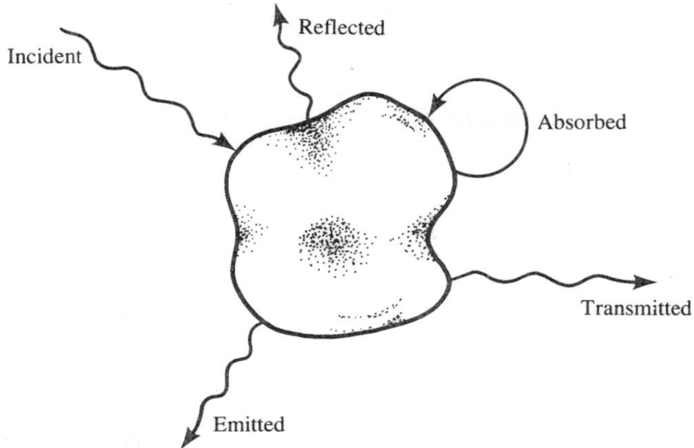

FIGURE 2.13

Energy flow during material interactions

while that for the entire band (see Problem 2.2) is

$$\langle \varepsilon_e \rangle = \frac{\int_0^\infty \varepsilon_{e\lambda} L_{e\lambda,b}\, d\lambda}{\dfrac{\sigma T^4}{\pi}} \tag{2.65}$$

A material is referred to as a graybody when its emissivity is independent of wavelength; otherwise the materials are generally termed as spectral or colored bodies.

The remaining three material radiation ratios are measured in terms of the incident radiation. Absorptance is the ratio of the absorbed radiant exitance and the incident radiant exitance. The absorptance α_e of a radiator is thus defined as

$$\alpha_e = \frac{E_{e,a}}{E_{e,i}} \tag{2.66}$$

where $E_{e,a}$ is the energy absorbed by the radiator and $E_{e,i}$ is the energy incident on the radiator. A material generally increases its internal energy by absorbing radiation. Thus, at equilibrium, the material counteracts the increase in its internal energy by emitting an equivalent amount of radiation. **Kirchhoff's law** governing such an energy transfer states that all thermal radiators must absorb and emit radiation at the same rate irrespective of wavelength and its material composition. Thus

$$\varepsilon_e = \alpha_e \tag{2.67}$$

If this were not so, a thermal gradient would develop and the body would cease to be isothermal. Thus, if a certain power is absorbed by a material, there must be an equal amount of emission so as to maintain the isotropic nature of the energy flow. Equation (2.67) applies to total emission over all wavelengths as well as to the emission over any finite bandwidth. Note that the emissivity of a blackbody is unity. Thus the radiation given out of a blackbody consists exclusively of emitted radiation and not reflected radiation.

Reflectance of a radiating surface is defined as the ratio of the reflected flux and the incident flux. For an opaque material, the reflectance ρ_e is given by

$$\rho_e = \frac{M_{e,r}}{E_{e,i}} \tag{2.68}$$

where $M_{e,r}$ is the radiant exitance due to reflection. Highly reflecting radiators are generally poor blackbodies. The transmittance of a medium is defined likewise as the ratio of the transmitted flux to the incident flux. The transmittance τ_e of a radiator is given by

$$\tau_e = \frac{M_{e,t}}{E_{e,i}} \tag{2.69}$$

where $M_{e,t}$ is the transmitted radiant exitance of the radiator. Note that it is also possible to obtain spectral radiation ratios. Again, for each of the four radiometric ratios, we could have used ratios of radiant energy, or radiant flux, or radiant intensity, or radiance.

When a flux of irradiation is incident upon a radiator, the input irradiance $E_{e,i}$ is related to the absorbed, reflected, and transmitted components by

$$E_{e,i} = E_{e,a} + M_{e,r} + M_{e,t} \tag{2.70}$$

which gives us

$$\alpha_e + \rho_e + \tau_e = 1 \tag{2.71}$$

Oftentimes the total radiant flux of an optical radiator that may consist mainly of emitted radiant flux and reflected radiant flux is called the **optical signature**. The emitted portion of the signature depends on whether the radiator is passive or active. The reflected part of the signature, on the other hand, depends on the type of ambient source irradiating the radiator and also on the spectral reflectance of the radiator. Optical signatures corresponding to passive sources usually consist of both types of radiation (emitted and reflected) but, in the case of active sources, the reflected flux is insignificant when compared to the emitted flux.

There are several types of radiation (specifically light) sources. In the visible range, the spectral radiant exitance distribution of thermal radiators such as carbon, platinum, and tungsten are reasonably equivalent to that of a blackbody radiator. In particular, tungsten can be operated at high enough temperature, either in vacuum or in inert gas, for long periods of time before being completely evaporated. Note that tungsten's emissivity decreases only slightly with increasing wavelength and temperature. For most practical purposes, the spectral transmittance of the glass bulb of a tungsten lamp is considered uniform in the visible region of the spectrum. The addition of a small quantity of halogen (chlorine, fluorine, bromine, and iodine) usually provides for a regenerative halogen cycle. When this added compound comes in contact with the hotter filament, tungsten is redeposited on the filament.

In addition to thermal radiators, there is an important class of light sources e.g., carbon arc lamps, mercury vapor lamps, fluorescent lamps, and flashtubes—that are referred to as **electric discharge lamps**. Basically the electric discharge lamps involve passing electric current through either a gas or vapor. Upon collision, accelerated electrons give up their kinetic energies to the atoms, molecules, and ions that subsequently emit radiation as they return to their normal energy state. Finally, there are light-emitting diodes (LEDs) and lasers that are described in greater detail in one or more of the subsequent chapters.

Radiant flux, for most active sources, is determined by integrating either the spectral radiance or spectral exitance function over the wavelength region of interest. The passive sources are generally modeled as blackbody radiators; however, such simplification may introduce significant error in the calculation of the corresponding optical signatures. At a constant temperature, the total emitted radiant flux of a radiator, which is made of a single material and whose spectral radiant exitance is $M_{e\lambda}$, is obtained as

$$\Phi_e = \int\int M_{e\lambda}\varepsilon_{e\lambda}\,d\lambda\,dA_s \tag{2.72}$$

where $\varepsilon_{e\lambda}$ is generally a constant for graybodies and dA_s is the elemental area.

One approach to using Eq. (2.72) to suit the complexities of a realistic radiator is to treat the radiator as if it were made up of a number of radiating subareas. In like manner, the reflected part of the optical signature due to a number of ambient sources is obtained by adding the effects of each individual ambient source. The total reflected radiant flux is thus given by

$$\Phi_r = \sum_m \int\int E_{e\lambda,m}\rho_{e\lambda}\,d\lambda\,dA_s \tag{2.73}$$

where $E_{e\lambda,m}$ is the spectral irradiance of the mth ambient source that is irradiating the radiator and $\rho_{e\lambda}$ is the spectral reflectivity.

2.5 Temperature

The thermodynamic temperature of a radiator can be determined by any one of the four following radiometric schemes:

1. By measuring its total radiation
2. By measuring its radiation only at a particular wavelength
3. By determining the wavelength distribution of its radiation
4. By determining the apparent color of its radiation

The corresponding thermodynamic temperatures are respectively called radiation temperature, brightness temperature, difference (or ratio) temperature, and color temperature.

Radiation temperature T_R of a radiator is defined as the temperature of a blackbody that emits the same total radiance or radiant exitance. Thus, using Eq. (2.56), it is given by

$$T_R = \left(\frac{M_e}{\sigma}\right)^{1/4} \tag{2.74}$$

Using Eqs. (2.62) and (2.44), the radiation temperature of a radiator is often approximated by

$$T_R = \begin{cases} \varepsilon_e^{1/4} T & \text{gray body} \\ \left[\int_0^\infty \varepsilon_e \dfrac{2\pi hc^2}{\sigma\lambda^5} \dfrac{1}{e^{hc/k\lambda T} - 1} d\lambda\right]^{1/4} & \text{colored body} \end{cases} \tag{2.75}$$

where T is the absolute temperature of the body and ε_e is its emissivity. The relative errors involved in this approximation are, respectively,

$$\text{Error} = \begin{cases} 1 - (\varepsilon_e)^{1/4} & \text{gray body} \\ 1 - T\left[\int_0^\infty \varepsilon_e(\lambda) \dfrac{2\pi hc^2}{\sigma\lambda^5} \dfrac{1}{e^{hc/k\lambda T} - 1} d_\lambda\right]^{-1/4} & \text{colored body} \end{cases} \tag{2.76}$$

Brightness temperature T_B of a body is defined as the true temperature of a blackbody radiator that yields the same radiance within a narrow spectral band as the body itself. Thus, using Eq. (2.62), emissivity can be expressed in terms of T_B as

$$\varepsilon_e = \frac{e^{hc/k\lambda T} - 1}{e^{hc/k\lambda T_B} - 1} \tag{2.77}$$

Now solving Eq. (2.77) for T_B, one obtains

$$T_B = \left[\frac{k\lambda}{hc} \ln\left(\frac{e^{hc/k\lambda T} - 1}{\varepsilon_e} + 1\right)\right]^{-1} \tag{2.78}$$

When held at the same temperature T, a blackbody emits more radiation than a graybody at all wavelengths. Therefore, T_B is always less than T.

Difference temperature T_D of a radiating body is equal to the true temperature of the blackbody that has the same spectral distribution as the body in question. For two chosen wavelengths λ_1 and λ_2 the ratio of the spectral radiant exitances is given by

$$\frac{M_{e\lambda}|_{\lambda=\lambda_1}}{M_{e\lambda}|_{\lambda=\lambda_2}} = \frac{\dfrac{\varepsilon_{e1}}{\lambda_1^5}[e^{hc/k\lambda_2 T} - 1]}{\dfrac{\varepsilon_{e2}}{\lambda_2^5}[e^{hc/k\lambda_1 T} - 1]} \tag{2.79}$$

where ε_{e1} and ε_{e2} are the respective emissivities. The difference temperature T_D is usually determined by solving Eq. (2.79). Using Wien's approximation, we find that

$$T_D = \frac{hc\lambda_1\lambda_2}{k(\lambda_1 - \lambda_2)\ln\left[\dfrac{M_{e\lambda_1}}{M_{e\lambda_2}}\dfrac{\varepsilon_{e2}}{\varepsilon_{e1}}\left(\dfrac{\lambda_1}{\lambda_2}\right)^5\right]} \tag{2.80}$$

It is essential at this point to discuss the chromaticity curve (known also as the CIE diagram) because of its significance in the definition of color temperature. But, again, this discussion is not possible without defining the concept of the tristimulus values. The **tristimulus values** of a color correspond to the amounts of the three primary colors (matching stimuli) which, in combination, have the characteristics identical to those of the given color. The CIE chromaticity coordinates are simply the ratios of each tristimulus value of the color to their sum. The chromaticity of a color in turn is defined as the color quality of a stimulus expressed in terms of the chromaticity coordinates. It is quite possible that the spectral distribution of a source may not exactly conform to Planck's finding but nevertheless may have the same or nearly same chromaticity as that of a blackbody at some temperature of the radiator. This particular temperature is referred to as the color temperature.

The CIE standard of 1931 involves the average tristimulus values of an average human with an angular subtense at the eye of the observer limited to a maximum of 4°. However, the CIE standard of 1964 redefines the values in terms of an angular subtense of more than 4° at the eye of the observer. The two CIE diagrams are shown in Fig. 2.14, where x and y are the corresponding coordinates of the 1931 curve and x_{10} and y_{10} are the corresponding coordinates of 1964 curve.

A radiator having a set of normalized chromaticity coordinates also has a set of relative radiant exitances such that

$$\frac{M_{e\lambda}|_{\lambda=\lambda_1}}{M_{e\lambda}|_{\lambda=\lambda_2}} = \frac{\lambda_2^5[e^{hc/k\lambda_2 T_C} - 1]}{\lambda_1^5[e^{hc/k\lambda_1 T_C} - 1]} \tag{2.81}$$

where T_C is the **color temperature**. T_C in turn is determined from the solution of the equation:

$$\frac{e^{hc/k\lambda_2 T_C} - 1}{e^{hc/k\lambda_1 T_C} - 1} = \frac{\varepsilon_{e1}(e^{hc/k\lambda_2 T} - 1)}{\varepsilon_{e2}(e^{hc/k\lambda_1 T} - 1)} \tag{2.82}$$

where ε_{e1} and ε_{e2} are the emissivities, respectively, at λ_1 and λ_2. Fig. 2.15 shows the color-temperature curve that corresponds to the spectral distributions as given by Eq. (2.82). Table 2.4 lists the locus of the color-temperature curve from 1000 K to infinity. The color temperature of a blackbody (or graybody) is the same as its absolute temperature. For all other radiators, color temperature has no relation to its absolute temperature. The blue sky, for example, has a color temperature in the neighborhood of 20,000 K whereas an overcast sky may have a color temperature of about 7000 K.

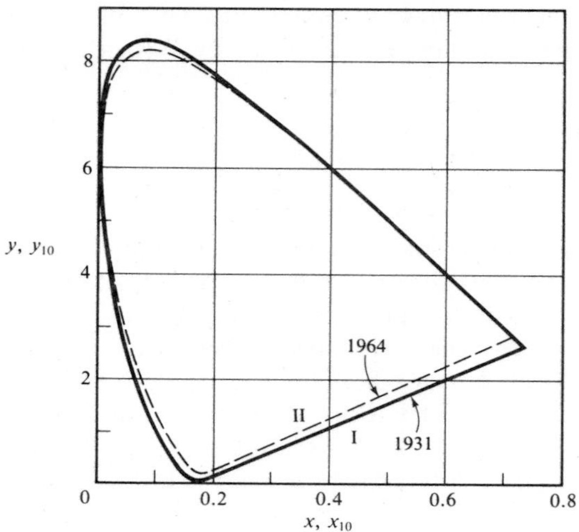

FIGURE 2.14

The CIE diagrams. [Adapted from D. B. Judd and G. Wyszecki, *Color in Business, Science & Industry*, 3rd ed. (New York: Wiley, 1975). © 1975 by John Wiley & Sons, Inc.]

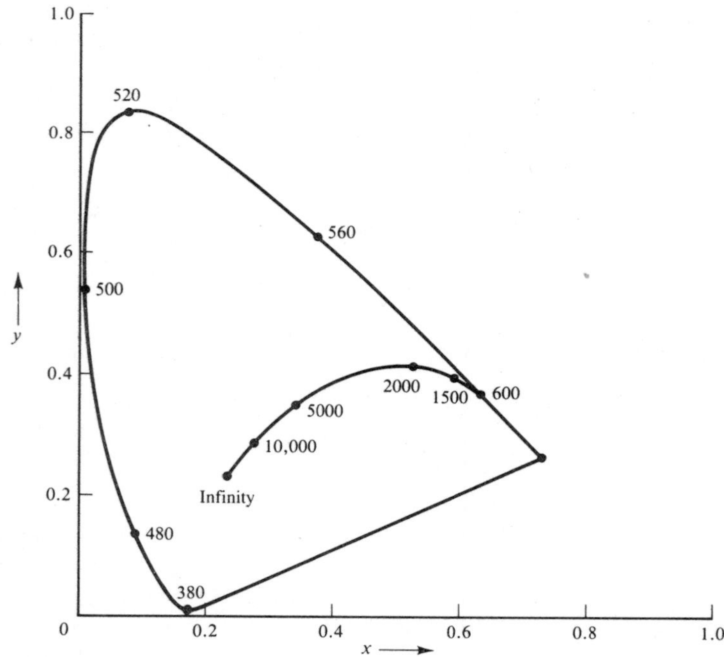

FIGURE 2.15

The color-temperature curve

TABLE 2.4

The Coordinates for the Color-Temperature Curve of Fig. 2.15

Temperature (K)	*Chromaticity Coordinates* x	y
1000	0.6528	0.3444
1500	0.5857	0.3931
2000	0.5267	0.4133
2500	0.4770	0.4137
3000	0.4369	0.4041
3500	0.4053	0.3907
4000	0.3805	0.3768
4500	0.3608	0.3636
5000	0.3451	0.3516
5500	0.3325	0.3411
6000	0.3221	0.3318
6500	0.3135	0.3237
7000	0.3064	0.3166
7500	0.3005	0.3103
8000	0.2952	0.3048
8500	0.2908	0.3000
9000	0.2869	0.2956
10000	0.2807	0.2884
20000	0.2560	0.2570
30000	0.2501	0.2489
Infinity	0.2392	0.2345

Problems

2.1 Show that the total flux emitted by an isotropic source, $\Phi_{e,\text{tot}}$, is given by $\Phi_{e,\text{tot}} = 4\pi I_e$, where I_e is the radiant intensity.

2.2 Show that for a Lambertian source, $M_e = \pi L_e$.

2.3 (a) Calculate the irradiance E_e for a Lambertian disk source of radius a at a point P, a distance R from the center of the source. (b) What happens if a is increased drastically?

2.4 (a) Calculate the irradiance E_e for a Lambertian spherical source of radius a at a point P at a distance R from the center of the source. (b) Show that at great distances, the observer cannot differentiate between a disk and a spherical source. [Use the answer of Problem 2.3(a).]

2.5 Express the monochromatic radiant exitance in terms of frequency.

2.6 Determine the relative error involved in Wien's approximation of radiation given by Eq. (2.48).

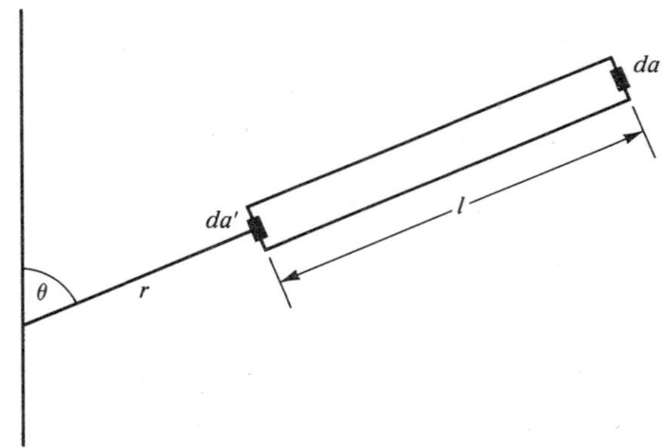

FIGURE 2.P1

2.7 Planck's law in photon flux units is given by

$$M_{e\lambda}(T) = \frac{2\pi c}{\lambda^4} \frac{1}{e^{hc/k\lambda T} - 1} \qquad \text{(photons/s-m}^2\text{-}\mu\text{m)}$$

Derive the Wien's displacement law for this photon flux blackbody curve. How does the wavelength corresponding to maximum photon flux differ from that corresponding to the maximum radiant power?

2.8 Calculate $F(\lambda, T)$ of a blackbody in the visible spectral band at a temperature of 7500 K.

2.9 A detector of area da is placed within the back of a housing of length l having a front opening of area da'. The detector housing is placed at a distance r from the source as shown in Fig. 2.P1. Show that the radiance is given by $(d^2\Phi_e/da\,da')l^2$.

2.10 The radiance of semiconductor lasers is proportional to $\cos^m \theta$ where $15 \leqslant m \leqslant 20$. Determine the power radiated into a cone whose apex angle is θ_a.

2.11 Write a computer program to plot:
(a) relative radiant exitance:

$$\int_0^\lambda \frac{M_{e\lambda}}{\sigma T^4}\, d\lambda$$

(b) relative photon flux density:

$$\int_0^\lambda \frac{M_{ep,\lambda}}{\sigma_p T^3}\, d\lambda$$

(c) integral of radiant exitance:

$$\int_0^\lambda \frac{\partial M_{e\lambda}}{\partial T}\, d\lambda$$

(d) integral of photon contrast:

$$\int_0^\lambda \frac{\partial M_{ep,\lambda}}{\partial T}\, d\lambda$$

as a function of λ (in micrometers) for $T = 200$ K, 300 K, 400 K, 500 K, and 600 K. State the physical significance of the curves where

$$M_{ep,\lambda} = \frac{2\pi c}{\lambda^4} \frac{1}{e^{hc/k\lambda T} - 1} \qquad \text{(photons/s-m}^2\text{-}\mu\text{m)}$$

and $\sigma_p = 1.5204 \times 10^{11}\ \text{sec}^{-1}\ \text{cm}^{-2}\ \text{K}^{-3}$.

2.12 Determine the brightness temperature when $hc/k\lambda T \gg 1$. Estimate the relative error for such a simplification.

2.13 Determine the equation equivalent to Eq. (2.80) when Wien's approximation is invalid.

2.14 For an object at infinity, show that Eq. (2.26) becomes

$$E_{ei} = \frac{\pi L_{eo}}{4} \left(\frac{n_i}{n_o}\right)^2 \frac{1}{F^2}$$

where F number is defined as the ratio of the focal length and lens diameter, and n_o and n_i are the refractive indices of the object and image space, respectively.

2.15 (a) The radiant exitance per unit frequency interval is given by $M_{ev} = cu_{ev}/4$. Show that M_{ev} has its maximum value at frequency ν_m provided that $\nu_m/T = 5.878887 \times 10^{10}$/s-K.
(b) Show that the maximum number of photons emitted per unit time per unit frequency interval in all directions in the solid angle 2π by a unit area of a blackbody at temperature T occurs at a frequency ν_p such that $\nu_p/T = 3.320552 \times 10^{10}$/s-K.
(c) Show that the total number of photons N of all frequencies is related to the temperature of the blackbody by

$$N/T^3 = 1.520431 \times 10^{15}/\text{m}^2\text{-s-K}^3$$

2.16 Consider Fig. 2.3 and the following cases: (a) The detector is rotated such that its normal coincides with the direction of propagation. (b) The detector is moved along the direction of propagation, parallel to itself, and is placed at a distance r from the source. (c) The detector of case (b) is rotated so that its normal coincides with the direction of propagation. Show that the radiant flux for the preceding cases has $\cos^3\theta$, $\cos^2\theta$, and $\cos\theta$ dependence, respectively. Comment on the cosine-dependence characteristics for each of the preceding cases when the normal to the source is made to coincide with the direction of propagation.

2.17 Show that the irradiance at the detector for an extended source, which is imaged by a biconvex lens onto the detector (placed at the focal plane), is $L_e \, dA_l \cos^4 \theta / f^2$, where θ is the angle between the optic axis and the line joining the lens and the detector, dA_l is the lens area and f is the focal length.

References and Suggested Readings

Boyd, R. W. *Radiometry and the Detection of Optical Radiation.* New York: Wiley, 1983.

Burkhard, D. G., and D. L. Shealy. "Simplified Formula for the Illuminance in an Optical System." *Appl. Opt.* Vol. 20 (1981): 897.

Carreras, C., and A. Corrons. "Absolute Spectroradiometric and Photometric Scales Based on an Electrically Calibrated Pyroelectric Radiometer." *Appl. Opt.* Vol. 20 (1981): 1174.

Carter, R. C., and E. C. Carter. "High Contrast Sets of Colors." *Appl. Opt.* Vol. 21 (1982): 2936.

Colclaser, R. A., and S. Diehl-Nagle. *Materials and Devices for Electrical Engineers and Physicists.* New York: McGraw-Hill, 1985.

Dereniak, E. L., and D. G. Crowe. *Optical Radiation Detectors.* New York: Wiley, 1984.

Farmer, A. J. D. "Fourier Approach to Absolute Spectroradiometry." *Appl. Opt.* Vol. 20 (1981): 3050.

Frieberg, A. T. "Energy Transport in Optical Systems with Partially Coherent Light." *Appl. Opt.* Vol. 25 (1986): 4547.

Hoover, H. L. "Sunglasses, Pupil Dilation, and Solar Ultraviolet Irradiation of the Human Lens and Retina." *Appl. Opt.* Vol. 26 (1987): 689.

Jannson, T. "Radiative Transfer Function." *J. Opt. Soc. Am.* Vol. 70 (1980): 1544.

Kessel, J. "Transmittance Measurement in the Integrating Sphere." *Appl. Opt.* Vol. 25 (1986): 2752.

Keyes, R. J., (Ed.). *Optical and Infrared Detectors* (2nd ed.). Berlin: Springer-Verlag, 1980.

Kingston, R. H. *Detection of Optical and Infrared Radiation.* Berlin: Springer-Verlag, 1979.

Klein, M. V. *Optics.* New York: Wiley, 1970.

Marion, J. B., and M. A. Heald. *Classical Electromagnetic Radiation* (2nd ed.). Orlando, FL: Academic Press, 1980.

Nicodemus, F. E. "Radiometry," in *Applied Optics and Optical Engineering* Vol. 9. Shannon, R. R., and J. C. Wyant (Eds.). New York: Academic Press, 1983.

Ohwada, Y. "Numerical Calculation of Effective Emissivities of Diffuse Cones with a Series Technique." *Appl. Opt.* Vol. 20 (1981): 3332.

Pancove, J. I. (Ed.). *Display Devices.* Berlin: Springer-Verlag, 1980.

Pinson, L. J. *Electro-Optics.* New York: Wiley, 1985.

Pradhan, M. M., and R. K. Grag. "Pyroelectric Null Detector for Absolute Radiometry." *Appl. Opt.* Vol. 21 (1982): 4456.

Tardy, H. L. "Flux Concentrators in Integrating Sphere Experiments: Potential for Increased Detector Signal." *Appl. Opt.* Vol. 24 (1985): 3914.

Taylor, J. H. "Radiation Exchange." *Appl. Opt.* Vol. 26 (1987): 619.

Taylor, J. H. "Key Ideas of Radiation Exchange: Energy Associated with Electromagnetic Radiation." *Opt. News.* Vol. 13 (January 1987): 26.

Taylor, J. H. "Key Ideas of Radiation Exchange: Planck's Radiation Law." *Opt. News.* Vol. 13 (February 1987): 26.

Taylor, J. H. "Radiant Emissivity." *Opt. News.* Vol. 13 (April 1987): 13.

Wolfe, W. L., and G. J. Zissis (Ed.). *The Infrared Handbook.* Washington, D.C.: Office of Naval Research, Department of Navy, 1978.

Wyatt, C. L. *Radiometric System Design.* New York: McMillan, 1987.

Wyszecki, G., and W. S. Stiles. *Color Science: Concepts and Methods, Quantitative Data and Formulae* (2nd ed.). New York: Wiley, 1982.

Young, M. *Optics and Lasers Including Fibers and Integrated Optics* (2nd rev. ed.). Berlin: Springer-Verlag, 1984.

Zerlaut, G. A., and T. E. Anderson. "Multiple Integrating Sphere Spectrophotometer for Measuring Absolute Spectral Reflectance and Transmittance." *Appl. Opt.* Vol. 20 (1981): 3797.

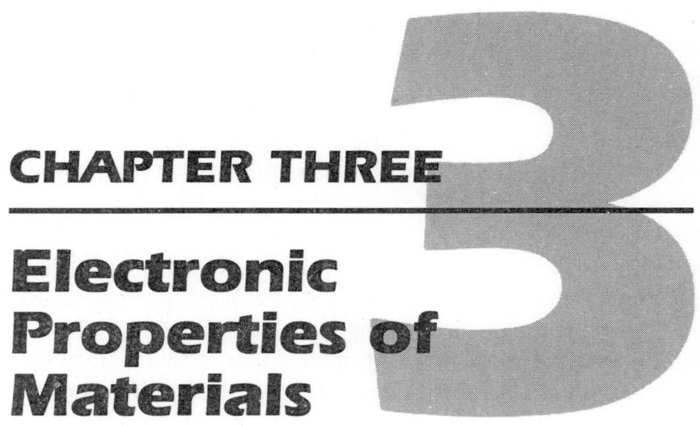

CHAPTER THREE

Electronic Properties of Materials

3.1 Introduction

In Chapter 1 we considered the behavior of a single electron when it was confined in a field of atoms. However, the overall material characteristics of a solid can be determined only by the complex behavior of multiple electrons. We have already seen that each and every electron has a specific set of quantum numbers associated with it and that no two electrons in the same atom can have the same set of quantum numbers. When atoms are isolated, as in a gas, we expect little or no interaction among the electron wave functions. But as soon as the interatomic spacing is decreased, the electrons become confined to only discrete energy levels.

In this chapter we investigate the behavior of electrons in materials suitable for fabricating electro-optic devices. The vast majority of all electro-optic devices are fabricated from a class of materials known as **semiconductors**. These semiconductors include elements such as silicon (Si) and germanium (Ge), compounds such as zinc selenium (ZnSe), and alloys such as mercury cadmium telluride ($HgCd_xTe_{1-x}$), where $0 \leqslant x \leqslant 1$. It will become evident that only a certain group of elements and alloys has semiconducting characteristics.

The geometric organization of atoms within a material plays a significant role in determining the exact characteristics of the material. Accordingly, solids are classed by the extent of their ordered regions into three types: amorphous, polycrystalline, and crystalline. Amorphous solids have no long-range atomic order, polycrystalline solids have a limited and localized atomic order, and crystalline solids have a very regular three-dimensional atomic order. However, understanding the geometric organization of atoms is not sufficient for understanding the relatively complex behavior of semiconductor materials.

The chemical elements in the widely known periodic table (see Appendix A) are normally classed into eight distinct groups. The periodic classification of elements is primarily based on the number of electrons available in the outermost subshells. In particular, atoms in elements like Si and Ge have four electrons in their outermost subshells and, therefore, they belong to a group listed in column IV of the periodic table. The compound semiconductors are formed basically by combining elements either from column III (those having three electrons in their outermost shell) and column V (those having five electrons in their outermost shell), or from column II (those having two electrons in their outermost shell) and column VI (those having six electrons in their outermost shell), or from column I (those having one electron in their outermost shell) and column VII (those having seven electrons in their outermost shell). Examples of the first type (referred to as zinc-blende structure) are GaAs, GaP, and InSb; examples of the second type are CuO and CdS; an example of the third type is AgCl.

The basic criterion in selecting a combination of elements for producing a semiconductor involves having exactly four electrons in the outermost shell of each element in the combination. For example, in the case of an isolated silicon atom (atomic number: 14; electronic configuration: $1s^2 2s^2 2p^6 3s^2 3p^2$), there are two electrons in its innermost shell (for $n = 1$), eight electrons in the intermediate shell (for $n = 2$), and four electrons in the outermost shell (for $n = 3$). Note that there can be up to eight energy levels in the outermost shell, comprising two at the s-level and six at the p-level. Consequently, in an isolated silicon atom, only half of the eight allowed states are filled. The same is true of the isolated Ge atom (atomic number: 32) whose electronic configuration is $1s^2 2s^2 2p^6 3s^2 3p^6 3d^{10} 4s^2 4p^2$. In crystalline form, the outer electrons (four from each atom) participate in a chemical bond whereby the outermost electrons are shared between the neighboring atoms. Such a chemical bond, where electrons are shared by the atoms, is referred to as the **covalent bond**. Each of the crystalline Si (or Ge) atoms has four nearest neighbors that provide a stable bonding configuration.

Interestingly, the overwhelming majority of semiconductors form a special crystalline order called the **diamond lattice**. The diamond lattice structure shown in Fig. 3.1 is formed of cube-shaped unit cells, each of which has eight corner atoms, six edge-centered atoms, and four interior atoms located in such a way that each atom in the lattice always has four near neighbors.

The four column IV elements whose atoms form diamond lattices are C, Si, Ge, and Sn. Each of the four valence electrons of any atom in the diamond lattice structures can be considered as shared between it and one of its four near neighbors, with each valence electron belonging simultaneously to two of the atoms. In effect, each atom in its outer subshell has eight shared electrons that form a very stable low-energy covalent structure. In comparison, a zinc-blende structure is mostly a covalent type but partially ionic. At room temperature, the unit cell of a crystalline Si lattice has its cubic sides equal to about

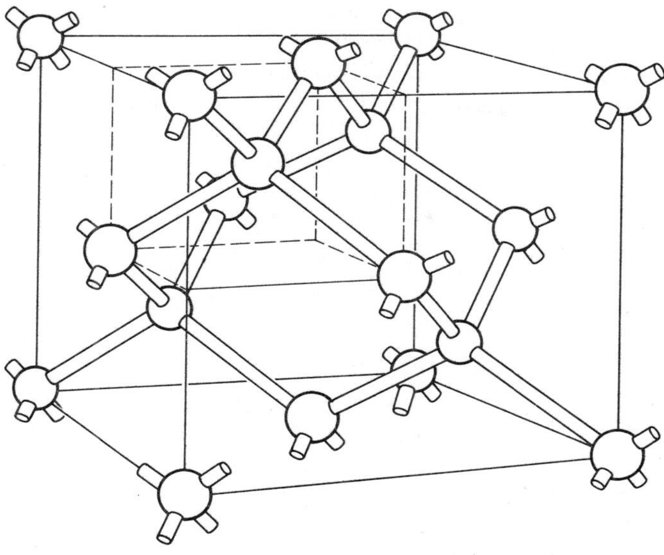

FIGURE 3.1
A unit cell in a diamond lattice

5.43×10^{-10} m, which amounts to an atomic density of about 5×10^{28} atoms/m^3 for the Si lattice.

In the following section, we investigate several important aspects of semiconductors: energy-band models, carrier propagation, doping, carrier distribution, and equations of state. Subsequent sections are devoted to junction diodes and bipolar transistors. The corresponding switching transient responses and frequency responses are not covered here since these phenomena are not critical to the understanding of electro-optic devices. However, any of the books listed at the end of the chapter can be used for reference.

3.2 Semiconductors

From the discussion of the Krönig-Penney model in Section 1.5.4, it is obvious that the energy of an electron, moving in a periodically varying potential field, is limited to only certain ranges. Electron energies outside of these allowed ranges are strictly prohibited. An isolated Si atom, for example, has four valence electrons in its outermost shell. But when they are in a group, Si atoms combine to form a diamond crystal lattice structure, in such a way that each of these atoms has exactly four near neighbors. This lattice arrangement is a consequence of the fact that of the eight allowed energy states in the outermost

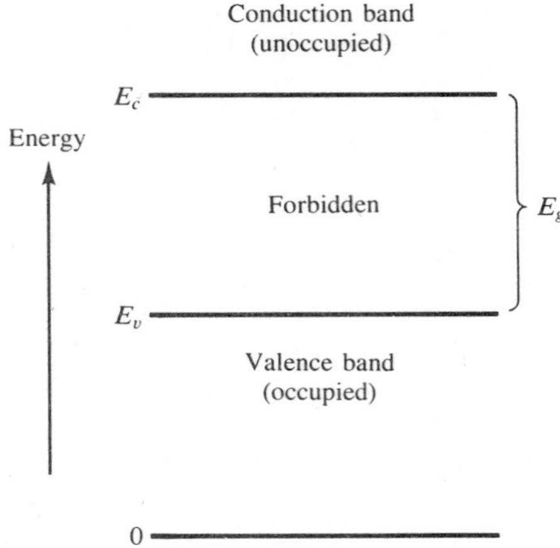

FIGURE 3.2
The energy-band diagram of a semiconductor

shell of an isolated Si atom, only four are occupied. With four neighboring silicon atoms, it becomes possible for the central atom to share its outermost electrons with the neighboring atoms, thus resulting in a stable atomic system. This simplistic chemical-bond model is, however, inadequate to describe the energy-related aspects of the semiconductors. In the next subsection we examine the energy-band model to describe important semiconductor characteristics.

3.2.1 Energy-Band Diagrams

When, for example, n isolated Si atoms are brought together to form an n atom Si crystal, the resulting crystal allows for a total of $8n$ allowed states. However, the distribution of energy states within a crystalline structure is governed by the Krönig-Penney model. Accordingly, exactly half of these energy states ($4n$) remain depressed in energy while the other half increase in energy. This phenomenon allows for the introduction of a forbidden energy region between the two allowable bands. The upper band of the allowed states forms the **conduction band**, while the lower band of the allowed states forms the **valence band**, and the disallowed band forms the forbidden gap, or **band gap**, as shown in Fig. 3.2.

It is natural for the electrons to first occupy the lowest possible energy states. Consequently, in an n-atom crystal, the outermost $4n$ electrons are expected to occupy the lowest $4n$ valence states. Thus, at temperatures approaching $T = 0$ K, the conduction band remains mostly empty and the valence band remains mostly occupied. The numerical difference between the lowest conduction-band energy value E_c and the highest valence-band energy value E_v is referred to as the **band-gap energy** E_g. As the temperature increases, the concentration of electrons in the conduction band increases and that in the valence band decreases. The introduction of an energy exceeding E_g into the crystal allows the electrons to move from the valence band to the conduction band.

When an amount of energy exceeding E_g is introduced into the crystalline structure, the atomic bonds of the crystal are reorganized whereby one Si–Si bond per atom is broken for one or more atoms. The breaking of each Si–Si bond in turn causes a valence-band electron to migrate into the conduction band, thus introducing in it a free carrier. As soon as the free carrier of charge $-e$ is produced in the conduction band, a void (absence of an electron) with a net positive charge $+e$ is generated in the valence band. The **void**, known also as a **hole**, is also able to move freely within the valence band. Any one of the neighboring valence band electrons can drift easily into the hole location because such a drift requires very little additional energy. This movement of a valence-band electron introduces a new hole at the location of its origin. A hole is thus conceptualized as another subatomic particle that can move around. Such random movement of free electrons and holes results in a zero mean current. But as soon as an electric field is applied, the movement of the charge carriers is affected, resulting in a drift of the carriers in a particular direction.

The energy-band model, introduced in this subsection, enables us to characterize material properties as well. In the case of insulators, E_g is very large, and that generally prohibits the excitation of electrons from the valence band to the conduction band. In the case of conductors, however, either there is no gap or the gap is very minimal, and that results in an abundance of carriers in the conduction band. Semiconductors fall in between these two types—insulators and conductors—and possess an E_g value that is neither too large nor too small. Table 3.1 lists some of the sample values of E_g for a few elements and compounds. It is interesting to note that while Si and Ge are semiconductors, carbon (C) is an insulator and tin (Sn) is a conductor, even though all four of these are column IV elements.

TABLE 3.1

Band-Gap Energies

Material	*Band-Gap Energy (in eV)*	*Material*	*Band-Gap Energy (in eV)*
Si	1.12	GaAs	1.42
Ge	0.66	InAs	0.36
C (diamond)	5.47	InSb	0.17
Sn	0.08	GaP	2.26
ZnO	3.35	CdSe	1.70
AlAs	2.16	InP	1.35

A semiconductor is termed **intrinsic** when it contains few or no impurity atoms. Practically, it may be impossible to realize an absolutely pure semiconductor. Thus, as long as the number of impurity atoms is absolutely minimal so as not to affect the general characteristics of the host atoms, the semiconductor is referred to as intrinsic. All of the conduction-band free electrons of an intrinsic semiconductor are the products of thermal excitation. For every free electron present in the conduction band there is a hole present in the valence band. Thus, in an intrinsic semiconductor, the number of electrons per volume n is exactly equal to the number of holes per volume p. Thus

$$n = p \equiv n_i \tag{3.1}$$

where n_i is 1.45×10^{16} electrons/m^3, 2.4×10^{19} electrons/m^3, and 1.79×10^{12} electrons/m^3 at room temperature in Si, Ge, and GaAs, respectively. Equation (3.1) is true as long as a hole is generated along with the generation of each free electron. Note that each cubic meter of an intrinsic Si crystal contains approximately 5×10^{28} atoms. Comparison of this number with the value of n_i convinces us that only a small fraction of Si–Si bonds are actually broken at room temperature due to thermal energy.

The concentration of carriers in an intrinsic semiconductor is rather too small to fabricate useful electro-optic devices out of intrinsic materials. We show later that devices such as photoconductors that are made using intrinsic material need to be relatively large in area to produce a significant device response. The number of charge carriers and holes in a semiconductor could be drastically increased by intentionally introducing appropriate impurities into the intrinsic crystal lattice. This technique, called **doping**, involves the addition of dopants in controlled amounts; this addition changes the intrinsic semiconductor into an **extrinsic** semiconductor.

Dopants belonging to column V of the periodic table cause an increase of free electrons, whereas those of column III cause an increase of holes. The resulting extrinsic semiconductors are referred to respectively as n-type (donor-doped material having more electrons than holes) and p-type (acceptor-doped material having more holes than electrons) semiconductors. Consider, for example, that a column IV atom (having four valence electrons) of an intrinsic semiconductor is being replaced by a column V dopant atom (having five valence electrons) of about the same size. Four of the valence electrons of the dopant will take part in the covalent bonding with the neighboring column IV atoms, while the fifth electron will be only weakly attached to the dopant atom. When the intrinsic semiconductor is suitably doped, the weakly attached electrons of the dopants contribute to the pool of free electrons. Some of the more common donors are phosphorus (P), arsenic (As), and antimony (Sb), while boron (B), indium (In), aluminium (Al), and gallium (Ga) are the more common acceptors. A column V dopant has five electrons in its outermost shell; thus while four of the five electrons fit snugly into the bonding structure, the fifth electron wanders freely. Again, since a column III dopant has three electrons in its outermost shell, it participates in the crystal lattice with

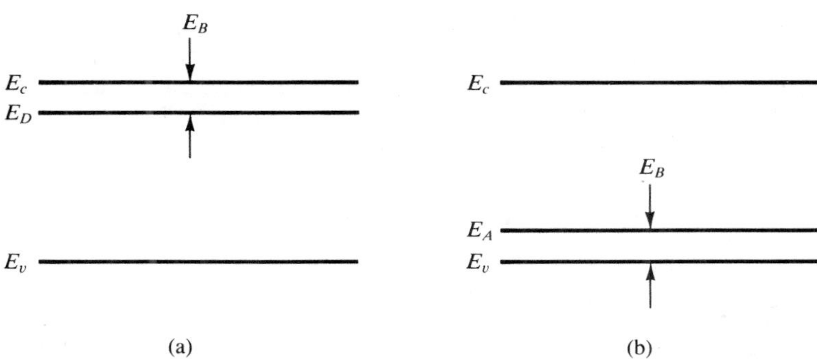

FIGURE 3.3

Energy-band diagram for extrinsic semiconductor: (a) *n*-type and (b) *p*-type

one of the four bonds broken due to the presence (absence) of a (an) hole (electron). Free electrons in the *n*-type and holes in the *p*-type semiconductors are called **majority carriers**, whereas holes in the *n*-type and free electrons in the *p*-type semiconductors are called **minority carriers**.

The addition of dopants does not violate the overall charge neutrality. An *n*-type material, for example, has excess free electrons, but it also has an equal number of ionized, positively charged donor atoms. However, excessive doping densities may result in undesirable stresses in the semiconductor crystals. More importantly, the generated carriers can start to influence each other, thus damaging the desirable electrical properties of the semiconductor. Such an excessively doped semiconductor, often referred to as **degenerate**, violates the constraints of Pauli's exclusion principle.

The doping mechanism is also explainable in terms of the energy-band model. The donor energy level E_D lies just below E_c such that the difference of the two levels equals the corresponding donor binding energy E_B. At room temperature, the donor electrons acquire sufficient energy to transform themselves to conduction-band carrier electrons. Likewise, the presence of acceptors introduces an acceptor energy level E_A near the bottom of the band gap but slightly above E_v, so that the difference between E_v and E_A is equal to the acceptor binding energy. At room temperature, the valence-band electrons jump onto the acceptor levels, leaving behind excess holes in the valence band. Consequently, the intrinsic semiconductor becomes *p*-type. Figure 3.3 illustrates the energy band in a doped material, while Table 3.2 lists dopant-site binding energies for a number of familiar acceptors and donors.

Thermal generation is not the only means by which carriers are generated in semiconductors. Absorption of optical energy is often a cause for generating carriers. In an intrinsic semiconductor, a valence-band electron can absorb the energy (if equivalent to E_g) of a photon and, as a result, move to an allowed state in the conduction band, generating a free electron and a hole. Likewise, a bound electron and a hole are generated in an acceptor-doped semiconductor if the optical energy is equivalent to the acceptor binding energy. The photogenerated excess carriers increase the conductivity of the semiconductor. This particular type of conductivity is referred to as **photoconductivity**. Some of the more commonly used photoconductive materials are PbSe, PbTe, PbS, and CdS, which have band-gap energies of 0.27, 0.31, 0.37, and 2.40 eV, respectively.

3.2.2 Carrier Distribution

It is important to determine the distribution characteristics of energy levels in both conduction and valence bands. The energy scenario of the semiconductor is somewhat similar to the energy distribution of multiple free electrons confined in an infinite square potential well. We have already considered the case of a single electron confined in an infinite one-dimensional potential well in Section 1.5.3. By extending this one-dimensional problem to three dimensions, we find the quantized energy of an electron using Eq. (1.125) to be

$$E_q = \frac{\pi^2 \hbar^2}{2m_n^* a^2} q^2 \tag{3.2}$$

where m_n^* is the effective mass of the electron as defined in Eq. (1.148), a is the well width, and q is the quantum number such that

$$q^2 = q_x^2 + q_y^2 + q_z^2 \tag{3.3}$$

where q_x, q_y, and q_z are quantum numbers corresponding to the three Cartesian coordinates.

The set of values of q_x, q_y, and q_z determines a specific energy state E_q. Equal values of E_q generally lie on the surface of a sphere whose radius is q. The

TABLE 3.2

Binding Energies for the Dopant-Sites

Host Semiconductor	*Si (eV)*	*Ge (eV)*
Donor		
Antimony (Sb)	0.039	0.0096
Arsenic (As)	0.049	0.0127
Phosphorus (P)	0.045	0.012
Acceptor		
Aluminium (Al)	0.057	0.0102
Boron (B)	0.045	0.0104
Gallium (Ga)	0.065	0.0108
Indium (In)	0.16	0.0112

number of quantum states η with energy equal to or less than E_q is proportional to the volume of this sphere. However, only one-eighth of the sphere's volume is to be considered since the quantum numbers are all nonnegative. Thus, using Eq. (3.2),

$$\begin{aligned}\eta &= \left(\frac{1}{8}\right)\left(\frac{4}{3}\right)\pi q^3 \\ &= \frac{\pi}{6}\left[\frac{2m_n^* a^2 E}{\pi^2 \hbar^2}\right]^{3/2}\end{aligned} \tag{3.4}$$

The density of energy states, defined as the number of energy states per unit volume a^3 in the energy interval E and $E + dE$, is thus given by

$$\frac{1}{a^3}\frac{d\eta}{dE}dE = \frac{1}{a^3}\frac{\pi}{4}\left[\frac{2m_n^* a^2 E}{\pi^2 \hbar^2}\right]^{1/2}\left(\frac{2m_n^* a^2}{\pi^2 \hbar^2}\right)dE = \frac{m_n^*}{\pi^2 \hbar^3}\sqrt{\frac{m_n^* E}{2}}\,dE \tag{3.5}$$

Like the effective electron mass, the effective mass of a hole m_p^* is also different from the hole rest mass m_e. Electrons and holes can be treated as classical Newtonian particles as long as their effective masses are used in place of the corresponding rest masses. The effective carrier masses for a few of the important semiconductors are listed in Table 3.3.

The density of states at an energy E not too distant from the band edges, denoted by $\eta_c(E)$ and $\eta_v(E)$, respectively, for the conduction and valence bands, can be given by

$$\eta_c(E) = \frac{m_n^*}{\pi^2 \hbar^3}\sqrt{\frac{m_n^*(E - E_c)}{2}} \qquad E \geqslant E_c \tag{3.6}$$

and

$$\eta_v(E) = \frac{m_p^*}{\pi^2 \hbar^3}\sqrt{\frac{m_p^*(E_v - E)}{2}} \qquad E \leqslant E_v \tag{3.7}$$

TABLE 3.3

Effective Masses

Semiconductor	m_p^*/m_e	m_n^*/m_e
CdSe	0.45	0.130
GaAs	0.082	0.068
GaP	0.60	0.82
Ge	0.04–0.28	0.082–1.64
InAs	0.4	0.023
InSb	0.4	0.015
PbS	0.25	0.25
PbTe	0.2	0.17
Si	0.16–0.49	0.19–0.98

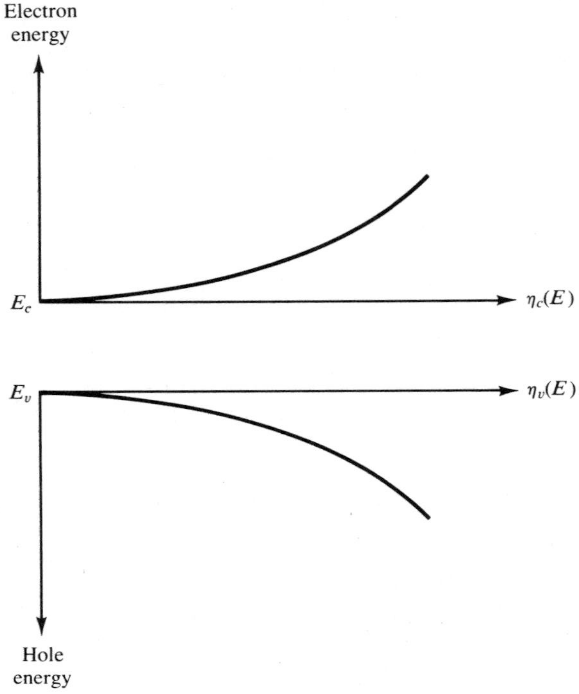

FIGURE 3.4

Density of states within the bands

The density of states is exactly zero at the band edges, but it increases as the square root of the energy as one goes farther away from the band-edge energies. The characteristics of the density of states is illustrated in Fig. 3.4.

Equations (3.6) and (3.7) describe the distribution of states at a given energy E, but they do not identify which of these energy states are actually occupied by carriers. It is, therefore, imperative to look for a distribution function for electrons and holes. Provided the number of the carriers and their total energy are known, the distribution function will provide the probability of finding these particles at any specific energy. We have already encountered the quantum-mechanical probability distribution expression for photons, referred to as the Bose-Einstein distribution function, in Eq. (2.42). But photons are quite different from either electrons or holes because, while an unlimited number of photons can be accommodated in an allowed state, no two electrons (or holes) can ever have the same energy state (as per Pauli's exclusion principle).

A particular probability distribution function, known as the Fermi-Dirac function, $f_{\text{FD}}(E)$, is generally used to quantify the probability that an available energy state is occupied by a Fermion (i.e., electrons and holes). The

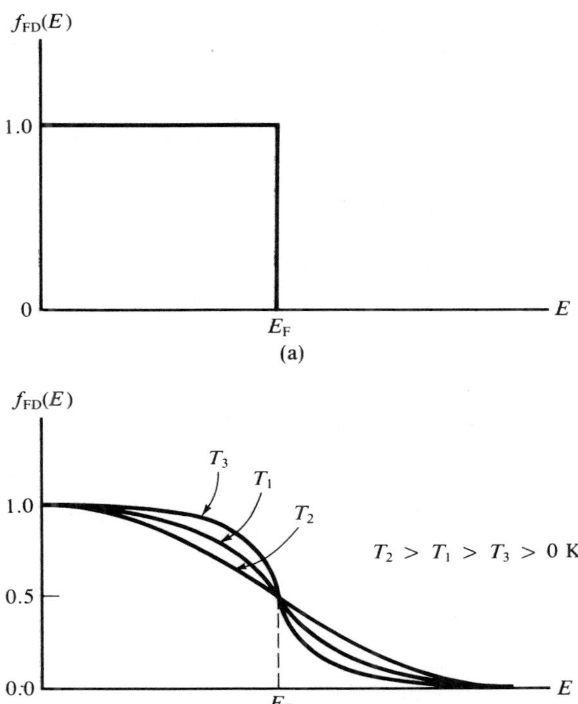

FIGURE 3.5

Fermi function at (a) absolute zero and (b) $T > 0$ K

Fermi-Dirac function, at thermodynamic equilibrium, is given by

$$f_{FD}(E) = \frac{1}{1 + e^{(E - E_F)/kT}} \tag{3.8}$$

in terms of Fermi energy E_F which is the particular energy at which the Fermi-Dirac function is 0.5. At temperature $T \to 0$ K, $f_{FD}(E)$ is unity when $E < E_F$ and is zero when $E > E_F$. Figure 3.5 shows the Fermi function at both absolute zero and also at $T > 0$ K. Note that Eq. (3.8) reduces to the classical Maxwell-Boltzmann distribution function (for particles that are distinguishable) when $E \gg E_F$.

The distribution of carriers in the respective energy bands may now be derived from $f_{FD}(E)$ and the respective density of states. The product $\eta_c(E) f_{FD}(E)$ corresponds to the distribution of conduction-band electrons, and $\eta_v(E)[1 - f_{FD}(E)]$ corresponds to the distribution of valence-band holes. Consequently, the number of carriers is small at the band edges and also at energies far different from those of the band edges.

The exact value of E_F is an important factor in determining the exact nature of the carrier distribution. In the intrinsic semiconductor, E_F appears more or less at the middle of the band gap. In the case of the extrinsic semiconductor, when E_F is closer to E_c, the material is n-type, and when E_F is closer to E_v, the material is p-type. Consequently, a material can be classified as either intrinsic, n-type, or p-type by observing the E_F value. More importantly, Pauli's exclusion principle also affects the carrier distribution characteristics. Pauli's principle states that each energy state (described by the quantum numbers n, l, and m_l) can be occupied by only two electrons (one having a positive spin and the other having a negative spin). The total number of conduction-band electrons is thus obtained from

$$n = \int_{E_c}^{\infty} 2\eta_c(E) f_{FD}(E)\, dE \tag{3.9}$$

and similarly the total number of valence-band holes is given by

$$p = \int_{0}^{E_v} 2\eta_v(E)[1 - f_{FD}(E)]\, dE \tag{3.10}$$

Equation (3.9) is evaluated as follows:

$$n = \frac{m_n^* \sqrt{2m_n^*}}{\pi^2 \hbar^3} \int_{E_c}^{\infty} \frac{\sqrt{E - E_c}}{1 + e^{(E - E_F)/kT}}\, dE$$

$$= \frac{m_n^* \sqrt{2m_n^*}}{\pi^2 \hbar^3} (kT)^{3/2} \int_{0}^{\infty} \frac{x^{1/2}}{1 + e^{x - x_0}}\, dx \tag{3.11}$$

where

$$x = \frac{E - E_c}{kT} \tag{3.12}$$

and

$$x_0 = \frac{E_F - E_c}{kT} \tag{3.13}$$

Equation (3.11) is not an easy equation to solve. However, the tabulated value of the associated integral is generally available in mathematical handbooks. If E_F is considered to have values less than or equal to $E_c - 3kT$, then we find that

$$n = N_c e^{(E_F - E_c)/kT} \tag{3.14}$$

since

$$\int_{0}^{\infty} s^{1/2} e^{-s}\, ds = \frac{\sqrt{\pi}}{2} \tag{3.15}$$

where

$$N_c \equiv \frac{1}{\sqrt{2}} \left[\frac{m_n^* kT}{\pi \hbar^2} \right]^{3/2} \tag{3.16}$$

Similarly, Eq. (3.10) can be evaluated readily if we consider E_F to be equal to or larger than $E_v + 3kT$, yielding

$$p = N_v e^{(E_v - E_F)/kT} \tag{3.17}$$

where

$$N_v \equiv \frac{1}{\sqrt{2}} \left[\frac{m_p^* kT}{\pi \hbar^2} \right]^{3/2} \tag{3.18}$$

Whenever E_F is confined within $3kT$ of either E_c or E_v, the semiconductor is referred to as **degenerate**; otherwise the semiconductor is called **nondegenerate**. A degenerate state of the semiconductor usually corresponds to an excessive amount of doping.

A semiconductor is intrinsic when $n = p \equiv n_i$ or, in other words, when E_F is approximately equal to $E_i = (E_c + E_v)/2$. Thus, for an intrinsic semiconductor,

$$n_i = N_c e^{(E_i - E_c)/kT} = N_v e^{(E_v - E_i)/kT} \tag{3.19}$$

Accordingly, we can rewrite Eqs. (3.14) and (3.17) as

$$n = n_i e^{(E_F - E_i)/kT} \tag{3.20}$$

and

$$p = n_i e^{(E_i - E_F)/kT} \tag{3.21}$$

respectively. These two equations are valid only as long as the semiconductor is nondegenerate and when it is also true that

$$np = n_i^2 = N_c N_v e^{-E_g/kT} \tag{3.22}$$

Equation (3.22) refers to the very important fact that either hole or electron concentration in a nondegenerate semiconductor can be varied as long as the product np is a constant.

For the case of extrinsic semiconductors, we can write the general charge neutrality equation as

$$p - n + N_D - N_A = 0 \tag{3.23}$$

where N_D and N_A are numbers of donor atoms/volume and acceptor atoms/volume, respectively. At room temperature, Eq. (3.23) is valid only if the semiconductor has had sufficient thermal energy to ionize all of its donor and acceptor sites.

In an n-type material, $N_D \gg N_A$ and, thus, Eq. (3.23) can be solved using Eq. (3.22) to give

$$n = \frac{N_D}{2} + \left[\left(\frac{N_D}{2} \right)^2 + n_i^2 \right]^{1/2} \tag{3.24}$$

In practice, in the n-type semiconductor $N_D \gg n_i$ which yields

$$n \simeq N_D \tag{3.25a}$$

$$p \simeq \frac{n_i^2}{N_D} \tag{3.25b}$$

Similarly, for the p-type semiconductor, we obtain

$$p \simeq N_A \tag{3.26a}$$

$$n \simeq \frac{n_i^2}{N_A} \tag{3.26b}$$

It must be noted that n_i increases as temperature increases, and at high enough temperature n_i far exceeds both N_A and N_D. Consequently, at elevated temperature, the semiconductor becomes intrinsic since both n and p approach n_i.

Equation (3.19) may be solved using Eqs. (3.16) and (3.18) for the intrinsic energy level E_i to give

$$\begin{aligned} E_i &= \frac{E_c + E_v}{2} + \frac{kT}{2}\ln\left(\frac{N_v}{N_c}\right) \\ &= \frac{E_c + E_v}{2} + \frac{3}{4}kT\ln\left(\frac{m_p^*}{m_n^*}\right) \end{aligned} \tag{3.27}$$

Therefore, E_i (and thus Fermi energy) is not quite at the mid-point of the band gap except when $T = 0$ K. Using the values listed in Table 3.3, we find that at room temperature the Fermi energy is located slightly below the mid-point of the band gap of both Si and Ge, whereas it is located slightly above the mid-point of the band gap of GaAs. In the extrinsic materials, E_F is not a constant but changes with the doping level. With the increase in doping, E_F gradually shifts toward the band edges, and at significantly high doping the material becomes completely degenerate.

3.2.3 Current and Continuity Equations

While the last subsection described the carrier distribution in semiconductors under equilibrium conditions, this subsection explores the consequences of having perturbing forces in the semiconductor. For example, when an electric field is applied, a current is expected to flow through the semiconductor. There are three major types of carrier activity that may occur within semiconductors. These are **drift**, **diffusion**, and **recombination-generation**. All of these processes may take place simultaneously—but independently—inside the semiconductor.

The first type—the current that is caused by the movement of carriers in response to an applied electric field alone—is referred to as **drift current**. When an external field of magnitude $\mathscr{E}$ is applied across the semiconductor, the holes—

each with charge $+e$—move along the direction of the field, and the electrons—each with charge $-e$—move counter to the direction of the field. The velocity of a carrier in a semiconductor is different from that in a vacuum. In a material, the carrier velocity is often affected by repeated acceleration and retardation due to multiple collisions. The effective velocity of the carriers is often described in terms of an average drift velocity v_d. The hole drift current density is thus given by the product epv_d. In the presence of an electric field, the randomly moving hole is accelerated to a value given by $e\mathscr{E}/m_e$. The drift velocity of a hole is obtained in terms of its mean free time (i.e., the duration t_r between collisions) as

$$v_d = \frac{et_r}{m_p^*}\mathscr{E} \equiv \mu_p \mathscr{E} \tag{3.28}$$

where μ_p is a proportionality constant referred to as the mobility of holes. Over most of the useful ranges of the electric field, the drift velocity varies directly as the applied electric field. As per definition, the velocity per unit applied electric field is used to determine the mobility of the carrier.

The hole current density is given by

$$J_p|_{\text{drift}} = e\mu_p p \mathscr{E} \tag{3.29}$$

Similarly, the electron current density becomes

$$J_n|_{\text{drift}} = e\mu_n n \mathscr{E} \tag{3.30}$$

where μ_n is the mobility of the electron. It should be realized that the mobility of a carrier is not an absolute constant but a function of both doping and temperature. The characteristics of mobility for both electrons and holes are shown in Fig. 3.6. Mobilities for some of the typical semiconductors at 300 K are listed in Table 3.4.

The **total drift current** under steady state is given by

$$J_{\text{drift}} = \sigma \mathscr{E} \equiv e(\mu_n n + \mu_p p)\mathscr{E} \tag{3.31}$$

where σ is the conductivity. Alternatively, we can determine the resistivity ρ of the semiconductor as

$$\rho = \frac{\mathscr{E}}{J_{\text{drft}}} = \frac{1}{e(\mu_n n + \mu_p p)} \tag{3.32}$$

Whenever there exists a carrier gradient in the semiconductor, carriers redistribute themselves by moving from regions of higher concentrations to regions of lower concentrations, as shown in Fig. 3.7. Such movement of carriers results in a **diffusion current**—the second type of carrier activity. Random thermal motion generally works as a catalyst to this process until the carriers are uniformly distributed throughout the space. Diffusion current requires no electric field. The diffusion current density is directly proportional to the concentration gradient. For simplicity, consider that the variation of carrier

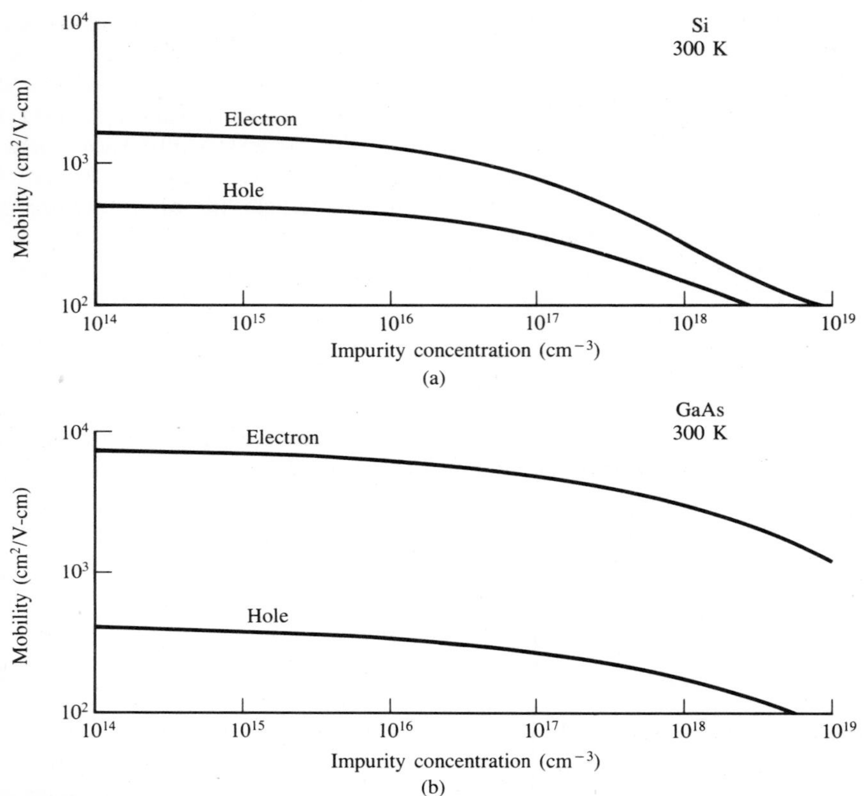

FIGURE 3.6

Carrier mobilities as functions of doping and temperature

TABLE 3.4

The Mobility for a Number of Semiconductors at 300 K

Semiconductors	μ_n	μ_p
Ge	3900	1900
Si	1500	450
GaAs	8500	400
GaP	110	75
InSb	80000	1250
InAs	33000	460
InP	4600	150
CdS	340	50
CdTe	1050	100
PbS	600	700
PbTe	6000	4000

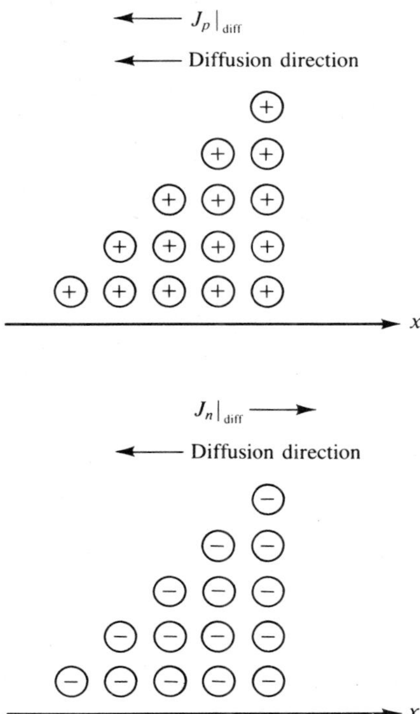

FIGURE 3.7

Diffusion of holes and electrons

concentration is dominant only along the x axis. The diffusion current densities are then given by

$$J_p|_{\text{diff}} = -eD_p\frac{dp}{dx} \tag{3.33a}$$

$$J_n|_{\text{diff}} = eD_n\frac{dn}{dx} \tag{3.33b}$$

where D_p and D_n are the respective diffusion constants. The negative sign in Eq. (3.33a) accounts for the fact that for the same type of gradient, the resulting hole and electron currents flow in opposite directions. The total current density due to both drift and diffusion of carriers can be written as

$$J_p = e\mu_p p\mathscr{E} - eD_p\frac{dp}{dx} \tag{3.34a}$$

$$J_n = e\mu_n n\mathscr{E} + eD_n\frac{dn}{dx} \tag{3.34b}$$

At thermal equilibrium, the Fermi energy is invariant throughout the semiconductor. Consequently, the gradient of n can be derived from Eq. (3.20) as follows:

$$\frac{dn}{dx} = -\frac{n}{kT}\frac{dE_i}{dx} \tag{3.35}$$

But the applied electric field $\mathscr{E}$ is related to the intrinsic energy level E_i by

$$e\mathscr{E} = \frac{dE_i}{dx} \tag{3.36}$$

Under equilibrium conditions, $J_n = 0$. Thus Eq. (3.34b) can be expressed as

$$e\mu_n n\mathscr{E} + eD_n\left(-\frac{n}{kT}\frac{dE_i}{dx}\right) = 0 \tag{3.37}$$

which in turn reduces to

$$en\mathscr{E}\left(\mu_n - \frac{eD_n}{kT}\right) = 0 \tag{3.38}$$

But since $\mathscr{E}$ is not equal to 0,

$$\frac{D_n}{\mu_n} = \frac{kT}{e} \tag{3.39}$$

Likewise, for the holes, we obtain

$$\frac{D_p}{\mu_p} = \frac{kT}{e} \tag{3.40}$$

The last two equations are both known as Einstein's relations for electrons and holes. In many calculations of interest, these equations are used for expressing diffusion current characteristics in terms of drift current characteristics and vice versa.

The third form of carrier activity includes a number of processes whereby carriers are either created or annihilated. For example, when a photodetecting semiconductor is subjected to light, excess charge carriers are generated, increasing the electrical conductivity of the detector. When light is taken away, the semiconductor returns to equilibrium by means of **electron-hole recombination**. At times the recombination takes place by means of a recombination center whereby a hole jumps up to the recombination energy level and an electron jumps down to the same energy level, resulting in the annihilation of the pair. This recombination concept is utilized to design light-emitting diodes (LEDs) as well as diode lasers. The rate of change of carrier concentration resulting from the recombination process is proportional to the deviation of the respective carrier concentrations from their equilibrium values. Thus the net rate of change of carrier concentration due to both recombination and generation is given by

$$\left.\frac{dp}{dt}\right|_{R\text{-}G} = -\frac{\Delta p}{\tau_p} + \frac{dn_L}{dt} \tag{3.41a}$$

$$\left.\frac{dn}{dt}\right|_{R\text{-}G} = -\frac{\Delta n}{\tau_n} + \frac{dn_L}{dt} \tag{3.41b}$$

where τ_n and τ_p are the respective time constants for the minority carriers, Δp and Δn are the respective excess carrier densities, and n_L is the number per unit volume of electron-hole pairs generated by light absorption. Physically the time constant is interpreted as the average time that an excess minority carrier survives in the midst of the majority carriers before being annihilated. Note that under low levels of injection, the majority carrier concentration remains essentially unaffected, whereas the minority carrier concentration undergoes a drastic change by several orders of magnitude.

The overall effect of all three carrier activities causes a change in the carrier concentration. The resulting rate of change in the carrier concentration is given by

$$\frac{\partial n}{\partial t} = \frac{1}{e}\frac{\partial J_n}{\partial x} + \left.\frac{\partial n}{\partial t}\right|_{R\text{-}G} \tag{3.42a}$$

$$\frac{\partial p}{\partial t} = -\frac{1}{e}\frac{\partial J_p}{\partial x} + \left.\frac{\partial p}{\partial t}\right|_{R\text{-}G} \tag{3.42b}$$

where J_p and J_n are respectively given by Eqs. (3.34a) and (3.34b) that account for both drift and diffusion. Equations (3.42a and b) are commonly known as the **continuity equations**. By limiting these equations to minority carriers and low-level injections, we can come up with the minority carrier diffusion equations. The assumptions that $\mathscr{E} = 0$ in the semiconductor bulk region and that the equilibrium minority carrier concentration is invariant all throughout the semiconductor are both necessary in deriving the minority carrier diffusion equations. Using Eq. (3.34a) we find that

$$-\frac{1}{e}\frac{\partial J_p}{\partial x} = -\frac{1}{e}\frac{\partial}{\partial x}\left(e\mu_p p\mathscr{E} - eD_p\frac{\partial p}{\partial x}\right) = D_p\frac{\partial^2 p}{\partial x^2} \tag{3.43}$$

Since $p = p_0 + \Delta p$, where p_0 is the equilibrium minority hole concentration, Eq. (3.42b) can be rewritten as

$$\frac{\partial \Delta p_n}{\partial t} = D_p\frac{\partial^2 \Delta p_n}{\partial x^2} - \frac{\Delta p_n}{\tau_p} + \frac{dn_L}{dt} \tag{3.44}$$

Similarly, Eq. (3.42a) can be reduced to

$$\frac{\partial \Delta n_p}{\partial t} = D_n\frac{\partial^2 \Delta n_p}{\partial x^2} - \frac{\Delta n_p}{\tau_n} + \frac{dn_L}{dt} \tag{3.45}$$

In both of these minority carrier diffusion equations, subscripts of the excess carrier density variables indicate the type of the majority carrier region.

3.3 Junction Diode

While the last section dealt with only uniformly doped semiconductors, we now consider a step junction of two uniformly doped extrinsic semiconductors having different doping on the two sides of the junction. In particular, when a part of the crystal is doped with acceptors while the other part is doped with donors, the resulting step junction is referred to as *p-n* junction diode. In the beginning it will be appropriate to restrict our discussion to the unbiased case when no applied voltage or light is allowed to impinge on the diode.

It is reasonable to assume that charge neutrality exists everywhere in the *p-n* junction of Fig. 3.8. This neutrality is not accomplished simply by a balance of free electron flow and hole flow. The flow of one in the direction opposite to that of the other actually amounts to a net current in only one of these two directions. Thus, to achieve equilibrium, it is necessary for both free electron current and hole current to be zero separately. For the purposes of our discussion, the electron and hole densities in the *p*-side are designated as n_p and p_p, respectively. Similarly, in the *n*-side these carrier densities are denoted by n_n and p_n, respectively. Thus while n_n and p_p represent the majority carrier concentrations, n_p and p_n represent the minority carrier concentrations. Oftentimes a subscript of 0 is also used alongside to indicate the corresponding equilibrium values. At equilibrium, the majority hole density in the *p*-side is

$$p_{p0} = N_A \tag{3.46}$$

while the minority electron density is given by

$$n_{p0} = \frac{n_i^2}{N_A} \tag{3.47}$$

Similarly, the carrier concentrations in the *n*-side are

$$n_{n0} = N_D \tag{3.48a}$$

$$p_{n0} = \frac{n_i^2}{N_D} \tag{3.48b}$$

Since the extrinsic carrier densities are much larger than the intrinsic carrier density, Eqs. (3.46) through (3.48) indicate the presence of a strong concentration gradient in regions in the immediate neighborhood of the junction. Once a contact is established between the donor-doped and acceptor-doped semiconductors in the form of a *p-n* junction, holes will diffuse from the *p*-side to the *n*-side and electrons will diffuse from the *n*-side to the *p*-side in response to their respective concentration gradients. Ideally this diffusion process should continue until a complete uniformity of concentrations has been achieved. But as holes continue to diffuse from the *p*-side to the *n*-side, they leave behind ionized acceptor atoms. Similarly, ionized donor atoms are left behind in the *n*-side due to the diffusion of electrons. The build-up of ionized donor atoms and ionized

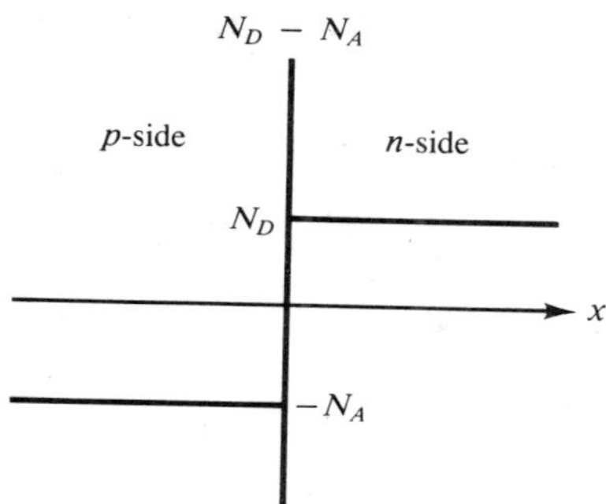

FIGURE 3.8

A step *p-n* junction diode

acceptor atoms results in an electric field that is directed along the negative x axis. The resulting electric field ultimately stops further diffusion of carriers from one side to the other. At thermal equilibrium, the net flow of charges thus becomes zero.

The narrow region near the junction where the mobile carriers are depleted is called the **depletion region** and is shown in Fig. 3.9(a) where x_p is the penetration length of the depletion region into the p-side and x_n is the penetration length of the depletion region into the n-side. The accumulation of ionized charge densities in the depletion region gives rise to an electric field as shown in Fig. 3.9(c). The exact value of the field $\mathscr{E}$ is, however, given by the Poisson's equation

$$\frac{d\mathscr{E}}{dx} = \frac{\rho(x)}{K_s \varepsilon_0} \tag{3.49}$$

where K_s is the relative dielectric constant of the semiconductor material and $\rho(x)$ is the total charge density given by

$$\rho(x) = e(p - n + N_D - N_A) \tag{3.50}$$

The dielectric constants for some of the important semiconductors are listed in Table 3.5.

The solution of Eq. (3.49), in turn, can be used for determining the potential difference in terms of the electric field since

$$V(x) = -\int_{-\infty}^{x} \mathscr{E}(x)\, dx \tag{3.51}$$

where we have arbitrarily assumed that potential is zero at $x = -\infty$. The net amount of potential difference that exists across the depletion region in the absence of any bias at thermal equilibrium is called **built-in potential** V_B. Figure 3.9(d) illustrates the potential distribution across the unbiased diode and identifies the built-in potential.

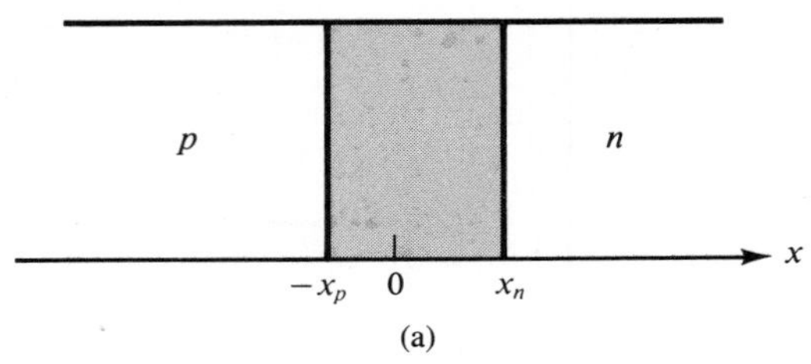

(a)

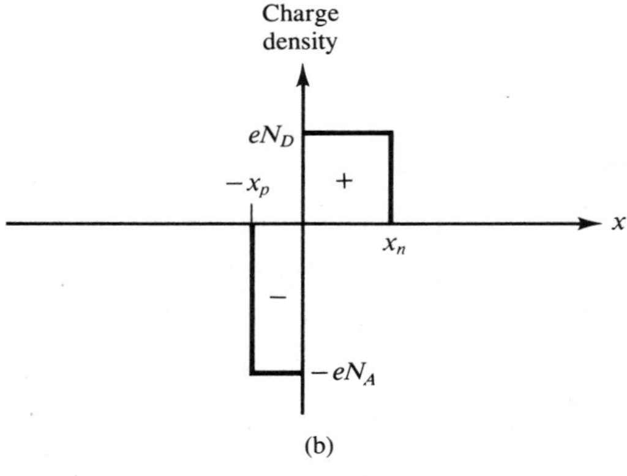

(b)

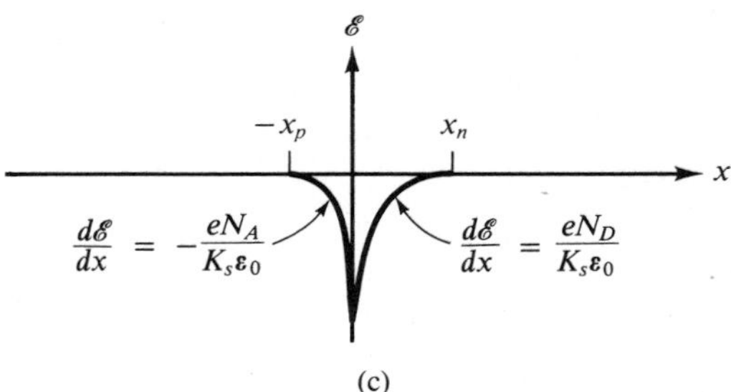

(c)

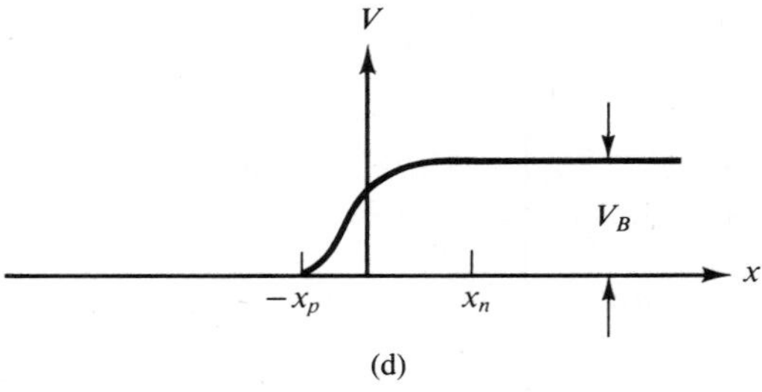

(d)

FIGURE 3.9

Depletion region characteristics in a *p-n* junction when applied voltage is zero: (a) depletion region, (b) charge density, (c) field distribution, and (d) potential distribution

TABLE 3.5

Dielectric Constants of Semiconductors

Semiconductor	K_s
Ge	16.0
Si	11.9
CdS	5.4
CdTe	10.7
GaAs	13.1
GaP	11.1
InSb	17.7
PbS	17.0
ZnO	9.0

At thermal equilibrium, the net current density equals zero. Thus using Eq. (3.40), Eq. (3.34a) gives us

$$\mathscr{E} = \frac{D_p}{\mu_p}\frac{1}{p}\frac{dp}{dx} = \frac{kT}{e}\frac{1}{p}\frac{dp}{dx} \tag{3.52}$$

Equation (3.52) can be used next to evaluate the built-in (or barrier) potential as

$$V_B = -\int_{-\infty}^{\infty} \mathscr{E}\,dx = -\frac{kT}{e}\int_{-\infty}^{\infty}\frac{1}{p}\,dp = -\frac{kT}{e}\ln(p)\Big|_{p(-\infty)}^{p(+\infty)} \tag{3.53}$$

Now using Eqs. (3.46) and (3.48b), we get

$$V_B = -\frac{kT}{e}\left[\ln\left(\frac{n_i^2}{N_D}\right) - \ln(N_A)\right] = \frac{kT}{e}\ln\frac{N_D N_A}{n_i^2} \tag{3.54}$$

Having obtained the built-in potential across the junction, we can next determine the variation of the field within the depletion region. Equation (3.49) can be employed to evaluate the field variation, but it turns out to be difficult since the carrier concentrations (and consequently charge density too) are functions of both V and $\mathscr{E}$. This hurdle is overcome by making certain reasonable approximations that are commonly known as depletion approximations. The depletion approximations are (a) charge density in the bulk region ($x > x_n$ and $x < -x_p$) is zero and (b) charge density ρ in the depletion region is made up of only dopants, i.e., $\rho \simeq eN_A$ in $-x_p \leqslant x \leqslant 0$ and $\rho \simeq eN_D$ in the $0 \leqslant x \leqslant x_n$ region. These assumptions are almost true in most practical cases. Assuming complete ionization of the impurities, Eq. (3.49) thus reduces to

$$\frac{d\mathscr{E}}{dx} = \begin{cases} \dfrac{eN_D}{K_s\varepsilon_0} & 0 \leqslant x \leqslant x_n \\[2ex] -\dfrac{eN_A}{K_s\varepsilon_0} & -x_p \leqslant x \leqslant 0 \end{cases} \tag{3.55}$$

such that $\mathscr{E}(x > x_n) = \mathscr{E}(x < -x_p) = 0$. Consequently, we can visualize that $\mathscr{E}(x)$ has a negative slope on the p-side and a positive slope on the n-side. The maximum value of the field occurs at the junction, but it is negative everywhere in the depletion region, as shown in Fig. 3.9(c). Using Eq. (3.55), we obtain for the depletion region

$$\mathscr{E}(x) = \begin{cases} -\dfrac{eN_D}{K_s\varepsilon_0}(x_n - x) & 0 \leqslant x \leqslant x_n \\ -\dfrac{eN_A}{K_s\varepsilon_0}(x_p + x) & -x_p \leqslant x \leqslant 0 \end{cases} \tag{3.56}$$

But since the field has to be continuous at $x = 0$,

$$-\frac{eN_A}{K_s\varepsilon_0}x_p = -\frac{eN_D}{K_s\varepsilon_0}x_n \tag{3.57}$$

or

$$N_A x_p = N_D x_n \tag{3.58}$$

Equation (3.58) confirms the fact that, within the depletion region, the total negative charge must be equal to the total positive charge. The areas under the charge density curve due to positive and negative charges are therefore expected to be exactly equal in Fig. 3.9(b).

Having determined the field distribution within the depletion region we can use Eq. (3.51) to evaluate the potential distribution within the depletion region. In the p-side of the depletion region,

$$\begin{aligned} V(x) &= \frac{eN_A}{K_s\varepsilon_0}\int_{-x_p}^{x}(x_p + x)\,dx \\ &= \frac{eN_A}{2K_s\varepsilon_0}(x_p + x)^2 \end{aligned} \tag{3.59}$$

And similarly in the n-side of the depletion region, we obtain

$$\begin{aligned} V(x) &= \frac{eN_D}{K_s\varepsilon_0}\int_{x}^{x_n}(x_n - x)\,dx \\ &= -\frac{eN_D}{2K_s\varepsilon_0}(x_n - x)^2 + V_B \end{aligned} \tag{3.60}$$

since $V(x_n) = V_B$. Again, $V(x)$ must be continuous at $x = 0$ and thus

$$-\frac{eN_D}{2K_s\varepsilon_0}x_n^2 + V_B = \frac{eN_A}{2K_s\varepsilon_0}x_p^2 \tag{3.61}$$

which can be manipulated using Eq. (3.58) as follows:

$$x_p^2\frac{eN_A}{2K_s\varepsilon_0}\left(1 + \frac{N_A}{N_D}\right) = V_B \tag{3.62}$$

or

$$x_p = \left[\frac{2K_s\varepsilon_0}{e} V_B \frac{N_D}{N_A(N_A + N_D)}\right]^{1/2} \tag{3.63}$$

Now using Eqs. (3.58) and (3.63) we obtain

$$x_n = \left[\frac{2K_s\varepsilon_0}{e} V_B \frac{N_A}{N_D(N_A + N_D)}\right]^{1/2} \tag{3.64}$$

Finally, the width W of the depletion region is obtained by adding the absolute values of x_n and x_p to give

$$W = \left[\frac{2K_s\varepsilon_0}{e} V_B \frac{N_A + N_D}{N_A N_D}\right]^{1/2} \tag{3.65}$$

As expected, the depletion region penetrates less into the side with the heavier doping. This assertion also agrees with the conclusion of Eq. (3.58). The depletion width varies as the square root of the built-in potential. As we shall see later, an application of bias voltage may change the value of this junction potential, thus changing the value of the depletion width.

Consider now the effects of having an external bias source of voltage V_A across the junction diode with its positive terminal connected to the p-side and its negative terminal connected to the n-side. A portion of the voltage drop will appear across the bulk regions. However, in most junction diodes, the length of each region is small compared with the cross-sectional area. Thus, since resistance is negligible in the bulk region, we may safely assume that the bias voltage is most effective only in the depletion region. It becomes readily obvious that the junction voltage V_j across the depletion region will now change from V_B to $V_B - V_A$. All of the equations derived thus far are still valid except that each occurrence of V_B must be replaced by $V_B - V_A$. Thus in the p-side of the depletion region:

$$\mathscr{E}(x) = -\frac{eN_A}{K_s\varepsilon_0}(x_p + x) \tag{3.66a}$$

$$V(x) = \frac{eN_A}{2K_s\varepsilon_0}(x_p + x)^2 \tag{3.66b}$$

and

$$x_p = \left[\frac{2k_s\varepsilon_0}{e}(V_B - V_A)\frac{N_D}{N_A(N_A + N_D)}\right]^{1/2} \tag{3.66c}$$

and, similarly, in the n-side of the depletion region:

$$\mathscr{E}(x) = -\frac{eN_D}{K_s\varepsilon_0}(x_n - x) \tag{3.67a}$$

$$V(x) = -\frac{eN_D}{2K_s\varepsilon_0}(x_n - x)^2 + V_B - V_A \tag{3.67b}$$

and

$$x_n = \left[\frac{2K_s\varepsilon_0}{e}(V_B - V_A)\frac{N_A}{N_D(N_A + N_D)}\right]^{1/2} \tag{3.67c}$$

and, finally, the depletion width of the biased junction is given by

$$W = \left[\frac{2K_s\varepsilon_0}{e}(V_B - V_A)\frac{N_A + N_D}{N_A N_D}\right]^{1/2} \tag{3.68}$$

It can be seen that for the case of the forward bias ($V_A > 0$), the junction width decreases, whereas for the case of the reverse bias ($V_A < 0$), the junction width increases. Accordingly, the electric field decreases with forward bias and increases with reverse bias. When $V_A = 0$, the difference between the E_c (or E_v) levels on both sides of the junction is eV_B so that the drift and diffusion components of the current cancel each other out. As soon as $V_A > 0$ is applied, the difference between the corresponding energy levels on both sides is reduced by eV_A, as shown in Fig. 3.10(a), and the diffusion currents from the corresponding majority carrier sides become dominant over the corresponding drift currents. As the potential barrier is lowered, many more electrons (from the *n*-side conduction band) will have sufficient energy to diffuse to the *p*-side, just as many more holes (from the *p*-side valence band) will have sufficient energy to diffuse to the *n*-side. These flows are illustrated in Fig. 3.10(a). Alternatively, when $V_A < 0$ is applied, the difference between the corresponding energy levels on either side of the junction is increased by eV_A, depleting the amount of diffusion current. In the latter case, therefore, the drift current is found to be dominant over the corresponding diffusion current, as shown in Fig. 3.10(b). In any event, the drift current is generally very small even though we may expect it to be proportional to the applied voltage because of the fact that there are very few minority electrons and minority holes to contribute to the drift current. Consequently, we expect to see a large current (due to diffusion) when a diode is forward biased and a very small reverse current (due to drift) when it is reverse biased.

In order to quantitatively determine the current caused by an applied voltage, we are required to solve for the hole current I_p, due to minority holes in the *n*-side, and electron current I_n, due to minority electrons in the *p*-side. These currents are given by Eqs. (3.34a) and (3.34b), respectively:

$$I_p \simeq -eD_p A\frac{d\Delta p_n}{dx} \tag{3.69a}$$

$$I_n \simeq eD_n A\frac{d\Delta n_p}{dx} \tag{3.69b}$$

where A is the cross-sectional area of the junction and the electric field $\mathscr{E} = 0$ in the bulk region. For all practical purposes, the depletion region is generally assumed to have no generation or recombination of carriers. Accordingly, the

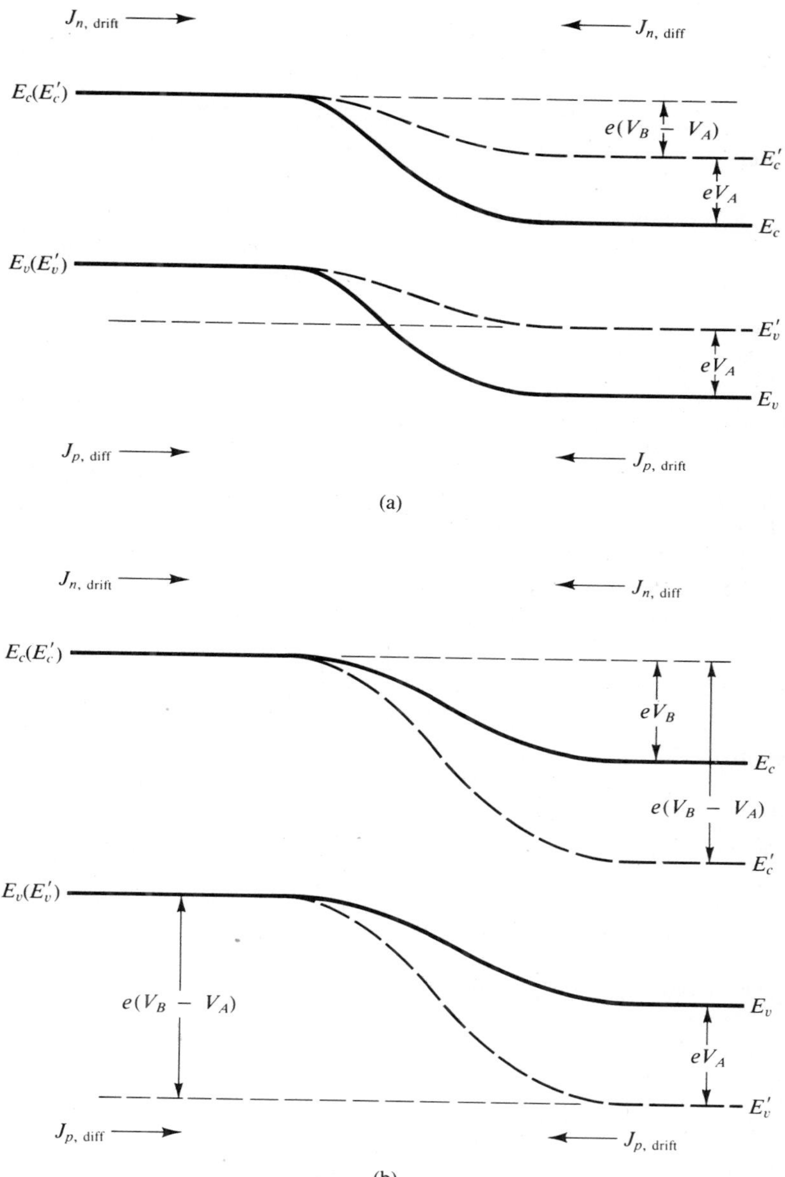

FIGURE 3.10

Current in a *p-n* diode under (a) forward bias and (b) reverse bias

total current is given by

$$I = I_p(x)|_{x=x_n} + I_n(x)|_{x=-x_p} \tag{3.70}$$

However, the solutions of Eqs. (3.69a) and (3.69b) cannot be determined without knowing the excess minority carrier concentrations, Δp_n and Δn_p, in the bulk region.

The excess minority carrier concentrations can be determined by solving the corresponding steady-state minority carrier equations that are given by Eqs. (3.44) and (3.45). These equations reduce to

$$\frac{d^2\Delta p_n}{dx^2} = \frac{\Delta p_n}{\tau_p D_p} \tag{3.71a}$$

$$\frac{d^2\Delta n_p}{dx^2} = \frac{\Delta n_p}{\tau_n D_n} \tag{3.71b}$$

for n- and p-regions, respectively. But again, to solve Eqs. (3.71a) and (3.71b) we need to know the boundary conditions (the values of excess minority carrier concentrations at the depletion edges).

In the depletion region, the field is nonzero. Accordingly, using Eq. (3.52), the junction voltage V_j in the presence of the applied voltage V_A is given by

$$V_j = V_B - V_A = -\frac{kT}{e}\ln(p)\Big|_{x=-\infty}^{x=\infty} = \frac{kT}{e}\ln\left(\frac{p(x)|_{-x_p}}{p(x)|_{x_n}}\right) \tag{3.72}$$

Therefore,

$$\frac{p(x)|_{-x_p}}{p(x)|_{x_n}} = e^{e(V_B - V_A)/kT} \tag{3.73}$$

or

$$p(x)|_{x_n} = p(x)|_{-x_p} e^{-eV_B/kT} e^{eV_A/kT} \tag{3.74}$$

However, from Eq. (3.54), we obtain

$$e^{-eV_B/kT} = \frac{n_i^2}{n_{n0} p_{p0}} \tag{3.75}$$

such that at equilibrium $p_{p0} = N_D$ and $n_{n0} = N_A$. Then Eq. (3.74) can be expressed as

$$\begin{aligned} p(x)|_{x_n} &= p(x)|_{-x_p}\left(\frac{n_i^2}{n_{n0} p_{p0}}\right) e^{eV_A/kT} \\ &= \frac{n_i^2}{n_{n0}} e^{eV_A/kT} = p_{n0} e^{eV_A/kT} \end{aligned} \tag{3.76}$$

since $p_{p0} \simeq p(x)$ at $x = -x_p$. The excess hole concentration at $x = x_n$ is, therefore, given by

$$\Delta p(x)|_{x_n} = p(x)|_{x_n} - p_{n0} = p_{n0}[e^{eV_A/kT} - 1] \quad (3.77)$$

In the same manner, the excess electron concentration at $x = -x_p$ is found to be

$$\Delta n(x)|_{-x_p} = n_{p0}[e^{eV_A/kT} - 1] \quad (3.78)$$

Equation (3.71) can now be solved separately in the n-bulk and then in the p-bulk regions. For example, Eq. (3.71a) can be solved in the n-bulk region $(x \geqslant x_n)$ to yield

$$\Delta p_n(x) = C_1 e^{x/L_p} + C_2 e^{-x/L_p} \quad (3.79)$$

where $L_p = (\tau_p D_p)^{1/2}$ and C_1 and C_2 are two arbitrary constants. C_1 becomes zero since $\Delta p_n(x) = 0$ at $x = \infty$. C_2 can be determined by making use of Eq. (3.77) as follows:

$$C_2 = \Delta p_n(x)|_{x_n} e^{x_n/L_p} = p_{n0}[e^{eV_A/kT} - 1]e^{x_n/L_p} \quad (3.80)$$

Therefore,

$$\Delta p_n(x) = p_{n0}[e^{eV_A/kT} - 1]e^{(x_n - x)/L_p} \quad (3.81)$$

The diffusion length L_p is equivalent to a distance from the junction at which the excess hole distribution is reduced to $1/e$th of its value at the injection point. The probability that a hole injected at $x = 0$ will survive at an arbitrary x without being recombined is $\exp(-x/L_p)$, and the probability that the hole will recombine within the differential length dx is dx/L_p. Thus the average distance through which the hole survives before being recombined is

$$\langle x \rangle = \int_0^\infty \frac{1}{L_p} e^{-x/L_p} dx = L_p \quad (3.82)$$

Equation (3.82) reasserts our earlier observation that L_p is identical to the average distance through which a hole diffuses before being recombined.

In like manner, Eq. (3.71b) can also be solved for the p-bulk region $(x \leqslant -x_p)$ to give

$$\Delta n_p(x) = n_{p0}[e^{eV_A/kT} - 1]e^{(x_p + x)/L_n} \quad (3.83)$$

where

$$L_n = [D_n \tau_n]^{1/2} \quad (3.84)$$

Equations (3.69a) and (3.69b) can now be evaluated at the depletion edges to yield

$$I_p(x)|_{x_n} = -eD_p A \left.\frac{d\Delta p_n(x)}{dx}\right|_{x = x_n}$$

$$= \frac{eD_p A p_{n0}}{L_p}[e^{eV_A/kT} - 1] \quad (3.85)$$

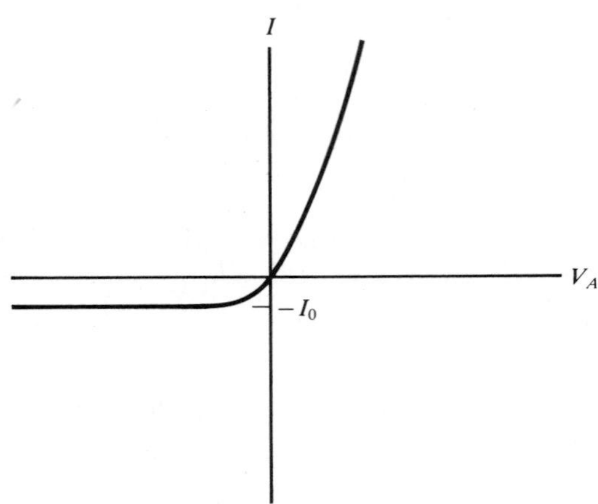

FIGURE 3.11

The ideal diode characteristics

and

$$I_n(x)|_{-x_p} = eD_n A \left.\frac{d\Delta n_p(x)}{dx}\right|_{x=-x_p}$$

$$= \frac{eD_n A n_{p0}}{L_n}[e^{eV_A/kT} - 1] \tag{3.86}$$

Equation (3.70) can therefore be evaluated to give for the total current

$$I = eA\left[\frac{D_p p_{n0}}{L_p} + \frac{D_n n_{p0}}{L_n}\right](e^{eV_A/kT} - 1)$$

$$= I_0[e^{eV_A/kT} - 1] \tag{3.87}$$

where I_0, referred to as the saturation current, is given by

$$I_0 \equiv eA\left[\frac{D_p p_{n0}}{L_p} + \frac{D_n n_{p0}}{L_n}\right]$$

$$= eAn_i^2\left[\frac{D_p}{L_p N_D} + \frac{D_n}{L_n N_A}\right] \tag{3.88}$$

The ideal diode characteristics, as described by Eq. (3.87), are plotted in Fig. 3.11. When it is forward biased, the junction current depends strongly on the bias voltage V_A. In particular, the dependence is essentially exponential when $V_A \gg kT/e$. The error introduced if we neglect -1 in Eq. (3.87) is only about 5% when $eV_A = 3kT$. Again, the reverse-biased current is essentially independent of the bias voltage and is equal to $-I_0$. Note also that the saturation current is dependent on n_i^2, which in turn strongly depends on both temperature and band-gap energy [as per Eqs. (3.16), (3.18), and (3.22)]. Thus, for only a 6 K increase in temperature in Si or an 11 K increase in temperature in Ge the saturation current is doubled. But since the saturation current in Si is much less than that in Ge, Si diodes are preferable over Ge diodes for operations at higher temperatures.

3.4 Bipolar Junctions

Bipolar junctions (or transistors) are formed by coupling two *p-n* junctions in the same host crystal. It may be either of the form n^+-*p*-*n* or p^+-*n*-*p*. The three extrinsic regions of the transistor, emitter (E), base (B), and collector (C), as identified in Fig. 3.12 (page 110), form two distinct junctions. The base is common to both of the junctions. The emitter region is more heavily doped than the collector region. Consequently, the terminals between collector and emitter cannot be interchanged without the device characteristics being changed drastically. The doping magnitude of the base region lies somewhere between that of the other two regions.

In a transistor under normal operating conditions, the junction between emitter and base is forward biased and that between the collector and base is reverse biased. This most common operating condition is referred to as the **active region operation**. Most of the linear signal amplifiers, such as operational amplifiers, are operated in this region since only then do they have their largest signal gain and least signal distortion. Besides this active region operation, there are three more operating conditions, namely, saturation, cut-off, and inverted. Figure 3.13 (on page 110) shows the biasing requirements for all of these operating conditions where V_{xy} implies that x has a higher potential than y. When we need to have a low output at the logic device (corresponding to binary 0), we tend to operate in the saturation region because it's possible to get the least voltage drop across the collector and emitter in this region, even with a large collector current flow. On the other hand, to have a high output at the logic device (corresponding to binary 1), we tend to operate in the cut-off region so that we can realize a large voltage drop across the collector and emitter with almost nonexistent collector current. The inverted region represents the scenario where the collector acts like an emitter and the emitter acts like a collector. When signal gain is not an objective, such an operation region is highly sought after.

In the more common active region operation of a p^+-*n*-*p* transistor, the forward-biased, emitter-base junction permits the holes to move freely from the emitter into the base. This flow of holes is also accompanied by a flow of electrons from the base into the emitter, as shown in Fig. 3.14 (page 110). The two currents I_{Ep} and I_{En} combine to produce a positive emitter current I_E. Similarly, the collector current I_C at the collector-base junction is determined by adding the hole current I_{Cp} entering the depletion region from the base side and the electron current I_{Cn} entering the depletion region from the collector side. The difference

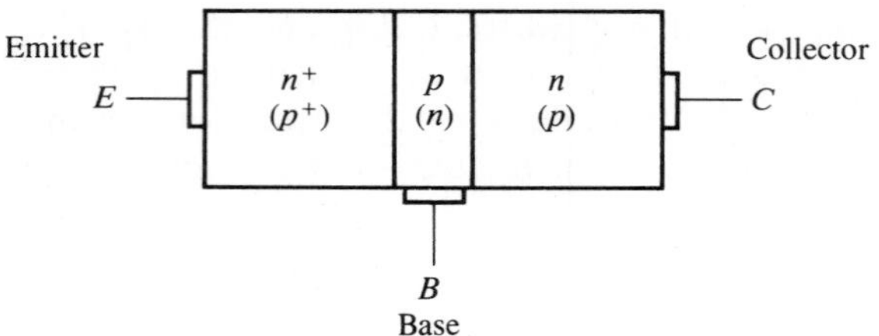

FIGURE 3.12

***n*⁺-*p*-*n* (*p*⁺-*n*-*p*) bipolar transistor**

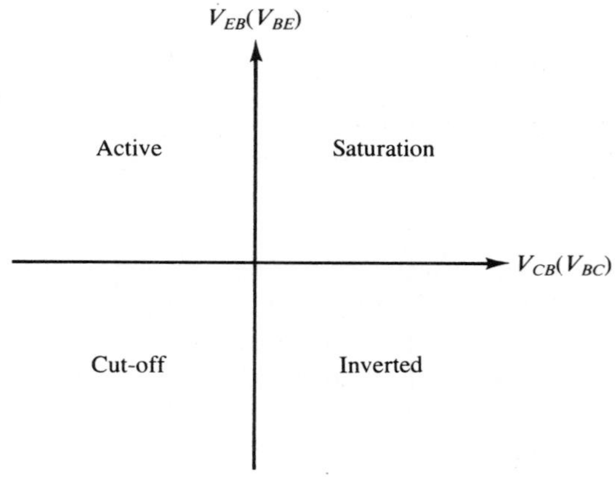

FIGURE 3.13

Regions of operation for p^+-n-p (n^+-p-n) bipolar junctions

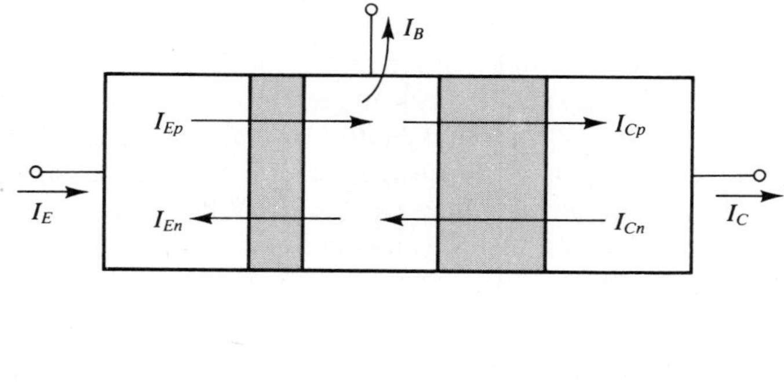

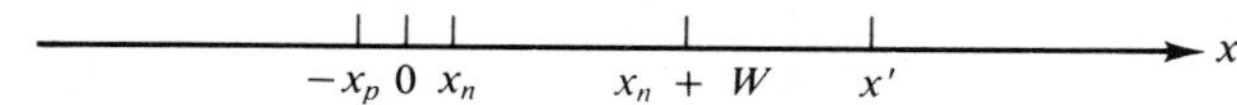

FIGURE 3.14

A p^+-n-p transistor and its associated parameters

between the emitter current and the collector current shows up in the form of base current I_B.

No generation-recombination is expected to occur in the depletion regions of an ideal transistor. To avoid confusion, the minority carrier terminologies such as diffusion length, diffusion constant, and doping level, for the three regions—emitter, base, and collector—are denoted by the subscripts E, B, and C, respectively. The emitter current crossing the emitter-base junction can be obtained using Eq. (3.34) as follows:

$$I_E = I_{Ep}(x)|_{x_n} + I_{En}(x)|_{-x_p}$$

$$= eAD_E \frac{d\Delta n_E}{dx}\bigg|_{-x_p} - eAD_B \frac{d\Delta p_B}{dx}\bigg|_{x_n} \tag{3.89}$$

where $\mathscr{E} = 0$ in the bulk regions and A is the junction cross-sectional area. Similarly, the collector current crossing the collector-base junction is found from

$$I_C = I_{Cp}(x)|_{x_n+W} + I_{Cn}(x)|_{x'}$$

$$= -eA\left[D_B \frac{d\Delta p_B}{dx}\bigg|_{x_n+W} - D_c \frac{d\Delta n_c}{dx}\bigg|_{x'}\right] \tag{3.90}$$

where W is the base width. Equations (3.89) and (3.90) involve the excess minority carrier concentrations, Δp_B, Δn_E, and Δn_C, respectively, in the base, emitter, and collector regions. These three quantities are evaluated by solving the three minority carrier diffusion equations:

$$\frac{d^2\Delta p_B(x)}{dx^2} = \frac{\Delta p_B(x)}{L_B^2} \tag{3.91a}$$

$$\frac{d^2\Delta n_E(x)}{dx^2} = \frac{\Delta n_E(x)}{L_E^2} \tag{3.91b}$$

$$\frac{d^2\Delta n_c(x)}{dx^2} = \frac{\Delta n_c(x)}{L_C^2} \tag{3.91c}$$

in their respective regions. To solve them, however, we need to know the appropriate boundary values for the excess minority carrier concentrations at the edges. But these boundary values are determined by the junction voltages V_{EB} and V_{CB}. Using Eqs. (3.77) and (3.78), we find

$$\Delta n_E(x)|_{-\infty} = 0 \tag{3.92a}$$

$$\Delta n_E(x)|_{-x_p} = n_{EO}[e^{eV_{EB}/kT} - 1] \tag{3.92b}$$

$$\Delta p_B(x)|_{x_n} = p_{BO}[e^{eV_{EB}/kT} - 1] \tag{3.92c}$$

$$\Delta p_B(x)|_{x_n+W} = p_{BO}[e^{eV_{CB}/kT} - 1] \tag{3.92d}$$

$$\Delta n_C(x)|_{x'} = n_{CO}[e^{eV_{CB}/kT} - 1] \tag{3.92e}$$

$$\Delta n_C(x)|_{+\infty} = 0 \tag{3.92f}$$

It is reasonable to assume that, for most practical transistors, the base width W is infinitesimally small and, therefore, the diffusion length L_B is considered significantly large. Thus Eq. (3.91a) reduces to

$$\frac{d^2\Delta p_B(x)}{dx^2} \simeq 0 \tag{3.93}$$

Using the boundary conditions of Eqs. (3.92c) and (3.92d), Eq. (3.93) readily yields

$$\Delta p_B(x) = \Delta p_B(x)|_{x_n} - \frac{\Delta p_B(x)|_{x_n} - \Delta p_B(x)|_{x_n+W}}{W} x \tag{3.94}$$

On the other hand, the solution of Eq. (3.91b) turns out to be

$$\Delta n_E(x) = A_1 e^{x/L_E} + A_2 e^{-x/L_E} \tag{3.95}$$

where A_2 is set to zero to satisfy the boundary condition of Eq. (3.92a). Then, by incorporating the boundary condition of Eq. (3.92b), we find that

$$\Delta n_E(x) = n_{EO}[e^{eV_{EB}/kT} - 1]e^{(x+x_p)/L_E} \tag{3.96}$$

Eqs. (3.94) and (3.96) can next be introduced in Eq. (3.89) to give the emitter current as

$$I_E = eA\left[\frac{D_E n_{EO}}{L_E} + \frac{D_B p_{BO}}{W}\right][e^{eV_{EB}/kT} - 1] - \frac{eAD_B}{W} p_{BO}(e^{eV_{CB}/kT} - 1) \tag{3.97}$$

Equation (3.90) can be determined by taking note of the fact that, in the absence of recombination-generation in the base,

$$I_{Ep}(x)|_{x_n} = I_{Cp}(x)|_{x_n+W} \tag{3.98}$$

The collector current therefore becomes

$$I_C = \frac{eAD_B}{W} p_{BO}[e^{eV_{EB}/kT} - 1] - eA\left[\frac{D_C n_{CO}}{L_C} + \frac{D_B p_{BO}}{W}\right][e^{eV_{CB}/kT} - 1] \tag{3.99}$$

while the corresponding base current becomes

$$I_B = \frac{eAD_E}{L_E} n_{EO}(e^{eV_{EB}/kT} - 1) + \frac{eAD_C}{L_C} n_{CO}(e^{eV_{CB}/kT} - 1) \tag{3.100}$$

for the ideal bipolar transistor.

The emitter and collector currents described so far are, however, based on the fact that the base diffusion length approaches infinity. Assuming a small base width, the solution of $\Delta p_B(x)$ resulted in a linear equation given by Eq. (3.94).

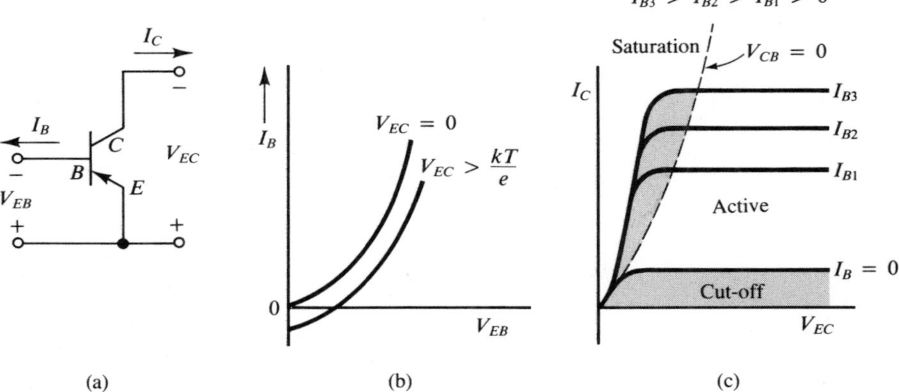

FIGURE 3.15

Common emitter p^+-n-p characteristics

A rigorous calculation shows that for a finite value of L_B, the expressions for I_E and I_C, respectively, become:

$$I_E = \frac{eAD_E}{L_E} n_{E0}(e^{eV_{EB}/kT} - 1) + \frac{eAD_B}{L_B} p_{B0}\left\{\coth\left(\frac{W}{L_B}\right)\right\} \times \left[e^{eV_{EB}/kT} - 1 - \left\{\text{sech}\left(\frac{W}{L_B}\right)\right\}\{e^{eV_{CB}/kT} - 1\}\right] \quad \textbf{(3.101)}$$

and

$$I_C = \frac{eAD_B}{L_B} p_{B0}\left\{\text{cosech}\left(\frac{W}{L_B}\right)\right\}\left[e^{eV_{EB}/kT} - 1 - \cosh\left(\frac{W}{L_B}\right)\right] - \frac{eAD_C}{L_C} n_{C0}(e^{eV_{CB}/kT} - 1) \quad \textbf{(3.102)}$$

The transistors are used mostly in circuits, either in the common-emitter or common-base format. Figure 3.15(a) illustrates the common-emitter configuration, whereas Figs. 3.15(b) and 3.15(c) respectively illustrate the corresponding ideal input-output characteristics. The two input variables are I_B and V_{EB}, whereas the output variables are I_C and V_{EC}. Note the similarity between the forward-biased diode characteristics of Fig. 3.11 and the transistor input characteristics, I_B versus V_{EB} when $V_{EC} = 0$. When V_{EC} becomes greater than a couple of kT/e volts, the transistor input characteristics become independent of V_{EC}. It can be seen that at large enough values of V_{EC}, additional voltage fails to influence the injection of carriers at the emitter-base junction.

Figure 3.15(c) illustrates the active region where, for a fixed I_B, I_C is constant and is relatively independent of V_{EC}. A large I_C/I_B (i.e., gain) is present all

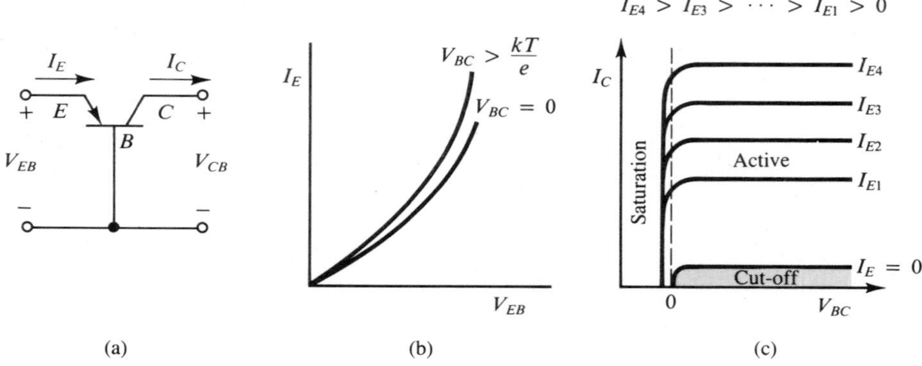

FIGURE 3.16

Common base p^+-n-p characteristics

throughout the active region when $V_{CB} < 0$. In the saturation region, however, the flow of holes (from the collector to the base) increases as the V_{CB} becomes more positive. This particular flow of holes opposes the other flow of holes, which is arriving from the emitter, thus resulting in a net decrement of collector current.

Figure 3.16 illustrates the characteristics of a p^+-n-p transistor in a common-base configuration. The input characteristics, I_E versus V_{EB}, at $V_{CB} = 0$ are similar to the forward-based diode characteristics shown in Fig. 3.11. For all V_{CB} values larger than a few kT/e volts, the input characteristics become independent of V_{BC}. As far as the output is concerned, I_C is independent of V_{BC} when I_E is a constant. In the saturation region (when $V_{BC} < 0$), the collector begins to inject holes into the base, which therefore opposes the hole flux injected by the emitter (when $I_E > 0$). Consequently, the collector current decreases severely in the saturation region.

3.5 Deviations from Ideal Device Characteristics

Although the theory developed in the last few sections is in reasonable agreement with what is observed in practice, there are several important points of difference between an ideal diode and a real diode, as well as between an ideal transistor and a real transistor. It may be beyond the scope of this chapter to cover all of these deviations and their causes. The device deviation phenomena that are of interest include the Zener effect, avalanche breakdown, and the recombination-generation effect in diodes and the base-width modulation effect in the bipolar transistors.

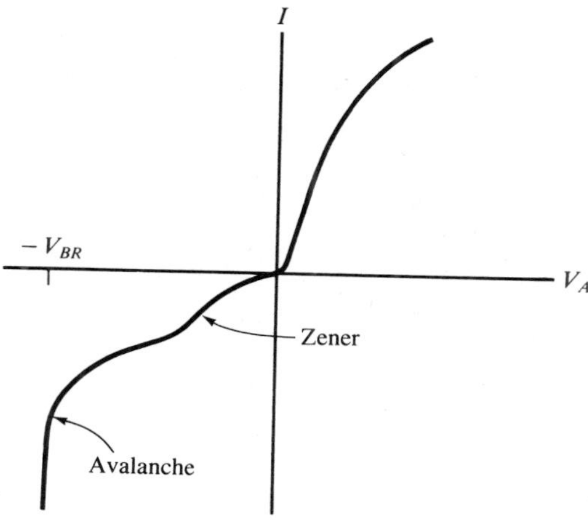

FIGURE 3.17

Real diode characteristics

Zener breakdown generally occurs in diodes that are heavily doped on both sides of the junction and subjected to breakdown voltages of approximately 4 volts or less. With the increase in doping, the depletion width decreases [per Eq. (3.65)] until the depletion width approaches approximately 100 Å or less. Figure 3.10 shows that with larger reverse bias, the valence band (with filled states) on the *p*-side appears to have almost the same energy as the conduction band (with empty states) on the *n*-side. This scenario is equivalent to that of electrons on one side of a potential barrier while the other side of the well is completely empty. It becomes quantum mechanically possible, therefore, for the electrons to tunnel from one side to the other. This tunneling results in increasing the reverse current. Once this tunneling effect sets in, the additional reverse voltage causes a further increase in the reverse current, resulting in a diode characteristic such as that shown in Fig. 3.17.

Avalanche breakdown is also responsible for reverse bias breakdown in the diodes but at very large reverse voltage. With increasing reverse bias, the electric field at the middle of the depletion width increases continually (per Fig. 3.9). At a critical value of this electric field, the electron will acquire enough energy to accelerate and collide with a crystal lattice atom. This collision may free an electron-hole pair. When electron-hole pairs are repeatedly freed, the result may be an avalanche of additional carriers. At very large values of reverse bias, the magnitude of current increases very rapidly, approaching infinity as the voltage approaches the breakdown voltage.

While deriving the ideal diode equation, we assumed that the net generation-recombination rate in the depletion region was insignificant. When a reverse bias is applied, the generation of carriers actually dominates in the depletion region because the carrier concentrations fall far short of their

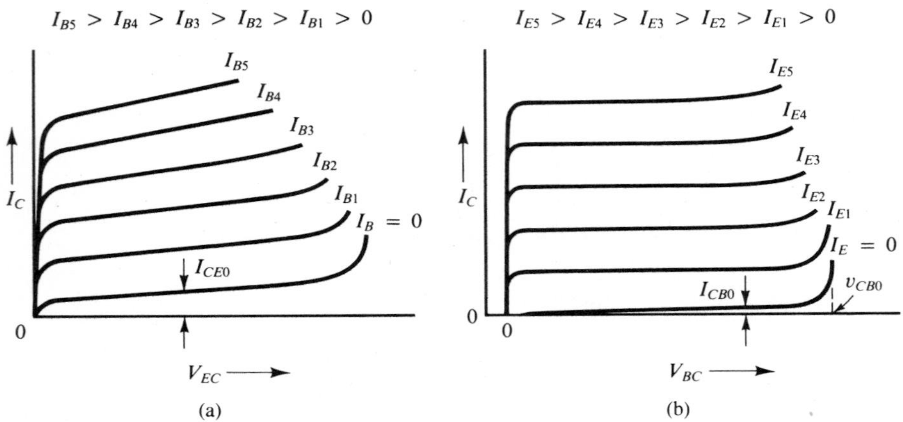

FIGURE 3.18

Real p^+-n-p transistor output characteristics for a (a) common emitter and (b) a common base

equilibrium values. Similarly, when the diode is forward biased, recombination of carriers dominates in the depletion region. Each recombination event removes an excess electron-hole pair. At very low levels of the forward current, the recombination current component dominates the diode. This and a few other factors result in an I-V characteristic of the form

$$I = I_0[\exp(eV_A/nkT) - 1] \tag{3.103}$$

where n is generally referred to as an ideality factor and is a measure of how close to ideal the conditions under which the diode was fabricated were.

Base-width modulation happens to be only one of the many effects that mold the real transistors to deviate from the expected ideal responses. The common-emitter output response, as shown in Fig. 3.18(a), shows a significant effect of base-width modulation. Since W decreases with increasing V_{EC} when V_{EB} is a constant, I_C increases with increasing V_{EC} according to Eq. (3.99). In the common-base active region, for example, it can be easily seen from Eq. (3.97) that I_E varies inversely as the emitter-base junction width. As V_{BC} becomes more forward biased, W decreases drastically, which increases I_E when V_{EB} is a constant. As I_E increases with the increase of V_{BC}, and since V_{EB} decreases to keep I_E fixed, I_C does not increase significantly. I_C will not, however, remain constant for all V_{BC} but will increase slightly at higher bias voltage. In real transistors, junctions are not abrupt but graded. This causes the base transit time to decrease, often increasing the high-frequency performance of the transistor. Again, high current operating conditions may cause the rate of minority carrier injection to go beyond the limits of low-level injection. Such an operating condition would thus invalidate the basic assumption that no generation and recombination occur in the depletion regions.

Problems

3.1 Determine the carrier distribution when E_F is (a) closer to the conduction band, (b) at the middle of the band gap, and (c) closer to the valence band. Sketch the plots as functions of energy.

3.2 Derive Eqs. (3.15) and (3.17).

3.3 Show that the built-in potential V_B for a diode having an equal level of doping on either side of the junction is given by

$$V_B = \frac{e}{2K_s\varepsilon_0} \frac{N_A N_D}{N_A + N_D} (x_n + x_p)^2$$

3.4 Obtain E_i as a function of E_F and doping level for both *p*-type and *n*-type semiconductors. At what levels of doping in Si does the semiconductor become degenerate? Plot E_F against doping levels.

3.5 Determine the resistivity of an extrinsic semiconductor. Plot both resistivity and conductivity for both *p*-type and *n*-type semiconductors as functions of doping.

3.6 Obtain the general expression for the carrier concentration under the following conditions: (a) steady-state, (b) no concentration gradient, (c) steady-state and no concentration gradient, and (d) steady-state and no recombination-generation.

3.7 Repeat the plots of Fig. 3.9 for the cases of (a) forward-biased and (b) reverse-biased diodes.

3.8 For a linearly graded symmetric junction, $N_A - N_D = -mx$ and $x_n = x_p$, where m is any positive constant. Determine $\mathscr{E}$ and V as a function of x for such a junction.

3.9 Plot the minority carrier concentrations in the semiconductor bulk regions as a function of position for both forward and reverse bias.

3.10 Derive Eqs. (3.97), (3.99), and (3.100).

3.11 Derive Eqs. (3.101) and (3.102).

3.12 Assume a quasi-ideal p^+-n-p transistor where $\Delta p_B(x)$ is linear as described by Eq. (3.94) but recombination is allowed in the base. Show that this base current due to recombination only is

$$I_B|_{\text{recomb}} = \frac{eAW}{2\tau_B} p_{B0}[(e^{eV_{EB}/kT} - 1) + (e^{eV_{CB}/kT} - 1)]$$

where τ_B is the base time constant. Then determine the total base current in the quasi-ideal transistor. What is the effect on I_C and I_E?

3.13 Calculate the depletion layer junction capacitance of the *p*-*n* diode using $C_j = dQ_j/dV$, where Q_j is the stored charge on either side of the step junction and V is the bias voltage.

3.14 Repeat Problem 3.13 for the graded symmetric junction diode (see Problem 3.8). How does your answer differ from that of Problem 3.13?

3.15 Show that the current flowing through a narrow-base p^+-n diode, whose n-side length l is significantly smaller than L_p, is given by

$$I = \frac{eAD_p p_{n0}}{L_p} \coth\left(\frac{l}{L_p}\right)[e^{eV_A/kT} - 1]$$

Note: Employ the condition $\Delta p_n = 0$ at $x = l$ in Eq. (3.79).

References and Suggested Readings

Arora, N. D., J. R. Hauser, and D. J. Roulston. "Electron and Hole Mobilities in Silicon as a Function of Concentration and Temperature." *IEEE Trans. Elect. Dev.* Vol. ED-29 (1982): 292.

Bar-lev, A. *Semiconductor and Electronic Devices.* (2nd ed.). Englewood Cliffs, NJ: Prentice-Hall, 1980.

Barber, H. D. "Effective Mass and Intrinsic Concentration in Silicon." *Solid State Electr.* Vol. 10 (1967): 1039.

Blakemore, J. S. "Semiconducting and Other Major Properties of Gallium Arsenide." *J. Appl. Phys.* Vol. 53 (1982): R123.

Blakemore, J. S. "Approximations for Fermi-Dirac Integrals, Especially the Function $F_{1/2}(\eta)$ Used to Describe Electron Density in a Semiconductor." *Solid State Electr.* Vol. 25 (1982): 1067.

Colclaser, R. A., and S. Diehl-Nagle. *Materials and Devices for Electrical Engineers and Physicists.* Reading, MA: Addison-Wesley, 1987.

Hess, K. "Ballistic Electron Transport in Semiconductors." *IEEE Trans. Electr. Dev.* Vol. ED-28 (1981): 937.

Hobgood, H. M., G. W. Eldridge, D. L. Barrett, and R. N. Thomas. "High-Purity Semi-Insulating GaAs Material for Monolithic Microwave Integrated Circuits." *IEEE Trans. Elect. Dev.* Vol. ED-28 (1981): 140.

Hummel, R. E. *Electronic Properties of Materials: An Introduction for Engineers.* Berlin: Springer-Verlag, 1985.

Lee, D. S., and J. G. Fossum. "Energy-Band Distortion in Highly Doped Silicon." *IEEE Trans. Elect. Dev.* Vol. ED-30 (1983): 626.

Li, S. S., and W. R. Thurber. "The Dopant Density and Temperature Dependence of Electron Mobility and Resistivity in *n*-Type Silicon." *Solid-State Electr.* Vol. 20 (1977): 609

Linares, L. C., and S. S. Li. "An Improved Model for Analyzing Hole Mobility and Resistivity in *p*-Type Silicon Doped with Boron, Gallium, and Indium." *J. Electrochem. Soc.* Vol. 128 (1981): 601.

Marshak, A. H., and C. M. Van Vliet. "Electrical Current and Carrier Density in Degenerate Materials with Nonuniform Band Structure." *Proc. IEEE* Vol. 72 (1984): 148.

Mott, N. F. "Recombination: A Survey." *Solid-State Electr.* Vol. 21 (1980): 1275.

Navon, D. H. *Semiconductor Microdevices and Materials.* New York: Holt, Rinehart, & Winston, 1986.

Pierret, R. F. *Advanced Semiconductor Fundamentals* (Vol. VI of Modular Series on Solid State Devices). Reading, MA: Addison-Wesley, 1987.

Priu, J., and W. Budde. "Compliance of Si Photodiodes with Talbot's Law." *Appl. Opt.* Vol. 20 (1981): 2676.

Queisser, H. J. "Recombination at Deep Traps." *Solid-State Electr.* Vol. 21 (1980): 1495.

Richou, F., G. Pelous, and D. Lecrosnier. "Thermal Generation of Carriers in Gold-Doped Silicon." *J. Appl. Phys.* Vol. 51 (1980): 6252.

Seeger, K. *Semiconductor and Electronic Devices: An Introduction.* Berlin: Springer-Verlag, 1985.

Smith, P. M., J. Frey, and P. Chatterjee. "High-Field Transport of Holes in Silicon." *Appl. Phys. Letts.* Vol. 39 (1981): 332.

Streetman, B. G. *Solid State Electronic Devices* (2nd ed.). Englewood Cliffs, NJ: Prentice-Hall, 1980.

Sze, S. M. *Physics of Semiconductor Devices* (2nd ed.). New York: John Wiley & Sons, 1981.

Sze, S. M. (Ed.). *VLSI Technology.* New York: McGraw-Hill, 1983.

CHAPTER FOUR

Photoemitters and Photodetectors

4.1 Introduction

Optoelectronic devices are designed to generate or intercept efficiently with optical energy—i.e., with visible, ultraviolet, or infrared light. Accordingly, optoelectronic devices can be either light emitters or light sensors. In principle, light emitters and light sensors utilize light in exactly the reverse manner from one another.

Light emitters can be either incoherent or coherent. Incoherent light emitters, such as light-emitting diodes (LEDs) and incandescent lights, radiate optical energy over a broad range of wavelengths. In comparison, coherent light emitters, such as a gas laser, radiate optical energy over a very narrow band. Interestingly, compound semiconductors, such as GaAs, make better light emitters than do the simple single semiconductors, Si or Ge.

Light sensors, also referred to as light detectors, convert incoming optical energy to an equivalent response that is easily measurable. Light sensors comprise all photodetectors, including the solar cells that convert solar energy to an equivalent amount of electrical energy. Note that from among the semiconductors, either Si or Ge may be used to make light sensors, even though these semiconductors emit more heat than light.

The photoemission process, often referred to as **luminescence**, can be classified into three types—photoluminescence, cathodoluminescence, and electroluminescence—according to the nature of the excitation mechanisms utilized. Note, however, that luminescence does not include incandescence, which is purely a result of a material's temperature. When carriers are excited by means of photoabsorption, the process is called **photoluminescence**. When high-energy electron bombardment is used for the excitation of carriers, the process is referred to as the **cathodoluminescence**, and when an electric field is employed for excitation of the carrier, the process is called **electroluminescence**. In the case of LEDs, for example, electroluminescence plays the most effective role.

Light sensors are of two types: thermal and quantum. **Thermal sensors** are those in which the absorbed light is converted to heat, thus producing a detectable change in temperature. **Quantum sensors** generate carriers as a result of the absorption of photons. Quantum sensors, or detectors, are classified according to three types: photoemissive, photoconductive, and photovoltaic. In **photoemissive detectors**, absorbed photons generate free electrons that escape from materials, typically enclosed within a vacuum container. The generated electrons are subsequently collected and measured. In **photoconductive sensors**, generated free charge carriers increase the conductivity of the semiconductor material. The conductivity is then measured by estimating the resultant change in the current flow. A **photovoltaic detector** consists of a *p-n* junction whose barrier potential changes with the absorption of incident photons. The resulting change in the corresponding open-circuit output voltage is used to measure the amount of absorbed radiation.

Quantum light sensors are available as solid-state devices or as photoemissive tubes. While the solid-state devices are compact, the photoemissive light sensor tubes are bulky. The spectral responsivity of photoemission-related detectors is mostly limited to ultraviolet and the visible part of the spectrum, while that of the solid-state photodetectors ranges from near ultraviolet through the visible and into the infrared regions of the spectrum. These photoemissive sensors or detectors utilize either an external or an internal photoelectric effect. In the former, the energy of the incoming photon is large enough to free an electron from the surface of a photosensitive material. In the latter case, a small amount of photon energy raises an electron from a valence band or an impurity band to the conduction band.

There are a variety of thermal detectors: calorimeters, thermopiles, bolometers, pyroelectric detectors, and pneumatic detectors. Pyroelectric detectors, for example, utilize the phenomenon whereby the spontaneous polarization of a piezoelectric crystal is changed when its temperature is changed. The differential pyroelectric effect generates current that is in proportion to the rate of change in temperature rather than to the magnitude of the change.

In comparison to quantum sensors, thermal detectors are far slower and less uniform in their responses. While quantum sensors are mostly narrow-band devices, thermal detectors are generally broadband in character. However, thermal sensors are simpler devices in that they require no cooling accessories.

Light sensors can produce meaningful optoelectronic displays according to two basic approaches. In the first method, an image is directed onto a large-area detector within a vacuum chamber referred to as a tube, and the projected image is read by means of electron scanning. Such camera tubes are frequently used in the television industry. The second method consists of display devices formed of individual sensors (each corresponding to a single pixel) that are organized in a two-dimensional array. The arrays of such sensors can be organized so that the resulting image exceeds the resolution of commercial television. While intrinsic silicon semiconductor materials can sense visible and near-infrared optical signals, extrinsic silicon materials respond to longer wavelengths and are, thus, desirable in the detector arrays.

In this chapter, we concentrate on light emitters and quantum sensors. A section in this chapter is devoted to the physics of photographic media. Photographic film is a somewhat different type of detector that is often used to record the incoming light pattern. We leave the discussion of lasers to Chapter 5 because many issues pertaining to lasers need an in-depth investigation, and we return to the discussion of noise characteristics of detectors in a subsequent chapter.

4.2 Light-Emitting Diodes

A light-emitting diode (LED) is a solid-state *p-n* junction diode that gives out light upon the application of a bias voltage. The luminescence process in this case is electroluminescence, which is associated with emission wavelengths in the visible and infrared regions of the spectrum. When a forward bias is applied to the *p-n* junction diode, carriers are injected into the depletion region in large numbers. Because of their closeness, the electron-hole pairs undergo a recombination that is associated with the emission of energy. Depending on the semiconductor band-gap characteristics, this energy can be in the form of heat (as phonons) or light (as photons).

The solution of the Schrödinger equation for a crystal (see Fig. 1.16) has already revealed a relationship between the energy E of an electron in a solid and its wave vector **k**. It may be noted that the lattice structure affects the motion of an electron only when k is close to $n\pi/l$ (where n is any integer and l is the crystal periodicity) and the effect of this constraint is to introduce an energy band gap between the allowed energy bands. Figure 4.1(a) shows portions of two E versus k curves for neighboring energy bands within the regions $k = \pi/l$ and $k = -\pi/l$ (also known as the reduced zone).

While the upper band of Fig. 4.1 represents the energy of conduction band electrons, the curvature of the lower band can be associated with electrons having negative effective mass (Eq. 1.148). The concept of negative effective mass can readily be identified with the concept of holes in the valence band. While the majority of the electrons are identified with the minima of the upper E-k curve, the majority of the holes are identified with the maxima of the lower E-k curve. The minimum value of the conduction band and the maximum value of the valence band in Fig. 4.1(a) both have identical k values. A semiconductor having such a characteristic is said to have a direct band gap, and the associated recombination in such a semiconductor is referred to as direct.

The **direct recombination** of an electron-hole pair always results in the emission of a photon. In a direct band-gap semiconductor, the emitted photon is

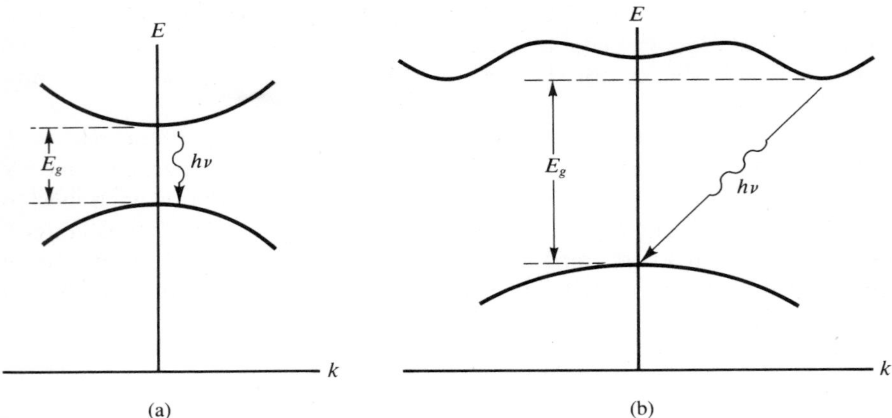

FIGURE 4.1

***E* versus *k* for semiconductors having (a) a direct band gap and (b) an indirect band gap**

not associated with any change in momentum (given by $\hbar k$) since $\Delta k = 0$. However, for some of the semiconducting materials, the E versus k curve may be somewhat different, as shown in Fig. 4.1(b). While the minimum conduction-band energy can have a nonzero k, the maximum valence-band energy can have $k = 0$. The electron-hole recombination in such a semiconductor is referred to as indirect.

An **indirect recombination** process involves a momentum adjustment. Most of the emission energy is thus expended in the form of heat (as phonons). Very little energy is left for the purpose of photon emission, which in most cases is a very slow process. Furthermore, since both photons and phonons are involved in this energy exchange, such transitions are less likely to occur. The interband recombination rate is basically given by

$$\frac{dn}{dt} = B_r np \tag{4.1}$$

where B_r is a recombination-dependent constant which for a direct band-gap semiconductor is $\sim 10^6$ times larger than that for an indirect band-gap semiconductor. For direct recombination, B_r value ranges from 0.46×10^{-10} to 7.2×10^{-10} cm^3/s.

All semiconductor crystal lattices are reasonably identical, being dissimilar only in terms of their band characteristics. Si and Ge both have indirect band transitions, whereas GaAs, for example, is a semiconductor that has direct band transition. Thus, while Si and Ge are preferred for fabricating transistors and integrated circuits, GaAs is preferred for the fabrication of LEDs.

The direct recombination (when $k =$ constant) results in a photon emission whose wavelength (in μm) is given by

$$\lambda = \frac{hc}{E_g} = \frac{1.24}{E_g(\text{eV})} \tag{4.2}$$

where E_g is the band-gap energy. The LEDs under proper forward-biased conditions can operate in the ultraviolet, visible, and infrared regions. For the visible region, however, we need to be concerned with the spectral luminous efficiency curves of Fig. 2.6, which account for the fact that the visual response to any emission is a function of wavelength. It is unfortunate that there is not a single element semiconductor suitable for fabricating LEDs, but there are many binary and ternary compounds that can be used for fabricating LEDs. Table 4.1 lists some of these binary semiconductor materials. The ternary semiconductors include GaAlAs, $CdGeP_2$, and $ZnGeP_2$ for the infrared region operation, $CuGaS_2$ and $AgInS_2$ for the visible region operation, and $CuAlS_2$ for the ultraviolet region operation. Ternary semiconductors are used because their energy gaps can be tuned to a desired emission wavelength by picking appropriate composition.

Of the ternary compounds, gallium arsenide-phosphide (written as $GaAs_{1-x}P_x$) is an example that is basically a combination of two binary semiconductors, namely, GaAs and GaP. The corresponding band-gap energy of the semiconductor can be varied by changing the value of x. For example, when $x = 0$, $E_g = 1.43$ eV. E_g increases with increasing x until $x = 0.44$ and $E_g = 1.977$ eV, as shown in Fig. 4.2. However for $x \geqslant 0.45$, the band gap is indirect.

TABLE 4.1

Binary Semiconductors Suitable for LED Fabrication

	Material	E_g(eV)	*Emission Type*
III–V	GaN	3.5	UV
II–VI	ZnS	3.8	UV
II–VI	SnO_2	3.5	UV
II–VI	ZnO	3.2	UV
III–VII	CuCl	3.1	UV
II–VI	BeTe	2.8	UV
III–VII	CuBr	2.9	UV–Visible
II–VI	ZnSe	2.7	Visible
III–VI	In_2O_3	2.7	Visible
II–VI	CdS	2.52	Visible
II–VI	ZnTe	2.3	Visible
III–V	GaAs	1.45	IR
II–VI	CdSe	1.75	IR–Visible
II–VI	CdTe	1.5	IR
III–VI	GaSe	2.1	Visible

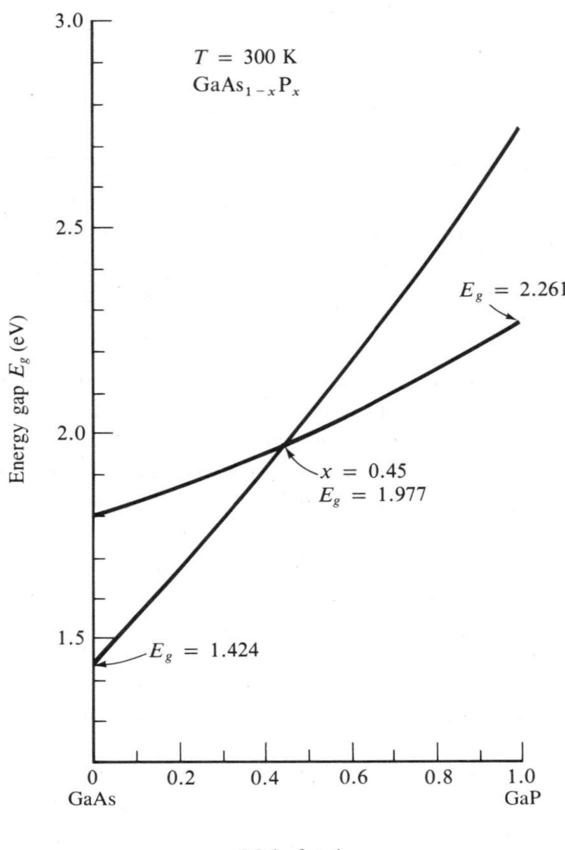

FIGURE 4.2

Band-gap energy versus x in $GaAs_{1-x}P_x$. (After H. J. Casey, Jr., and M. B. Panish. *Heterostructure Lasers.* NY: Academic Press, 1978. Courtesy of Academic Press.)

The most common composition of $GaAs_{1-x}P_x$ used in LEDs has $x = 0.4$ and $E_g \simeq 1.8$ eV. This band-gap energy corresponds to an emission of red light. The calculators and watches often use this particular composition of $GaAs_{1-x}P_x$.

Interestingly, the indirect band gap of $GaAs_{1-x}P_x$ (with $1 \geqslant x \geqslant 0.45$) can be used to output light ranging from yellow through green provided the semiconductor is doped with impurities such as nitrogen. The dopants introduced in the semiconductor replace phosphorus atoms which, in turn, introduce electron trap levels very near to the conduction band. For example, at $x = 0.5$, the doping of nitrogen increases the LED efficiency from 0.01% to 1%, as shown in Fig. 4.3. It must be noted, however, that nitrogen doping shifts the peak

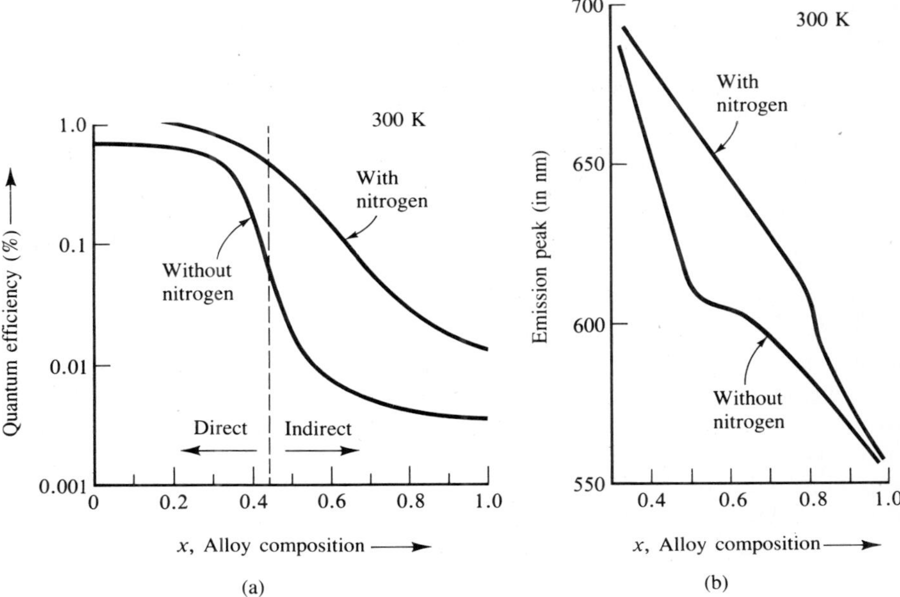

FIGURE 4.3

The effects of nitrogen doping in $GaAs_{1-x}P_x$: (a) quantum efficiency versus x and (b) peak emission wavelength versus x

emission wavelength toward the red. The shift is comparatively larger at and around $x = 0.5$ than at $x = 1.0$. The energy emission in nitrogen-doped $GaAs_{1-x}P_x$ devices is a function of both x as well as nitrogen concentration.

Nitrogen is a different type of impurity from those commonly encountered in extrinsic semiconductors. Nitrogen, like arsenic and phosphorus, has five valence electrons, but it introduces no net charge carriers in the lattice. It provides active radiative recombination centers in the indirect band-gap materials. For an electron, a recombination center is an empty state in the band gap into which an electron falls and, then, thereafter, falls into the valence band by recombining with a hole. For example, while a GaP LED emits green light (2.23 eV), a nitrogen-doped GaP LED emits yellowish-green light (2.19 eV), and a heavily nitrogen-doped GaP LED emits yellow light (2.1 eV).

The injected excess carriers in a semiconductor may recombine either radiatively or nonradiatively. Whereas nonradiative recombination generates phonons, radiative recombination produces photons. Consequently, the internal quantum efficiency η defined as the ratio of the radiative recombination rate R_r to the total recombination rate, is given by

$$\eta = \frac{R_r}{R_r + R_{nr}} \tag{4.3}$$

where R_{nr} is the nonradiative recombination rate. However, the injected excess carrier densities, Δn and Δp, return to their equilibrium value exponentially as

$$\Delta p \equiv \Delta n = \Delta n_0 e^{-t/\tau} \tag{4.4}$$

where τ is the carrier lifetime and Δn_0 is the excess electron density at equilibrium. Since $\Delta n/R_r$ and $\Delta n/R_{nr}$ are respectively equivalent to the radiative recombination lifetime τ_r and the nonradiative recombination lifetime τ_{nr}, we can obtain effective minority carrier bulk recombination time τ as

$$\frac{1}{\tau} = \frac{1}{\tau_r} + \frac{1}{\tau_{nr}} \tag{4.5}$$

such that $\eta = \tau/\tau_r$. The reason that a fast recombination time is crucial is that the longer the carrier remains in an excited state, the larger the probability that it will give out energy nonradiatively. In order for the internal quantum efficiency to be high, the radiative lifetime τ_r needs to be small. For indirect band-gap semiconductors, $\tau_r \gg \tau_{nr}$ so that very little light is generated, and for direct band-gap semiconductors, τ_r increases with temperature so that the internal quantum efficiency deteriorates with the temperature.

As long as the LEDs are used as display devices, it is not too important to have fast response characteristics also. However, the LEDs are also used for the purpose of optical communications, and for those applications it is appropriate to study their time response characteristics. For example, an LED can be used in conjunction with a photodetector for transmitting optical information between two points. The LED light output can be modulated to convey optical information by varying the diode current. Most often the transmission of optical signals is facilitated by introducing an optical fiber between the LED and the photodetector.

There can be two different types of capacitances in diodes that can influence the behavior of the minority carriers. One of these is the **junction capacitance** (see Problem 3.13), which is caused by the variation of majority charge in the depletion layer. While it is inversely proportional to the square root of bias voltage in the case of an abrupt junction, it is inversely proportional to the cube root of bias voltage in the case of a linearly graded junction. The second type of capacitance, known as the **diffusion capacitance**, is caused by the minority carriers.

Consider an LED that is forward biased with a dc voltage. Consider further that the bias is perturbed by a small sinusoidal signal. When the bias is withdrawn or reduced, charge begins to diffuse from the junction as a result of recombination until an equilibrium condition is achieved. Consequently, as a response to the signal voltage, the minority carrier distribution contributes to a signal current.

Consider a one-dimensional p-type semiconducting material of cross-sectional area A whose excess minority carrier density is given by Eq. (3.45) as

$$\frac{\partial \Delta n_p}{\partial t} = D_n \frac{\partial^2 \Delta n_p}{\partial x^2} - \frac{\Delta n_p}{\tau} \tag{4.6}$$

As a direct consequence of the applied sinusoidal signal, the excess electron distribution fluctuates about its dc value. In fact, we may assume excess minority carrier density to have a time-varying component as described by

$$\Delta n_p(x, t) = \langle \Delta n_p(x) \rangle + \Delta n'_p(x) e^{j\omega t} \tag{4.7}$$

where $\langle \Delta n_p(x) \rangle$ is a time-invariant quantity. By introducing Eq. (4.7) into Eq. (4.6), we get two separate differential equations:

$$\frac{\partial^2}{\partial x^2} \langle \Delta n_p(x) \rangle = \frac{\langle \Delta n_p(x) \rangle}{L_n^2} \tag{4.8a}$$

and

$$\frac{\partial^2}{\partial x^2} \Delta n'_p(x) = \frac{\Delta n'_p(x)}{(L_n^*)^2} \tag{4.8b}$$

where

$$L_n^* = \frac{L_n}{\sqrt{1 + j\omega\tau}} \tag{4.9a}$$

and

$$L_n = [D_n \tau]^{1/2} \tag{4.9b}$$

Equation (4.8a) is identical to Eq. (3.71b), and its dc solution is well known. Again, the form of Eq. (4.8b) is similar to that of Eq. (4.8a) and, therefore, its solution is given by

$$\Delta n'_p(x) = \Delta n'_p(0) e^{-x/L_n^*} \tag{4.10}$$

Since the frequency dependent current $I(\omega)$ is simply a product of eAD_n and the concentration gradient, we find that

$$I(\omega) = \left| eAD_n \frac{d\Delta n'_p(x)}{dx} \right|_{x=0} = \frac{I(0)}{\sqrt{1 + \omega^2 \tau^2}} \tag{4.11}$$

where $I(0)$ is the intensity emitted at zero modulation frequency. We can determine the admittance next by dividing the current by the perturbing voltage. The real part of the admittance, in this case, will be equivalent to the diode conductance, whereas its imaginary part will correspond to the diffusion capacitive susceptance.

The modulation response as given by Eq. (4.11) is, however, limited by the carrier recombination time. Often an LED is characterized by its modulation bandwidth, which is defined as the frequency band over which signal power (proportional to $I^2(w)$) is half of that at $\omega = 0$. Using Eq. (4.11), the 3-dB modulation bandwidth is given by

$$\Delta\omega \simeq \frac{1}{\tau_r} \tag{4.12}$$

where the bulk lifetime has been approximated by the radiative lifetime. Some times the 3-dB bandwidth of the LED is given by $I(\omega) = \frac{1}{2}I(0)$, but this simplification contributes to an erroneous increase in the bandwidth by a factor of 1.732.

Under conditions of thermal equilibrium, the recombination rate is proportional to the product of initial carrier concentrations, n_0 and p_0. Then, under nonequilibrium conditions, additional carriers $\Delta n = \Delta p$ are injected into the material. Consequently, the recombination rate of injected excess carrier densities is given by initial carrier concentrations and injected carrier densities as

$$\begin{aligned} R_{\Delta n} &= [B_r(n_0 + \Delta n)(p_0 + \Delta p) - B_r n_0 p_0] \\ &= B_r(n_0 + p_0 + \Delta n)\Delta n \end{aligned} \tag{4.13}$$

where B_r is the same constant introduced in Eq. (4.1). For p-type GaAs, for example, $B_r = 1.7 \times 10^{-10}$ cm^3/s when $p_0 = 2.4 \times 10^{18}$ holes/cm^3. Eq. (4.13) is used to define the radiative carrier recombination lifetime by

$$\tau_r = \frac{\Delta n}{R_{\Delta n}} = \frac{1}{B_r(n_0 + p_0 + \Delta n)} \tag{4.14}$$

In the steady-state condition, the excess carrier density can be calculated in terms of the active region width d by

$$\Delta n = \frac{J\tau_r}{ed} \tag{4.15}$$

where J is the injection current density.

The radiative recombination lifetime is found by solving Eq. (4.14) after having eliminated Δn from it using Eq. (4.15):

$$\tau_r = \frac{\left[(n_0 + p_0)^2 + \dfrac{4J}{B_r ed}\right]^{1/2} - (n_0 + p_0)}{\dfrac{2J}{ed}} \tag{4.16}$$

Thus, while for the low carrier injection (i.e., $n_0 + p_0 \gg \Delta n$), Eq. (4.16) reduces to

$$\tau_r \simeq \frac{1}{B_r(n_0 + p_0)} \tag{4.17a}$$

for the high carrier injection (i.e., $n_0 + p_0 \ll \Delta n$), it reduces to

$$\tau_r \simeq \sqrt{\frac{ed}{JB_r}} \tag{4.17b}$$

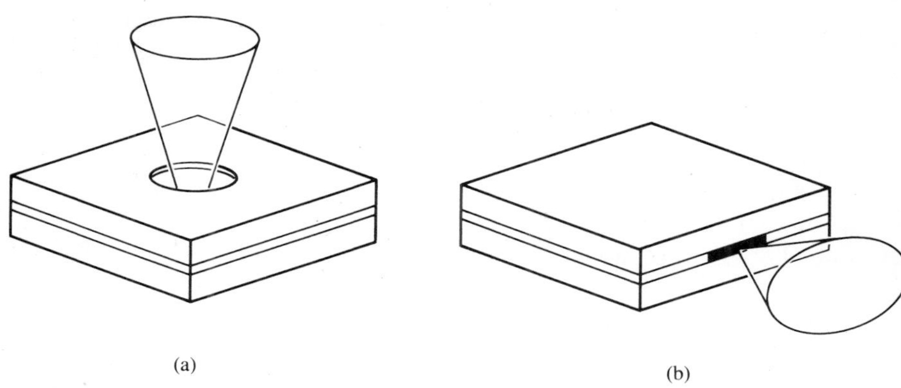

FIGURE 4.4

LED type: (a) surface emitter and (b) edge emitter

Equation (4.17a) indicates that in highly doped semiconductors, τ_r is small. But the doping process has its own problem, since in many of the binary LED compounds higher doping may introduce nonradiative traps just below the conduction band, thus nullifying Eq. (4.12). In comparison to Eq. (4.17a), Eq. (4.17b) provides a better alternative whereby τ_r can be reduced by decreasing the active region width or by increasing the current density. For the case of *p*-type GaAs, the radiative lifetimes vary between 2.6 ns and 0.35 ns, respectively, when p_0 varies between 1.0×10^{18} holes/cm^3 and 1.5×10^{19} holes/cm^3.

Usually, LEDs are operated at low current ($\simeq 10$ mA) and low voltages ($\simeq 1.5$ V), and they can be switched on and off in the order of 10 ns. In addition, because of their small sizes, they can be reasonably treated as point sources. It is, therefore, not surprising that they are highly preferred over other light sources for applications in fiber optic data links.

Two particular LED designs are popular: **surface emitters** and **edge emitters**. They are shown in Fig. 4.4. In the former, the direction of major emission is normal to the plane of the active region, whereas in the latter the direction of major emission is in the plane of the active region. The emission pattern of the surface emitters is very much isotropic, whereas that of the edge emitters is highly directional.

As the LED light originating from a medium of refractive index n_1 goes to another medium of refractive index n_2 ($n_2 < n_1$), only a portion of incident light is transmitted. This behavior follows from Fig. 1.4. In particular, the portion of the emitted light corresponds to only that which originates from within a cone of semi-apex angle θ_c, such that

$$\theta_c = \sin^{-1}\left(\frac{n_2}{n_1}\right) \quad \textbf{(4.18)}$$

In the case of an LED, n_1 corresponds to the refractive index of the LED medium and n_2 corresponds to that of air (or vacuum). Light originating from *beyond* angle θ_c undergoes a total internal reflection. However, the light directed from *within* the cone of the semi-apex angle θ_c will be subjected to Fresnel's loss as described in Section 1.3. Thus the overall transmittance T is given by

$$T = 1 - \left(\frac{n_1 - n_2}{n_1 + n_2}\right)^2 \tag{4.19}$$

Accordingly, the total electrical-to-optical conversion efficiency in LEDs is given by

$$\begin{aligned}
\eta_{\text{LED}} &= T[(\text{solid angle within the cone})/(4\pi)] \\
&= \frac{1}{2} T \int_0^{\theta_c} \sin\theta \, d\theta \\
&= \frac{1}{2} T(1 - \cos\theta_c) \\
&\simeq \frac{1}{4} T \sin^2\theta_c \\
&= \frac{1}{4}\left(\frac{n_2}{n_1}\right)^2 \left[1 - \left(\frac{n_1 - n_2}{n_1 + n_2}\right)^2\right]
\end{aligned} \tag{4.20}$$

Only two schemes increase the electrical-to-optical conversion efficiency in an LED. The first technique involves guaranteeing that most of the incident rays strike the glass-to-air interface at angles less than θ_c. It is accomplished by making the semiconductor-air interface hemispherical. The second method involves schemes whereby the LED is encapsulated in an almost transparent medium of high refractive index. The latter means is comparatively less expensive. If a glass of refractive index 1.5 is used for encapsulation, the LED efficiency can be increased by a factor of 3. Two of the possible encapsulation arrangements and the corresponding radiation patterns are illustrated in Fig. 4.5.

LEDs are often used in conjunction with a phototransistor to function as an optocoupler. The optocouplers are used in circumstances when it is desirable to have a transmission of signals between electrically isolated circuits. They are used to achieve noise separation by eliminating the necessity of having a common ground between the two systems. Depending on the type of coupling material, these miniature devices can provide both noise isolation as well as high voltage isolation. Figure 4.6 shows a typical case where two optocouplers are used to attain a chopper circuit. The two optocouplers chop either the positive or the negative portion of the input signals with a frequency of one-half that of the control signal that is introduced at the T flip-flop. The operational amplifier provides an amplified version of the chopped output waveform. In comparison, a chopper circuit that uses simple bipolar transistors produces noise spikes in the output because of its inherent capacitive coupling.

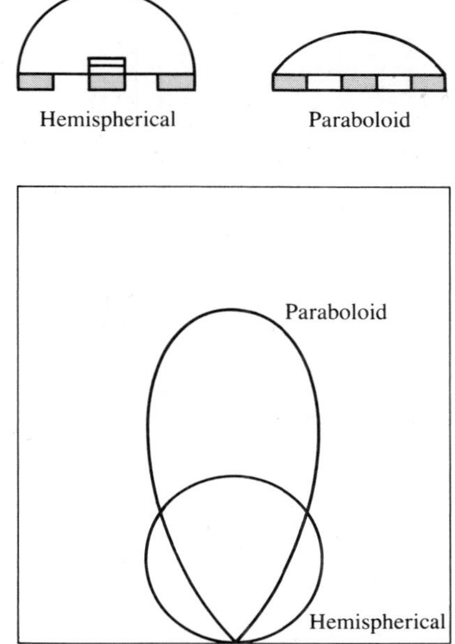

FIGURE 4.5

LED encapsulation geometries and their radiation patterns

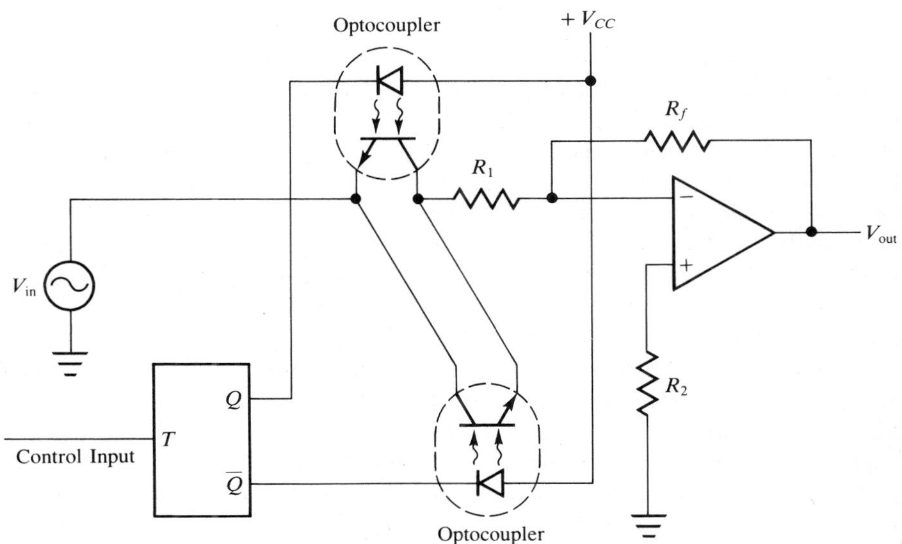

FIGURE 4.6

A chopping circuit with an amplifier

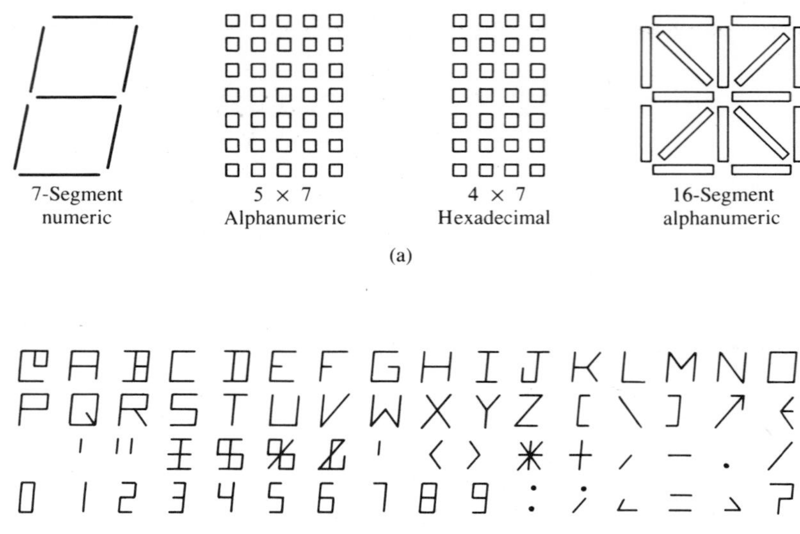

(b)

FIGURE 4.7

(a) LED display formats and (b) displayed alphanumeric characters using 16-segment displays

The visible LEDs are best known for their uses in displays and indicator lamps. In applications where more than a single source of light is required, an LED array can be utilized. An LED array is a device consisting of a row of discrete LEDs connected together within or without a common reflector cavity. Figure 4.7(a) shows different LED arrangements for displaying hexadecimal numeric and alphanumeric characters, whereas Fig. 4.7(b) shows, for example, the possible alphanumeric characters using 16-segment displays. In digital systems, the binary codes equivalent to these characters are usually decoded and, consequently, a specific combination of LED segments are turned on to display the desired alphanumeric character.

The dot matrix display provides the most desirable display font. It gives more flexibility in shaping characters and has a lower probability of being misinterpreted in case of a display failure. However, these displays involve a large number of wires and increased circuit complexity. LED displays, in general, have excellent viewing angle, high resonance speed ($\simeq$ 10 ns), long life, and super interface capability with electronics with almost no duty cycle limitation. LEDs with blue emission are not available commercially. When compared with passive displays, LED displays consume more power and involve complicated wiring with at least one wire per display element.

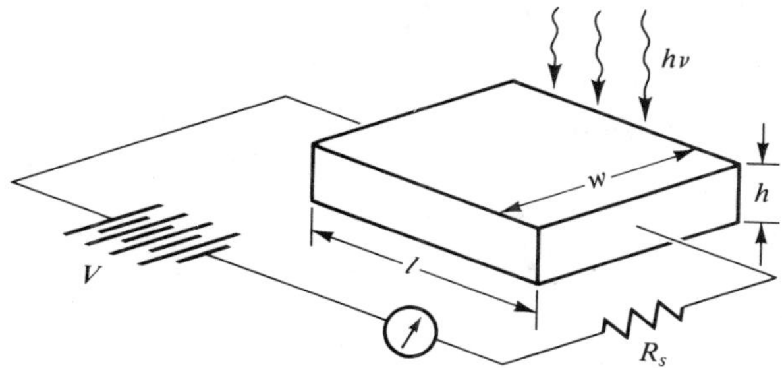

FIGURE 4.8

A photoconductor circuit

4.3 Photoconductors

Almost all semiconductors exhibit certain degrees of photoconductivity. Accordingly, a photoconductor is the simplest of all photodetection devices, built exclusively of only one type of semiconductor that has a large surface area and two ohmic contact points. In the presence of an energized incident photon, the excited valence-band electron of the photoconductor leaves behind a hole in the valence band. Often an extrinsic semiconductor is better suited for the purpose of photoconduction. For example, a far infrared sensitive photoconductor can be designed by introducing an acceptor level very close to the valence band or by introducing a donor level very close to the conduction band. Consequently, photoconduction has two causes. It is caused either by the absorption of photons at the impurity levels in an extrinsic semiconductor or due to the band-gap transition in an intrinsic semiconductor. Obviously, photoconductors have to be cooled in order to avoid excessive thermal excitation of carriers.

Figure 4.8 shows a typical photoconductor circuit where R_s is the series resistance. Assume further that the resistance of the photoconductor is larger than R_s so that most of the bias voltage appears across the photoconductor surface. To guarantee that in the absence of incoming light the number of carriers is a minimum, the working temperature is maintained sufficiently low. Incident light continues to affect both generation and recombination of carriers until the photoconductor has reached a new equilibrium at a higher carrier concentration. The change in the carrier density causes a reduction in the photoconductor's resistance. In fact, there are a great many commercial applications of photoconductors where the fractional change in resistance is quite large. In the presence of an electric field, the generated excess majority carriers drift away from the appropriate terminals.

The absorbed portion of the incoming monochromatic light that falls normally onto the photoconductor is determined in terms of the absorption coefficient α. In the case of an extrinsic semiconductor, α is very small (1/cm to 10/cm) since the number of available impurity levels is relatively small. But in the case of an intrinsic photoconductor, α is relatively large ($\simeq 10^4$/cm) as the number of available electron states is very large. The absorbed optical power $P_{abs}(y)$ is given by

$$P_{abs}(y) = P_{in}(1 - R)e^{-\alpha y} \tag{4.21}$$

where P_{in} represents the incoming optical power and R is the surface reflectance of the photoconductor. At steady state, the generation and recombination rates are equal to each other. Consequently,

$$\begin{aligned} \frac{\alpha P_{abs}(y)}{h\nu lw} &= \frac{\alpha P_{in}(1-R)e^{-\alpha y}}{h\nu lw} \\ &= \frac{n(y)}{\tau_n} = \frac{p(y)}{\tau_p} \end{aligned} \tag{4.22}$$

where τ_n and τ_p are the mean life times, respectively, of electrons and holes, $n(y)$ and $p(y)$ are the carrier densities, respectively, of electrons and holes, and the product lw is the surface area of the photoconductor.

The total drift current passing through the intrinsic photoconductor is determined using Eqs. (3.31) and (4.22):

$$\begin{aligned} i_s &= \int_0^h e\mathscr{E}w[\mu_n n(y) + \mu_p p(y)]\, dy \\ &= \frac{e\mathscr{E}}{h\nu l}\int_0^h [\mu_n\tau_n + \mu_p\tau_p] P_{abs}(y)\, dy \\ &= \frac{\eta_{pc} e P_{in}}{h\nu l}\mathscr{E}(\mu_n\tau_n + \mu_p\tau_p) \end{aligned} \tag{4.23}$$

where the quantum efficiency η_{pc} is defined as

$$\eta_{pc} = \alpha(1 - R)\int_0^h \exp(-\alpha y)\, dy \tag{4.24}$$

and $\mathscr{E}$ is the electric field. In the case of an extrinsic pohotoconductor, the signal current of Eq. (4.23) reduces to

$$i_s = \begin{cases} \dfrac{\eta_{pc} e P_{in}}{h\nu}\left[\dfrac{\mu_n \mathscr{E}\tau_n}{l}\right] & n\text{-type} \\ \dfrac{\eta_{pc} e P_{in}}{h\nu}\left[\dfrac{\mu_p \mathscr{E}\tau_p}{l}\right] & p\text{-type} \end{cases} \tag{4.25}$$

In either case, the quantity within the square bracket is generally referred to as the **photoconductive gain** G given by

$$G = \begin{cases} \dfrac{\tau_n}{\tau_d} & n\text{-type} \\[2ex] \dfrac{\tau_p}{\tau_d} & p\text{-type} \end{cases} \tag{4.26}$$

where τ_d is the average carrier drift time or transit time between the two metal contacts. This follows from Eq. (3.28) since drift velocity is given by the product of electric field and carrier mobility. The photoconductive gain generally measures the effective charge transferred through the external circuit due to each of the photoinduced carriers. A high gain is attained by reducing τ_d. This can be accomplished by increasing the volume of the photoconductor and decreasing the separation between the metal contacts. Accordingly, photoconductive detectors are often prepared in the shape of a long ribbon with metal contacts along its edges. However, it should be noted that a longer carrier lifetime will affect the device response. The current diminishes at a faster rate if light is withdrawn at any instant. Consequently, the device is not sufficiently effective unless the duration of exposure exceeds the carrier lifetime.

Photoconductors are relatively easy to construct, but they are relatively slow in their operation. They require external voltage sources and in most cases must be cryogenically cooled to minimize the effect of thermally generated charge carriers. Thus what appears to be a less-expensive detector in the beginning becomes quite expensive when all of the peripherals are taken into account. Some of the common photoconductor materials are PbS, CdS, CdSe, InSb, and $Hg_xCd_{1-x}Te$. While InSb has a good response ($\simeq$ 50 ns), CdS and CdSe have poor responses ($\simeq$ 50 ms). CdS and CdSe are used for detecting visible light, and both have very high photoconductive gain ($\simeq 10^4$).

4.4 *p–n* and *p–i–n* Photodiodes

In general, a **photovoltaic detector** consists of a semiconductor junction so that the equilibrium energy bands on the two sides of the junction are shifted relative to one another as shown in Fig. 3.10. If a sufficiently energized photon is incident in the neighborhood of the junction, it produces an electron-hole pair which in turn causes a current to flow through the wire that connects the two components of the junction. Such a mode of operation which requires no external bias is said to be **photovoltaic**. Photovoltaic detectors have large surface areas so that they can generate a large photocurrent in the absence of a bias. However, they are nonlinear in their responses. Light-powered meters and solar cells are common examples of this type of detector. Interestingly, we may use photovoltaic detectors in the so-called photoconductive mode by applying a reverse bias. When used in this mode, the detector has a remarkable linear response.

Photodiodes are examples of bipolar semiconductor junctions that are operated in reverse bias. These photodetectors are generally sensitive, easily biased, small in area, and compatible with integrated optics components. Consequently, they are suitable for systems like those of fiber optic communication links. The basics of the reverse-biased *p-n* junction photodiodes were discussed in Chapter 3. Beyond a certain bias voltage, the detector response is relatively invariant to further changes in bias voltage. However, the frequency response is generally improved at higher bias values. The frequency response is often limited by two factors: carrier diffusion time across the depletion layer and junction capacitance of the diode. The carrier diffusion time is generally reduced by increasing the bias voltage but without exceeding the value of the breakdown voltage, whereas the junction capacitance is improved by incorporating an intrinsic layer between the *p* and *n* regions as in a *p-i-n* photodiode. Our attention later in this section is geared toward the details of such semiconductor devices.

As soon as a semiconductor junction is established, electrons start flowing from the *n*-region to the *p*-region, leaving behind donor ions, and holes start flowing from the *p*-region to the *n*-region, leaving behind acceptor ions. This flow of electrons and holes builds up a depletion layer at the junction. In the absence of any bias, however, the drift and diffusion components of the total current balance each other out. A reverse bias, on the other hand, greatly reduces the diffusion current across the junction but leaves the drift component relatively unaltered.

The photodiode is reverse biased such that a current (generated by incoming photons) proportional to the number of absorbed photons can be generated. With an optical energy in excess of the band-gap energy, electron and hole pairs are generated in photodiodes. Those pairs that are generated in the depletion region are driven by the electric field through the junction, thus contributing to a reverse current. In addition, those pairs that are generated in the bulk regions, but within the diffusion length of the depletion region, diffuse into the depletion region and also contribute to the reverse current. If we neglect the amount of recombination loss in the depletion region, we can estimate the photocurrent by

$$I_\lambda = \frac{e\eta_{pn}P_{\text{abs}}}{h\nu} \tag{4.27}$$

where P_{abs} is the absorbed optical power and η_{pn} is the conversion efficiency. The effective conversion efficiency is reduced by the fact that some of the electron-hole pairs of the bulk areas diffuse into the depletion region.

The number of minority holes generated in the *n*-side but within the diffusion length of the depletion region is AL_pg where g is the generation rate and A is the cross-sectional area of the junction. Similarly, the number of minority electrons generated in the *p*-side but within the diffusion length of the depletion region is AL_ng. Using Eq. (3.88), the net photocurrent in the reverse-biased photodiode is thus given by

$$I = eA\left[\frac{L_p p_{n0}}{\tau_p} + \frac{L_n n_{p0}}{\tau_n}\right][e^{eV_A/kT} - 1] - eAg(L_p + L_n) \tag{4.28}$$

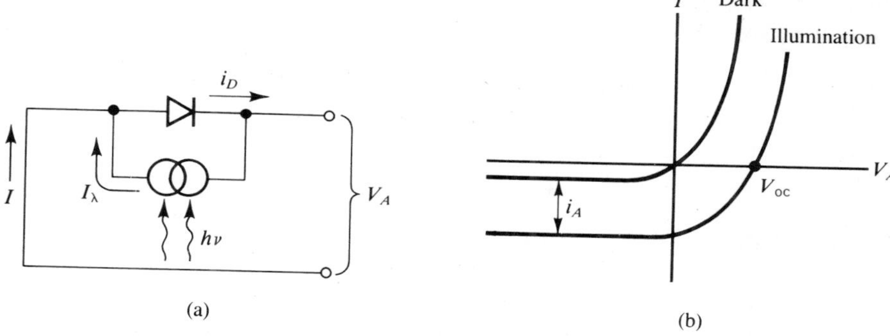

FIGURE 4.9

(a) Equivalent photodiode circuit and (b) its *V-I* characteristics

where the first term refers to photodiode dark current i_d and the second term accounts for the oppositely directed diffusion photocurrent.

When the photodiode is short-circuited (i.e., when $V_A = 0$), the photocurrent is not zero since the current caused solely by the collection of optically generated carriers in the depletion region is nonzero. The equivalent circuit of the photodiode and the corresponding *V-I* characteristics are shown in Fig. 4.9. If the photodiode is open-circuited (i.e., when $I = 0$) in the presence of illumination, an open-circuit photovoltage $V_A = V_{oc}$ appears across the photodiode terminals.

The magnitude of the open-circuit photovoltage, V_{oc} is found from Eq. (4.28):

$$V_{oc} = \frac{kT}{e}\ln\left[\frac{g(L_p + L_n)}{L_p\dfrac{p_{n0}}{\tau_p} + L_n\dfrac{n_{p0}}{\tau_n}} + 1\right] = \frac{kT}{e}\ln\left[\frac{I_\lambda}{I_0} + 1\right] \tag{4.29}$$

where $-I_0$ is the peak reverse dark current. The open-circuit voltage is thus a logarithmic function of the incident optical power P_{abs}. In a symmetrical *p-n* photodiode, I_λ/I_0 approaches the value g/g_{th}, where $g_{th} = p_{n0}/\tau_p$ is the equilibrium thermal generation-recombination rate. Thus, as the minority carrier concentration is increased, g_{th} increases due to the decrease in the carrier lifetime. Consequently, the increase of the minority carrier concentration does not allow V_{oc} to grow indefinitely. In fact. V_{oc} is limited by the equilibrium junction potential.

The power delivered to the load is given by

$$P_L = IV_A = I_0V_A[e^{eV_A/kT} - 1] - I_\lambda V_A \tag{4.30}$$

Thus the particular voltage V_{Am}, corresponding to the maximum power transfer, is found by setting the derivative of P_L to zero. Consequently, we obtain

$$\left[1+\frac{eV_{Am}}{kT}\right]e^{eV_{Am}/kT}=1+\frac{I_\lambda}{I_0} \tag{4.31}$$

Since a *p-n* photodiode is also used as a solar cell for converting the sunlight to electrical energy, we may increase the value of V_{Am} as well as the corresponding photocurrent I_m. Note that the photodiode can achieve a maximum current of I_λ and a maximum voltage of V_{oc}. Often, therefore, the efficiency of a photodiode is measured in terms of the ratio, $(V_{Am}I_m/V_{oc}I_\lambda)$, also known as the fill-factor. The present thrust of solar-cell research is thus directed toward increasing this ratio. By cascading thousands of individual solar cells, we can generate an enormous amount of power that is sufficient for energizing orbiting satellites.

The mode of operation where the photodiode circuit of Fig. 4.9(a) is applied across a simple load is photovoltaic. The voltage across the load R_L can be used to evaluate the current flowing through it. However, if the photodiode in conjunction with a load is subjected to a relatively large external bias, the operation will be referred to as **photoconductive**. The latter mode is preferred over the photovoltaic because the current flowing through the load is generally large enough and, therefore, approaches I_λ. Thus, while the current-to-optical power relationship in the photovoltaic mode is logarithmic, it is linear in the photoconductive mode. Since the depletion-layer junction-capacitance C_j (see Problem 3.13) in an abrupt junction is proportional to $A[(N_DN_A)/\{V_A(N_D+N_A)\}]^{1/2}$, the photovoltaic mode contributes to a larger capacitance and, therefore, to a slower operation. In comparison, the photoconductive photodiode has a faster response.

A cut-off frequency, f_c, is generally defined as the frequency when the capacitive impedance of the photodiode equals the value of the load resistance. Therefore,

$$f_c=\frac{1}{2\pi R_LC_j} \tag{4.32}$$

Thus the junction capacitance has to be decreased to increase the frequency response. This is achieved by decreasing the junction area, by reducing the doping, or by increasing the bias voltage. While there is a physical limit to the smallest junction area, the other two requirements in effect tend to increase depletion width (per Eq. 3.68), drift transit time, and bulk resistance, none of which is truly desirable.

The *p-n* photodiode discussed so far has one weakness, in that the incident optical power is not fully utilized in the optical-to-electric conversion process because the depletion width of a *p-n* junction is extremely small. Because of this

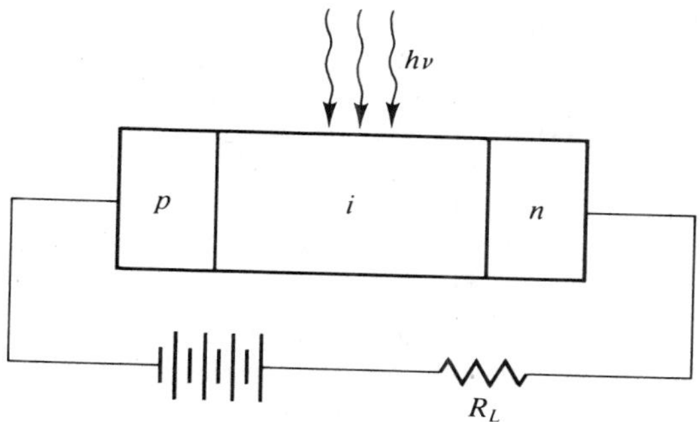

FIGURE 4.10

Reverse-biased *p-i-n* photodiode

physical limitation, the *p-n* photodiodes do not have a desirable response time. This obstacle is overcome by introducing a semi-insulating thick intrinsic (lightly doped) semiconductor layer between its *p*-layer and its *n*-layer, as shown in Fig. 4.10. Such especially organized photodiodes are referred to as *p-i-n* photodiodes.

In *p-i-n* photodiodes, the separating electric field occupies a large fraction of the device. The wider the thickness of the intrinsic layer, the higher the quantum efficiency. High field strength in the intrinsic layer allows the electron-hole pairs to be driven rapidly toward the respective extrinsic regions. However, the carrier transit time is generally proportional to the thickness of the intrinsic layer. Accordingly, there is a design compromise between the expected quantum efficiency and the desirable response time. For a typical *p-n* photodiode, the response time is in the order of 10^{-11} s, whereas that for a *p-i-n* photodiode is about 10^{-9} s. The quantum efficiency of a *p-i-n* photodiode can be anywhere in the range of 50% through 90%. Usually, indirect band-gap semiconductors are preferred over direct band-gap semiconductors as photodiode materials because otherwise there is a significant conversion loss due to surface recombination. Indirect band-gap materials engage phonons to conserve momentum during the transfer. A *p-i-n* configuration eliminates part of this problem because of its longer absorption length.

Si photodiodes (having a maximum quantum efficiency of 90% at 0.9 μm) are used mostly in the wavelength region below 1 μm, whereas Ge photodiodes (having a maximum quantum efficiency of 50% at 1.35 μm) are preferred in the ranges above 1 μm. In addition to the single-element semiconductor photodiodes, there are many ternary (e.g., InGaAs, HgCdTe, and AlGaSb) as well as quaternary (e.g., AlGaAsSb and InGaAsP) photodiodes that are commercially produced.

4.5 Avalanche Photodiode

The photodiodes introduced in the last section have proven to be very successful in their applications in the background-limited photodetection. However, the photodiodes lack internal gain and in many cases require an amplifier to provide noticeable signal currents. APD (avalanche photodiode) is a specific photodiode that makes use of the avalanche phenomenon that was introduced in Section 3.5. By adjusting the bias voltage to a level where it is on the verge of breakdown, we can accelerate the photogenerated carriers. The accelerated carriers, in turn, produce additional carriers by collision ionization.

Avalanche gain is generally dependent on impact ionizations encountered in the regions having sufficiently high electric field. This gain is achieved by subjecting the reverse-biased semiconductor junction to a voltage below its breakdown field ($\simeq 10^5$ V/cm). Electrons and holes thereby acquire sufficient kinetic energy to collide inelastically with a bound electron and ionize it generating an extra electron-hole pair. These extra carriers, in turn, may have sufficient energy to cause further ionization until an avalanche of carriers has resulted. Such a cumulative avalanche process is normally represented by a multiplication factor M that turns out to be an exponential function of the bias. Gains of up to 1000 can be realized in this way. This makes an APD that competes strongly with another high-gain photodetector device, known as a photomultiplier tube, in the red and near infrared.

The probability that a carrier ionization will occur depends primarily on the electric field in the depletion layer. Again, since the electric field in the depletion layer is a function of position (see Fig. 3.9), the ionization coefficients, α and β, respectively, for the electrons and the holes, turn out also to be functions of position. The ionization coefficients are particularly low at lower values of electric fields, as shown in Fig. 4.11, for the case of silicon.

Consider the reverse-biased p-n junction of depletion width W shown in Fig. 4.12. The entering hole current $I_p(0)$ increases as it travels toward the p-side, and the entering electron current $I_n(W)$ increases as it travels toward the n-side. In addition, hole and electron currents due to generation in the depletion layer also move in their respective directions. Thus for the total hole and electron currents we can write

$$\frac{dI_p(x)}{dx} = \alpha(x)I_n(x) + \beta(x)I_p(x) + g(x) \tag{4.33a}$$

$$-\frac{dI_n(x)}{dx} = \alpha(x)I_n(x) + \beta(x)I_p(x) + g(x) \tag{4.33b}$$

where $g(x)$ is the rate per unit length with which the pairs are generated thermally and/or optically. Equation (4.33a) is integrated from $x = 0$ to $x = x$ and, similarly, Eq. (4.33b) is integrated from $x = x$ to $x = W$ to give

$$I_p(x) - I_p(0) = \int_0^x [\alpha(x) - \beta(x)]I_n(x)\,dx + I\int_0^x \beta(x)\,dx + \int_0^x g(x)\,dx \tag{4.34a}$$

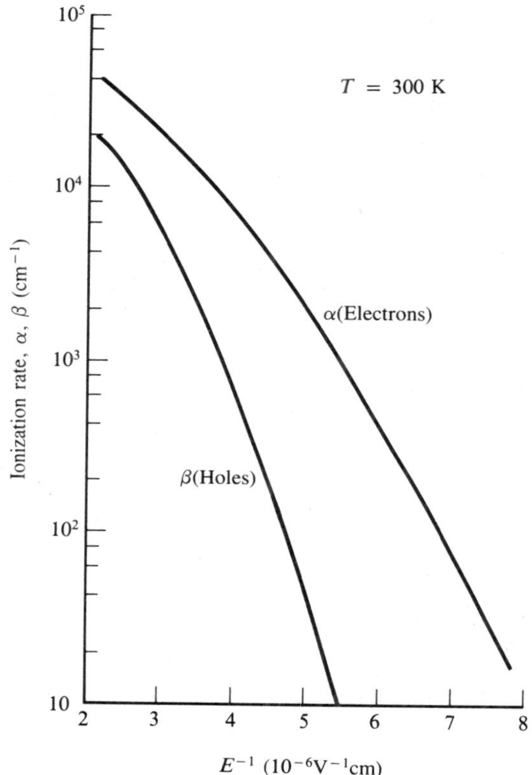

FIGURE 4.11

Ionization coefficients for silicon at 300 K. (Adapted from M. J. Howes and D. V. Morgan, eds., *Optical Fibre Communications: Device, Circuit, and Systems*. New York: Wiley, 1980. Copyright © 1980. Reprinted by permission of John Wiley & Sons, Ltd.)

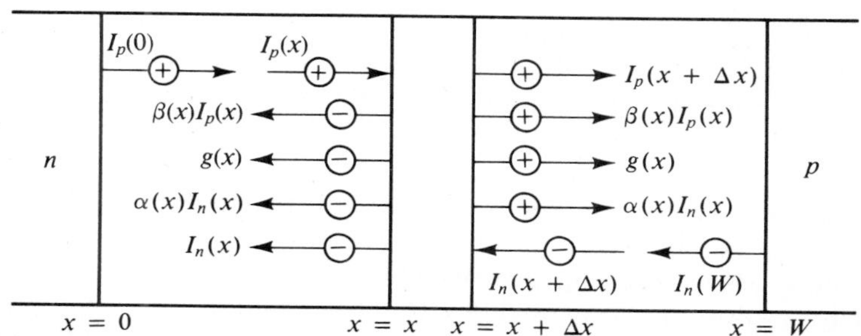

FIGURE 4.12

Avalanche in a reverse-biased *p–n* junction

and

$$-I_n(W) + I_n(x) = \int_x^W [\alpha(x) - \beta(x)]I_n(x)\,dx + I\int_x^W \beta(x)\,dx + \int_x^W g(x)\,dx \qquad \textbf{(4.34b)}$$

where the sum $I \equiv I_n(x) + I_p(x)$ is independent of position. Note, however, that I is equivalent to the saturation current I_0 as defined in Eq. (3.88). Adding Eqs. (4.34a) and (4.34b), we obtain

$$I = \frac{I_0 + I_g + \int_0^W [\alpha(x) - \beta(x)]I_n(x)\,dx}{1 - \int_0^W \beta(x)\,dx} \qquad \textbf{(4.35)}$$

where I_g represents the total generation current.

For a very special case when $\alpha(x) = \beta(x)$, the total current is given by

$$I = M(I_0 + I_g) \qquad \textbf{(4.36)}$$

where M is the avalanche multiplication factor as defined by

$$M = \frac{1}{1 - \int_0^W \alpha(x)\,dx} \equiv \frac{1}{1 - \delta} \qquad \textbf{(4.37)}$$

Ideally speaking, the avalanche condition is thus given by

$$\int_0^W \alpha(x)\,dx = 1 \qquad \textbf{(4.38)}$$

when M becomes infinite. In most practical cases, electron and hole coefficients are not equal and these coefficients vary with the electric field. A practical avalanche photodiode is thus referred to by its ionization rate ratio, $K(\equiv \beta/\alpha)$. We can then arrive at an expression for M after going through some extremely cumbersome mathematics and multiple assumptions:

$$M = \frac{K - 1}{K - e^{(K-1)\delta}} \qquad \textbf{(4.39)}$$

We may note from Fig. 4.11 that for most electrical fields of interest, K is negligible. Thus the avalanche multiplication factor reduces to

$$M \simeq e^{\delta} \qquad \textbf{(4.40)}$$

When $K = 0$, the gain increases exponentially with δ, but it does not necessarily become infinite. As shown in Fig. 4.13, with K approaching 1, the gain approaches infinity at a still smaller value of the field.

It is appropriate to note that the electric field can be easily altered by changing the level of doping. The ionization coefficient α is often given by

$$\alpha = Ae^{-B/|\mathscr{E}|} \qquad \textbf{(4.41)}$$

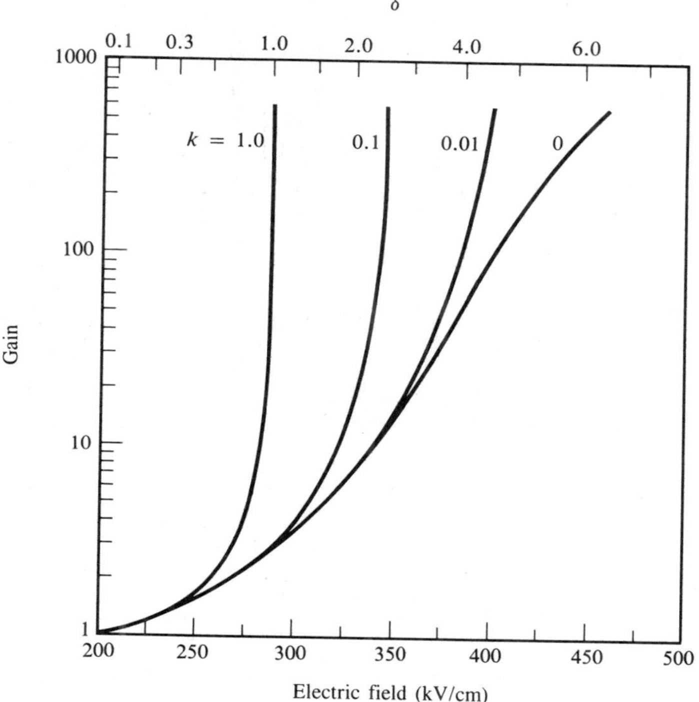

FIGURE 4.13

Gain versus electric field in an APD. (Adapted from P. P. Webb, R. J. McIntyre, and J. Conradi, "Properties of Avalanche Photodiodes." *RCA Rev*, Vol. 35 (1974): p. 234. Courtesy of General Electric Company.)

where A and B are material constants and $\mathscr{E}$ is the electric field described in Eq. (3.56) in terms of doping level. For silicon, A and B are, respectively, 9×10^5/cm and 1.8×10^6 V/cm. At a gain of 100 when $K = 0.01$, for example, a 0.5% alteration in the doping changes the gain by about 20%, whereas the gain would have changed by almost 320% if $\alpha = \beta$. The choice of K and doping is, therefore, critical in the design of an APD. APDs are primarily meant to be used with small signals, and they require special power suppies to maintain them in their avalanche mode.

The phototransistor, like APD, is a detector that exhibits current gain. It can be regarded as a combination of a simple photodiode and a transistor. Phototransistors are photoconductive bipolar transistors that may or may not have a base lead. Light is generally absorbed in the base region. A *p-n-p* phototransistor is shown in Fig. 4.14. When there is no light, no current flows because there is no base control current. Upon illumination, holes that are

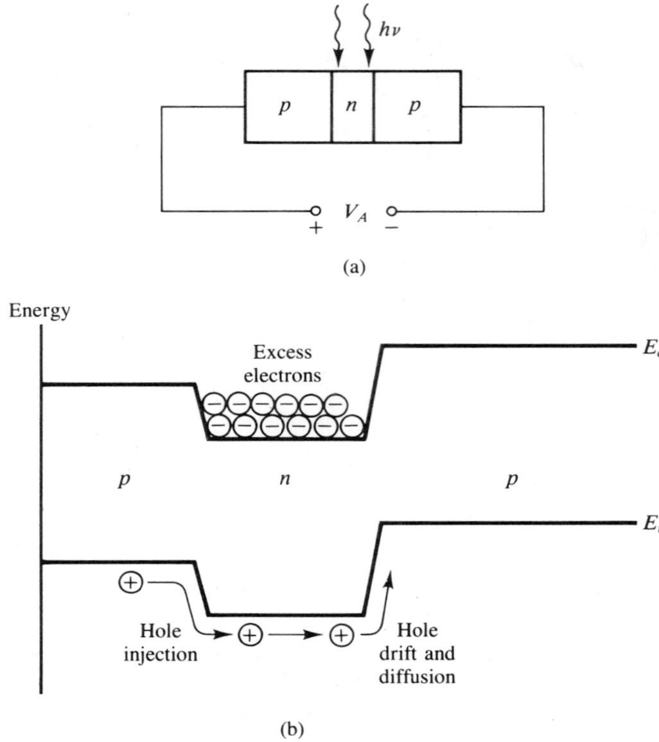

FIGURE 4.14

(a) A biased phototransistor and (b) its energy diagram

excited in the base diffuse out leaving behind an overall negative charge which in turn forward-biases the emitter-base junction. Such a bias results in a hole current that flows from the emitter to the collector until the negative charge of the excess base electrons are neutralized by recombination. In the phototransistor, unlike the photodiode, a much larger current flows through the device. Thus while the phototransistor works very much like the photodiode, it amplifies the photogenerated current. Also, a longer recombination time for the excess base electrons contributes to a higher gain.

For the phototransitors,

$$I_E = I_C + I_B \tag{4.42}$$

where I_E, I_C, and I_B are the quantities introduced in Section 3.4. In the presence of illumination, the base current is given by $\eta I_{\text{abs}} Ae\lambda/hc$, where A is the junction area, η is the internal quantum efficiency, and I_{abs} is the intensity of the absorbed light. The collector current I_C has two components: (a) the standard diode reverse saturation current, I_{CBO} and (b) the portion of the emitter current αI_E

that crosses into the collector where $\alpha \leqslant 1$. The leakage current I_{CBO} corresponds to the collector current at the edge of the cutoff when $I_E = 0$, as shown in Fig. 3.14(c). Thus

$$I_E = (I_B + I_{CBO})\left[1 + \frac{\alpha}{1-\alpha}\right] \tag{4.43}$$

The ratio $\alpha/(1-\alpha)$ is an active region performance parameter of a phototransistor. This ratio is usually in the order of $\simeq 10^2$. In the absence of light, the current flowing in a phototransistor is $I_{CBO}[1 + \{\alpha/(1-\alpha)\}]$, which is much larger than that in a photodiode under similar (dark) conditions. When illuminated, phototransistor current approaches $I_B[1 + \{\alpha/(1-\alpha)\}]$, thus contributing to a significant gain like that of an APD. The only limitation of the phototransistor happens to be its response time, which is about 5 μs, whereas that in a photodiode is on the order of 0.01 μs.

4.6 Vacuum Photodiode and Photomultiplier

Electrons may be emitted when light of an appropriate frequency ν strikes the surface of solids. Such light-emitting solids are called **photocathodes**. The minimum energy necessary for the emission of an electron is referred to as the **work function** ϕ of the solid. In the specific case of semiconductors, electron affinity (energy difference between the vacuum level and E_c) plays the role of work function. The behavior of an electron in solids is like that of an electron in a finite potential well (Section 1.4.2), where the difference between the highest occupied (bound) level and the vacuum (free) level is ϕ, as shown in Fig. 4.15(a). The Fermi level is equivalent to the highest possible bound energy level (Eq. 3.8). Thus the kinetic energy E of an emitted electron is given by

$$E = h\nu - \phi \tag{4.44}$$

Since electrons reside at or below the Fermi level, E corresponds to the maximum possible kinetic energy. The emission of an electron thus requires a minimum of ϕ $(= h\nu)$ photoenergy. However, in the case of a semiconductor, this minimum energy is equivalent to $E_g + E_a$ where E_a is the electron affinity energy. Often it may become necessary to reduce the value of E_a, which is accomplished by making the semiconductor surface p-type. The band bending at the surface results in a downward shift of the conduction band by an amount E_b, as shown in Fig. 4.15(b). Consequently, the effective electron affinity becomes

$$E_a' = E_a - E_b \tag{4.45}$$

In certain semiconductors—those referred to as having negative electron affinity—E_b exceeds E_a. Semiconductors such as these are used for infrared photocathodes.

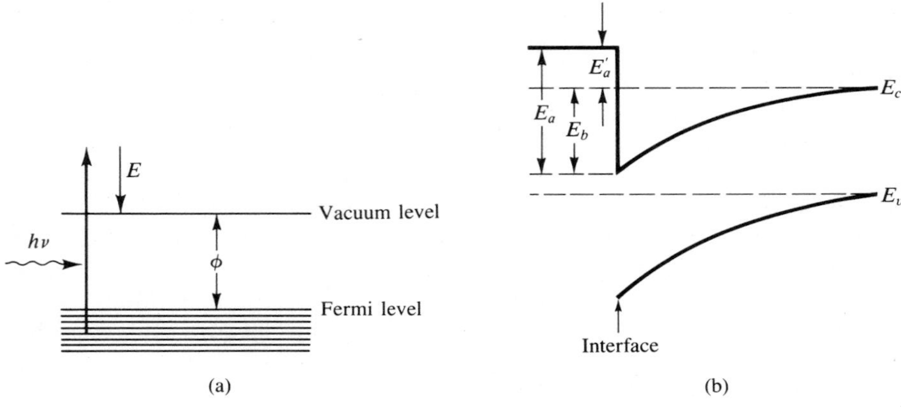

FIGURE 4.15

Energy level diagram in a (a) solid-vacuum interface and in a (b) band-bended semiconductor-vacuum interface

The vacuum photodiode shown in Fig. 4.16 is a quantum detector designed by placing a photocathode along with another electrode (referred to as an anode) within a vacuum tube. In practice, the photocathode consists of a semicylindrical surface while the anode is placed along the cylindrical axis. When an optical energy in excess of the work function illuminates the photocathode, current begins to flow in the circuit. When the bias voltage V_A is large enough ($\simeq 100$ V), the emitted electrons are collected by the anode. When optical energy falls below the work function level, current ceases to exist, irrespective of the bias voltage. For efficient collection of electrons, the distance between the anode and the photocathode is kept to a minimum by making sure that the associated capacitance value remains reasonable. Often the anode is made of highly transparent gridlike wires so as not to impede the incoming optical energy. In comparison, solid-state photodetectors are not only smaller, faster, and less power consuming but also more sensitive. Consequently, vacuum photodiodes are used only when the incoming optical energy is more than a certain maximum that may otherwise damage the solid-state photodetectors.

The characteristic curve of a vacuum phototube shows that the photocurrent for a given illumination is invariant above saturation voltage. The saturation voltage is mildly wavelength and illumination sensitive. Since the operating voltage of a phototube is usually larger than the saturation voltage, minor fluctuations in the supply voltage do not cause any discrepancy in the phototube's performance. An important feature of a phototube is the fact that photocurrent varies linearly as light flux. A slight departure from the linearity occurs at high enough flux values and is caused by the space-charge effects. This nonlinearity is avoided by using a large anode-to-photocathode voltage. In practice, the flux level sets a lower limit on the value of the load

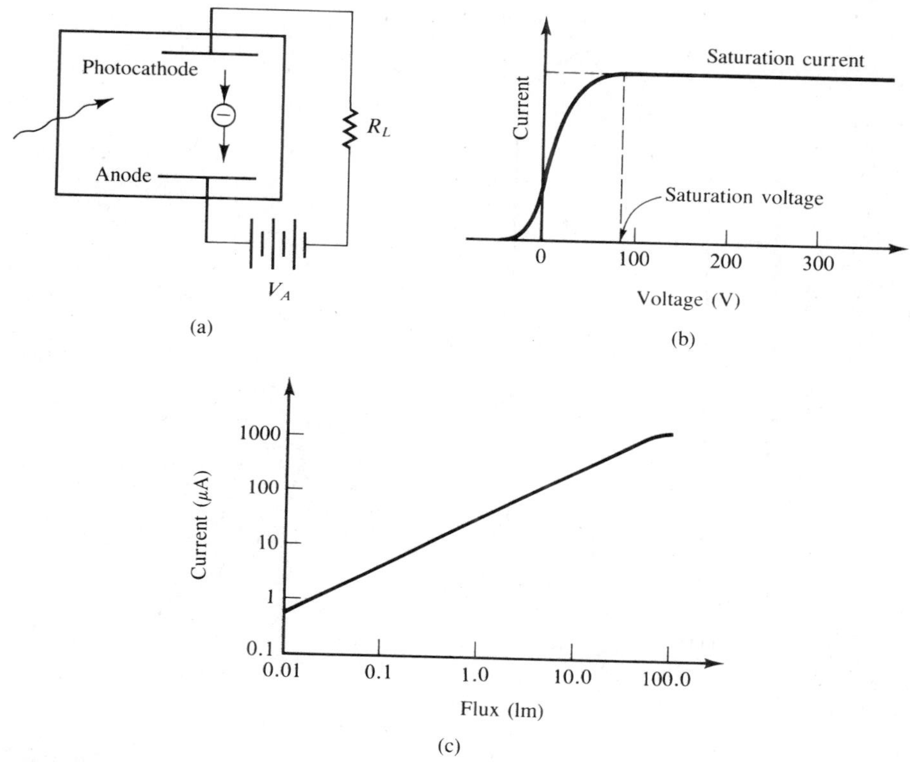

FIGURE 4.16

A vacuum photodiode: (a) circuit, (b) characteristic curve, and (c) current versus light flux

resistance. The load, used to produce a usable signal voltage, in turn, sets a lower limit on the time constant.

Gas-filled phototubes are identical to vacuum phototubes except that they contain approximately 0.1 mm of an inert gas. The inert gas provides a noise-free amplification ($\simeq$5–10) by means of ionization of the gas molecules. However, inert gases have poor frequency responses. Again the response, which is basically nonlinear, is a function of the applied voltage. The phototubes are, therefore, used in applications where the frequency response is not critical.

Photoemissive tube technology is used to develop an alternate but quite popular high-gain device known as a **photomultiplier**. In a photomultiplier tube (PMT), the photoelectrons are accelerated through a series of anodes (referred to as **dynodes**) housed in the same envelope; these dynodes are maintained at successively higher potentials. A photoelectron emitted from the photocathode is attracted to the first dynode because of the potential difference. Depending on the energy of the incident electron and the nature of the dynode surface,

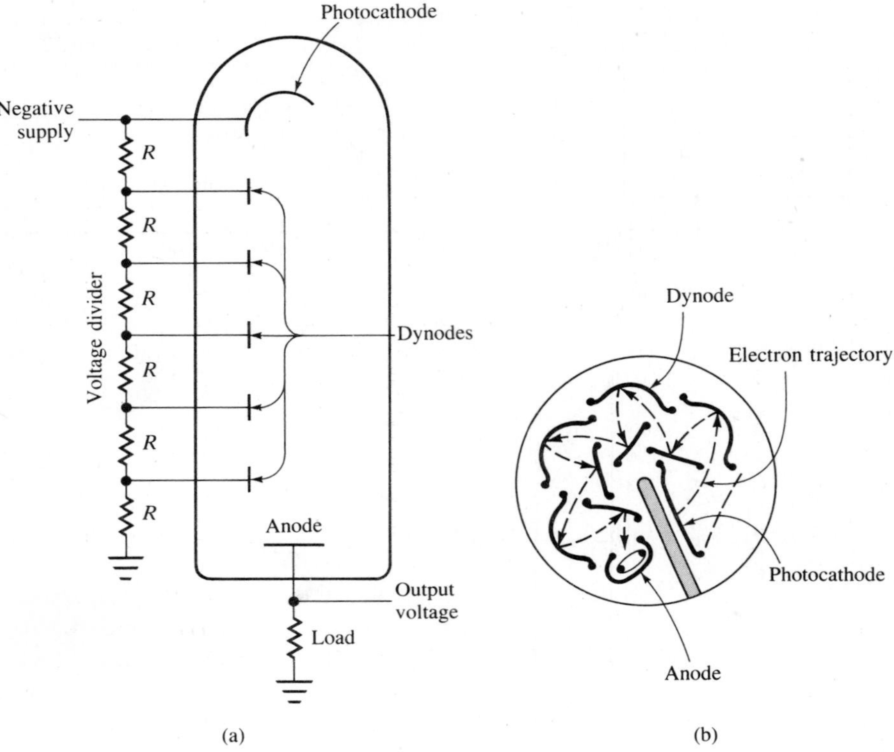

FIGURE 4.17

A photomultiplier tube: (a) schematic of a 5-stage PMT and (b) focusing dynode structure

secondary electrons are emitted upon impact at the first dynode. Each of these secondary electrons produces more electrons at the next dynode, and so on until the electrons from the last dynode are collected at the anode. The dynodes are constructed from materials that, on the average, emit δ (larger than 1) electrons for each of the incident electrons. One such photomultiplier is shown in Fig. 4.17(a), where δ is a function of the interdynode voltage. When the PMT is provided with N such dynodes, the total current amplification factor is given by

$$G = \frac{i_{\text{out}}}{i_{\text{in}}} \delta^N \tag{4.46}$$

Thus, with fewer than ten dynodes and δ less than 5, the gain can easily approach 10^6.

The problems of a PMT are quite the same as those of a vacuum photodiode. However, a PMT is undoubtedly more sensitive. The response of a

PMT is comparatively slower since electrons have to move through a longer distance. In addition, there is a finite spread in the transit time because all of the electrons may not have identical velocities and trajectories. This transit time spread is often reduced, not by reducing the number of dynodes but by increasing the value of δ. However, it must be noted that for most photocathode materials, the maximum wavelength of incoming light is permitted to be about 1200 nm. Thus, for the detection of longer wavelength radiation, a solid-state detector is preferred. PMTs are commonly operated with $\simeq 10^2$ V between the dynodes, which is advantageous because the overall gain of the tube may be varied over a wide range by means of a relatively small voltage adjustment. But, at the same time, it is also disadvantageous because the voltage supply for the PMT must be extremely stable for the calibration to be reliable. We can show that in an N-stage PMT operating at an overall voltage V, a fluctuation ΔV in the voltage produces a change ΔG in the gain G such that

$$\Delta G = GN\frac{\Delta V}{V} \tag{4.47}$$

Consequently, a 1% voltage fluctuation in a 10-stage PMT will cause a 10% change in the gain.

Different types of PMTs are distinguishable by their geometric arrangement of dynodes. In particular, the focusing-type PMT, as shown in Fig. 4.17(b), employs electrostatic focusing between the adjacent dynodes and thereby reduces the spread in the transit time. These PMTs are, however, somewhat more noisy and unstable than the unfocused types. Like phototubes, PMTs have an exceptionally linear response.

It is appropriate to introduce a solid-state-equivalent of PMT, known as the staircase avalanche photodiode (SAPD), that has been added lately to the list of photodetectors. The noise in an APD increases with the increase in the ratio of the ionization coefficient $k\ (=\beta/\alpha)$. On the other hand, a high k is required for a higher gain. An SAPD provides a suitable solution to this apparent anomaly by incorporating PMT-like stages in the solid-state APDs. An unbiased SAPD consists of a graded-gap multilayer material (almost intrinsic) as shown in Fig. 4.18(a). Each dynode-like stage is linearly graded in composition from a low band-gap value E_{gl} to a high band-gap value, E_{gh}. The materials are chosen so that the conduction band drop ΔE_c at the end of each stage equals or just exceeds the electron ionization energy. Note, however, that ΔE_c is much larger than the valence-band rise ΔE_v. Consequently, only electrons contribute to the impact ionization provided the SAPD is biased as shown in Fig. 4.18(b).

A photoelectron generated next to p^+-contact drifts toward the first conduction band under the influence of the bias field and the grading field (given by $\Delta E_c/l$ where l is the width of each step). But this field value is not large enough for the electrons to impact ionize. In this device, only the bias field is responsible for the hole-initiated ionization. The actual impact ionization process in each stage occurs at the very end of the step when the conduction-band discontinuity undergoes a ΔE_c change. The total SAPD gain becomes $(2-f)^N$ where N is the

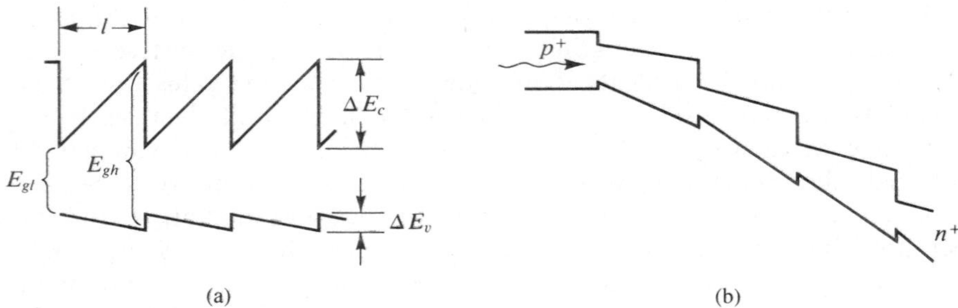

FIGURE 4.18

Staircase APD: (a) unbiased and (b) biased

number of stages and f is the average fraction of electrons that do not impact ionize in each of the stages. The critical bias SAPD field just exceeds $\Delta E_c/l$ so as to provide the electrons with necessary drift through l but not impact ionize.

4.7 Charge-Coupled Imagers

An important solid-state photodetecting device is the **charge-coupled imager**, which is composed of a closely spaced array of charge-coupled devices (CCD) arranged in the form of a register. Each of the CCD units is provided with a storage potential well and is, therefore, able to collect photogenerated minority carriers. The collected charges are shifted down the array and converted into equivalent current or voltage at the output terminal. To understand the overall function of such repetitive storage and transfer of charge packets, consider the metal-oxide semiconductor (MOS) structure of Fig. 4.19, where the metal electrode and the p-type semiconductor are separated by a thin SiO_2 layer of width x_0 and dielectric constant K_0. Silicon nitride, Si_3N_4, is also used for the insulating layer. The capacitance of such a structure depends on the voltage between the metal plate and the semiconductor.

In thermal equilibrium, the Fermi level is constant all throughout the device, and carrier densities are given by Eqs. (3.20) and (3.21), respectively. For simplicity, we may assume (a) that the work function difference between the metal and the semiconductor is zero and (b) that there is no charge accumulated in the insulator or at the junction between the insulator and the semiconductor. Consequently, the device may be considered to have no built-in potential.

A biased MOS capacitor results in two space-charge regions by the displacement of mobile carriers, as shown in Fig. 4.20. The total bias voltage V_G applied at the gate input G is shared between the oxide layer and the semiconductor surface, whereas only a negligible voltage appears across the metal plate. Under reverse bias, the surface potential gives rise to an upward

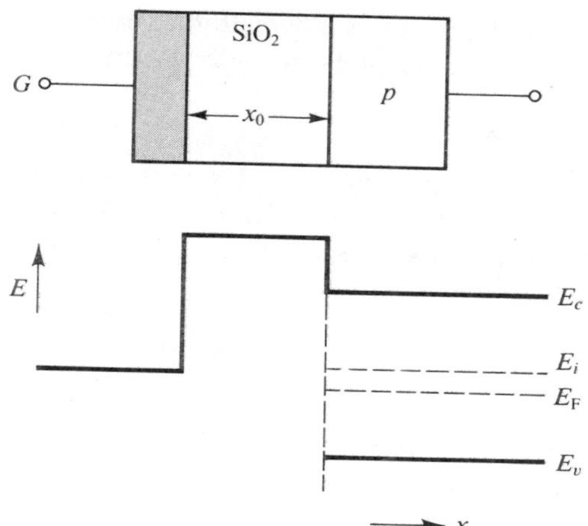

FIGURE 4.19

MOS capacitor and its unbiased energy-band diagram

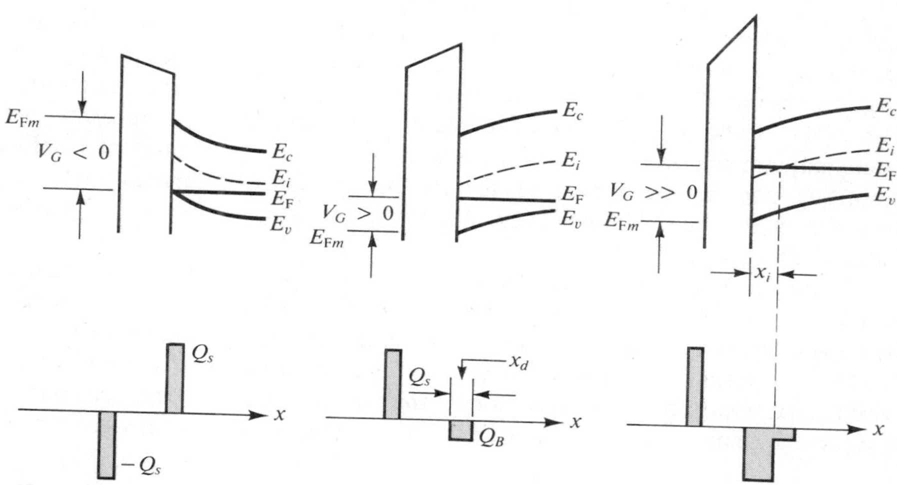

FIGURE 4.20

Energy-band diagram and charge distribution of an unbiased MOS capacitor: (a) $V_G < 0$, (b) $V_G > 0$, and (c) $V_G \gg 0$

bend in the energy diagram, as shown in Fig. 4.20(a). At the edges $E_i - E_F$ becomes comparatively larger, thus resulting in a higher hole density at the surface than that within the bulk region. This condition generally increases the surface conductivity. Again Fig. 4.20(b) shows the forward-biased case, where a decrease of E_i–E_F at the edges causes a depletion of holes at the semiconductor surface. The total charge per unit area in the bulk semiconductor is given by

$$Q_B = -eN_A x_d \tag{4.48}$$

where x_d is the width of the depletion layer. Using the depletion approximation in Poisson's equation (as used in the derivation of Eq. 3.59), we can arrive at the potential within the semiconductor as

$$V_s(x) = V_s(0)\left[1 - \frac{x}{x_d}\right]^2 \tag{4.49}$$

where

$$V_s(0) = \frac{eN_A}{2K_s\varepsilon_0}x_d^2 \tag{4.50}$$

is often referred to as the **surface potential**. The voltage characteristic is similar to that of a step junction having a highly doped *p*-side. Note, however, that as the bias voltage V_G is increased further, the band bending could result in a crossover of E_i and E_F within the semiconductor, as shown in Fig. 4.20(c). Consequently, the carrier depletion gives rise to an extreme case of carrier inversion whereby electrons are generated at the junction and holes are generated inside the semiconductor with two regions being separated by the crossover point. Therefore, a *p*–*n* junction is induced under the metal electrode. The effect of the gate voltage is to remove the majority carriers from the semiconductor region that is closest to the gate and introduce a potential well. Absorbed photons contribute to the freeing of minority carriers that are collected in the well. The resulting output signal corresponds to the photo-induced charge.

If the semiconductor were approximated as a borderline conductor, the metal-semiconductor structure could be envisioned as a parallel-plate capacitor with the oxide layer working as its dielectric material. However, in forward bias, the MOS structure is modeled by incorporating an additional capacitor in series with the oxide capacitor to accommodate the presence of a surface space-charge layer in the semiconductor. The overall MOS capacitance C is thus given by

$$\frac{1}{C} = \frac{1}{C_0} + \frac{1}{C_s} \tag{4.51}$$

where

$$C_0 = \frac{K_0\varepsilon_0}{x_0} \tag{4.52a}$$

and

$$C_s = \frac{K_s\varepsilon_0}{x_d} \tag{4.52b}$$

Neglecting the voltage drop in the metal plate, the forward bias can be expressed as

$$V_G = V_s(0) - \frac{Q_s}{C_0} \tag{4.53}$$

where Q_s is the density of induced charge in the semiconductor surface region and $V_s(0)$ is the surface potential. The gradient of the surface potential generally determines the minority carrier movements. The depth of the potential well is often decreased either by decreasing the oxide capacitance—i.e., by increasing the oxide thickness—or by increasing the doping level of the p-type material.

Sufficient forward bias may eventually induce an inversion layer. With the passing of time, electrons accumulate at the oxide-semiconductor junction, and a saturation condition is reached when the electron diffusion current leaving the junction is counterbalanced by the electron drift current arriving at the junction. The time required to reach this saturation condition is referred to as the **thermal-relaxation time**. The net flow of electrons is directed toward the junction prior to the thermal-relaxation time, whereas the net flow of electrons is directed away from the junction after the thermal-relaxation time has elapsed. Since there was no inversion layer prior to the saturation, the induced charge Q_s is obtained by summing Q_B and externally introduced charge Q_e. Equations (4.48) and (4.49) can be incorporated into Eq. (4.53) to give the surface potential as

$$V_s(0) = V_G - \frac{Q_e}{C_0} + \frac{eK_s\varepsilon_0 N_A}{C_0^2}\left[1 - \left\{1 + \frac{2C_0^2\left(V_G - \dfrac{Q_e}{C_0}\right)}{eK_s\varepsilon_0 N_A}\right\}^{1/2}\right] \tag{4.54}$$

The depth of the potential well x_d is often evaluated using Eqs. (4.54) and (4.49). The value of x_d is used in turn to evaluate C_s using Eq. (4.52b) and, consequently, we can determine the overall MOS capacitance as

$$C = \frac{C_0}{\left[1 + \dfrac{2C_0^2}{eN_AK_s\varepsilon_0}V_G\right]^{1/2}} \tag{4.55}$$

The MOS capacitor in effect serves as a storage element for some period of time prior to reaching the saturation point.

A CCD structure formed by cascading an array of MOS capacitors, as shown in Fig. 4.21(a), is often referred to as the **surface channel charge-coupled device** (SCCD). Basically, the voltage pulses are supplied in three lines, each connected to every third gate input (and consequently this CCD is called a three-phase CCD). In the beginning, G_1 gates are turned on, resulting in an accumulation and storage of charge under the gates. This step is followed by a step whereby G_2 is turned on, thus resulting in a charge equalization step across two-thirds of each cell. Subsequently G_1 is turned off, resulting in a complete transfer of all charges to the middle one-third of each cell. This process is repeated to transfer charge to the last one-third of the CCD cell. Consequently,

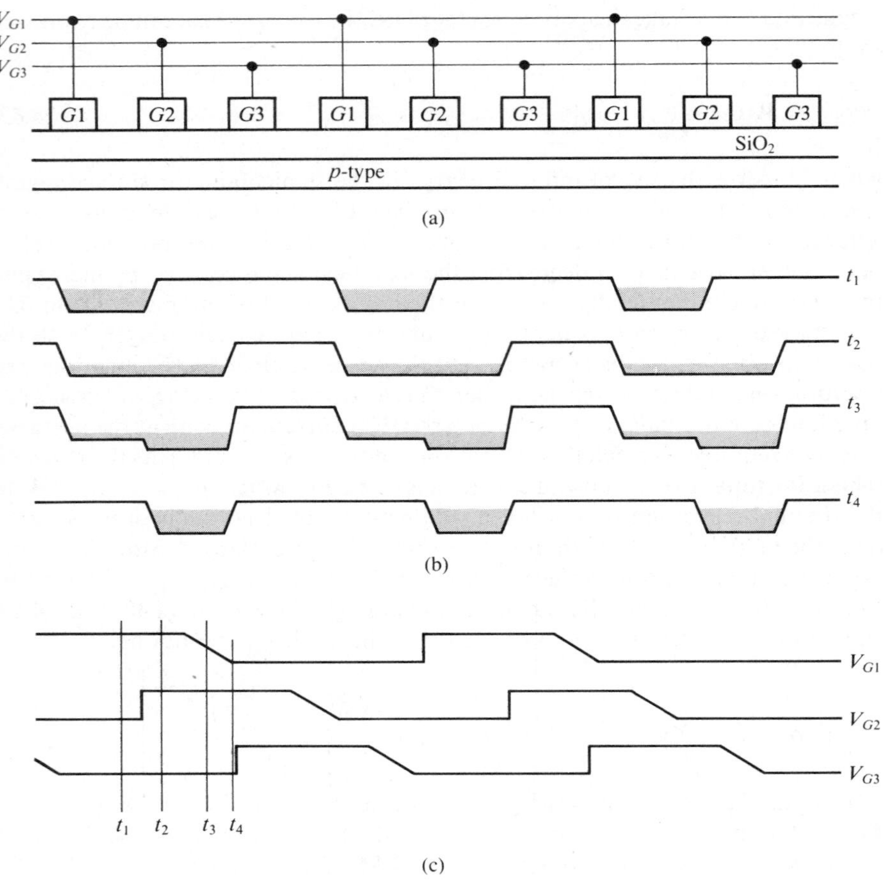

FIGURE 4.21

(a) CCD structure, (b) potential wells at different times, and (c) timing diagram

after a full cycle of clock voltages has been completed, the charge packets shift to the right by one cell, as illustrated in Fig. 4.21(b)–(c). When the CCD structure is formed using an array of photosensors, charge packets proportional to light intensity are formed and these packets are shifted to a detector for the final readout.

The CCD signal readout is also accomplished by using either two-phase or four-phase clocking schemes. In each of the cases, however, the transfer of charges is implemented by the sequentially applied clock pulses. There are three phenomena that enhance the transfer of charges in the SCCD. They are (a) self-induced drift, (b) thermal diffusion, and (c) fringe-field drift. The **self-induced drift**, responsible for most of the transfer, is essentially a repulsion effect between the like charges. The **thermal diffusion** component makes up for most of the remaining signal charge. It can be shown that for most materials, the thermal time constant is longer than the self-induced time constant. The upper frequency limit for switching operations is thus determined by the thermal time constant. For the SCCDs, this upper limit can be in the order of 10 MHz. The **fringe-field drift** is determined by the spacing of the electrodes and results in a smoothing out of the transitional potential fields. This third effect is responsible for the transfer of the final few signal electrons.

Figure 4.22(a) shows a system of MOS transistors along with a photodiode array, both of which can be embedded under the same monolithic structure. The system is able to perform sequential readout. A voltage pattern can be generated from the shift register so as to turn on only one transistor at a time. The switching voltage is shifted serially around all diodes. This scheme can also be extended to two dimensions, as shown in Fig. 4.22(b), where one row is switched on and all columns are then scanned serially. The process is repeated for the remaining rows until all photodiodes have experienced the scanning.

The primary item which hinders the charge transfer is surface state trapping that occurs along the semiconductor-oxide interface. These trapping energy levels are introduced by nonuniformities in the interface. These energy levels tend to trap and re-emit electrons, the rate of which is a function of clocking frequency and their positions relative to the Fermi level. This hindering in the charge transfer is reduced by transferring a constant amount of charge in each of the CCD wells. This charge fills most of the trapping states, as a result of which interaction of trapping levels and signal charge is minimized.

A popular method used to control the problem of surface trapping is accomplished by having what is known as a **buried channel CCD** (BCCD). It involves implementing a thin secondary layer that is of opposite polarity to that of the substrate material along the oxide surface. The fringe fields are much smoother in a BCCD than in an SCCD, but the well depth is smaller. Hence, a BCCD can switch information at a faster rate, but it cannot hold as much signal. Switching speeds of up to 350 MHz are not uncommon for BCCDs.

In applications where a semiconductor depletion region is formed, there is a constant generation of electron-hole pairs due to the thermal vibration of the crystal lattice at any temperature above 0 K. This generation of carriers constitutes a dark current level and determines the minimum frequency with which the transfer mechanism can occur. The time taken by a potential well to fill up with dark electrons can be quite long in some of the CCDs. There are two basic types of CCD imagers: the **line imager** and the **area imager**. In a line imager, charge packets are accumulated and shifted in one direction via one or two parallel CCD shift registers, as shown in Fig. 4.23(a), where the CCD register is indicated by the shaded regions. The two basic types of CCD area imagers are shown in Figs. 4.23(b) and 4.23(c).

In the **interline transfer CCD** (ITCCD), photocells are introduced between the vertical CCD shift registers. Polysilicon strips are placed vertically over each line of the photocells to provide shielding. During an integration period

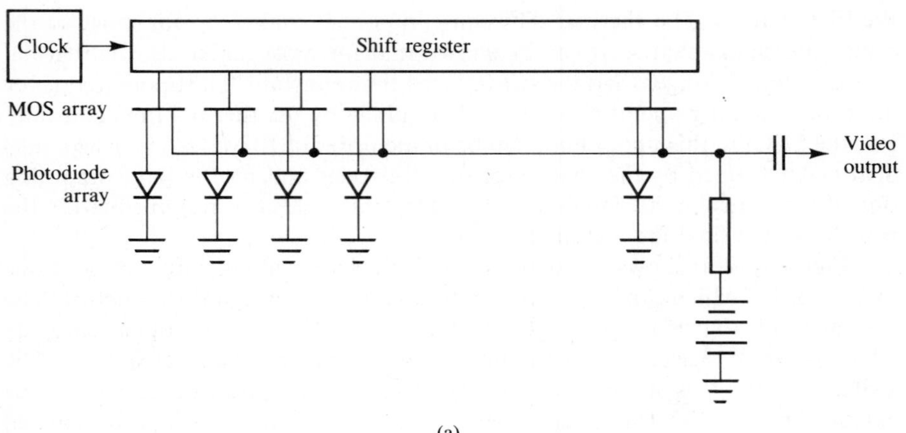

(a)

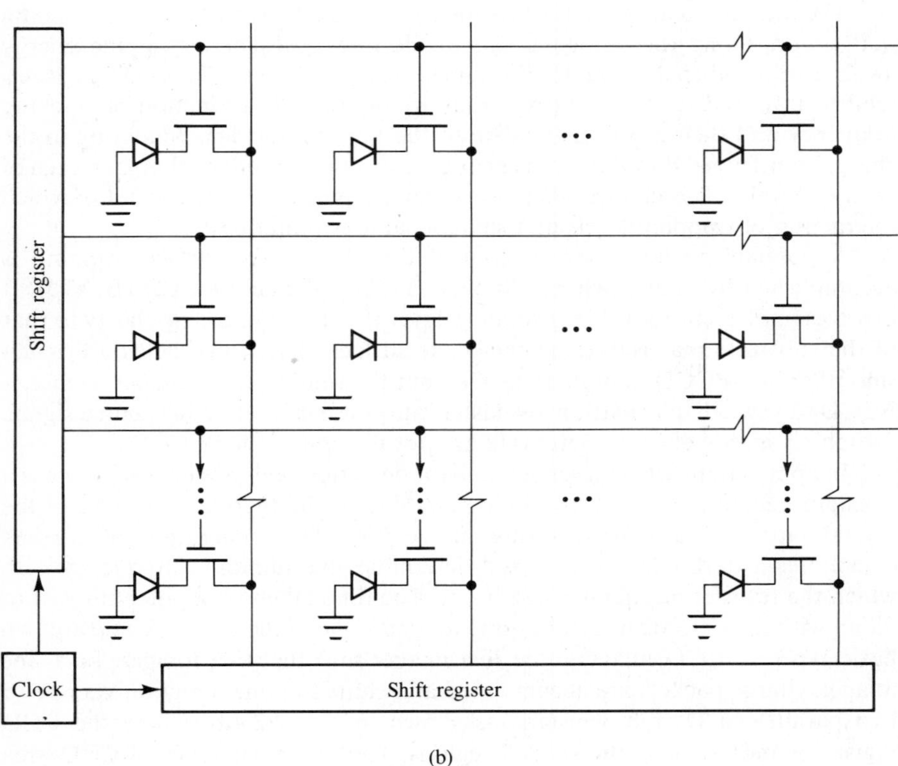

(b)

FIGURE 4.22

MOS readout scanner: (a) one-dimensional and (b) two-dimensional

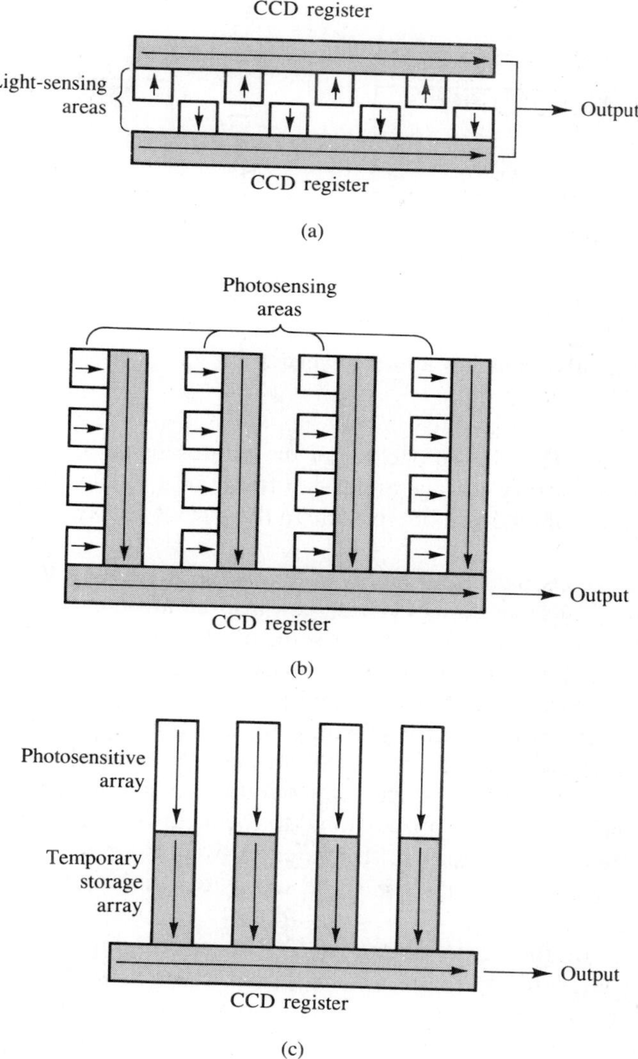

FIGURE 4.23

CCD imager: (a) line, (b) interline transfer, and (c) frame transfer

(referred to as one frame), all of the cells are switched with a positive voltage. The ITCCD is obtained by extending the line imager to a two-dimensional matrix of line imagers in parallel. The outputs of the line imagers are fed into a single output CCD register. For the case of the **frame transfer CCD** (FTCCD), the sensor is divided into two halves: a photosensitive section and a temporary

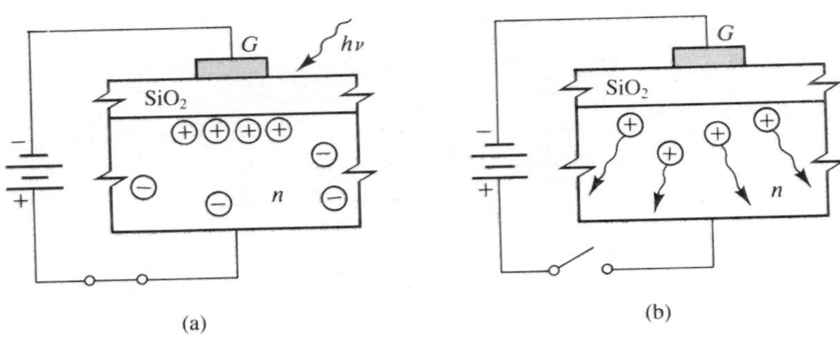

FIGURE 4.24

CID charge (a) accumulation and (b) injection

storage section. The charge packets of the photosensing array are transferred over to the temporary storage array as a frame of a picture. Subsequently, the information is shifted down one-by-one to the output register and is then shifted horizontally.

Besides CCDs, a different family of MOS charge transfer devices, referred to as the **charge-injection devices** (CID), can also be an integral part of the focal plane array. CIDs involve exactly the same mechanism for detecting photons as CCDs. They differ only in the methods used for reading out the photo-induced charges. The basic mechanism of charge injection is illustrated in Fig. 4.24, where the CID consists of an *n*-substrate, for example. Application of a negative gate voltage causes the collection of photo-induced minority charge carriers in the potential well adjacent to the semiconductor-oxide interface. This accumulation of charge is directly proportional to the incident optical irradiance. Once the gate voltage is withdrawn, the potential well dissipates and the minority carriers are injected promptly into the substrate, resulting in a current flow in the circuit.

Because of the serial nature of CCDs, optical input signals cause the resulting charge to spill over into adjacent cells. This effect, referred to as **blooming**, causes the image to appear larger than its actual magnitude. In comparison, CIDs are basically x–y addressable such that any one of their pixels can be randomly accessed with little or no blooming. However, the CID capacitance (sum of the capacitances of a row and a column) is much larger than the CCD capacitance and, therefore, the CID images tend to be noisier.

4.8 Cathode-Ray Tubes

We have already discussed LEDs and their uses as display devices. LEDs are relatively inexpensive solid-state devices that are small in size and reasonably reliable. In this section, we introduce a non-solid-state display device known as

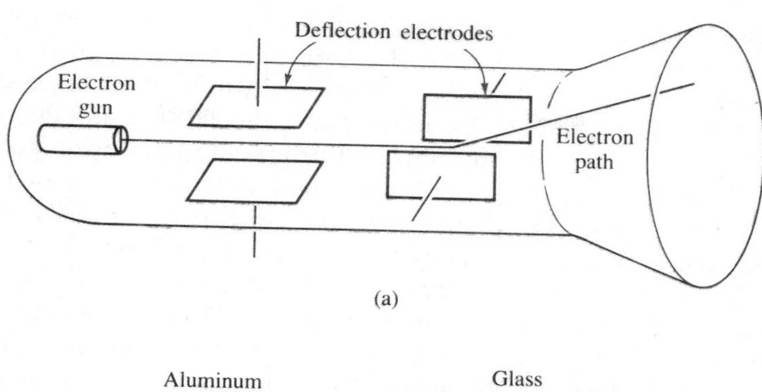

(a)

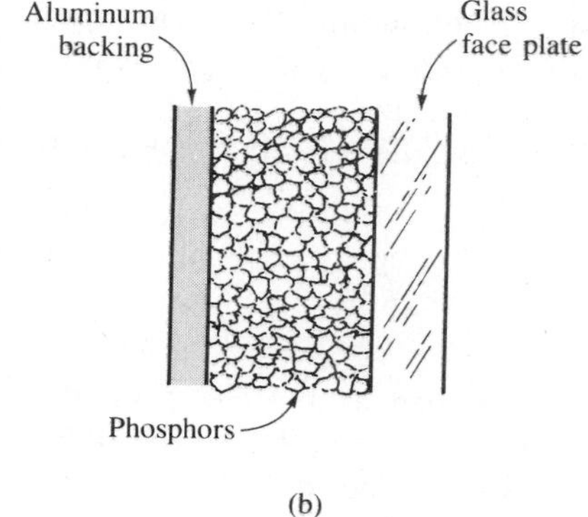

(b)

FIGURE 4.25

(a) A CRT schematic and (b) a CRT screen

the **cathode-ray tube** (CRT) that is deeply entrenched in our day-to-day video world. In spite of competing technologies such as flat-panel display, it is unlikely that CRTs will be totally replaced. In spite of its large power consumption and bulky size, it is by far the most common display device found in both general and special purpose usage, aside from displaying small alphanumerics. The cost factor and the trend toward using more and more high-resolution color displays are the key factors that guarantee the CRT's longevity. The CRTs have satisfactory response speed, resolution, design, and life. Besides, there are very few electrical connections and CRTs can present more information per unit time at a lower cost than any other display technology. A CRT display is generally subdivided into categories such as having electrostatic or magnetic deflection, monochromatic or color video, and single or multiple beams.

Figure 4.25(a) shows the schematic of a CRT display where the cathodoluminescent phosphors are used at the output screen. **Cathodoluminescence** refers to the emission of radiation from a solid when it is bombarded by a beam of electrons. The electrons in this case are generated by thermionic emission of a cathode and are directed onto the screen by means of a series of deflection plates held at varying potentials. The electron beam is sequentially scanned across the screen in a series of lines by means of electrostatic or electromagnetic fields (introduced by deflection plates) acting orthogonally to the direction of electron trajectory. The bulb enclosing the electron gun, the deflectors, and the screen is made air-free for the purpose of having an electron beam and a display area. The to-be-displayed video signals are applied to both the electron gun and deflectors in synchronization with the scanning signals. The display is usually refreshed sixty times a second to avoid having a flickering image. While in the United States CRT displays consist of 525 scan lines, the number is about 625 overseas. The phosphor screen is often treated as being split into two interlaced halves. Thus, if a complete refreshing cycle takes t_r time, only odd-numbered lines are scanned during the first $t_r/2$ period, and the even-numbered lines are scanned during the remaining half. Consequently, our eyes treat the refreshing rate as if it were $2/t_r$ Hz instead of only $1/t_r$ Hz.

Using Newton's second law, the velocity of the bombarded electron is given by

$$v = \sqrt{\frac{2eV}{m_e}} \tag{4.56}$$

where V is the potential applied at the electron gun. If this electron now encounters an electric field **E** at an angle θ to the potential gradient, then the horizontal displacement x of the electron is given as a function of time t by

$$x = (v \sin\theta)t \tag{4.57}$$

while its corresponding vertical displacement is given by

$$\begin{aligned} y &= (v \cos\theta)t + \tfrac{1}{2}at^2 \\ &= (v \cos\theta)t - \frac{e|\mathbf{E}|}{2m_e}t^2 \end{aligned} \tag{4.58}$$

where a is the acceleration due to the electric field. Finally, t can be eliminated to yield the trajectory of the electron as

$$y = \frac{x}{\tan\theta} - \frac{|\mathbf{E}|}{4V\sin^2\theta}x^2 \tag{4.59}$$

Equation (4.58) corresponds to that of a parabola whose slope α is given by

$$\tan\alpha = \frac{1}{\tan\theta} - \frac{|\mathbf{E}|}{2V\sin^2\theta}x \tag{4.60}$$

However, if instead a magnetic field **B** is applied, the resultant force on the electron would act in a direction perpendicular to the plane of **B** and v. The electron is, therefore, subjected to a helical path whose radius r is given by

$$r = \frac{m_e v \cos\theta}{e|\mathbf{B}|}$$

$$= \frac{\sin\theta}{|\mathbf{B}|}\sqrt{\frac{2m_e V}{e}} \tag{4.61}$$

while its angular velocity is given by $\omega = e|\mathbf{B}|/m_e$. Consequently, the pitch of the resulting helical path is

$$p = \frac{2\pi}{\omega} v \cos\theta$$

$$= \frac{\cos\theta}{|\mathbf{B}|}\sqrt{\frac{8m_e\pi^2 V}{e}} \tag{4.62}$$

As described earlier, the deflected electron strikes the CRT phosphor screen causing the phosphors of that CRT location to emit light. It is interesting to note that both absorption and emission distributions of phosphors are bell shaped, but the distribution peaks are relatively displaced in wavelength. When compared with the absorption distribution, the emission distribution peaks at a higher wavelength. This shift toward the red end of the spectrum is referred to as the **Stokes' shift**. This phenomenon is utilized to convert ultraviolet radiation to useful visible radiation. It is used often in fluorescent lamps to increase their luminous efficiencies. In particular, the CRT illumination, caused by the cathodoluminescent phosphors, is a strong function of both current and accelerating voltage and is given by

$$L_e = Kf(i)V^n \tag{4.63}$$

where K is a constant, $f(i)$ is a function of current, and n ranges between 1.5 and 2.

TABLE 4.2

Selected Phosphors and Their Characteristics

Phosphor	*Peak Wavelength*	*Color*	*Efficiency*	*Time for 90% Decay*
Zn_2SiO_4:Mn	525 nm	Yellowish-green	4.7	24 ms
ZnS:Cu	542 nm	Yellowish-green	12.4	0.1 ms
(Zn, Cd)S:Ag	530 nm	Yellowish-green	18.5	0.06 ms
$CaWO_4$	430 nm	Blue	3.4	0.03 ms
La_2O_2S:Tb	543 nm	Green	11.8	1 ms
Gd_2O_2S:Tb	544 nm	Green	10.2	1 ms
Y_2O_2S:Eu	627 nm	Red	13.1	0.9 ms
YVO_4:Eu	617 nm	Orange	12.0	9 ms
ZnS:Ag	460 nm	Blue	15.2	0.03 ms

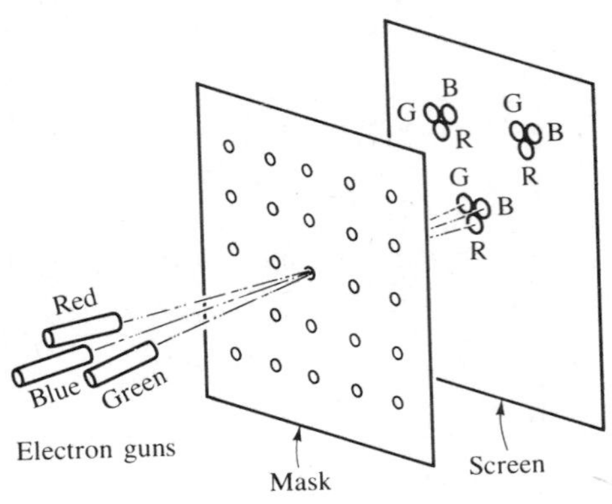

FIGURE 4.26

Shadow masking in a color CRT

With larger accelerating voltage, electrons penetrate further into the phosphor layer, causing more phosphor cells to irradiate.

The factors that are taken into consideration in selecting a particular phosphor are, namely, *decay time*, *color*, and *luminous efficiency*. Note that even though a phosphor may have a higher radiant efficiency in the red while another may have a lower radiant efficiency in the green, the latter may have a more desirable luminous efficiency curve. Table 4.2 lists some of the more common phosphors and their characteristics. Usually, the phosphor screen consists of a thin layer ($\simeq 5\ \mu$m) of phosphor powder placed between the external glass face plate and a very thin layer ($\simeq 0.1\ \mu$m) of aluminum backing, as shown in Fig. 4.24(b). The aluminum backing prevents charge build-up and helps in redirecting light back toward the glass plate. The aluminum backing is thin enough so that most of the electron beam energy can get through it. A substantial amount of light that reaches the glass at normal or nearly normal angles is transmitted. But a portion of the incident light (beyond the critical angle of incidence) may get totally internally reflected at the glass-air interface, some of which may again get totally internally reflected at the phosphor-glass interface. Such physical circumstances produce a series of concentric circles of reduced brightness instead of producing one bright display spot. The combination of diffused display spots results in a display spot that has a Gaussian distribution profile.

Of the many available methods, the most common one for introducing color in a CRT display involves the use of a metal mask and three electron guns, each corresponding to a primary phosphor granule (red, blue, and green), as shown in Fig. 4.26. The three electron guns are positioned at different angles, so that while each of the electron beams is passing through a particular mask-hole strikes a particular primary phosphor dot. All three beams are deflected simultaneously. In addition, the focus elements for the three guns are connected in parallel so that a single focus control is sufficient to manipulate all beams. The three primary dots are closely packed in the screen so that proper color can be generated for each signal. Misalignment of the three beams causes a loss of purity for the colors. In any event, when compared to the monochrome display, the CRT color reproduction process involves a loss of resolution to a certain degree because the primary phosphor cells are physically disjointed.

4.9 Imaging Tubes

Imaging tubes convert a visual image into equivalent electrical signals that are used thereafter for viewing the image on display devices such as CRTs. They are used as in a television camera tube in which a single multilayer structure serves both as an image sensor and as a charge storage device. The single multilayer structure is subsequently scanned by a beam of low-velocity electrons to produce a video signal. In particular, when the characteristics of the photosensor layer during the optical-to-electric conversion depend on photosensor's photoconductive property, the imaging tube is referred to as the **vidicon**. Figure 4.27 shows a typical vidicon whose thin target material consists of a photoconductive layer, such as selenium (Se) or antimony trisulphide (Sb_2S_3), placed immediately behind a transparent conductive film of SnO_2. Its charge retention quality is very good since the resistivity of the photoconductor material is very high ($\simeq 10^{12}$ Ω-cm). The conducting layer is connected to a positive potential V_B via a load resistor. The other side of the target is scanned with an electron beam almost in the same way as CRT scanning. The output video signal is generally capacitively coupled, and the majority of vidicons employ magnetic focusing and deflection schemes.

The vidicon target is often modeled as a set of leaking capacitors, each of which corresponds to a minute area of the target. One side of these capacitors is tied together by means of a transparent conductive layer and is connected to a bias voltage V_B. The low-velocity scanning electron beam makes momentary contact with each of the miniature areas, charging them negatively. The target has a high resistance in the dark, but when it is photoirradiated its resistance drops significantly. In absence of illumination, the scanning beam drives the target to a potential value close to that of the cathode, which allows for a small amount of dark current to flow when the beam is removed. The decrease in resistivity due to photoirradiance causes the capacitorlike target to discharge itself in the absence of the electron beam. However, when the electron beam scans this discharged area, it will recharge the target. More current is being taken away from the illuminated area than from the unilluminated areas, thus generating a video signal at the output. The video signal is found to be proportional to the nth power of illumination, where n is a positive number less than unity.

The dark current is rather large in vidicons. Again, the spectral response of Sb_2S_3 is very poor at wavelengths greater than 0.6 μm. An imaging tube referred

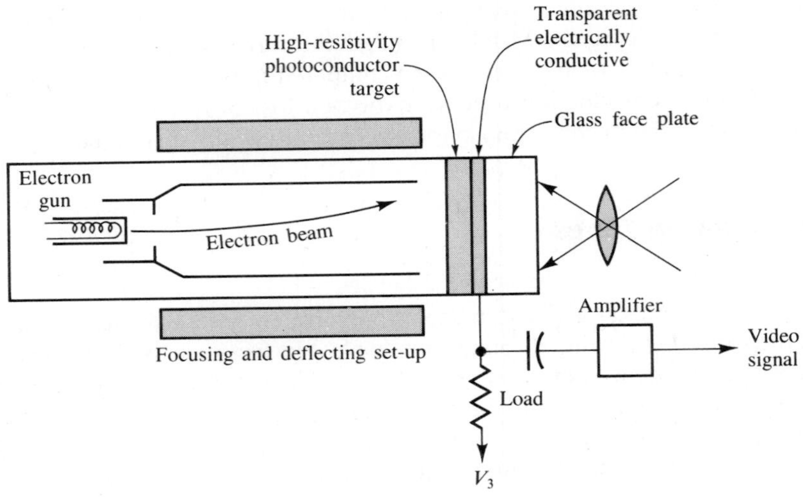

FIGURE 4.27

A vidicon structure

to as a **plumbicon** is often used to overcome the aforementioned shortcomings of a vidicon. The plumbicon is essentially identical to a vidicon, except that the photoconductive target is replaced by a layer of lead oxide (PbO) that behaves like a *p–i–n* diode and not like a photoconductor.

The band-gap energy of lead oxide is 2 eV and, therefore, it is not too sensitive to red. However, the introduction of a thin layer of lead sulphide (with a band gap of about 0.4 eV), along with lead oxide, eliminates the problem of red insensitivity in the plumbicon. The transparent conductive film acts like an *n*-type region, and the PbO layer acts like a *p*-type region, whereas the region between the two behaves like an intrinsic semiconductor. Photoirradiance of this *p–i–n* structure generates carriers in the plumbicon. But the flow of carriers in the opposite direction generally reduces the amount of stored charge. However, in the absence of photoirradiance, the reverse bias gives rise to a dark current that is negligible in comparison to that encountered in the vidicon. There are two serious disadvantages of using plumbicons because their resolution (fewer than 1000 lines) is often limited by the thick lead oxide layer and because the change in target voltage cannot be used to control their sensitivity. In spite of these demerits, they are used widely in color-TV studios. However, the lead oxide layer can be replaced by an array formed of thousands of silicon diodes to increase the sensitivity of plumbicons.

Another important imaging tube, referred to as an **image intensifier**, is of significant importance in the transmittal of images. In principle, it is a vacuum photodiode equipped with a photocathode on the input window and a phosphor layer on the output window. Image intensifiers are devices in which

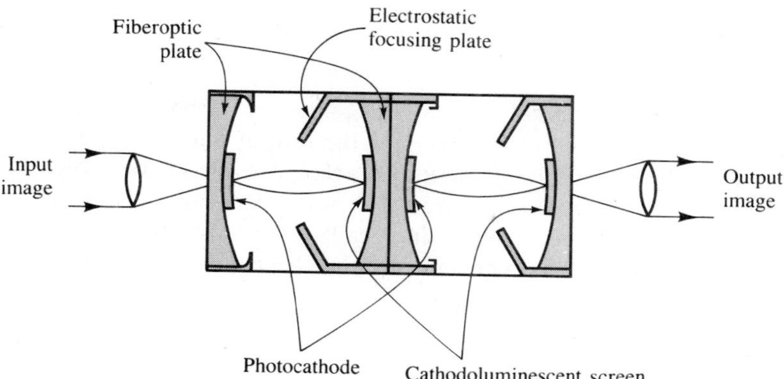

FIGURE 4.28

Two cascaded image intensifiers

the primary optical image is formed on a photocathode surface (with S20 phosphor layer backing), and the resulting photocurrent from each of the image points is intensified by increasing the energy of the electrons as shown in Fig. 4.28. The windows are made of fiber optic plates, so that the plane image surface of the input can be transformed to the curved object and image surfaces of a simple electrostatic lens. The electrons strike a luminescent screen and the intensified image is produced by means of cathodoluminescence. It is possible to cascade more than one such intensifier with fiber-optic coupling between them, making sure that an accelerating potential is applied between the photocathode and the screen. Such a cascade device, along with an objective lens and an eyepiece, is used in the direct-view image intensifier. In any event, it is possible to achieve luminance gains of up to 1000 with each image intensifier. An image intensifier such as this can also be designed by increasing the number of electrons (as in a photomultiplier).

4.10 Photographic Recording Medium

The photographic recording medium consists of a base, made of either acetate film or glass plate, and a layer of photographic emulsion, where photosensitive silver halide is suspended in a gelatin support. When exposed to light, the silver halide particles undergo a physical change, reducing to tiny metallic silver particles. The process of reduction is completed (or *developed*) and then *fixed* by a set of prescribed chemical processes that remove the remaining silver halide, leaving behind the silver grains. The transmittance of the developed film depends strictly on the presence of silver grains that are basically opaque at the optical frequencies. The intensity transmittance T of the developed photographic transparency is defined as

$$T(x, y) = \left\langle \frac{I_{out}(x, y)}{I_{in}(x, y)} \right\rangle \tag{4.64}$$

where I_{out} and I_{in}, respectively, represent the output and the input irradiance and $\langle\rangle$ denotes the localized ensemble average over an area that is large compared to the size of the film grain but small compared to an area within which the output intensity changes drastically. The density of the transparency D, defined as the silver mass per unit area, has been determined by Hurter and Driffield to be

$$D = -\log(T) \tag{4.65}$$

The most commonly used photographic recording characteristic is thus known as the Hurter-Driffield (H-D) curve, which relates the density with the logarithm of the energy used to expose the photographic media. Figure 4.29 shows a typical H-D curve having three distinct regions: shoulder, linear, and toe regions. In both the toe and the shoulder regions, density is quite independent of the exposure. However, in the linear region, the density is proportional to the logarithm of the energy. The linear region is described by

$$D = \gamma \log E - D_0 \tag{4.66}$$

where γ, referred to as the **film gamma**, is the slope of the linear region of the H-D curve and $-D_0$ is the intercept of this line with the density axis. A photographic medium with large value γ is referred to as a high-contrast recording medium, while that with a small value of γ is called a low-contrast recording medium. The photographic gamma is a function of not only the type of emulsion but also of the developing time and developing chemicals.

Since $E = It$ where t is the time of exposure, the intensity transmittance can be determined from Eq. (4.66) using Eq. (4.65) as

$$T = \delta I^{-\gamma} \tag{4.67}$$

where $\delta = 10^{D_0} t^{-\gamma}$ is a positive constant. It is apparent that a one-step photographic process provides an intensity transmittance that is nonlinear with respect to the input irradiance. However, a two-step process, consisting of taking an intermediate transparency at first and then passing the light through the negative to obtain the final transparency (also known as the positive transparency), yields a different option. Using Eq. (4.67) repeatedly, we find the intensity transmittance of the positive transparency to be

$$T_p = \delta_p (I)^{\gamma_p} \tag{4.68}$$

where δ_p is a positive constant, $\gamma_p = \gamma_1 \gamma_2$, and γ_n is the film gamma corresponding to the nth step involved. By choosing $\gamma_1 \gamma_2 = 1$, it is possible to obtain a linear relationship between the intensity transmittance and the input irradiance.

Whenever the photographic media is used as an element in the coherent optical processors (discussed in Chapter 8), it is more reasonable to use

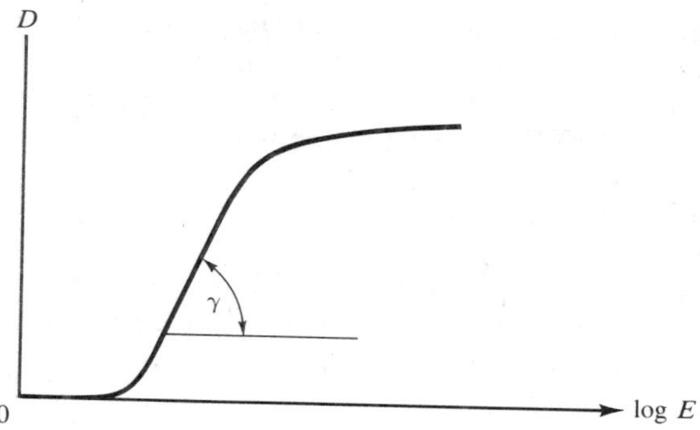

FIGURE 4.29

A typical Hurter-Driffield curve

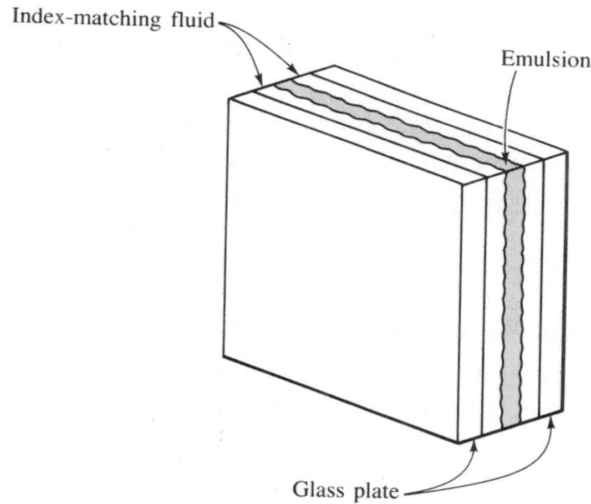

FIGURE 4.30

An index-matching liquid gate

amplitude transmittance rather than intensity transmittance. It is important to realize that neither the base thickness nor the emulsion thickness is uniform. In fact, emulsion thickness varies in accordance with the spatial density, which in turn depends on the exposure variations. The amplitude transmittance T_a is given by

$$T_a(x, y) = [T_i(x, y)]^{1/2} \exp[j\phi(x, y)] \tag{4.69}$$

where T_i is the intensity transmittance and $\phi(x, y)$ accounts for the random phase shifts. However, film thickness variations that are mostly uncontrollable are usually not desirable in most practical applications. Such variations are often avoided by means of an index-matching liquid gate, as shown in Fig. 4.30, where the film emulsion is immersed in index-matching fluid (having a refractive index very similar to that of the emulsion) placed between two parallel glass plates. The amplitude transmittance of the developed film when submerged in the index-matching liquid gate can thus be written as

$$T_a(x, y) = [T_i(x, y)]^{1/2} \tag{4.70}$$

A two-step contact printing process would reveal as before that a linear relationship exists between the amplitude transmittance and the amplitude of the recording optical field when $\gamma_1\gamma_2 = 1$.

Problems

4.1 Determine the diffusion capacitance C_D of an LED in terms of its forward bias V_A. Plot G/G_0 and $2C_D/\tau G_0$ against $\omega\tau$, where G_0 is the conductance at $\omega = 0$. Comment on the physical meaning of the plots and justify your conclusions.

4.2 Show that for the photoconductor circuit of Fig. 4.8, the power across R_s becomes maximum when R_s equals the photoconductor resistance.

4.3 Show that the gain of a photoconductor is given by

$$G = [\tau'(\tau_n + \tau_p)]/(\tau_n\tau_p)$$

where τ' is the minority carrier lifetime.

4.4 Using Eq. (4.41), determine the breakdown voltage in a silicon APD in terms of its doping. Plot the breakdown voltage versus doping level when $A = 9 \times 10^5$ V/cm and $B = 1.8 \times 10^6$ V/cm.

4.5 (a) The multiplication factor of an APD (shown in Fig. 4.12) is also given by

$$M(x) = 1 + \int_0^x \alpha(x)M(x)\,dx + \int_x^W \beta(x)M(x)\,dx$$

where α and β are the ionization rates. Starting with a differential equation for $M(x)$, show that

$$M(x) = \frac{e^{-\int_x^W [\alpha(x) - \beta(x)]\,dx}}{1 - \int_0^W \alpha(x)e^{-\int_x^W [\alpha(x) - \beta(x)]\,dx}\,dx}$$

(b) Show that the solution for $M(x)$ from part (a) reduces to Eq. (4.39) where $k = \beta/\alpha$.

4.6 Show for a MOS capacitor that its capacitance is given by

$$C = \frac{C_0}{\left[1 + \frac{2C_0^2}{eN_A K_s \varepsilon_0} V_G\right]^{1/2}}$$

where V_G is the bias voltage and C_0 is the oxide capacitance.

4.7 Plot C/C_0 versus V_G for the MOS capacitor where V_G is the bias voltage and C_0 is the oxide capacitance.

4.8 A varied number of phenomena in CCDs can amount to retaining a small quantity of charge during each of the transfers. In a CCD scheme having n sensing elements, each of which involves m transfers, show that the fraction of the charge transferred is $1 - mn\varepsilon$ where ε is the fractional charge per elemental transfer.

4.9 Show that the vertical displacement due to electrostatic deflection in a CRT is given by

$$y \simeq \frac{Ll_d V_d}{2Va_1} \frac{\ln\left(\frac{a_2}{a_1}\right)}{\left(\frac{a_2}{a_1}\right) - 1}$$

where L is the distance between the deflectors and the screen, l_d is the length of the deflector ($l_d \ll L$), a_1 is the entrance deflector separation, a_2 is the exit deflector separation, and V_d is the deflector potential.

4.10 Find the vertical displacement for the magnetic deflection in a CRT in terms of the magnetic flux density and other pertinent parameters.

References and Suggested Readings

Awwal, A. A. S., A. K. Cherri, M. A. Karim, and D. L. Moon. "Dynamic Response of an Electro-Optic Imaging System." *SPIE*, Vol. 1116 (1989): 185.

Aukerman, L. W., F. L. Vernon, and Y. Song. "Radiation Threshold Levels for Noise Degradation of Photodiodes." *Opt. Engn.* Vol. 23 (1984): 678.

Boivin, L. P. "Diffusers in Silicon-Photodiode Radiometers." *Appl. Opt.* Vol. 21 (1982): 918.

Burnham, R., and D. Scarl. "Photomultiplier Dark Pulses." *Appl. Opt.* Vol. 24 (1985): 293.

Carter, R. C., and E. C. Carter. "High Contrast Sets of Colors." *Appl. Opt.* Vol. 21 (1982): 2936.

Cheo, P. K. *Fiber Optics Devices and Systems.* Englewood Cliffs, NJ: Prentice-Hall, 1985.

Cherri, A. K., A. A. S. Awwal, M. A. Karim, and D. L. Moon. "Restoration of Motion-Degraded Images in Electro-Optical Displays." *SPIE*, Vol. 1116 (1989): 198.

Coleman, C. I. "Imaging Detectors for the Ultraviolet." *Appl. Opt.* Vol. 20 (1981): 3693.

Dallas, W. J. "The Correlogram: A Two-Transmission Layer Display Device." *Appl. Opt.* Vol. 24 (1985): 3945.

Dewey, A. G. "Laser-Addressed Liquid Crystal Displays." *Opt. Engn.* Vol. 23 (1984): 230.

Emmons, R. B., S. R. Hawkins, and K. F. Cubb. "Infrared Detectors: An Overview." *Opt. Engn.* Vol. 14 (1975): 21.

Garbardi, D. R., and D. L. Shealy. "Coupling of Domed Light-Emitting Diodes with a Multimode Step-Index Optical Fiber." *Appl. Opt.* Vol. 25 (1986): 3435.

Gardner, J. L., and F. J. Wilkinson. "Response Time and Linearity of Inversion Layer Silicon Photodiodes." *Appl. Opt.* Vol. 24 (1985): 1531.

Itakura, Y., A. Sugimura, and S. Tsutsumi. "Amplitude-Modulated Reticle Constructed by a LC Cell Array." *Appl. Opt.* Vol. 20 (1981): 2819.

Jarominski, J. "Optical Crosstalk in LED Displays with Light Scattering Optical Cavities." *Appl. Opt.* Vol. 21 (1982): 3256.

Jarominski, J. "Optical Efficiency of LEDs with Hemispherical Lenses." *Appl. Opt.* Vol. 21 (1982): 2461.

Johnson, E. J., and M. A. Karim. *Digital Design: A Pragmatic Approach.* Boston: PWS Publishers, 1987.

Keiser, G. *Optical Fiber Communications.* New York: McGraw-Hill, 1981.

Liesegang, G. W., and P. R. Smith. "Vidicon Characteristics under Continuous and Pulsed Illumination." *Appl. Opt.* Vol. 21 (1982): 1437.

Lou, D. Y. "Characterization of Optical Disks." *Appl. Opt.* Vol. 21 (1982): 1602.

McCreight, C. R., and J. H. Goebal. "Infrared Charge-Injection-Device Array Performance at Low Background." *Appl. Opt.* Vol. 20 (1981): 3189.

McIntyre, R. J. "Multiplicative Noise in Avalanche Diodes." *IEEE Trans. on Electr. Dev.* Vol. ED-13, No. 1 (1966): 164.

Melchior, K., M. Rueff, and E. Schmidberger. "Sensors and Flexible Production." *Opt. Engn.* Vol. 23 (1984): 507.

Perez, O., and M. A. Karim. "An Efficient Implementation of Joint Fourier Transform Correlation Using a Modified LCTV." *Microwave & Opt. Tech. Lett.* Vol. 2 (1989): 193.

Pike, J. N. "Fluorescence of Mixed Powder Samples: A Six-Flux Theory." *Appl. Opt.* Vol. 20 (1981): 1167.

Priu, J., and W. Budde. "Compliance of Si Photodiodes with Talbot's Law." *Appl. Opt.* Vol. 20 (1981): 2676.

Sandel, B. R., and A. L. Broadfoot. "Statistical Performance of the Intensified Charge-Coupled Device." *Appl. Opt.* Vol. 25 (1986): 4135.

Schmieder, D. E., and B. Wolfe. "Empirical Television Sensor Performance Model." *Opt. Engn.* Vol. 24 (1985): 882.

Shimomura, T., and S. Kobayashi. "Color Contrast Criteria in a Guest-Host Mode Liquid Crystal Display." *Appl. Opt.* Vol. 20 (1981): 819.

Sze, S. M. *Physics of Semiconductor Devices* (2nd ed.). New York: Wiley, 1981.

Tannas, Jr., L. E., ed. *Flat-Panel Displays and CRTs*. New York: Van Nostrand Reinhold, 1985.

Trakalo, M., P. P. Webb, P. Poirier, and R. J. McIntyre. "Avalanche Photodiode Thirty-Two-Element Linear Array with Minimal Dead Space." *Appl. Opt.* Vol. 26 (1987): 3594.

Uchida, T. "Multicolored Liquid Crystal Displays." *Opt. Engn.* Vol. 23 (1984): 247.

Wang, J. Q., P. L. Richards, J. W. Beeman, N. M. Haegel, and E. E. Haller. "Optical Efficiency of Far-Infrared Photoconductors." *Appl. Opt.* Vol. 25 (1986): 4127.

Williams, G. F., F. Capasso, and W. T. Tsang. "The Graded Bandgap Multilayer Avalanche Photodiode: A New Low-Noise Detector." *IEEE Trans. on Electr. Dev. Lett.* Vol. EDL-3 (1982): 71.

Wilson, J., and J. F. B. Hawkes. *Optoelectronics: An Introduction*. Englewood Cliffs, NJ: Prentice-Hall International, 1985.

Wolfe, W. L., and G. J. Zissis, ed. *The Infrared Handbook*. Washington, DC: Office of Naval Research, Department of the Navy, 1978.

Yu, F. T. S., S. Jutamulia, and D. A. Gregory. "Real-Time Liquid Crystal TV XOR- and XNOR-Gate Binary Image Subtraction Technique." *Appl. Opt.* Vol. 26 (1987): 2738.

CHAPTER FIVE

Lasers

5.1 Introduction

The word **laser** is an acronym for **l**ight **a**mplification by **s**timulated **e**mission of **r**adiation. It refers to a unique photo-emitting device in which electromagnetic radiation can stimulate a medium (referred to as **lasing medium**) as a result of which the medium emits optical radiation of a single frequency. In addition, the light output of a laser is in phase with the input electromagnetic radiation. Lasers are equipped with an optical resonator, consisting of two precisely aligned mirrors, to provide positive internal feedback to the system. Repeated energy feedback amplifies the radiation beam and converts it into an intense output beam. So that it will have an effective output beam of light, one of the two resonator mirrors is made partially transmitting. Not surprisingly, lasers are analogous to electrical positive feedback oscillators, except for the fact that an optical oscillator is able to operate in more than one mode. Depending on the type of output, a laser can be either **continuous-wave** (CW) type or **pulse** type.

Laser radiation is not too different from any other form of electromagnetic radiation. However, several special properties of laser output make this device unique. Accordingly, the laser plays an important role in electro-optic applications. Lasers are used in a wide variety of practical applications, such as optical metrology; welding; spectroscopy; industrial, medical, and military applications; optical communications; optical computing; material processing; optical information processing; and many more. The output beam of laser light is highly monochromatic (or discrete as referred to in Section 2.3), coherent, directional, bright (owing to small beam divergence), and, in many cases, polarized. In comparison, the LEDs encountered in Chapter 4 are incoherent and non-narrow-band. Apart from that of a semiconductor laser, the output beam of laser light is highly directional and has a small beam divergence angle. While it is quite impossible to collect the total emitted radiation of conventional sources efficiently because their radiations spread into the 4π solid angle, it is possible to collect and make use of laser emission very efficiently. Typical emission values of the most common gas laser, e.g., the He–Ne laser, is about 10^{10} W/m^2-sr, whereas that of the sun is only about 1.3×10^6 W/m^2-sr. In addition, laser light can be focused to a very small diffraction-limited spot size. It is important to note that while CW lasers produce very low ($\simeq 1$ μW) to very high ($\simeq 1$ kW) power, pulsed lasers can produce power in the order of a terawatt (TW). Yet the power conversion efficiency for most lasers is very poor, ranging from less than 1% to a few percent. In practice, they require large power supplies and are, therefore, very bulky.

In a lasing medium (which may at times be dispersed in a host medium), photons are emitted spontaneously as atoms from higher energy states undergo transitions to lower energy states. However, such emissions are not only insignificant but are too random to produce a highly directional laser output. To develop a working laser, therefore, we are expected to cause a nonequilibrium condition in the lasing medium. This imbalance causes more lasing atoms or molecules to occupy an excited state, while comparatively fewer atoms or molecules occupy the less-excited states. Such a rearrangement of atoms or molecules is referred to as the **population inversion**. This nonequilibrium condition is achieved by an acceptable pumping mechanism, such as electric discharge, *RF* excitation, injection of high current densities, chemical reactions, nuclear excitation, and flashing gas-discharge light. Population inversion allows the atoms or molecules, when they are stimulated by an incoming electromagnetic radiation, to undergo transitions to lower energy states. Each such stimulated transition is also associated with an emission of a photon. Note, however, that the stimulated emission probability of a photon is extremely high when it is compared to the spontaneous emission probability of a photon. In effect, it is this stimulated emission of photons that is instrumental in producing a laser.

The light produced from the transition of atoms or molecules to a lower state is subjected to a positive feedback by means of a properly designed optical resonator. A positive feedback ensures a large buildup of stimulated radiation, producing an extremely intense beam of light output. In effect, the excitation process drives the lasing medium to the threshold of population inversion, exciting atoms or molecules to undergo radiative transitions. Photons traveling along the resonator axis contribute to most of the power buildup. Often the geometry of the laser resonator is crucial in having a highly directional laser output. The combined effect of the resonator feedback and the stimulated emission produce an extremely bright light output that is strikingly different from those of other light sources.

In this chapter, we first dwell on the types of electromagnetic radiative processes and then on the conditions for emission of light. Next, we explore the

Gaussian properties of laser output, the feedback, the amplification process, and, finally, various types of lasers.

5.2 Radiative Processes

In thermal equilibrium, materials absorb and emit energy at the same rate. There are three fundamental radiative processes that may take place when an electromagnetic wave interacts with a lasing material. These are **spontaneous emission**, **stimulated emission**, and **absorption**. The spontaneous emission, because of its very nature, is distributed over a wide range of frequencies. As stated in the last section, spontaneous emission of photons does not directly contribute to a laser output. The dynamic behavior of a laser is often described with reasonable precision by a set of coupled rate equations involving the three radiative processes. In their simplest form, these are a pair of simultaneous differential equations describing the population inversion and the laser radiation field as functions of time.

Consider an atomic system that has two nondegenerate energy states E_2 and E_1 such that $\Delta E = E_2 - E_1$. From the discussion of Chapter 1, it is obvious that when an atom undergoes transitions between these two states, it either emits or absorbs a photon of frequency $\nu = \Delta E/h$. Under normal circumstances, the atom will be in the lower level because physical systems tend to be more stable at the lowest energy state. Given the presence of an incoming photon, there is a high probability that an atom will configure itself at the excited energy state, but it returns again to the ground state by giving up a photon of energy ΔE. We now examine the details of the three radiative processes for a proper understanding of the process of achieving population inversion.

1. *Spontaneous emission:* This is a random radiative process whereby an electron in the excited state E_2 decays to state E_1 and, in so doing, gives out ΔE energy in the form of a photon, as shown in Fig. 5.1(a). If the population density of the excited energy state is N_2 and that of the lower energy state is N_1, then the rate of spontaneous decay of these atoms is found proportional to N_2. Consequently,

$$\left.\frac{dN_2}{dt}\right|_{\mathrm{sp}} = \left.\frac{dN_1}{-\,dt}\right|_{\mathrm{sp}} = -A_{21}N_2 \tag{5.1}$$

where A_{21} is called the **spontaneous emission probability**. The spontaneous emission lifetime $\tau_{21,\mathrm{sp}}$, defined as the average time during which the electron survives in the excited state before contributing to a spontaneous emission, is equivalent to $1/A_{21}$. The probability that a particular atom will undergo a spontaneous transition in time dt is $A_{21}\,dt$. The population density of the excited state is given by

$$N_2 = N_2^0 e^{-t/\tau_{21,\mathrm{sp}}} \tag{5.2}$$

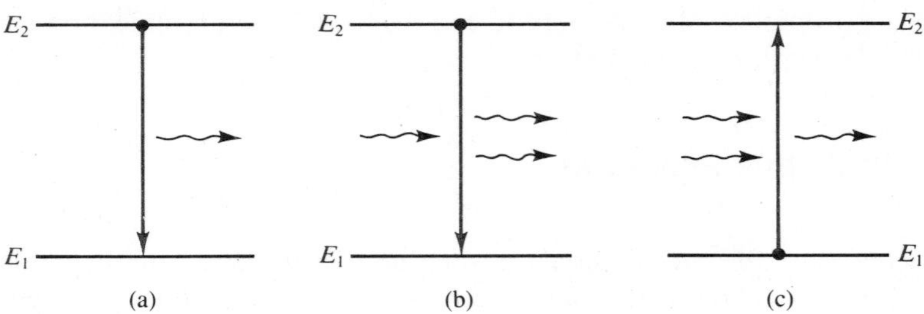

FIGURE 5.1

Radiative processes: (a) spontaneous emission, (b) stimulated emission, and (c) absorption

where N_2^0 is the value of N_2 at $t = 0$. Equation (5.2) justifies the fact that if an atomic system is subjected to only spontaneous emission, N_2 will deplete rapidly thus causing a rise in the population density N_1 of the lower energy state.

2. *Stimulated emission:* It is possible that an input photon can interact with an excited system and trigger emission of excess energy. A photon passing sufficiently close by an excited atom may cause the atom to undergo a radiative transfer before it would otherwise do so spontaneously. The excess photon energy output has the same frequency, phase, direction, and polarization as the stimulating photons. Consequently, the process involves an amplification of coherent light. Stimulated emission is associated with a specific energy level difference in a material, and the emitted photons show a high degree of spatial and temporal coherence with the incident photon. The fact that the stimulated and stimulating photons are in phase results in a high degree of coherence in lasers. In practice, a less than perfect temporal coherence is achieved because of the randomness present in radiative transfers due to vibrations and thermal variations among other things. In comparison, the spontaneous emission has no definite phase, and the resulting photon emission can be in any random direction. The stimulated emission rate is dependent on the incoming field as well as on the number of atoms to be stimulated and is given by

$$\left.\frac{dN_2}{dt}\right|_{\text{st}} = -\left.\frac{dN_1}{dt}\right|_{\text{st}} = -B_{21}N_2u_{e\nu} \tag{5.3}$$

where $u_{e\nu}$ is the average spectral energy density per unit frequency for blackbody radiation, and B_{21} is the stimulated emission probability per unit time per unit spectral energy density. Note that $u_{e\nu}\,d\nu$ provides the total photon density within the frequency interval ν and $\nu + d\nu$. The stimulated emission, as illustrated in Fig. 5.1(b), will be seen later to be the inverse of stimulated absorption.

3. *Absorption:* An electron in a lower energy state, after having absorbed an energy equivalent to ΔE, is raised to an excited energy state. In presence of a quasimonochromatic radiation field of frequency ν, the population of the less

excited state is depleted at a rate proportional to both the population of that state and radiation density. This deamplification process, called absorption and shown in Fig. 5.1(c), is associated with an absorption rate given by

$$\left.\frac{dN_2}{dt}\right|_{\text{abs}} = -\left.\frac{dN_1}{dt}\right|_{\text{abs}} = B_{12}N_1u_{ev} \tag{5.4}$$

where B_{12} is the absorption probability per unit spectral energy density. The input energy is accordingly decreased by an amount that was absorbed by the atomic system.

At thermal equilibrium, the photon absorption rate must equal the sum of the stimulated and spontaneous emission rates. An atomic system that is simultaneously subjected to all three radiative processes has an overall decay rate given by

$$\frac{dN_2}{dt} = -\frac{dN_1}{dt} = -A_{21}N_2 + B_{12}N_1u_{ev} - B_{21}N_2u_{ev} \tag{5.5}$$

However, at equilibrium, $dN_2/dt = dN_1/dt = 0$. Equation (5.5) yields the ratio of the population densities to be

$$\frac{N_2}{N_1} = \frac{B_{12}u_{ev}}{A_{21} + B_{21}u_{ev}} \tag{5.6}$$

There would be no absorption but only stimulated emission if N_1 were to be zero, but there would be only absorption if N_2 were to be zero. The ratio of the population densities at thermal equilibrium is also given by **Boltzmann distribution**:

$$\frac{N_2}{N_1} = \frac{g_2}{g_1}e^{-(E_2-E_1)/kT} \tag{5.7}$$

where g_i (referred to as the degeneracy) corresponds to the number of independent ways in which the atom can have the same energy E_i. Note that the ratio N_2/N_1 is always less than unity when $E_2 > E_1$ and $T > 0$ K. Normally, the energy level separation is on the order of 10^{-19} joules or more and, thus, at thermal equilibrium (at 300 K), N_1 exceeds N_2 by a factor on the order of 10^8. This implies that optical amplification is not realizable at thermal equilibrium. Otherwise, one gets something for nothing. The spectral energy density u_{ev} can be found by equating Eqs. (5.6) and (5.7) as

$$u_{ev} = \frac{A_{21}}{B_{21}}\left[\frac{g_1}{g_2}\frac{B_{12}}{B_{21}}e^{hv/kT} - 1\right]^{-1} \tag{5.8}$$

It may be noted that the spectral energy density is also given by Planck's radiation law using Eqs. (2.39) and (2.42):

$$\begin{aligned} u_{ev} &= \rho_{ev}f_{\text{BE}}hv \\ &= \frac{8\pi hv^3n^3}{c^3}\frac{1}{e^{hv/kT} - 1} \end{aligned} \tag{5.9}$$

where f_{BE} is the Bose-Einstein occupation number, n is the refractive index of the medium, and ρ_{ev} is the spectral mode density. The only way Eqs. (5.8) and (5.9) can then be equal is if and only if

$$\frac{g_1}{g_2} = \frac{B_{21}}{B_{12}} \tag{5.10a}$$

and

$$\frac{A_{21}}{B_{21}} = \frac{8\pi h\nu^3 n^3}{c^3} \tag{5.10b}$$

The last two equations are referred to as Einstein's relations. These equations interrelate the coefficients A_{21}, B_{12}, and B_{21}. The fact that $B_{21} = B_{12}$, when $g_1 = g_2$, for simplicity, implies that for a given radiation density, stimulated emission and absorption are equally probable. In addition, Eq. (5.10b) confirms that for most applicable frequencies, the rate of spontaneous emission is insignificant when compared to that of stimulated emission. Consequently, if favorable circumstances permit, the lasing within a material is expected to be dominated primarily by stimulated emission. Under conditions of thermal equilibrium, however, stimulated emission is not likely. The higher the frequency, the less likely the process.

We may conclude that all three radiative processes compete against each other according to Eq. (5.10). In building a laser, we expect to find stimulated emission to be the most dominating radiative process. This scenario can be achieved only by increasing energy density and by making N_2 larger than $(g_2/g_1)N_1$. This condition, referred to as the population inversion, appears to be unreachable according to Eq. (5.7) because $E_2 > E_1$. Before returning to the reason why this apparently impossible goal can be achieved, we must investigate absorption and amplification processes.

Consider that a collimated beam of light of unit cross-sectional area is passing through a lasing medium having an inverted population. For simplicity, let us assume that the system has only two energy states as assumed previously. We expect to find several aspects of energy transfer: a gain due to stimulated emission (from the inverted population); a loss due to absorption and scattering in the medium; absorption by the oscillator mirror; and transmission through the mirror. When a quasimonochromatic radiation of normal irradiance I_ν is incident on a lasing sample of thickness dx, the incremental change in the normal irradiance is found to be proportional to both incident irradiance and sample thickness dx. Consequently, the irradiance of the beam as a function of sample thickness is given by

$$I_\nu(x) = I_\nu(0)e^{(\beta-\alpha)x} \tag{5.11}$$

where $I_\nu(0)$ is the incident irradiance, β is the fractional gain per unit length, and α is the fractional loss per unit length. As described earlier, there are usually two mirrors with intensity reflectances R_1 and R_2 on the two sides of the oscillator.

Thus the roundtrip gain of the oscillator is given by

$$G = R_1 R_2 e^{2(\beta - \alpha)L} \tag{5.12}$$

where L is the length of the optical oscillator.

If the round-trip oscillator gain is less than unity, there will be only a deamplification in the system, and if it exceeds unity, the input light will be expected to be amplified. When β is high enough, the laser design can withstand alignment errors and unexpected losses, but when β is too small, none of these losses are tolerable. When the gain just balances out the losses, the laser reaches a steady-state condition, often referred to as a **threshold condition**. The threshold value of the gain per unit length is found by setting $G = 1$. Thus

$$\beta_{\text{th}} = \alpha - \frac{\ln(R_1 R_2)}{2L} \tag{5.13}$$

The gain per unit length must approach this threshold value in order for a continuous output to occur. In a pulsed laser, however, the gain β is at first increased to a value substantially higher than β_{th}. The pulsed laser begins with intense lasing but, as time progresses, the rate of stimulated emission decreases along with the decrease of population inversion. Thus, to sustain lasing, the gain has to be increased next. Such a process of increasing and decreasing β (corresponding to alternate lasing and no-lasing action) is characteristic of the pulsed lasers.

It is possible to realize population inversions in almost any medium—gas, liquid, or solid. The actual value of optical gain depends on both population inversion and the physical properties of the lasing medium. With a high value of gain, mirror alignment and cleanliness of the mirror may not be too critical. But in the case of a low gain medium, the mirrors must be accurately aligned and in perfect condition. A rigorous quantum mechanical treatment of the interaction between radiation and matter shows that the stimulated emission has the same directional properties, the same phase, the same polarization, and the same spectral characteristics as the stimulating emission.

A spectral analysis of emission and absorption, however, reveals that our energy level model needs a little modification. This arises from the quantum-mechanical uncertainty relation, in accordance to which the minimum value of the product of the spread in frequency and the corresponding spread in time is a finite positive value. Consequently, the emission of radiation is never strictly monochromatic but occupies a finite bandwidth of frequency. The function that relates frequency with the radiating intensity is often referred to as the **line shape function** $g(\nu)$. The quantity $g(\nu)\,d\nu$ is the probability for either emission or absorption of photons whose frequencies lie between ν and $\nu + d\nu$ such that

$$\int_0^\infty g(\nu)\,d\nu = 1 \tag{5.14}$$

The precise form of $g(\nu)$ depends on the particular mechanism that causes spectral broadening. The most noteworthy mechanisms are collision broadening, natural damping, and doppler broadening. Doppler broadening, for example, is introduced by apparent differences in frequency of emission as atoms tend to approach or move away from the observer. The induced transition rate per atom between the two energy states is given by

$$W_{ij}(v) = B_{ij}u_{ev} = \frac{c^3 u_{ev} g(v)}{8\pi h v^3 n^3 t_{21,\text{sp}}} = \frac{I_v c^2}{8\pi n^2 h v^3 t_{21,\text{sp}}} g(v) \tag{5.15}$$

where $I_v = cu_{ev}/n$ is the intensity of the optical wave, and the subscripts i and j refer respectively to the energy states 1 and 2.

It is possible to obtain the gain coefficient β in terms of either spontaneous or stimulated emission probability constants. The gain in irradiance per unit distance is equivalent to the product of depletion rate of atoms in the excited energy state and energy per photon. Because the contribution of spontaneous emission is relatively insignificant under the population inversion condition, using Eq. (5.10a) we can write

$$\frac{dI_v}{dx} = -\frac{dN_2}{dt} hv = u_{ev} g(v) hv [N_2 B_{21} - N_1 B_{12}] = \frac{I_v n h v g(v) B_{21}}{c} \left[N_2 - \frac{g_2}{g_1} N_1 \right] \tag{5.16}$$

since $I_v = u_{ev} c/n$, where n is the refractive index of the medium. We readily obtain irradiance from Eq. (5.16) as

$$I_v(x) = I_v(0) e^{\{[N_2 - (g_2/g_1)N_1] B_{21} n h v g(v) x/c\}} \tag{5.17}$$

By comparing Eq. (5.17) and the lossless form of Eq. (5.11), and using Eq. (5.10b), we get

$$\beta(v) = \frac{B_{21} n h v}{c} \left(N_2 - \frac{g_2}{g_1} N_1 \right) g(v) = \frac{A_{21} c^2}{8\pi v^2 n^2} \left(N_2 - \frac{g_2}{g_1} N_1 \right) g(v) \tag{5.18}$$

Consequently, at higher frequencies the gain is very small, which makes the construction of a laser very difficult, if not impossible. Because $A_{21} = 1/\tau_{21,\text{sp}}$, it is desirable to maintain as short a spontaneous lifetime as possible to have the largest possible gain. At the same time, note that the process of creating a population inversion may take a finite time. Thus $\tau_{21,\text{sp}}$ value is maintained just long enough to allow for the inversion of population.

The rate equations used in this section were arrived at through a number of simplifying assumptions, some of which are certainly unrealistic. For example, these equations have no positional or spectral dependencies. In addition, the pumping field has longitudinal as well as radial variations within the laser medium. In spite of these limitations, the rate equations remain a useful tool for providing a great deal of insight into the behavior of real laser systems. Details of the nature of laser emission are not obvious from a simple rate equation. Some of these detailed descriptions of a laser include spectral, temporal, and spatial distributions of emission. However, these details can often be determined independently. In general, the rate equations are useful for predicting the grosser features of laser output—for example, average and/or peak powers, threshold conditions, etc.

5.3 Laser Excitation

The laser is a combination of an optical amplifier and an optical resonator, which serve to provide a highly selective feedback mechanism. When the gain is sufficiently large to overcome the losses, regenerative oscillation begins and produces the laser beam. For a given transition, stimulated emission must exceed absorption for us to realize a desirable gain. The Boltzmann distribution of Eq. (5.7) refers to an equilibrium condition whereby only a few atoms are found in the excited energy state, whereas the majority of the atoms are found in the lower energy state. Clearly, therefore, a population inversion refers to a nonequilibrium condition in which we expect to find comparatively more atoms in the excited state. The condition $N_2 > N_1$ is sometimes called a **negative temperature condition** in accordance with the prediction of Eq. (5.7). In any event, for the population to be inverted, the number of atoms in the lower energy level must be decreased and the number in the upper energy level increased. This method of redistributing the population of the atoms is called **pumping**.

When two-level atoms are excited by irradiation, the number of upper-level atoms will increase but, at the same time, these excited atoms spontaneously decay to lower energy states by emitting energy. Equation (5.10a) predicts that no matter how intensely the atoms are excited, the population inversion is not realizable in a two-level system. At best, the two populations would be just equal to each other. Consequently, we examine other than two-level atomic systems to see if population inversion can be achieved. Notably, most atomic systems have a large number of energy states.

5.3.1 Three-Level Laser

Consider a three-level laser having a density of N_t active atoms, as shown in Fig. 5.2(a), where all three of the states are nondegenerate. At thermal equilibrium, the ground state E_1 is heavily populated, whereas state E_3 has the smallest of the three populations and E_2 has an intermediate-sized population. The exact distribution of the populations is, however, determined by the Boltzmann's distribution given in Eq. (5.7). It is fair to assume that in the beginning $N_1 \simeq N_t$.

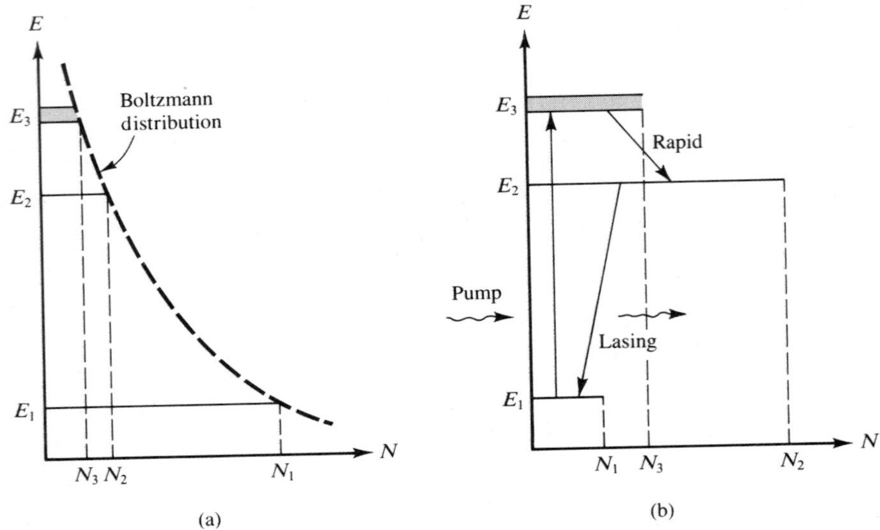

FIGURE 5.2

The three-level laser system: (a) prior to pumping and (b) after pumping

As the atoms are pumped from the lowest energy state to the top-most energy state, N_3 increases and N_1 decreases. Assume further that atoms from the top-most energy state decay rapidly to the intermediate state through some nonradiative process (i.e., by giving up energy to the neighboring molecules, thus increasing the lattice energy) and that the intermediate energy state is reasonably metastable. Consequently, pumping will be able to transfer atoms effectively from the lowest to the intermediate energy state via the top-most energy state. As this process continues, N_2 soon exceeds N_1, resulting in an inverted population.

Since the top-most energy level is not one of the lasing levels, it can be allowed to be broad enough to consist of a large number of closely spaced energy levels. A broadband flashlight can then be used as the pumping source, thus increasing the overall pumping efficiency. In practice, xenon, krypton, mercury, and tungsten lamps are used as pump sources. Since the lowest energy level is also the ground state, more than 50% of the ground-state atoms have to be pumped to the top-most energy level to attain population inversion. Again, the energy lost by the excited ions in making the transition from E_3 to E_2 is released as heat that has to be removed by a water-cooling system if the laser is to be operated continuously or at a high repetition rate.

Let a_{ij} represent the transition probability rate of an atom in transition from the energy state i to the energy state j. Let us assume further that the separation between the energy states is sufficiently greater than the thermal energy, such that $a_{ij} \gg a_{ji}$ when $E_i > E_j$. The rates at which the three population densities change are given, respectively, by

$$\frac{dN_1}{dt} = -W_p N_1 + W_l N_2 + a_{31} N_3 \tag{5.19a}$$

$$\frac{dN_2}{dt} = -W_l N_2 + a_{32} N_3 \tag{5.19b}$$

$$\frac{dN_3}{dt} = W_p N_1 - (a_{31} + a_{32}) N_3 \tag{5.19c}$$

where W_p is proportional to the pump beam energy density, W_l is proportional to the lasing energy density, and $N_t \equiv N_1 + N_2 + N_3$ is a constant for the laser system. The steady-state ($dN_1/dt = dN_2/dt = dN_3/dt = 0$) solutions of Eqs. (5.19) are given by

$$N_1 = \frac{W_l N_t (a_{31} + a_{32})}{W_l(a_{31} + a_{32}) + a_{32} W_p + W_p W_l} \tag{5.20a}$$

$$N_2 = \frac{a_{32} W_p N_t}{W_l(a_{31} + a_{32}) + a_{32} W_p + W_p W_l} \tag{5.20b}$$

$$N_3 = N_t - N_1 - N_2 \tag{5.20c}$$

By using Eqs. (5.20a) and (5.20b), we see that when

$$W_p > W_l \left[1 + \frac{a_{31}}{a_{32}} \right] \tag{5.21}$$

a population inversion is achievable. It is desirable that transition from the top-most energy level to the lowest energy level should be slow and that from the top-most energy level to the intermediate energy level must be very fast. Consequently, the inverted population, defined by $\Delta N = N_2 - N_1$, can be determined as follows:

$$\frac{\Delta N}{N_t} \simeq \frac{\dfrac{a_{32}}{W_l} \dfrac{W_p}{W_o}}{\left(\dfrac{a_{32}}{W_l} + 1 \right) \left(1 + \dfrac{W_p}{W_o} \right)} \tag{5.22}$$

where

$$W_o \equiv \frac{W_l(a_{31} + a_{32})}{W_l + a_{32}} \tag{5.23}$$

At very large value of the pump Eq. (5.22) reduces to

$$\mathrm{Lim}_{W_p \to \infty} \frac{\Delta N}{N_t} = \frac{1}{1 + \dfrac{W_l}{a_{32}}} \tag{5.24}$$

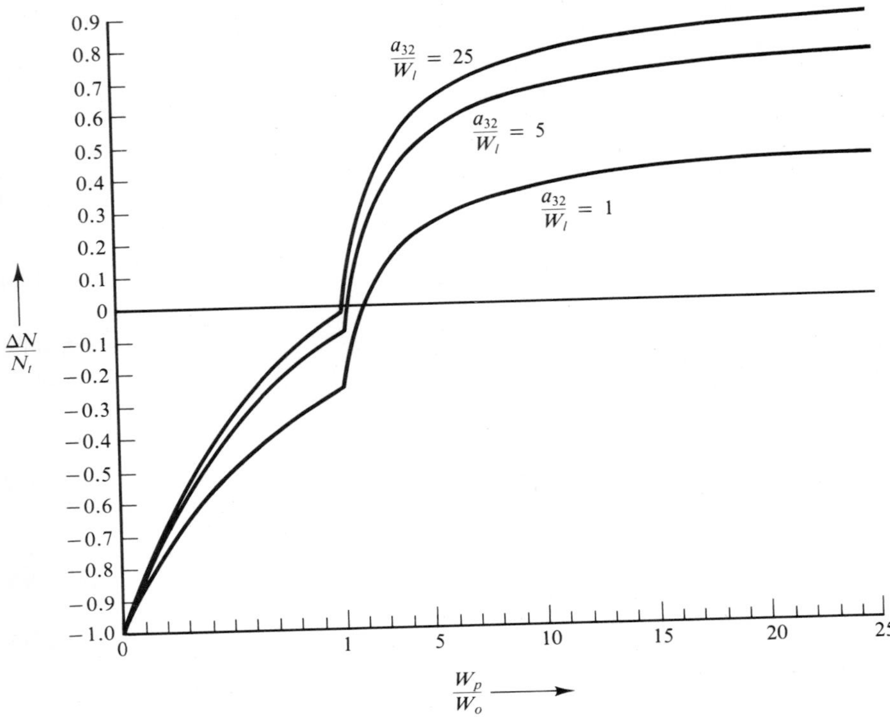

FIGURE 5.3

$\Delta N/N_t$ versus W_p/W_o plot for a three-level laser

Figure 5.3 shows a plot of $\Delta N/N_t$ versus W_p/W_o for different values of a_{32}/W_l. It can be seen that the larger values of a_{32}/W_l correspond to a larger population inversion, and hence to stronger lasing conditions.

The rate at which atoms decay from the intermediate energy state to the lowest energy state is given by $N_2 W_l$. However, for each atom present in the intermediate energy level, an energy equivalent to a minimum of $h\nu_p$ has to be pumped, where ν_p represents the average pumping frequency. Therefore, the minimum pumping power per unit volume of the lasing material is given by

$$P_{\text{3-level}} = N_2 W_l h\nu_p \tag{5.25}$$

But since the transition from the top-most to the intermediate energy state is very rapid, N_3 is negligible. At the threshold, therefore, we can assume that $N_2 \simeq \frac{1}{2} N_t$ and thus

$$P_{\text{3-level}} = \tfrac{1}{2} N_t W_l h\nu_p \tag{5.26}$$

5.3.2 Four-Level Laser

The four-level laser system, as shown in Fig. 5.4, has four energy states distributed in a manner such that the terminal energy state E_2 is well separated from the lowest energy state E_1. At thermal equilibrium, therefore, the population densities corresponding to the energy states E_2, E_3, and E_4 can be neglected in a four-level laser for all practical purposes. Again, since the population densities corresponding to the energy states E_2 and E_3 are negligibly small, it is easy to create a population inversion between E_2 and E_3. Consequently, a four-level laser has a smaller pumping requirement than a three-level laser.

For simplicity, the transition probability rate of an atom from energy state E_1 to energy state E_2 is not neglected in our treatment. The four rate equations for the laser of Fig. 5.4 then become

$$\frac{dN_1}{dt} = -\left(\frac{dN_2}{dt} + \frac{dN_3}{dt} + \frac{dN_4}{dt}\right) \tag{5.27a}$$

$$\frac{dN_2}{dt} = a_{12}N_1 - a_{21}N_2 + W_l N_3 + a_{42}N_4 \tag{5.27b}$$

$$\frac{dN_3}{dt} = -(W_l + a_{31})N_3 + a_{43}N_4 \tag{5.27c}$$

$$\frac{dN_4}{dt} = W_p N_1 - (a_{41} + a_{42} + a_{43})N_4 \tag{5.27d}$$

The steady-state solutions of Eq. (5.27) are obtained as follows:

$$N_1 = \frac{[a_{21}(a_{31} + W_l)(a_{41} + a_{42} + a_{43})N_t]}{[a_{12}(a_{31} + W_l)(a_{41} + a_{42} + a_{43}) + (W_l + a_{21})a_{43}W_p + (W_l + a_{31})(a_{42} + a_{21})W_p]} \tag{5.28a}$$

$$N_2 = \left[\frac{a_{12}}{a_{21}} + \frac{[a_{43}W_l + (a_{31} + W_l)a_{42}]W_p}{a_{21}(W_l + a_{31})(a_{41} + a_{42} + a_{43})}\right]N_1 \tag{5.28b}$$

$$N_3 = \frac{a_{43}W_p N_1}{(W_l + a_{31})(a_{41} + a_{42} + a_{43})} \tag{5.28c}$$

$$N_4 = N_t - N_1 - N_2 - N_3 \tag{5.28d}$$

The population inversion condition can now be obtained from Eqs. (5.28b) and (5.28c) as

$$W_p > a_{12}\left[\frac{a_{21}a_{43}}{(W_l + a_{31})(a_{41} + a_{42} + a_{43})} - \frac{a_{42}}{a_{41} + a_{42} + a_{43}} - \frac{a_{43}W_l}{(W_l + a_{31})(a_{41} + a_{42} + a_{43})}\right]^{-1} \tag{5.29}$$

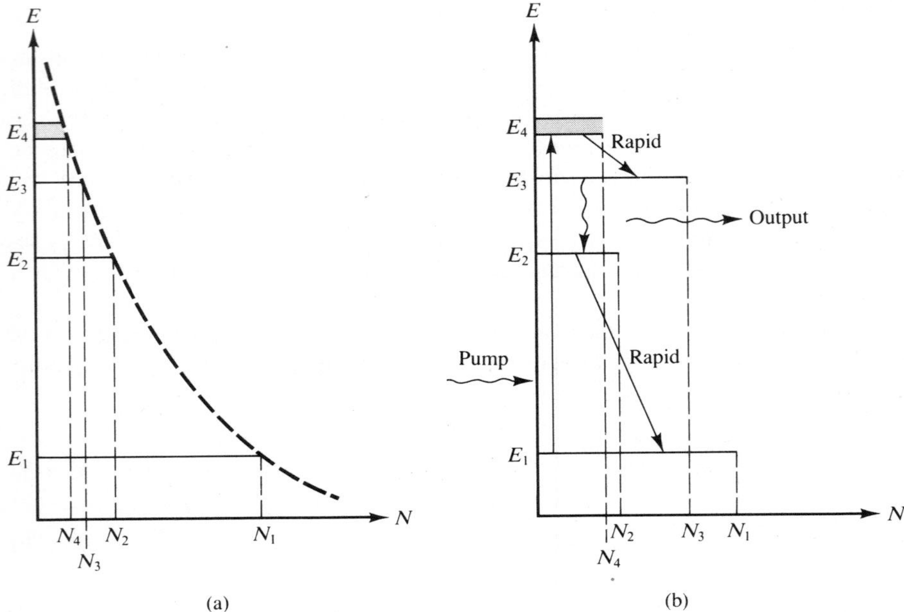

FIGURE 5.4

A four-level laser system at (a) thermal equilibrium and (b) after pumping

To have a satisfactory population inversion, we must have a large value for a_{21} and a small value for $W_l + a_{31}$. The relaxation rate of atoms at energy state E_2 (during transition from E_2 to E_1) has to be faster than the arrival rate of atoms into energy state E_2 to achieve population inversion for very small pump power. Since a_{42} is smaller than the algebraic sum of all downward transition probability rates at energy state E_4 and because $W_l < W_l + a_{31}$, Eq. (5.29) can be reduced to

$$W_p > (W_l + a_{31}) \frac{a_{41} + a_{42} + a_{43}}{a_{43}} e^{-(E_2 - E_1)/kT} \tag{5.30}$$

Comparing the pump energy densities of three-level and four-level lasers (see Eqs. 5.21 and 5.30), we find that they are identical except for the factor $\exp[-(E_2 - E_1)/kT]$. We may, therefore, readily conclude that the population inversion is more easily obtained in a four-level laser.

For a four-level laser, the pumping power per unit volume can be obtained from Eq. (5.25) by replacing N_2 with N_3. Again, since the atoms in E_2 level have shorter lifetimes than those in E_3 level at threshold, $N_3 - (g_3/g_2)N_2 \equiv N_{\text{th}} \simeq N_3$. Therefore, the minimum pumping power per unit volume in a four-level laser is

given by

$$P_{\text{4-level}} = N_{\text{th}} W_l h\nu_p \tag{5.31}$$

where the threshold population inversion N_{th} is given using Eq. (5.18) by

$$\begin{aligned} N_{\text{th}} &= N_3 - \frac{g_3}{g_2} N_2 \\ &= \frac{8\pi\nu^2 n^2 \beta_{\text{th}}}{A_{32} c^2 g(\nu)} \end{aligned} \tag{5.32}$$

5.4 Gaussian Characteristics of the Laser Beam

Having studied the condition for population inversion, we next focus our attention on the concepts of optical feedback by means of an optical oscillator. Usually, the optical oscillator also houses the lasing material. We could have depended simply on ray optics to determine the conditions for optical feedback, but ray optics fails to convey the extent of amplitude, phase, or spatial distribution of laser beam. Thus, to obtain a clearer description of the beams, we attempt in this section to analytically solve Maxwell's wave equation. Therefore, using Eqs. (1.10) and (1.14), we can begin with the equation for electric field

$$\nabla^2 E(x, y, z) + k^2 E(x, y, z) = 0 \tag{5.33}$$

where $k = \omega n/c$ is a very large number. The equation for the magnetic field intensity is identical in nature and, therefore, we shall limit ourselves to only Eq. (5.33).

The electromagnetic energy of an optical beam of light behaves like a nearly plane wave. For simplicity, we may assume that this beam of light is propagating along the z axis. Appropriately, we may assume a solution of the form

$$E(x, y, z) = \Psi(x, y, z) e^{-jkz} \tag{5.34}$$

for Eq. (5.33). The factor e^{-jkz} represents the plane wave characteristic of the beam of light, while the function $\Psi(x, y, z)$ is expected to vanish at large values of x or y. Further, $\Psi(x, y, z)$ should vary only slightly as z changes. Introducing Eq. (5.34) into Eq. (5.33), we get

$$\frac{\partial^2 \Psi}{\partial x^2} + \frac{\partial^2 \Psi}{\partial y^2} + \frac{\partial^2 \Psi}{\partial z^2} - 2jk \frac{\partial \Psi}{\partial z} = 0 \tag{5.35}$$

Both $\partial\Psi/\partial z$ and $\partial^2\Psi/\partial z^2$ are small but nonzero and $k\ (\equiv 2\pi n/\lambda)$ is a reasonably large number. Thus Eq. (5.35) reduces to

$$\frac{\partial^2 \Psi}{\partial x^2} + \frac{\partial^2 \Psi}{\partial y^2} - 2jk \frac{\partial \Psi}{\partial z} = 0 \tag{5.36}$$

We are already familiar with the method of separation of variables from our analysis of the hydrogen atom potential problem in Chapter 1. We can do the same for $\Psi(x, y, z)$. Accordingly, let us assume that

$$\Psi(x, y, z) = X(x)Y(y)Z(z)e^{-(x^2+y^2)/W(z)} \tag{5.37}$$

where both $Z(z)$ and $W(z)$ are slowly varying functions of z. The exponential factor guarantees that $\Psi(x, y, z)$ approaches zero at positions further from the propagation axis z.

Equation (5.36) can now be rewritten using Eq. (5.37) as

$$\left[\frac{1}{X(x)}\frac{d^2X}{dx^2} - \frac{4x}{W(z)X(x)}\frac{dX}{dx} + \frac{4m}{W(z)}\right] + \left[\frac{1}{Y}\frac{d^2Y}{dy^2} - \frac{4y}{W(z)Y(y)}\frac{dY}{dy} + \frac{4n}{W(z)}\right] - \frac{2jk}{Z(z)}\frac{dZ}{dz} - \frac{4(m+n+1)}{W(z)} + 2\left(\frac{\sqrt{x^2+y^2}}{W(z)}\right)^2\left[2 - jk\frac{dW}{dz}\right] = 0 \tag{5.38}$$

However, this equation is valid only as long as $X(x)$ is a function of x alone, $Y(y)$ is a function of y alone, and $Z(z)$ and $W(z)$ are functions of z alone. Consequently, the last term on the left-hand side must be zero, i.e.,

$$\frac{dW}{dz} = -j\frac{2}{k} \tag{5.39}$$

which yields

$$W(z) = -j\frac{2z}{k} + C_1 \tag{5.40}$$

where C_1 is an arbitrary constant. Defining $W(z) = \omega_0^2$ at $z = 0$, we see that $E(x, y, z)$ reduces to $(1/e)$th of its peak value at $r \equiv [x^2 + y^2]^{1/2} = \omega_0$. The radial distance ω_0 is often referred to as the **beam waist**. Equation (5.40) can be now written as

$$\frac{1}{W(z)} = \frac{1}{[\omega(z)]^2} + \frac{jk}{2R(z)} \tag{5.41}$$

where

$$\omega(z) = \omega_0\left[1 + \left(\frac{z}{z_0}\right)^2\right]^{1/2} \tag{5.42a}$$

$$R(z) = z\left[1 + \left(\frac{z_0}{z}\right)^2\right] \tag{5.42b}$$

and

$$z_0 = \frac{\pi\omega_0^2 n}{\lambda} \tag{5.42c}$$

Equation (5.38) can then be separated into three separate differential equations as follows:

$$\frac{1}{X}\frac{d^2X}{dx^2} - \frac{4x}{XW}\frac{dX}{dx} + \frac{4m}{W} = 0 \tag{5.43a}$$

$$\frac{1}{Y}\frac{d^2Y}{dy^2} - \frac{4y}{YW}\frac{dY}{dy} + \frac{4n}{W} = 0 \tag{5.43b}$$

$$\frac{1}{Z}\frac{dZ}{dz} = \frac{(m+n+1)\left(j\dfrac{k\omega_0^2}{2} - z\right)}{z^2 + \left(\dfrac{k\omega_0^2}{2}\right)^2} \tag{5.43c}$$

Equations (5.43a) and (5.43b) are standard differential equations that are solvable by means of Hermite polynomials defined as

$$X(x) = H_m\left(\sqrt{\frac{2}{W(z)}}\,x\right) \tag{5.44a}$$

$$Y(y) = H_n\left(\sqrt{\frac{2}{W(z)}}\,y\right) \tag{5.44b}$$

where the **Hermite polynomial** of order l and argument u, $H_l(u)$, is given by

$$H_l(u) = (-1)^l e^{u^2}\frac{d^l}{du^l}(e^{-u^2}) \tag{5.45}$$

Using Eqs. (5.40) and (5.41), Eq. (5.43c) can be integrated to give (after laborious manipulations) the remaining function $Z(z)$ as

$$\begin{aligned} Z(z) &= \frac{C_2}{\left(z + j\dfrac{k\omega_0^2}{2}\right)^{m+n+1}} \\ &= C_2\left[\frac{2}{jkW(z)}\right]^{m+n+1} \\ &= C_3\left[\frac{\omega_0}{\omega(z)}\right]^{m+n+1} e^{j(m+n+1)\tan^{-1}(z/z_0)} \end{aligned} \tag{5.46}$$

where C_2 is the new integration constant and $C_3 = C_2[2/(k\omega_0)]^{m+n+1}$. Therefore, the Hermite-Gaussian solution of Eq. (5.33) is given by

$$\begin{aligned} E(x, y, z) = {} & C_3 H_m\left(\sqrt{\frac{2}{W(z)}}\,x\right) H_n\left(\sqrt{\frac{2}{W(z)}}\,y\right)\left[\frac{\omega_0}{\omega(z)}\right]^{m+n+1} \\ & \times e^{-(x^2+y^2)/\omega^2(z)} e^{[-j\{kz + k(x^2+y^2)/\{2R(z)\} - (m+n+1)\tan^{-1}(z/z_0)\}]} \end{aligned} \tag{5.47}$$

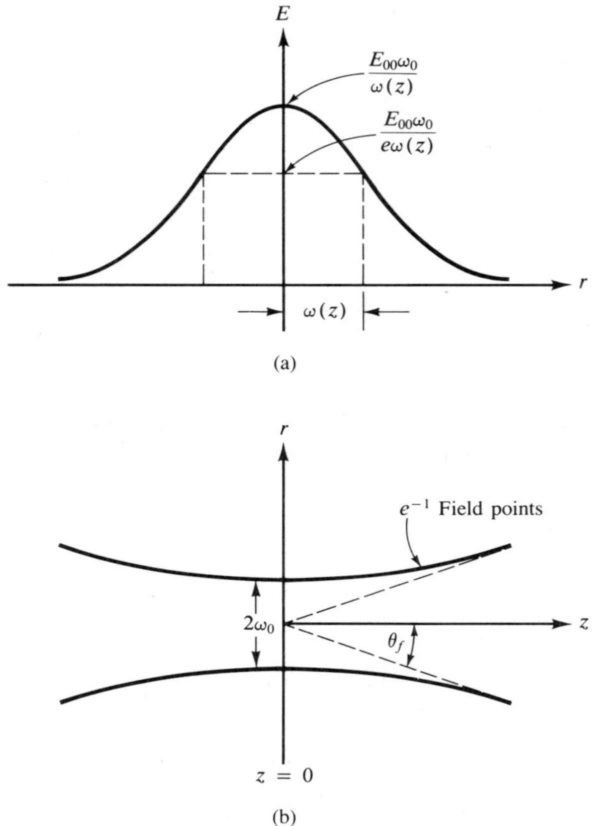

FIGURE 5.5

TEM_{00} characteristics: (a) $E(r)$ versus r and (b) e^{-1} field points versus z

We can see that for nonzero z, $\omega(z) > \omega_0$, which implies that the beam will never have a transverse width smaller than that corresponding to the minimum beam waist ω_0. Of particular interest to us is the TEM_{mn} (transverse electromagnetic) mode that corresponds to $m = n = 0$ (referred to as the fundamental mode). The field amplitude for the TEM_{00} mode is

$$E(x, y, z) = E_{00}\frac{\omega_0}{\omega(z)}e^{-[r/\omega(z)]^2} \tag{5.48}$$

where E_{00} is the peak value of the amplitude. When $r \equiv \sqrt{x^2 + y^2} = \omega(z)$, the field amplitude is only $1/e$th of what it is at $r = 0$. Figure 5.5(a) shows the amplitude of the field as a function of r, and Fig. 5.5(b) shows the $1/e$th field points as a function of propagation distance.

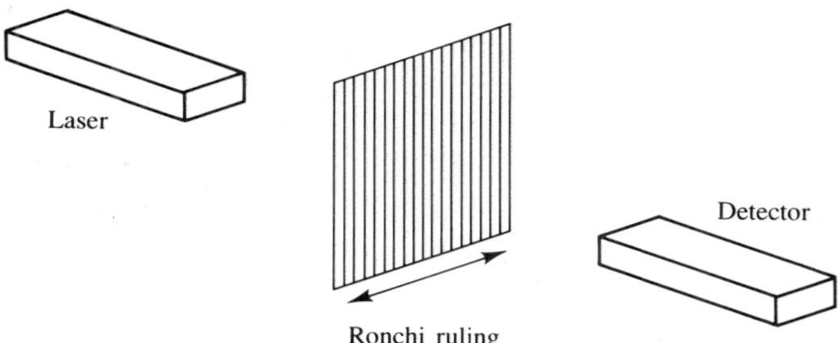

FIGURE 5.6

Setup for measuring Gaussian beam radius

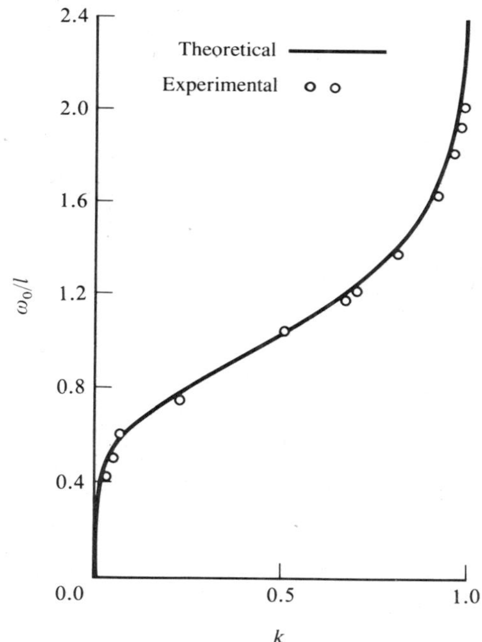

FIGURE 5.7

k versus ω_0/l for the setup of Fig. 5.6. [Adapted from M. A. Karim. "Measurement of Gaussian Beam Diameter Using Ronchi Rulings." *Electr. Lett.*, Vol. 21 (1985): p. 427.]

One of the ways to measure the Gaussian beam radius ω_0 accurately involves scanning a Ronchi ruling (having alternate transparent and opaque regions per period) of period l across the TEM_{00} laser beam. A detector placed directly on the other side of the Ronchi ruling, as shown in Fig. 5.6, is used to

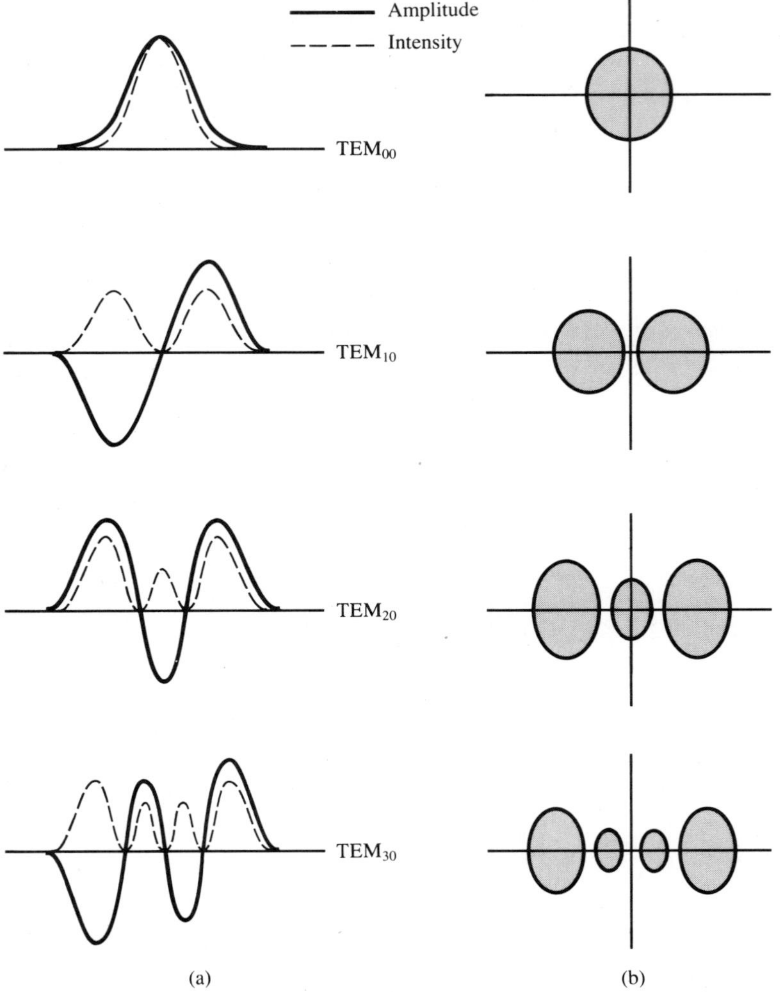

FIGURE 5.8

Higher order Gaussian beams: (a) amplitude and intensity and (b) beam cross section

detect the maximum and the minimum transmitted power. The detector power is a minimum when the center of the opaque region faces the Gaussian beam, and the detector power is a maximum when the center of the transparent region faces the Gaussian beam. The ratio of the minimum-to-maximum detected power k is found to be a function of the ratio ω_0/l as illustrated in Fig. 5.7. Consequently, the Gaussian beam radius is determined using the characteristics of Fig. 5.7 provided that k and l are known.

The far-field divergence angle θ_f is used to quantify the minimum spread of a Gaussian beam and is defined as the angle subtended by the asymptote (of e^{-1} field variation) with the direction of propagation. It is given by

$$\theta_f = \tan^{-1}\left(\left.\frac{d\omega}{dz}\right|_{z=\infty}\right) = \tan^{-1}\left(\frac{\lambda}{\pi\omega_0}\right) \tag{5.49}$$

The radial phase factor, $\exp[-j(kr^2)/(2R(z))]$, in the solution of Eq. (5.47) reflects the fact that the planes normal to the z axis are not equiphase surfaces. For positive $R(z)$, the phase of the field at nonzero r lags behind that at $r=0$. Similarly, for negative $R(z)$, the phase of the field at nonzero r leads that at $r=0$. Only at $z=0$ can we find a planar wavefront. We may infer that $R(z)$ provides the curvature of equiphase surfaces that are spherical in nature. According to Eq. (5.42b), the center of a spherical wavefront at $z \gg 0$ appears to be at $z = r = 0$, whereas the center of a spherical wavefront at $z \simeq 0$ appears to be at $z < 0$ and, finally, the center of a spherical wavefront at $z=0$ appears to be at $z=\infty$. Thus at $z=0$, the wavefront is planar and the beam waist is minimum.

It is appropriate now to comment on the behavior of a higher order Gaussian beam. The major difference between a higher order Gaussian beam and a TEM_{00} beam stems from the fact that the Hermite polynomials can force the beam intensity to diminish at a finite number of points. A few of the lower order Hermite polynomials, for example, are as follows:

$$H_0(u) = 1 \tag{5.50a}$$

$$H_1(u) = 2u \tag{5.50b}$$

$$H_2(u) = 4u^2 - 2 \tag{5.50c}$$

$$H_3(u) = 8u^3 - 12u \tag{5.50d}$$

$$H_4(u) = 16u^4 - 48u^2 - 12 \tag{5.50e}$$

Figure 5.8 shows examples of a few of the higher order laser modes. For large r, the intensity has an exponential decay characteristic, but at values closer to $r=0$, the intensity is drastically modified by the Hermite polynomials.

5.5 Optical Feedback

In an earlier section we discussed the fact that positive feedback is necessary in a laser if lasing action is to occur. This is achieved by placing the gain medium in an optical cavity (or resonator), which in turn is formed using a pair of mirrors on either side of the cavity. The resonator determines the frequency as well as the directional and spatial characteristics of output emission. The optical resonator serves to define one or more modes having significantly lower losses than all other modes. The initial ignition to the resonator is provided by any spontaneous photon emission if the photon is emitted along the cavity axis. The emitted energy is amplified as it passes through the gain medium and is fed back into the system by end mirrors to repeat the process. In a low-power laser, most of the amplified energy is reflected back by the end mirrors but, in the case of a high-power laser, a substantial portion of the amplified energy is transmitted out through the end mirror. For steady-state operation, the system losses—including those due to the power output—have to be equal to the gains in the system resonator.

In order to determine what happens to the Gaussian beam of light after it has gone through an arbitrary optical system, we need to relate the pertinent Gaussian beam characteristics to the parameters of the optical system. Consider a very simple system, such as the one shown in Fig. 5.9, that involves no mirrors or lenses but only a medium of length d. Since a ray is directed normal to the associated wavefront, an understanding of the ray gyration makes it possible to characterize the transformation of optical waves in an optical system.

A reasonable way to characterize a ray in a plane is to identify its position r from the axis and its slope r' with respect to the optic axis. For a paraxial ray, the angular deviation from the axis is small enough so that the sine and tangent of the angular deviation are estimated by the angle itself. Consequently, the output parameters of the ray are expressed as

$$r_{\text{out}} = r_{\text{in}} + d(r_{\text{in}})' \tag{5.51a}$$

$$(r_{\text{out}})' = (r_{\text{in}})' \tag{5.51b}$$

which can be expressed in a matrix format by

$$\begin{pmatrix} r_{\text{out}} \\ (r_{\text{out}})' \end{pmatrix} = \begin{pmatrix} 1 & d \\ 0 & 1 \end{pmatrix} \begin{pmatrix} r_{\text{in}} \\ (r_{\text{in}})' \end{pmatrix} \tag{5.52}$$

We observe from Eq. (5.52) that propagation of a ray through an optical system, although a trivial problem, can be described by a simple 2×2 matrix. In ray optics, the characteristic of this system is given by the corresponding ray matrix (referred to as an ABCD matrix), which is

$$\begin{bmatrix} A & B \\ C & D \end{bmatrix} = \begin{bmatrix} 1 & d \\ 0 & 1 \end{bmatrix} \tag{5.53}$$

Table 5.1 lists some of the more important optical system components and the corresponding 2×2 ray matrices. Note that for the system components having the same refractive index at both the entrance and exit planes, the determinant of the ray matrix is always unity.

It is appropriate now to define, for simplicity, a complex Gaussian beam parameter $q(z)$ in terms of $W(z)$ as follows:

$$\frac{1}{q(z)} \equiv -\frac{2j}{k}\frac{1}{W(z)} = \frac{1}{R(z)} - \frac{j\lambda}{\pi\omega^2(z)} \tag{5.54}$$

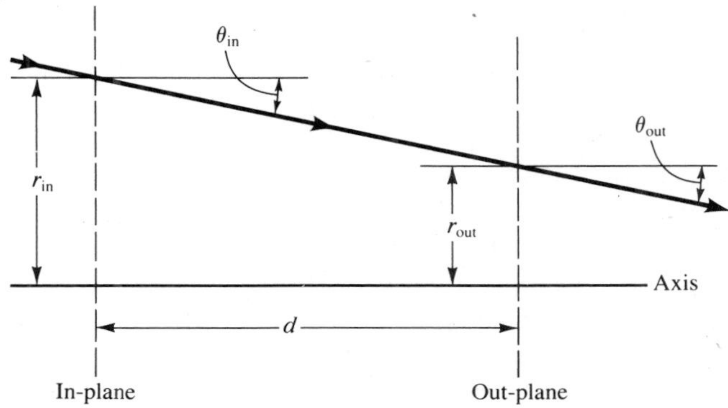

FIGURE 5.9

A ray propagating through a homogeneous medium

to characterize the Gaussian field distribution along the direction of propagation. Using Eq. (5.40), we obtain the q parameter at the system input, say at $z = z_1 \equiv z_{in}$, to be

$$q(z)|_{z=z_{in}} = z_1 + q(z)|_{z=0} \tag{5.55}$$

TABLE 5.1

Ray Matrices for a Few Optical Systems

(a) Thin lens $f > 0$ (biconvex); $f < 0$ (biconcave)	$\begin{pmatrix} 1 & 0 \\ -\frac{1}{f} & 1 \end{pmatrix}$
(b) Spherical interface $R =$ radius; $n_1 =$ input index; $n_2 =$ output index	$\begin{pmatrix} 1 & 0 \\ -\frac{n_2 - n_1}{n_2 R} & \frac{n_1}{n_2} \end{pmatrix}$
(c) Spherical mirror $R =$ radius	$\begin{pmatrix} 1 & 0 \\ -\frac{2}{R} & 1 \end{pmatrix}$
(d) A uniform medium of length d	$\begin{pmatrix} 1 & d \\ 0 & 1 \end{pmatrix}$

and, similarly, we find the q parameter at the system output, $z = z_1 + d \equiv z_{out}$, using Eq. (5.53) to be

$$q(z)|_{z=z_{out}} = z_1 + d + q(z)|_{z=0} = d + q(z)|_{z=z_{in}}$$
$$= \frac{Aq(z)|_{z=z_{in}} + B}{Cq(z)|_{z=z_{in}} + D} \tag{5.56}$$

Equation (5.56) is, incidentally, a unique relation, referred to as the ***ABCD* law**; it works amazingly well with all optical systems no matter how complex they are. The *ABCD* law allows us to determine the q parameter of an output Gaussian beam in terms of the q parameter of an input Gaussian beam and the elements of the 2×2 ray matrix of the optical system involved. For convenience, Eq. (5.54) is sometimes expressed alternatively as

$$\left.\frac{1}{q(z)}\right|_{z=z_{out}} = \frac{C + \left.\dfrac{D}{q(z)}\right|_{z=z_{in}}}{A + \left.\dfrac{B}{q(z)}\right|_{z=z_{in}}} \tag{5.57}$$

An optical resonator can be either stable or unstable. Most lasers, however, with the exception of a few high-power lasers, require a stable optical resonator. The resonator is said to be stable provided the complex q parameter of an incident Gaussian beam at any position remains unaltered after completion of one complete oscillation through the cavity. Using Eq. (5.57), the q parameter of a stable resonator at an arbitrary plane is thus given by

$$\frac{1}{q_r} = \frac{C + (D/q_r)}{A + (B/q_r)} \tag{5.58}$$

where A, B, C, and D are ray matrix elements for a complete round trip of that resonator. Equation (5.58) can now be solved for the steady-state complex parameter q_r as

$$\frac{1}{q_r} = \frac{D - A}{2B} \pm j\frac{\sqrt{4 - (A + D)^2}}{2B} \tag{5.59}$$

where $AD - BC = 1$.

Comparing Eqs. (5.54) and (5.59), we readily determine the radius of curvature and the spot size at any arbitrary resonator plane, respectively, as

$$R(z) = \frac{2B}{D - A} \tag{5.60a}$$

and

$$\omega(z) = \sqrt{\frac{2B\lambda}{\pi}}[4 - (A + D)^2]^{-1/4} \tag{5.60b}$$

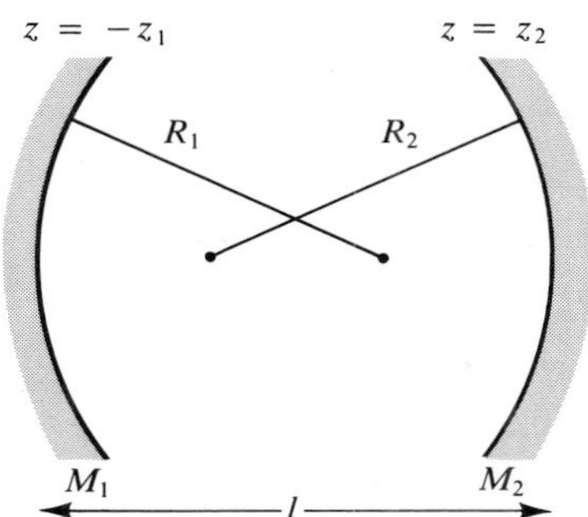

FIGURE 5.10

An optical feedback cavity

whereas the stability condition of the resonator is given by

$$4-(A+D)^2 \geqslant 0 \tag{5.61}$$

or

$$-1 \leqslant \frac{A+D}{2} \leqslant 1 \tag{5.62}$$

In order to relate Eq. (5.62) to the stability requirement of an optical feedback cavity, let us consider a generalized cavity as shown in Fig. 5.10. The two mirrors of radii of curvature R_1 and R_2 are allowed to face each other across a cavity of length l. The overall round trip ray matrix of the cavity is determined by multiplying the corresponding individual ray matrices. Accordingly we can begin from mirror M_1, for example, which gives us

$$\begin{aligned}\begin{pmatrix} A & B \\ C & D \end{pmatrix} &= \begin{pmatrix} 1 & 0 \\ -\dfrac{2}{R_1} & 1 \end{pmatrix}\begin{pmatrix} 1 & l \\ 0 & 1 \end{pmatrix}\begin{pmatrix} 1 & 0 \\ -\dfrac{2}{R_2} & 1 \end{pmatrix}\begin{pmatrix} 1 & l \\ 0 & 1 \end{pmatrix} \\ &= \begin{pmatrix} 1-\dfrac{2l}{R_2} & l+l\left(1-\dfrac{2l}{R_2}\right) \\ -\dfrac{2}{R_1}-\dfrac{2}{R_2}\left(1-\dfrac{2l}{R_1}\right) & \left(1-\dfrac{2l}{R_1}\right)\left(1-\dfrac{2l}{R_2}\right)-\dfrac{2l}{R_1} \end{pmatrix}\end{aligned} \tag{5.63}$$

We can then evaluate the stability requirement for the optical feedback oscillator using Eqs. (5.62) and (5.63) in terms of R_1, R_2, and l:

$$0 \leqslant g_1 g_2 \leqslant 1 \tag{5.64}$$

where

$$g_i = 1 - \frac{l}{R_i} \tag{5.65}$$

Figure 5.11 shows the generalized stability diagram in terms of g_1 and g_2. Note that the concentric ($R_1 = R_2 = l/2$), plane-parallel ($R_1 = R_2 = \infty$), and confocal ($R_1 = R_2 = l$) cavities are on the borderlines of the stability curves, suggesting that they may become very unstable should any of the system parameters malfunction. The concentric resonator, besides being sensitive to mirror misalignments, produces a very small spot size at the cavity center. Plane-parallel mirrors have a reasonable beam cross section, but they are also sensitive to mirror deformities. The line joining the concentric and plane-parallel resonator points corresponds to the class of symmetric resonators. The most useful of the stable resonators lie between Q and R, but closer to R. Beyond this stable region, the spot sizes become either infinite or imaginary. Very often the $g_1g_2 = 0.5$ condition is used to design a laser that is least sensitive to changes within the cavity.

Up to this point the optical cavity has been considered solely to support optical feedback. We shall find shortly that the cavity also plays an important role in determining the frequency characteristics of the laser. The resonating frequencies of the cavity are determined from the fact that they are supported by the resonator as long as the phase change encountered in one complete round trip is an integral multiple of 2π. Thus, in terms of transverse modes (m and n), the resonance condition for the cavity of Fig. 5.10 is given by

$$kl - (m + n + 1)\left[\tan^{-1}\left(-\frac{z_1}{z_0}\right) - \tan^{-1}\left(\frac{z_2}{z_0}\right)\right] = q\pi \tag{5.66}$$

using Eq. (5.47) where q is any integer. However, for the said cavity,

$$z_1 + z_2 = l \tag{5.67a}$$

$$R_1 = z_1\left[1 + \left(\frac{z_0}{z_1}\right)^2\right] \tag{5.67b}$$

$$R_2 = z_2\left[1 + \left(\frac{z_0}{z_2}\right)^2\right] \tag{5.67c}$$

Therefore, for the cavity of Fig. 5.10,

$$z_1 = \frac{g_2(1 - g_1)l}{g_1 + g_2 - 2g_1g_2} \tag{5.68a}$$

$$z_2 = \frac{g_1(1 - g_2)l}{g_1 + g_2 - 2g_1g_2} \tag{5.68b}$$

$$z_0 = \frac{l\sqrt{g_1g_2(1 - g_1g_2)}}{g_1 + g_2 - 2g_1g_2} \tag{5.68c}$$

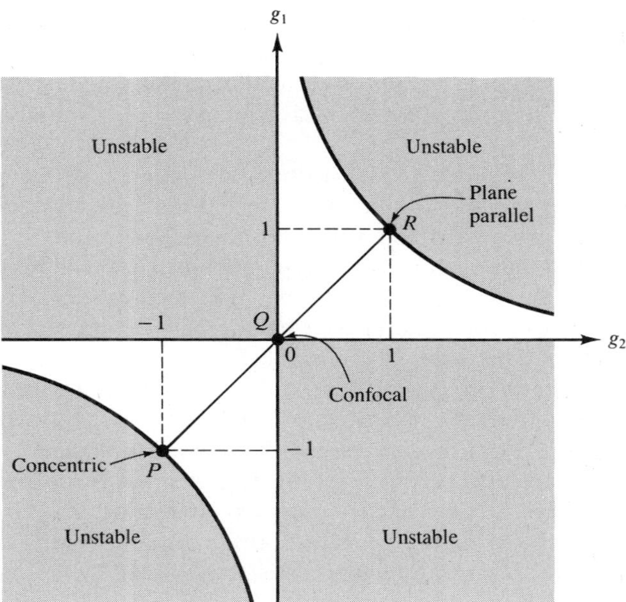

FIGURE 5.11

The stability diagram for optical oscillators

Since $k = 2\pi\nu/c$, Eq. (5.66) can also be expressed in terms of the resonant frequency using Eqs. (5.68a) through (5.68c) as

$$\nu_{mnq} = \frac{c}{2l}\left[q + \frac{m+n+1}{\pi}\cos^{-1}\sqrt{g_1 g_2}\right] \tag{5.69}$$

The longitudinal frequency spacing between the axial mode q is thus given by

$$\Delta\nu_{\text{long}} = \frac{c}{2l} \tag{5.70}$$

while that between the modes, due to a difference of one of the transverse modes (either m or n), is given by

$$\Delta\nu_{\text{trans}} = \frac{c}{2\pi l}\cos^{-1}\sqrt{g_1 g_2} \tag{5.71}$$

Equation (5.69) indicates that the transverse resonant frequencies depend on the sum $m + n$ and not on only m or only n. Thus the intermode frequency spacing, for example, in the case of a confocal resonator, is given by

$$\Delta\nu_{\text{trans,conf}} = \frac{c}{4l}\Delta(m+n) \tag{5.72}$$

Consequently, the transverse resonant frequencies in the case of a confocal resonator are positioned not only coincident to the axial resonant frequencies but also at the middle of the neighboring axial resonant frequencies. Figure 5.12 shows the corresponding frequency degeneracies of a confocal feedback resonator.

The domain of the stable optical feedback resonators is somewhat limited in the sense that the volume within the e^{-1} field contours is very small. For example, the spot size at the end mirrors of a confocal resonator is negligible when compared to that of either a plane-parallel or a concentric resonator (see Problem 5.14). Consequently, the effective gain medium that contributes to amplification is reasonably small. Lasers having such cavities therefore produce low-power outputs. This is why the unstable feedback cavities are important in producing high-power, high-gain lasers. In lasers having unstable cavities, comparatively less feedback is necessary to cause lasing. However, to sustain lasing, the single-pass gain has to exceed the single-pass loss.

Consider the unstable resonator of Fig. 5.13 where two convex mirrors face each other. A portion of the feedback energy is always lost in such a resonator. The energy that is lost can also be treated as the output of the system. Assume further that the two spherical wavefronts that are reflected by the mirrors appear to have originated from axial points z_1 and z_2, respectively. In fact, it can be argued that z_2 is the image of z_1 and vice versa. Therefore, we can use the object-image relationship of geometric optics to write

$$\frac{1}{L_1} - \frac{1}{L_2 + l} = 2l(g_1 - 1) \tag{5.73a}$$

and

$$\frac{1}{L_2} - \frac{1}{L_1 + l} = 2l(g_2 - 1) \tag{5.73b}$$

from which we get

$$L_1 = \frac{g_1 g_2 - g_2 \pm \sqrt{g_1 g_2 (g_1 g_2 - 1)}}{g_1 + g_2 - 2g_1 g_2} l \tag{5.74a}$$

$$L_2 = \frac{g_1 g_2 - g_1 \pm \sqrt{g_1 g_2 (g_1 g_2 - 1)}}{g_1 + g_2 - 2g_1 g_2} l \tag{5.74b}$$

Using simple geometric principles, the fraction of total power that survives each round trip can be obtained by multiplying the two reflected power fractions as

$$P_f = \pm \frac{\sqrt{g_1 g_2} - \sqrt{g_1 g_2 - 1}}{\sqrt{g_1 g_2} + \sqrt{g_1 g_2 - 1}} \tag{5.75}$$

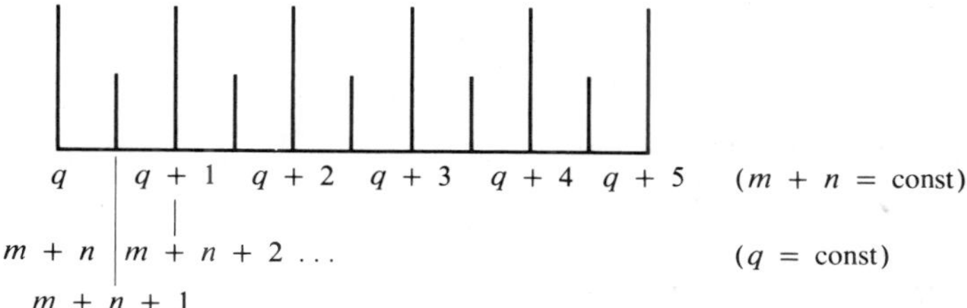

FIGURE 5.12

Frequency degeneracy in a confocal optical feedback cavity

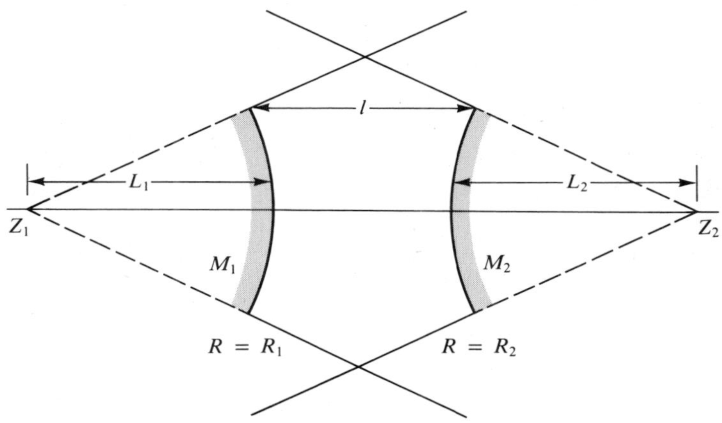

FIGURE 5.13

The parameters of an unstable optical feedback resonator

and, consequently, one can solve for $g_1 g_2$ in terms of P_f as

$$g_1 g_2 = \begin{cases} \dfrac{(1+P_f)^2}{4P_f} & \text{when } g_1 g_2 > 1 \\[2ex] -\dfrac{(1-P_f)^2}{4P_f} & \text{when } g_1 g_2 < 0 \end{cases} \tag{5.76}$$

Often the plot of Eq. (5.76) is superimposed on the stability diagram to characterize the equiloss contours.

An unstable optical feedback resonator satisfies the same resonance condition as that of Eq. (5.69). It is theoretically justifiable to have all kinds of transverse mode operations in an unstable cavity. In practice, however, all modes except the lowest order modes are avoided because the higher order modes involve excessive cavity losses. One such confocal annular laser cavity

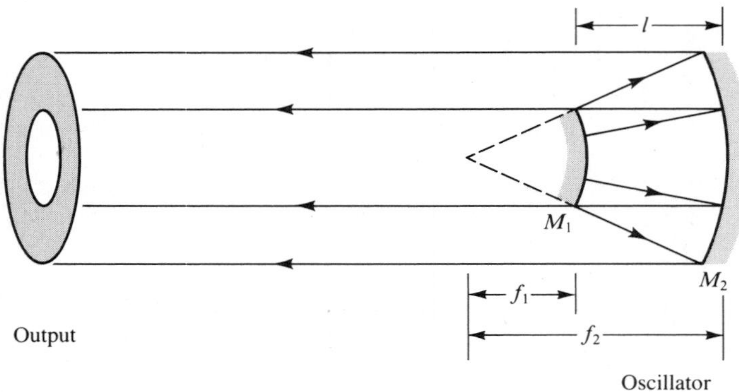

FIGURE 5.14

An annular laser with confocal cavity

and its corresponding output is shown in Fig. 5.14. Its laser mode behavior is quite similar to that of a Gaussian mode. At large distances, the annular irradiance distribution almost reduces to the familiar circular Gaussian irradiance distribution. However, a number of optical systems have already been designed to transform an annular beam to a uniformly distributed circular beam.

5.6 Q-Switching and Mode Locking

It is common to use Q-switching to produce high-powered lasers. Q-switching is a technique used to generate a short but very intense burst of light from the lasing material. Often the efficiency of a laser is characterized by its quality factor. The **quality factor** Q of a laser is inversely proportional to its cavity loss and is given by

$$Q = \frac{2\pi \nu E}{P_l} \tag{5.77}$$

where E is the energy stored in the resonator, P_l is the energy loss per unit time, and ν is the frequency of output light. Provided that the lifetime of a photon in the resonator is given by τ_c, Eq. (5.77) can be rewritten as a ratio of operating frequency and frequency bandwidth as

$$Q = \frac{2\pi \nu E}{E/\tau_c} = 2\pi \nu \tau_c \equiv \frac{\nu}{\Delta \nu} \tag{5.78}$$

where $\Delta \nu = (2\pi \tau_c)^{-1}$.

The resonator losses due to mirror reflectances can also be related to the Q

value. Assuming that the resonator of length L is made up of two end mirrors with reflectance values ρ_1 and ρ_2, respectively, the quality factor is given by

$$Q = \frac{2L}{c} \frac{2\pi v}{1 - \rho_1 \rho_2} \tag{5.79}$$

where the time necessary for a complete round trip is given by $2L/c$. It can be seen that Q values for lasers are very high, typically in the order of a million.

In the Q-switching mode, the laser resonator is subjected to a time-dependent loss. In the absence of lasing action, the gain due to population inversion can be made very large. In the initial phase of Q-switching, the Q is maintained low. When the population inversion has reached the desired peak value, the Q is switched to a high value, as a consequence of which all the accumulated energy is released together in a short period.

Consider a laser cavity in which a shutter is introduced in front of one mirror. With the shutter closed, if the active medium is pumped continually, population inversion keeps on growing even beyond the threshold value. When the shutter is subsequently opened, the energy stored in the cavity is released in the form of a short but intense pulse. If the shutter is opened in a much shorter time than the time required to build up the oscillation, the output consists of a single giant pulse. However, if the shutter is opened late, the output consists of a series of pulses, each having shorter peaks.

One of the oldest but most popular techniques involves rotating one of the laser mirrors about an axis normal to the resonator axis. For a typical laser resonator 30 to 40 cm in length, a rotational speed on the order of 25,000 revolutions per minute is needed to ensure that the laser output contains a giant pulse. When the mirrors are not parallel, the resonator losses are large. Under suitable conditions when the mirrors are just parallel, a giant pulse results at the output. An angular filter (where the transmission is a function of angle) is sometimes used in conjunction with a rotating prism or mirror to obtain effective Q-switching at lower mirror rotation speeds. In a subsequent chapter, we show that it is also possible to realize nonmechanical Q-switching using electro-optical effects. For example, an acousto-optic modulator can be used to generate laser pulses in the order of 10^{-6} s while Pockels and Kerr cells can be used to generate pulses on the order of 10^{-9} s.

In the previous section (see Eqs. 5.70 and 5.71) we saw that lasers are capable of operating in a variety of modes, both longitudinal and transverse. When a laser simultaneously operates in several such modes, the individual modes tend to oscillate independently, each with its own random phase. The same relative phase δ can be maintained between the modes by locking modes. The mode-locking technique can be used to generate pico-second laser pulses, each accompanied with high power. In an unlocked multimode laser, the total electric field is given by

$$E(t) = \sum_n E_{0n} e^{j[(\omega_0 + n\omega_c)t + \delta_n]} \tag{5.80}$$

where E_{0n}, $\omega_0 + n\omega_c$, and δ_n are, respectively, the amplitude, angular frequency, and phase of the nth mode, ω_0 ($\equiv q\pi c/L$) is the reference angular frequency, and q is any integer. Note that $E(t)$ is periodic in $T \equiv 2\pi/\omega_c = 2L/c$ since

$$
\begin{aligned}
E(t) &= \sum_i E_{0i} e^{[j\{(\omega_0 + n\omega_c)(t+T) + \delta_i\}]} \\
&= \sum_i E_{0i} e^{j[(\omega_0 + n\omega_c)t + (2\pi\omega_0/\omega_c) + 2\pi n + \delta_i]} \\
&= \sum_i E_{0i} e^{j[(\omega_0 + n\omega_c)t + \delta_i]} e^{j2\pi[\omega_0/\omega_c + n]} \\
&= E(t)
\end{aligned}
\tag{5.81}
$$

where ω_0/ω_c is an integer.

The most useful forms of mode locking are achieved when each of the δ_n's is zero. When such a condition is satisfied, the output power of the mode-locked laser is given by

$$
\begin{aligned}
P(t) &\propto \left| \sum_{n=-N}^{N} e^{j(\omega_0 + n\omega_c)t} \right|^2 \\
&\propto \frac{\sin^2\left(\dfrac{2N+1}{2}\omega_c t\right)}{\sin^2\left(\dfrac{\omega_c t}{2}\right)}
\end{aligned}
\tag{5.82}
$$

where, for simplicity, E_{0n} is assumed to be unity. Equation (5.82) shows that the resulting power train has peaks with a maximum value of $(2N+1)^2$, each with a pulse width of approximately $T/(2N+1)$. These peaks are distributed at periodic intervals of $2\pi/\omega_c$. Note also that the peak power is $2N+1$ times larger than the average power.

Mode-locking is achieved in lasers by several means. One of the more common ways to achieve passive mode-locking is to use a nonlinear absorber, such as an organic dye, in the cavity. The absorber absorbs all radiation in the cavity except when all modes are in perfect synchronization, in both position and phase.

5.7 Specific Lasers

Lasers are classified in a variety of ways. Sometimes they are classified according to the characteristics of their output as either continuous-wave or pulsed. Sometimes they are classified according to the characteristics of their lasing material—gas, solid, or liquid. Sometimes they are classified according to the characteristics of the pumping process as flashlamp-driven or current-driven. And finally they are sometimes classified according to their output wavelength—visible, infrared, or ultraviolet. There are hundreds of different types of laser sources available in the market today, each offering a wide variety of operating characteristics. Power level, wavelength, pulsewidth, tunability, coherence, and cost are just a few of the performance parameters considered in selecting the best laser for a given electro-optic application. Table 5.2 lists samples of the more common lasers. We do not discuss the design of each of these because that is beyond the purpose of this text. However, it may be worthwhile to discuss a few. For the purpose of discussion, we concentrate on only six of them. The first one to be covered is the He–Ne laser, which is possibly the most widely used coherent source in numerous electro-optic applications.

TABLE 5.2

A Few of the Common Lasers

Type	*λ(nm)*	*Power (W)*	*Specifications**	*Output Type*
Dye	500–900 (tunable)	0.02–2.0	BDa: 0.5; BDv: 1–1.5	CW; pumped by argon laser
	190–500 (tunable)	2M–15M	RR: 1–20; PL: 2–8 LW: 0.002	Pulsed; pumped by Nd-YAG
	200–980 (tunable)	0.1M–6M	RR: 0.1–50; PL: 200–700 LW: 0.4–0.001	Pulsed; pumped by coax. flash
Argon	488–514	0.002–25	BDa: 0.6–1.5; BDv: 0.9–1.1	CW
CO_2	10600	1–500K	BDa: 1.2–70; BDv: 1–10;	CW
		0.01–1200	RR: 1–8000; PL: 0.04–5000; BDa: 6–200; BDv: 1–50	Pulsed
He–Cd	325–442	0.001–0.05	BDa: 0.187–1.5; BDv: 0.3–3.1	CW
He–Ne	632.8	0.0003–25	BDa: 0.34–1.5; BDv: 0.9–4,1	CW
	1153	0.007–0.015	BDa: 1.6–2.3; BDv: 0.6–1	CW
ArF	193	0.03–35	RR: ≤250; PL: 0.0025–0.07; BDa: 110–30 × 30; BDv: 2 × 3–2 × 6	Pulsed
GaAs	750–900	0.001–0.2	Diverges rapidly BDv: 10 × 30–600 × 300	CW
	780–905	0.1–1150	RR: 5–50000;PL: 0.1–200; LW: 3.5–15; BDv: 10 × 30–400 × 280	Pulsed (available also in stack)
Nd–YAG	10640	0.5–50	BDa: 0.1–6; BDv: 1.5–15;	CW
solid	10600	0.0007–50	RR: 1–2500; PL: 0.01–20000; BDa: 3–10; BDv: 0.06–10	Pulsed
Ruby	694	0.02–20	RR: 0.033–120; PL: 0.3–2000; BDa: 2–19; BDv: 0.3–10	Pulsed

*BDa ≡ Beam diameter (in mm); BDv ≡ Beam divergence (in mrad); RR ≡ Rep. rate (in Hz); PL ≡ Pulse length (in ns); and LW ≡ Line width (in nm). BDa and BDv are measured at full angle, $1/e^2$ points, except for GaAs, which is measured at half energy half angle.

This is followed by discussions on the argon ion laser, the carbon dioxide laser, the neodymium laser, semiconductor lasers, and free electron lasers. Semiconductor lasers are cheap, efficient, and small in size, and consequently they are continually gaining prominence in optical communications applications.

5.7.1 Helium-Neon Laser

The helium-neon (He–Ne) laser is an electrically pumped low gain, four-level CW laser that is most famous for its TEM_{00} red emission at a wavelength of 632.8 nm. The available power level ranges from about 0.3 mW to 25 mW. The lasing medium of this laser consists of an He–Ne gas mixture with neon at a pressure of 0.1 torr and helium at a pressure of 1.0 torr. The gaseous mixture is enclosed in a long but narrow glass cavity having a set of two end mirrors (one completely reflecting and the other partially reflecting). The low gain nature of He–Ne laser requires that 99% or more of the generated light be fed back into the cavity to sustain the resonating process.

Figure 5.15 shows the energy level diagram of the He–Ne lasing gas. When an electrical discharge (using a dc voltage of 2 to 4 kV at about 50 mA) is introduced to the gaseous mixture, the accelerated electrons begin to collide with helium atoms, exciting them to higher energy states. These excited helium energy states are basically metastable. Coincidently, neon atoms have excited energy states approximately at the same levels as those of the helium atoms. Thus, when the excited helium atoms collide with the ground-level neon atoms, an energy exchange takes place, resulting in an excitation of neon atoms (to both E_4 and E_6) and a de-excitation of helium atoms (to the ground state). This energy exchange given by

$$\mathrm{He}^* + \mathrm{Ne} \rightarrow \mathrm{He} + \mathrm{Ne}^* \tag{5.83}$$

becomes highly probable because the excited helium energy states are metastable. Consequently, the population inversion is achieved between E_6 and E_5 (or E_3) and also between E_4 and E_3. These population inversions result in three unique emissions at 3390 nm (infrared), 632.8 nm (red), and 1150 nm (infrared), respectively. In reality, however, more than 150 different output transitions are possible, even though only three are dominant. Ideally it is possible to select any one of these wavelengths by choosing the characteristics of the resonator mirrors.

The neon atoms undergo spontaneous transition from the E_3 state to the E_2 state and then reach the E_1 state by colliding with the walls of the narrow tube. The probability of the final de-excitation of neon atoms from the E_2 state to the E_1 state decreases with increasing tube radius. Thus, for a tube with larger radius, it becomes more difficult to achieve population inversion. The laser gain is inversely proportional to the tube radius.

The most dominant neon emission (of wavelength 632.8 nm) has a spectral width of about 1500 MHz. Again, the length of the most common He–Ne laser resonator allows for either one or two modes to coexist. It is important to note

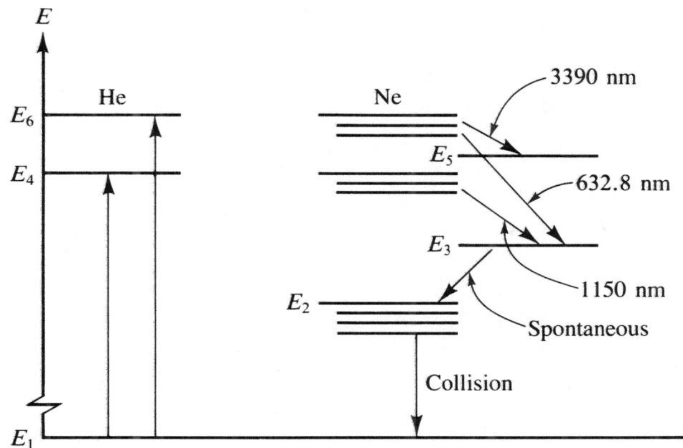

FIGURE 5.15

Energy level diagram for an He–Ne laser

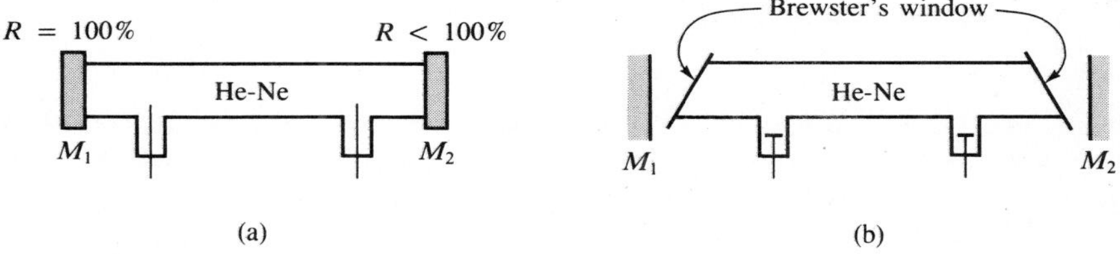

FIGURE 5.16

He–Ne laser with mirrors: (a) inside the tube and (b) outside the tube

that the thermal energy often changes the effective length of the cavity, which limits the stability of laser output to about 1%.

Figure 5.16 shows the two possible ways of designing the cavity: either by enclosing the mirrors within the tube or by putting the mirrors outside the tube. In the latter case, the tube ends are polished at Brewster's angle to guarantee pure polarized transmission. Internal mirrors are often susceptible to damage due to discharge in the cavity. The use of external mirrors provides flexibility, but it limits the laser modes to having only rectangular symmetry rather than having rotational symmetry.

5.7.2 Argon Ion Laser

The ion lasers use an ionized gas as the active medium. Basically they are four-level systems, and it is the excited states of the ions that are involved in the lasing. In the first step, collision provides the necessary excitation. Because of the

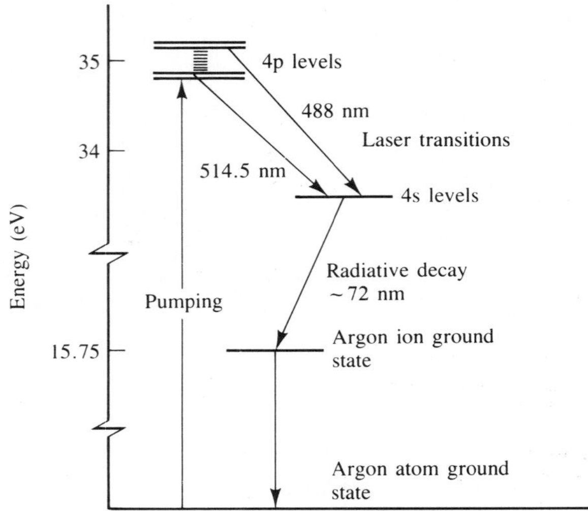

FIGURE 5.17

Energy levels for an argon ion laser

energy necessary to ionize an atom and then to excite it, the efficiencies of ion lasers are extremely low. Despite this fact, these lasers have a very high gain and can provide CW output. The common ion lasers are argon, krypton, and xenon. The argon laser has two strong outputs at 514.5 nm and 488 nm with powers in the range of 5 mW to 25 W. On the other hand, Kr (Xe) lasers provide output at 647.1 nm (995 to 539.5 nm) in the range of 5 mW to 6 W (200 W pulsed).

In the pumping process, first the neutral argon atom is ionized, which requires an energy of approximately 15.75 eV. Then the argon ion is further excited to a level approximately 19.68 eV above the argon ion ground state. Figure 5.17 shows the energy levels for the argon ion laser with most transition taking place between 454 and 529 nm, whereas the most intense ones lie at 488 and 514.5 nm. All lasing takes place between the 4p and 4s levels of the argon ion. The current densities required in argon ion lasers are about 1000 A/cm^2, which limits the tube bore diameter to a few millimeters. The ions drift toward the cathode and so a return path outside the discharge tube needs to be provided, usually by means of a tube made from a refractory material such as beryllium oxide. The use of Brewster windows and the use of external cavity mirrors help to protect mirror surfaces from damage. The efficiency is often increased by applying a magnetic field along the tube axis via a solenoid.

5.7.3 Carbon Dioxide Laser

The carbon dioxide (CO_2) laser is a molecular laser, and it is by far the most important and most efficient one of all. In a CO_2 laser, molecular vibrations,

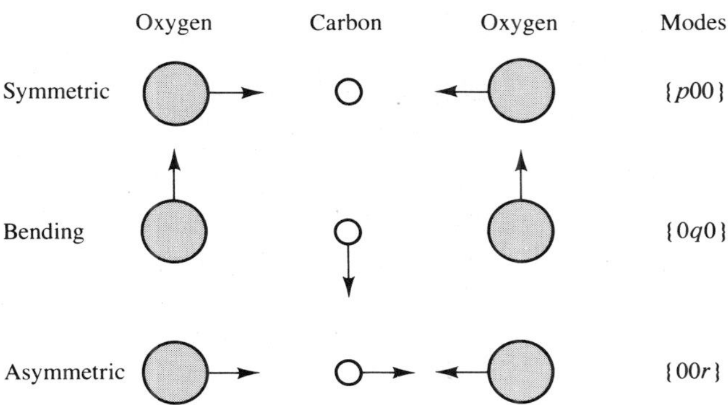

FIGURE 5.18

Vibrational modes of triatomic molecules

rather than electronic transitions, provide the mechanism for the lasing action. The CO_2 laser efficiency can go up to 30% while its power is reasonably high. It can easily provide several tens of watts and, by using a master oscillator power amplifier chain, one can generate terrawatts of pulsed power. Obviously, its applications are many; for example, welding and cutting of steel, weaponry, pattern cutting, and laser fusion, to name a few.

A CO_2 molecule involves quantization of the vibrational and rotational energy. The linear triatomic molecule has three distinct modes of vibration: symmetric, bending, and asymmetric, as shown in Fig. 5.18. The modes corresponding to these vibrations occur at 4.2×10^{13} Hz, 2.0×10^{13} Hz, and 7.0×10^{13} Hz, respectively. The bending mode, however, is degenerate in that it can occur in either of two perpendicular planes. One can refer to the particular state of the molecule by specifying a set of three vibrational quantum numbers (p, q, r), respectively, to designate symmetric, bending, and asymmetric quanta. Next, most of the vibrational levels are further split into rotational sublevels. However, rotational energy quanta are much smaller than the vibrational quanta and thus the rotational sublevels are very closely spaced.

Pure CO_2 can lase but very weakly; however, in combination with nitrogen (N_2) and helium (He), it can generate a larger power output. Helium transports waste heat to the tube walls by assisting in the deactivation of the lower laser levels. However, its primary purpose is to control the electron "temperature" of the discharge, which in turn regulates the relative division of the electrical power to the various processes. Often 10% oxygen is mixed with N_2 to improve electrode and optics life without affecting the power output. Note that there can be more than 200 vibrational-rotational transitions in the 8 to 18 μm wavelength range. The pertinent energy levels of CO_2, N_2, and He are shown in Fig. 5.19. Note that the first vibrational level of N_2 coincides very closely with the upper laser level $\{001\}$. The vibrational levels of N_2 are metastable and, thus, they can efficiently transfer energy selectively to the upper laser level.

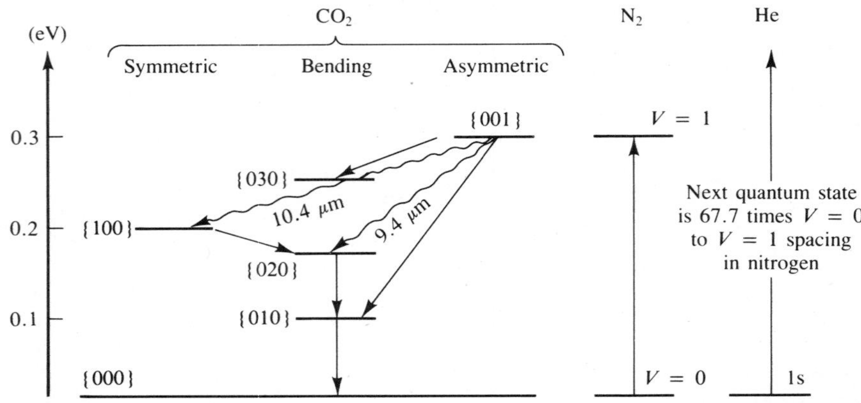

FIGURE 5.19

Energy levels of a CO_2 laser

Lasing occurs between {001} and {100} states at 10.6 μm and between {001} and {020} states at 9.6 μm. The probability for the latter transition is only about 5% of the 10.6 μm transition and, thus, it does not occur normally unless the 10.6 μm transition is deliberately eliminated. The lifetime of {001} of CO_2 in the presence of collision effects is on the order of milliseconds, whereas those of the lower states are on the order of microseconds. Selection rules require that the rotational state either increase or decrease by only one quantum on a transition from one vibrational state to another. That is when J is the rotation quantum number, $\Delta J = \pm 1$. In fact, a 10.2 μm transition occurs because of $\Delta J = 1$, whereas a 10.6 μm transition occurs because of $\Delta J = -1$.

The first excited N_2 state lies only 0.3 eV above the ground state, whereas the first excited state of helium (activated during the excitation of He in a He–Ne laser) lies approximately 20 eV above the ground state! This contrast explains why the CO_2 laser is so much more efficient than the He–Ne laser.

Electrons that are excited by electrical power transfer their power to the neutral gas atoms by collisions. A part of the energy is given up as the electrons elastically collide with the massive neutral atoms, raising the kinetic temperature of the gas. On the other hand, nitrogen atoms are raised from the ground state ($V = 0$) to higher vibrational states ($0 < V \leqslant 8$). Again, the excited nitrogen atoms, along with CO_2 at the (000) state, result in de-excited nitrogen that undergoes a transition from V to $V - 1$, while at the same time CO_2 moves from (000) to (001). In practice, about 60% of the electrical power contributes to pumping the upper laser level.

CO_2 lasers are usually equipped with systems, as shown in Fig. 5.20, to allow for a fresh supply of gas in the active region. The gas is often recirculated through a heat exchanger while a small amount of gas is continually removed and replaced to keep the gas mixture free of decomposition products such as

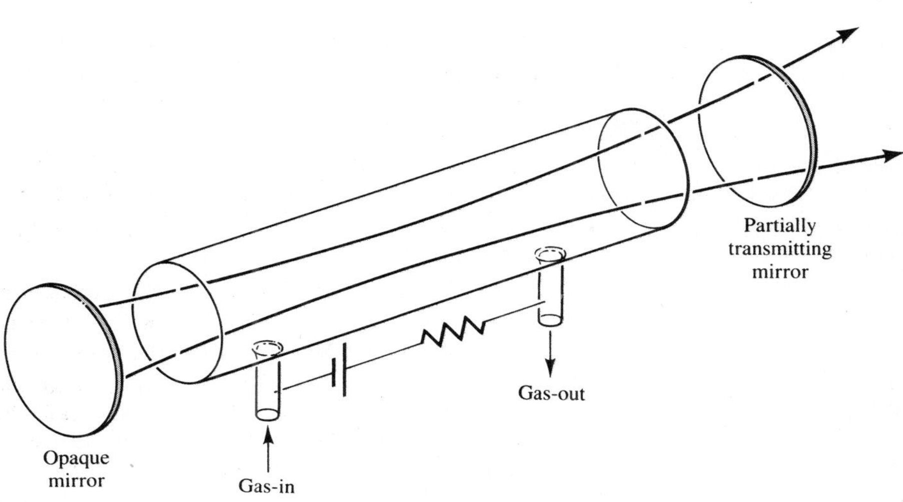

FIGURE 5.20

Axial discharge and axial flow in the CO_2 laser

carbon monoxide. The reaction CO_2-to-CO conversion is rapid. One way to overcome this is to add hydrogen or water vapor to the mixture to regenerate CO_2. Note that carbon monoxide absorbs strongly at 10.6 μm and would thus quench the lasing action if it were not removed in a timely fashion. In addition, CO is harmful for the optics as well as for the electrodes.

5.7.4 Neodymium Laser

The neodymium (Nd) laser is an example of a four-level laser where a solid material made by doping a rare earth into a variety of host materials is used as a lasant. The most common host materials are amorphous glass and crystalline yttrium aluminum garnate ($Y_3Al_5O_{12}$), or YAG. In such a laser, the lower laser level lies far above the ground state and is, thus, generally unpopulated so that the normalized population inversion is just equal to the number of atoms at the higher laser level.

In both Nd-glass and Nd-YAG lasers, the active atoms participate as if they were triply ionized in the form of Nd^{3+}. The Nd^{3+} ion surrounds itself with several oxygen atoms and thereby shields itself from its surroundings. It is no wonder that both Nd-glass and Nd-YAG lasers lase at 1.06 μm. In the former case, however, the dopant goes into the amorphous glass at random sides, and thus each dopant ion settles in a different environment. But in the latter case, Nd^{3+} substitutes for Y^{3+} in the YAG crystal, and thus each Nd^{3+} settles in almost identical environments.

The host material glass can be made very uniform and polished to unbelievable accuracy. However, it is true that because of the amorphous character of glass, its absorption bands are much broader than those in YAG. But on the other hand, glass permits high doping densities (up to 5%) and, therefore, glass-based amplifiers have a high figure of merit. The large bandwidth results in numerous simultaneously operated axial modes. The Q-switched glass laser is thus induced to operate in a mode-locked fashion, which puts all these modes in step, both in time and space, thereby resulting in a large bandwidth ($\simeq 10^{12}$ Hz), i.e., a narrow pulse width ($\simeq 10^{-12}$ s). But glass has a low thermal conductivity (about 1/40th that of YAG) and, thus, poses a problem for removing waste heat. Accordingly, Nd-glass lasers are limited to low duty cycle operations ($\simeq 1$ pulse/s). It is possible to make laser rods up to 1 meter long and several centimeters in diameter. The energy per pulse can be as high as 5000 J or more. One of the factors that limits power arises from the nonuniform heating of the rod. Temperature differences within the rod lead to variations in refractive indices, which in turn may cause beam distortion.

Nd-YAG lasers are capable of an average power output of the order of kilowatts. Since YAG has a relatively high thermal conductivity, the unwanted energy can be removed by conduction. With proper cooling, both CW and high repetition pulsed operation are possible. With acousto-optic modulation (see Chapter 7), Q-switched Nd-YAG lasers can provide pulse rates of up to many thousand pulses per second (pps). However, power begins to fall above 2000 pps. A single pulse may be isolated and amplified. Pulses with peak power in excess of $\simeq 10^6$ MW are possible. In CW operation, quartz-halogen lamps are used for pumping. The most effective pumping band lies between 700 nm and 900 nm. CW outputs of up to kilowatts are possible, but such outputs necessitate water cooling of the laser rod. By comparison, YAG crystals are quite expensive and difficult to grow in large sizes.

5.7.5 Semiconductor Laser

We explored the generation of incoherent light using LEDs in an earlier chapter. The emission from the LEDs results basically from the spontaneous recombination of electrons and holes that are injected across the semiconductor junction. The physics of semiconductor lasers is not too different in principle from that of the LEDs. Basically, the *p-n* junction serves as the lasing medium of a semiconductor laser. In this section, we concentrate on the mechanisms for obtaining population inversion in a region where there are ample numbers of injected electrons and holes. There exist about 25 known compound semiconductor diodes that lase over a whole range of wavelengths. The most common of these are the GaAs and $Al_xGa_{1-x}As$ lasers. Semiconductor lasers are particularly noteworthy for their remarkably small dimensions ($\simeq 0.001$ mm^3), high efficiency, and excellent controllability by means of junction current. The power output of a semiconductor laser is high enough to be considered suitable for the signal source of a fiber-optic system. The semiconductor laser feedback cavity is basically formed by the cleaved ends of

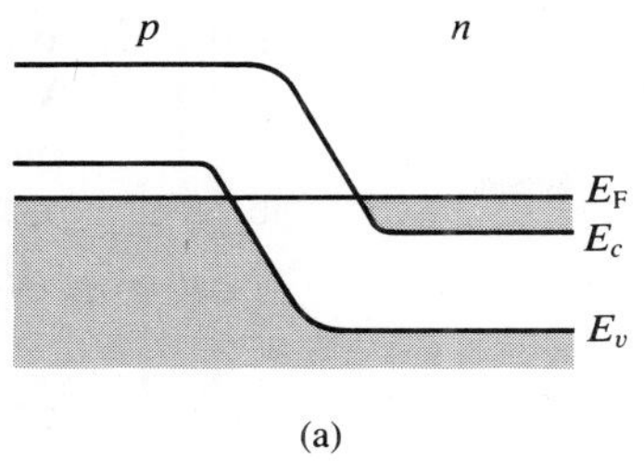

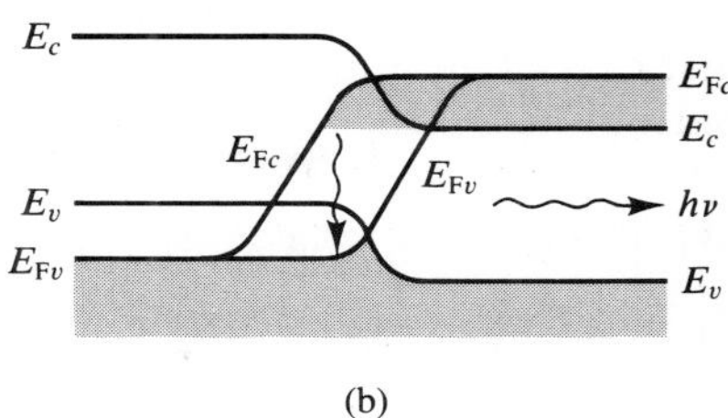

FIGURE 5.21

Semiconductor laser energy levels: (a) at no bias and (b) under forward bias

the integrated chip. In practice, a gold substrate is used as the metal contacts on both anode and cathode.

The *p*- and *n*-sides of a diode laser are both degenerate. This implies that both of the bulk regions are doped heavily, as a result of which the Fermi level lies well within the conduction and valence bands as shown in Fig. 5.21(a). In the absence of an applied voltage, the Fermi energy E_F remains invariant. With a forward bias exceeding E_g/e, however, electrons from the degenerate *n*-side are injected into the *p*-side, while holes are injected from the *p*-side to the *n*-side. The injected carriers, while diffusing through an average distance equal to diffusion length, recombine with the majority carriers. Thus, over a significant region at the junction (called the active region), the Fermi levels are separated by an amount more than the band-gap energy, as shown in Fig 5.21(b). Notably this narrow region is doubly degenerate. The carrier concentrations in this region are given by

$$n = n_i e^{(E_{Fc} - E_i)/kT} \tag{5.84a}$$

$$p = n_i e^{(E_i - E_{Fv})/kT} \tag{5.84b}$$

where E_{Fc} and E_{Fv} are the quasi-Fermi energy levels. The amplification condition, therefore, is given by

$$E_{Fc} - E_{Fv} > h\nu > E_g \tag{5.85}$$

where ν is the frequency of the stimulated emission.

The semiconductor region having an inverted population is generally confined within a thickness *t* on the order of 1 μm. However, the electromagnetic mode is spread out a little bit more beyond this thickness and is confined within a thickness *d* shown in Fig. 5.22 which is on the order of 5 μm. It is obvious that some of the active region recombination will be spontaneous in nature; but with a large enough injection, the stimulated emission exceeds the amount of spontaneous emission.

No external mirrors are necessary, in practice, for sustaining oscillation in the semiconductor laser. The high refractive index of the active region (that

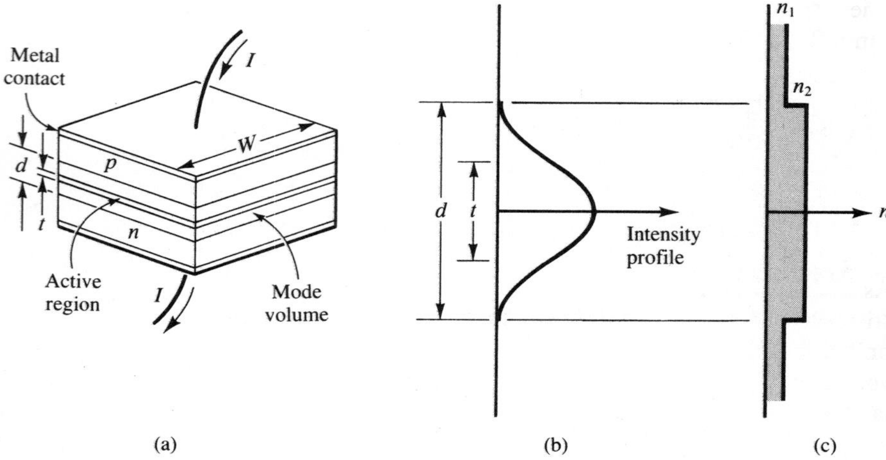

FIGURE 5.22

Semiconductor laser: (a) geometry, (b) intensity profile, and (c) refractive index profile

amounts to a reflectance of about 0.3 at the semiconductor-air interface) ensures that the active region functions as an optical feedback cavity. The diode is cleaved along the crystal planes normal to the junction plane, which forms a plane-parallel cavity. In such a cavity, the main losses are caused by carrier absorption and by scattering that takes place at the optical imperfections of the semiconductor. The *laser threshold* is defined as *the condition for which the round trip gain just equals the total losses incurred over the same distance in accordance to Eq. (5.13).*

A semiconductor laser is different from all other lasers in that the mode volume lWd extends beyond the active medium. Consequently, using Eq. (5.18), the associated gain per unit length is given by

$$\beta = \frac{c^2}{8\pi\nu^2 n^2 \tau_{21,\mathrm{sp}}}\frac{t}{d}\left(N_2 - \frac{g_2}{g_1}N_1\right)g(\nu) \tag{5.86}$$

where the effective inverted population is reduced by a factor t/d as long as $t < d$ and N_2 and N_1, respectively, are the total number of inverted electrons per unit volume at E_2 and E_1. When $d > t$, the gain constant β is inversely proportional to d. But if $d < t$, the factor (t/d) in Eq. (5.86) is replaced by unity.

N_1 can be assumed to be zero in the active region. Since in equilibrium the number of electrons injected within an interval is equal to that of spontaneous recombinations during the same interval, the threshold current density is given by

$$J_{\mathrm{th}} = \frac{(N_2)_{\mathrm{th}}}{\eta}\frac{et}{\tau_{21,\mathrm{sp}}} \tag{5.87}$$

where η is the internal quantum efficiency. Using Eq. (5.86) and the threshold condition of Eq. (5.13), the current density can be written as

$$J_{th} = \frac{8\beta_{th} ed\pi v^2 n^2}{c^2 \eta g(v)}$$

$$= \frac{8\pi edv^2 n^2}{c^2 \eta g(v)}\left[\alpha - \frac{\ln(R_1 R_2)}{2L}\right] \quad (5.88)$$

As the injection current begins to exceed the threshold value, it gives rise to a stimulated emission which in turn depletes the population inversion. In practice, only a fraction of the total generated power appears at the output. For a bias voltage of V_A, the total power input is $JWlV_A$. Consequently, the efficiency of the laser diode is given by

$$\text{efficiency} = \frac{\eta hv}{eV_A}\left(\frac{J - J_{th}}{J}\right)\left[\frac{\frac{\ln(R_1 R_2)}{2L}}{\frac{\ln(R_1 R_2)}{2L} - \alpha}\right] \quad (5.89)$$

For most diode lasers, the efficiency is very high as it approaches η.

The semiconductor laser described thus far is also known as a **homojunction laser** because there is only one junction in a single semiconductor material. In general, such a laser requires a larger amount of current density because of the relatively large values of d and α. This problem is often eliminated by making use of heterojunctions where, for example, an ultra-thin ($\simeq 0.5$ μm) layer of GaAs is sandwiched between a p-type $Ga_{1-x}Al_xAs$ layer and an n-type $Ga_{1-x}Al_xAs$ layer as shown in Fig. 5.23. The wider band gap of $Ga_{1-x}Al_xAs$ effectively confines the injected carriers within the GaAs region. Basically the heterojunction structure introduces a barrier before the electrons that are injected into the p-region at the left junction. Similarly a barrier is introduced before the holes that are injected into the n-region at the right junction. In addition, the lower refractive index value of the $Ga_{1-x}Al_xAs$ regions confines the oscillating light within the active region. Note that the change in the refractive index Δn, relative to GaAs, is approximately $-0.4x$ where x refers to the concentration fraction of aluminium. The use of heterojunctions reduces the value of both d and α, which in turn reduces the value of the current density much below that required in the homojunction lasers. Consequently, at room temperature, the heterojunction laser efficiency may increase to about 5%.

A diode laser is particularly noteworthy because of the great speed with which it can be turned on and off and because of the compatibility that it provides for interfacing it with semiconductor electronics. In addition, the diode lasers have a narrow bandwidth and, consequently, they are superb as transmitters in the fiber-optic communication systems.

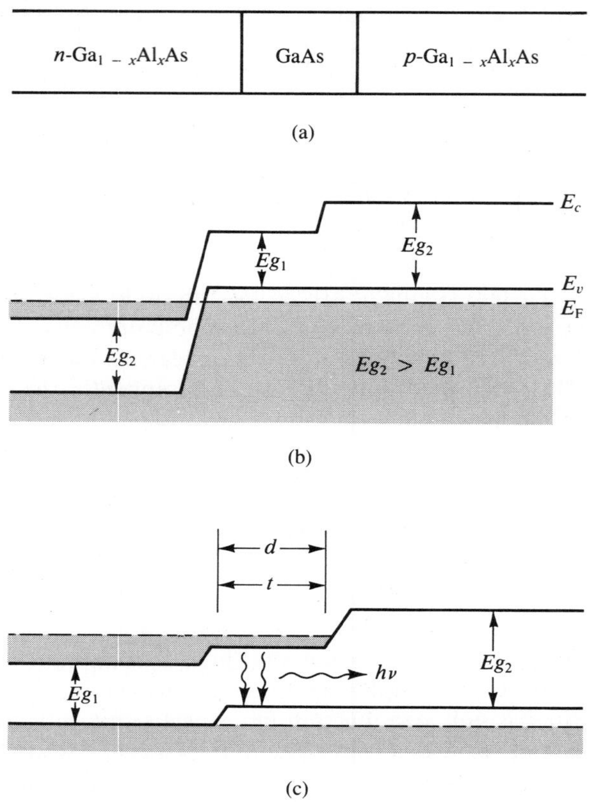

FIGURE 5.23

A double heterojunction: (a) schematic and energy level diagrams, (b) at equilibrium, and (c) under forward bias

5.7.6 Free Electron Laser

The free electron laser (FEL) is the newest class of oscillators that generates power output with tunability over a wide range from extreme ultraviolet to millimeter wavelengths. The operating principles of FEL are remarkably different from all other lasers considered so far in that they use beams of relativistic (traveling at a speed at or near the speed of light) electrons as their basic source of energy. You may recall from Chapter 1 that accelerating electrons do emit radiation. The accelerating electrons interact with their own fields, losing a part of their kinetic energy. Under certain conditions, however, these electrons can be made to give up their energy to a beam of photons traveling along the same direction.

The radiation given up by the accelerated charge carriers does not have an

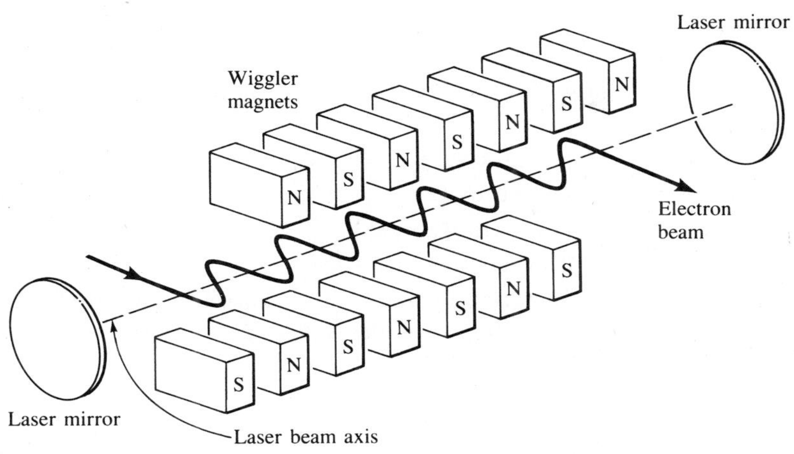

FIGURE 5.24

The free electron laser

isotropic distribution (like that of a spontaneous emission), but it is heavily weighted in the direction of the motion of electrons. Again, since the electric field of the optical beam is at right angles to its direction of propagation, the charge carriers are not able to transfer their energy to the photons because the charge carriers do not have a component of velocity perpendicular to the direction of photon flow. Thus, to achieve effective energy transfer between the charge carriers and photons, electrons are made to travel through a periodic array of permanent magnets with alternate polarities perpendicular to the direction of propagation of light. This arrangement of magnets is referred to as a **wiggler** and is shown in Fig. 5.24. The electrons passing through the wiggler field undergo transverse oscillations, as a result of which they spontaneously emit radiation, provided there is a proper feedback that is a maximum at a wavelength λ given by $\frac{1}{2}\Lambda/\gamma^2$ where Λ is the wiggler period and $m_0c^2(\gamma - 1)$ is the kinetic energy of the electrons. The parameter γ is the ratio of the total electron energy to its rest energy. One can thus control the wavelength by changing the kinetic energy of the electrons as well as by changing the wiggler period. The bandwidth of the FEL radiation is often given by $\frac{1}{2}\lambda/N$, where N is the number of wiggler periods.

Often it is not easy to transfer energy from accelerated charge carriers to photons. To overcome that hurdle, one may inject the charge carriers at an energy slightly higher than the resonant energy. The electrons then tend to return to the resonant condition, but this allows for transfer of a small amount of energy. Alternative methods may involve changing the wiggler period so that the final resonant electron energy is far less than that at the start. However, it must be borne in mind that the FEL is still under development and, therefore, general operating principles and systems are yet to be optimized.

Problems

5.1 Derive Eq. (5.38) from Eq. (5.36).

5.2 Verify Eq. (5.46).

5.3 Show that the total TEM_{00} power passing through free space at an arbitrary plane z is given by $(E_{00}^2 z_0 \lambda/2)(\varepsilon_0/\mu_0)^{1/2}$.

5.4 Show that the Gaussian wavefront at $z \gg (x^2 + y^2)^{1/2}$ is similar to that of a spherical wavefront originating at $z = 0$.

5.5 Plot the radius of curvature for the Gausian wavefront as a function of z and determine the location and the value of the minimum radius of curvature.

5.6 Derive Eqs. (5.68a–c) from Eqs. (5.67a–c).

5.7 Derive Eq. (5.69) from Eqs. (5.68a–c).

5.8 Determine the frequency degeneracy of an optical resonator for which $R_1 = R_2 \gg l$.

5.9 Verify Eqs. (5.74a–b).

5.10 Verify Eqs. (5.75) and 5.76).

5.11 Show the equiloss parameters in a g_1 versus g_2 diagram where the power loss per pass is 20%. Also identify the types of unstable cavities for each of the equiloss contours.

5.12 Show that for the annular laser, the fraction of the power emitted is given by $1 - (r_1/r_2)^2$ where r_i is the radius of curvature of the confocal mirrors.

5.13 Show that the cavity mode volume $V_{mn} = \pi\omega_0^2 l(m!n!)2^{m+n}$ where $V_{mn} = [\int\int\int |E_{mn}|^2 dx\, dy\, dz]/E_p^2$ and E_p is the peak value of the electric field within the cavity.

5.14 (a) Plot the ratio of the beam spot size at the mirrors and also at the center of a symmetrical resonator to its confocal value against l/R. (b) Plot the spot sizes (ω^2) at the end mirrors of a hemispherical resonator against l/R. Discuss the consequences of the plots obtained.

5.15 Show that for a Hermite-Gaussian beam, the ratio of power transmitted through a circular aperture of radius a placed transverse to the direction of propagation is $[1 - \exp\{-2(a/\omega)^2\}]$ for the TEM_{00} mode and $1 - [1 + 2(a/\omega)^2]\exp[-2(a/\omega)^2]$ for either the TEM_{10} or the TEM_{01} mode.

5.16 For a GaAs injection laser operating at a wavelength of 0.85 μm out of a cavity 500 μm long, determine the mode separation of the emitted light beam. *Note:* You may assume that the refractive index of GaAs does not change substantially with respect to wavelength.

5.17 Show that the longitudinal mode spacing in a semiconductor laser is given by $\Delta\lambda = \lambda^2[2n_s l\{1 - (\lambda/n_s)(dn_s/d\lambda)\}]^{-1}$ where l is the cavity length and n_s is the refractive index of the semiconductor.

References and Suggested Readings

Carter, W. H. "Spot Size and Divergence for Hermite-Gaussian Beams of Any Order." *Appl. Opt.* Vol. 19 (1980): 1027.

Casperson, L. W. "Synthesis of Gaussian Beam Optical Systems." *Appl. Opt.* Vol. 20 (1981): 2243.

Feit, M. D., and J. A. Fleck, Jr. "Spectral Approach to Optical Resonator Theory." *Appl. Opt.* Vol. 20 (1981): 2843.

Jahan, R., and M. A. Karim. "Refracting Systems for Gaussian-to-Uniform Beam Transformation." *Opt. & Laser Technol.* Vol. 21 (1989): 27.

Jones, K. A. *Introduction to Optical Electronics.* New York: Harper & Row, 1987.

Karim, M. A. "Measurement of Gaussian Beam Diameter Using Ronchi Rulings." *Electr. Lett.* Vol. 21 (1985): 427.

Karim, M. A., A. A. S. Awwal, A. M. Nasiruddin, A. Basit, D. S. Vedak, C. C. Smith, and G. D. Miller. "Gaussian Laser-Beam Diameter Measurement Using Sinusoidal and Triangular Rulings." *Opt. Lett.* Vol. 12, p. 93, 1987.

Karim, M. A., A. K. Cherri, A. A. S. Awwal, and A. Basit. "Refracting System for Annular Laser Beam Transformation." *Appl. Opt.* Vol. 26 (1987): 2446.

Muys, P., and H. Vanherzeele. "Pitfalls in the Use of Equivalent Resonators." *Appl. Opt.* Vol. 21 (1982): 3983.

Luxon, J. T., and D. E. Parker. "Practical Spot Size Definition for Single Higher-Order Rectangular-Mode Laser Beams." *Appl. Opt.* Vol. 20 (1981): 1728.

Luxon, J. T., and D. E. Parker. *Industrial Lasers and Their Applications.* Englewood Cliffs, NJ: Prentice-Hall, 1985.

Phillips, R. L., and L. C. Andrews. "Spot Size and Divergence for Laguerre Gaussian Beams of Any Order." *Appl. Opt.* Vol. 22 (1983): 643.

Self, S. N. "Focusing of Spherical Gaussian Beams." *Appl. Opt.* Vol. 22 (1983): 658.

Shimoda, K. *Introduction to Laser Physics.* Berlin: Springer-Verlag, 1984.

Smithers, M. E. "Laser Resonator with Nonuniform Gain and Magnification." *Appl. Opt.* Vol. 25 (1986): 118.

Svelto, O., and D. C. Hanna. *Principles of Lasers.* New York: Plenum Press, 1982.

Tache, J. P. "Derivation of ABCD Law for Laguerre-Gaussian Beams." *Appl. Opt.* Vol. 26 (1987): 2698.

Thyagarajan, K., and A. K. Ghatak. *Lasers Theory and Applications.* New York: Plenum Press, 1981.

Verdeyen, J. T. *Laser Electronics.* Englewood Cliffs, NJ: Prentice-Hall, 1981.

Wilson, J., and J. F. B. Hawkes. *Lasers Principles and Applications.* United Kingdom: Prentice-Hall International (UK) Ltd., 1987.

Yariv, A. *Optical Electronics* (3rd ed.). New York: Holt, Rinehart & Winston, 1985.

Zhou, G. S., and L. W. Casperson. "Modes of a Laser Resonator with a Retroreflective Mirror." *Appl. Opt.*, Vol. 20 (1981): 1621.

CHAPTER SIX

Noise in Optical Detection

6.1 Introduction

In the last two chapters we discussed the essential concepts for designing both optical sources and detectors. In our treatment, we deliberately ignored the fact that the performance of an electro-optic system consisting of a source, a detector, or a combination thereof has its own limitations. It turns out that the signal generated, processed, or detected in such a system is anything but ideal. The signals are usually corrupted by noise of many kinds.

The noise accompanying a signal is generally characterized by the random electromagnetic field that shares the same spectral region as that of the signal itself. If perfectly noiseless channels, amplifiers, and detectors were available, it would be possible to communicate over infinite distances with only a little power. Practical detection schemes, however, are both noisy and lossy. The detection of weak signals poses a very difficult problem because both noise and signal remain intertwined. Repeated amplification fails to recover a signal whenever the signal level is substantially below the noise level.

The detector noise can be classified into three distinct groups: the photon noise, the detector noise, and the post-detector noise. The signal source as well as the background sources that are present during the emission of the signal are responsible for introducing photon noise in a detector. In comparison, detector noise is caused by the physical processes that are employed in the detector in question for sensing the incoming emission. Finally, post-detector noises are introduced by the electronic amplification and transmission devices that are used to process the detected signal.

Detectivity is used to characterize the minimum amount of radiation a detector can measure with a reasonable certainty. The detectivity of a detector is generally dependent on the noise inherent in the detection process. Again, the responsivity of a detector, defined as the ratio of the output and the input, is used in determining how much of the input is actually translated by the detector into a meaningful output. In the subsequent sections we describe different aspects of noise so you will have a clearer understanding of both detectors and detection processes.

6.2 Mathematical Aspects of Noise

A time-varying voltage signal $v(t)$ can be described by its time-averaged value $\langle v \rangle$ and by its root-mean-square (rms) deviation from its time-averaged value $\langle [v - \langle v \rangle]^2 \rangle^{1/2}$ where $\langle x \rangle$ represents the average or expected value of x. Another important measure is the variance which is defined as the square of root-mean-square deviation. However, in most circumstances none of these parameters is enough to differentiate between two or more signals having different frequency components.

The spectral content of a voltage signal $V(\omega)$ is related to its real time response $v(t)$ by the **Fourier transform** relation

$$V(\omega) = \int_{-\infty}^{\infty} v(t) e^{-j\omega t} \, dt \tag{6.1}$$

where $\omega = 2\pi\nu$ is the angular frequency. In like manner, when $V(\omega)$ is known, $v(t)$ can be obtained by the **inverse Fourier transform** relation

$$v(t) = \frac{1}{2\pi} \int_{-\infty}^{\infty} V(\omega) e^{j\omega t} \, d\omega \tag{6.2}$$

However, in practice, it is not always possible to evaluate the Fourier transform integral over an infinite interval. Thus the signal $v(t)$ is often defined over a finite interval (also known as the resolution time of the system) T such that

$$v_T(t) = \begin{cases} v(t) & -T/2 \leqslant t \leqslant T/2 \\ 0 & \text{otherwise} \end{cases} \tag{6.3}$$

Consequently, the limited spectral content of the voltage signal represented by $V_T(\omega)$ can be expressed in terms of the truncated signal by the following pairs of equations:

$$V_T(\omega) = \int_{-T/2}^{T/2} v_T(t) e^{-j\omega t} \, dt \tag{6.4}$$

and

$$v_T(t) = \frac{1}{2\pi} \int_{-\infty}^{\infty} V_T(\omega) e^{j\omega t} \, d\omega \tag{6.5}$$

Since $v_T(t)$ is real, it can be shown that

$$V_T(\omega) = V_T^*(-\omega) \tag{6.6}$$

where * indicates complex conjugate operation.

An input signal such as the one just described may now be used to drive an amplifier whose input impedance is given by Z. The average power $\langle P \rangle$ associated with the input signal can be obtained by integrating the instantaneous power $v_T(t)i(t)$ and then dividing the result by T where $i(t) = v_T(t)/Z$ is the current flowing through the impedance. Thus

$$\begin{aligned}\langle P \rangle &= \frac{1}{T}\int_{-T/2}^{T/2} v_T(t)i(t)\,dt \\ &= \frac{1}{ZT}\int_{-T/2}^{T/2} v_T(t)v_T(t)\,dt \end{aligned} \tag{6.7}$$

Now, making use of Eqs. (6.4) and (6.5), we can rewrite Eq. (6.7) as

$$\begin{aligned}\langle P \rangle &= \frac{1}{ZT}\int_{-T/2}^{T/2} v_T(t)\left[\frac{1}{2\pi}\int_{-\infty}^{\infty} V_T(\omega)e^{j\omega t}\,d\omega\right]dt \\ &= \frac{1}{2\pi ZT}\int_{-\infty}^{\infty} V_T(\omega)\left[\int_{-T/2}^{T/2} v_T(t)e^{j\omega t}\,dt\right]d\omega \\ &= \frac{1}{2\pi ZT}\int_{-\infty}^{\infty} V_T(\omega)V_T^*(\omega)\,d\omega \\ &= \frac{1}{2\pi ZT}\int_{-\infty}^{\infty} |V_T(\omega)|^2\,d\omega \end{aligned} \tag{6.8}$$

by interchanging the order of the integrals. However, according to Eq. (6.6), $|V_T(\omega)|^2$ is even with respect to the frequency and, therefore, Eq. (6.8) may be reduced to

$$\langle P \rangle = \frac{1}{\pi ZT}\int_{0}^{\infty} |V_T(\omega)|^2\,d\omega \tag{6.9}$$

The integrand of Eq. (6.9) can be interpreted as the total energy delivered to a load of impedance Z. Accordingly, the **spectral density function** $S_T(\omega)$ can be defined as the energy delivered by a voltage signal $v(t)$ to a 1-ohm load per angular frequency interval:

$$S_T(\omega) = \frac{|V_T(\omega)|^2}{\pi T} \tag{6.10}$$

$S_T(\omega)\,d\omega$ describes the average power per unit impedance in the angular frequency interval between ω and $\omega + d\omega$. Equation (6.10) can also be written in terms of the frequency by

$$S_T(\nu) = 2\pi S_T(\omega) = \frac{2}{T}|V_T(\omega)|^2 \tag{6.11}$$

since

$$\int S_T(\nu)\, d\nu = \int S_T(\omega)\, d\omega \qquad \textbf{(6.12)}$$

An intuitive explanation of Eqs. (6.9) and (6.10) can be given as follows. The energy delivered by the truncated function increases or at least does not decrease with increasing T. Therefore, the quantity $|V_T(\omega)|^2$ increases or at least does not decrease with increasing T. As T becomes large, the fluctuations due to end effects in the integration become minimal and, thus, for all physically realizable random signals, $|V_T(\omega)|^2/\pi T$ is a converging function.

Equations (6.9) and (6.10) suggest a method that is used to determine the power spectral density of a signal. First, the data are collected over a record length of T units. The corresponding Fourier transform $V_T(\omega)$ is then calculated from this data using computational techniques. Taking its magnitude squared, the ratio $|V_T(\omega)|^2/\pi T$ is determined next.

Consider that $v(t)$ is periodic and that it is represented by a Fourier series as

$$v(t) = \sum_{p=-\infty}^{\infty} v_p e^{jp\omega_0 t} \qquad \textbf{(6.13)}$$

Accordingly, the total average power across a Z-ohm load due to $v(t)$ can be calculated using Parseval's theorem as follows:

$$\begin{aligned} \langle P \rangle &= \frac{1}{ZT} \int_{-T/2}^{T/2} v(t) v^*(t)\, dt \\ &= \sum_{q=-\infty}^{\infty} v_q \sum_{p=-\infty}^{\infty} v_p^* \frac{1}{ZT} \int_{-T/2}^{T/2} e^{j(q-p)\omega_0 t}\, dt \\ &= \frac{1}{Z} \sum_{p=-\infty}^{\infty} |v_p|^2 \end{aligned} \qquad \textbf{(6.14)}$$

Note that there is no contribution to the average power when p is not equal to q. For the case of a periodic signal, a line power spectrum can be plotted as shown in Fig. 6.1(a). Further, the cumulative power spectrum $S'_T(\omega)$ for this signal can be obtained simply by adding the terms of Eq. (6.14) over all harmonic numbers q up to and including ω such that

$$S'_T(\omega) = \int_{-\infty}^{\omega} S_T(\omega)\, d\omega = \frac{1}{2\pi ZT} \int_{-\infty}^{\omega} |V_T(\omega)|^2\, d\omega \qquad \textbf{(6.15)}$$

The result is plotted in Fig. 6.1(b).

We can also express the cumulative power spectrum as

$$S'_T(\omega) = \frac{1}{Z} \sum_{p=-\infty}^{\omega/\omega_0} |v_p|^2 u(\omega - p\omega_0) \qquad \textbf{(6.16)}$$

where u represents a unit step function. Consequently, the spectral power density

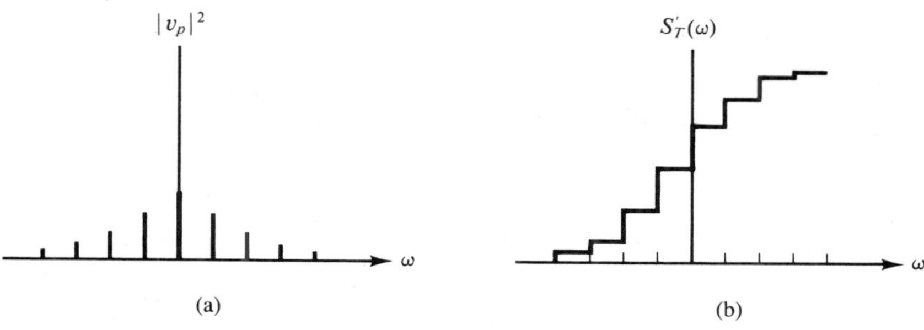

FIGURE 6.1

(a) Power spectrum and (b) cumulative power spectrum of a periodic function

of the periodic function introduced in Eq. (6.13) is given by

$$S_T(\omega) = \frac{dS'_T(\omega)}{d\omega} = \frac{1}{Z}\sum_p |v_p|^2 \delta(\omega - p\omega_0) \tag{6.17}$$

where $\delta(h)$ is the familiar impulse function in h. The spectral power density of a periodic function is, therefore, equivalent to a series of weighted impulse functions. Each of these weights corresponds to the squares of the respective Fourier coefficients. In case of an aperiodic function, however, $S'_T(\omega)$ is a smoothly varying function of frequency. According to the physical interpretation of $S_T(\nu)$, the signal can also be separated into various frequency groups by means of bandpass filters. Consequently, $S_T(\nu)\,d\nu$ can be evaluated to provide the average power per unit impedance in the frequency interval $\nu_p - d\nu/2$ and $\nu_p + d\nu/2$ where $d\nu$ is the filter bandwidth and ν_p is the central frequency of the filter.

The analysis of this section could have begun using a current signal $i(t)$. In that case, the corresponding equation for the average power would have resulted in

$$\langle P \rangle = \frac{Z}{\pi T} \int_0^\infty |I_T(\omega)|^2\, d\omega \tag{6.18}$$

Equation (6.18) provides a new definition of $S_T(\nu)$ because it is apparent that the mathematical quantity $S_T(\nu)\,d\nu$ can be represented either by an equivalent voltage generator or by an equivalent current generator, as shown in Fig. 6.2. These circuits are the well-known Thévenin-Norton conjugate pair of circuits. The power generating capacity of such signal generators depends on both the impedance and bandwidth characteristics of the circuit elements introduced at the output.

Consider the example of a detector that responds to an emission of photoelectrons. Consider further that the photoelectrons of charge e each had to move through a distance d with free electron velocity v before being finally

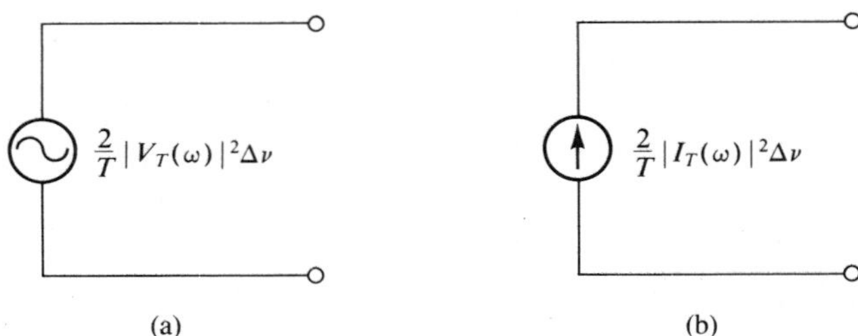

FIGURE 6.2

Equivalent circuit representation of $S_T(\nu)\,d\nu$

detected by the detector. The resulting transform $I_T(\omega)$ of the current signal $i(t)$ is given by

$$I_T(\omega) = \int_{-\tau/2}^{\tau/2} Gi(t)e^{-j\omega t}\,dt$$

$$= \frac{Gev}{d}\int_{-\tau/2}^{\tau/2} e^{-j\omega t}\,dt = Ge\frac{\sin(\omega\tau/2)}{\omega\tau/2} \qquad \textbf{(6.19)}$$

where G is the detector amplification factor and $\tau \equiv d/v$ is the photoelectron lifetime, which happens to be extremely small (a nanosecond or less). The photoelectrons contribute to a current pulse whose frequency components are spread out as predicted by Eq. (6.11). Thus

$$S_T(\nu) = \frac{2Ge}{T}\frac{\sin(\omega\tau/2)}{\omega\tau/2} \qquad \textbf{(6.20)}$$

which for very small values of τ reduces to

$$S_T(\nu) = \frac{2Ge}{T} \qquad \textbf{(6.21)}$$

Thus the process of photoelectron detection generates power over the whole spectrum of frequencies.

The spectral density function is useful in describing the relative amount of power of a given signal in terms of frequency. The total area under the spectral density curve generally represents the power associated with the signal. In addition, the spectral energy density accounts for the relative spectral energy attenuations in linear systems. To comprehend this, let us apply a signal voltage $v(t)$ to the input of a linear time-invariant system whose frequency transfer function is given by $H(\omega)$. The amplitude spectral density of the system output is, therefore, described by

$$G(\omega) = V_T(\omega)H(\omega) \qquad \textbf{(6.22)}$$

where $V_T(\omega)$ has already been described by Eq. (6.4). The normalized energy

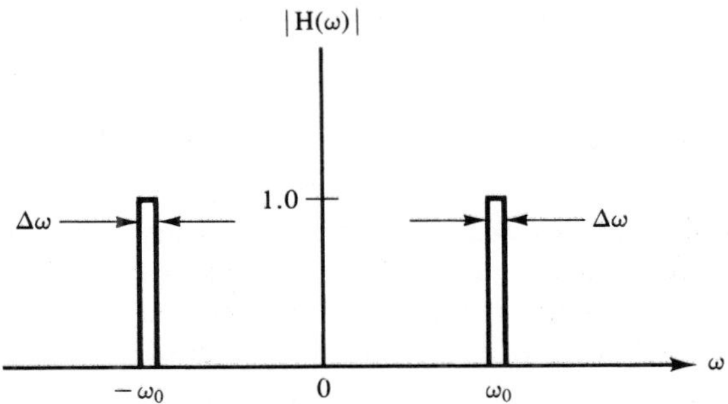

FIGURE 6.3

An ideal bandpass filter

density of $G(\omega)$ is thus given by

$$|G(\omega)|^2 = |V_T(\omega)|^2 |H(\omega)|^2 \tag{6.23}$$

provided the normalized average energy in the output signal per unit impedance is given by

$$P_0 = \frac{1}{2\pi T} \int_{-\infty}^{\infty} |V_T(\omega)|^2 |H(\omega)|^2 \, d\omega \tag{6.24}$$

It is to be noted that all phase information present in both the input signal and the system transfer function is lost in the calculation of power and energy density.

A useful interpretation of density function is obtained by considering an ideal (very narrow) bandpass filter whose characteristics are shown in Fig. 6.3. Consider applying $v(t)$ to the input of this bandpass filter whose frequency transfer function $H(\omega)$ is unity only over a small width of angular frequency $\Delta\omega$ at $\pm\omega_0$ while $H(\omega)$ is zero elsewhere. Using this narrow-band filter, the power per unit impedance is given by

$$\begin{aligned} P_0 &= \frac{1}{2\pi T} \int_{-\infty}^{\infty} |G(\omega)|^2 \, d\omega \\ &= \frac{1}{2\pi T} \left[\int_{-\omega_0-(1/2)\Delta\omega}^{-\omega_0+(1/2)\Delta\omega} |V_T(\omega)|^2 |H(\omega)|^2 \, d\omega \right. \\ &\qquad \left. + \int_{\omega_0-(1/2)\Delta\omega}^{\omega_0+(1/2)\Delta\omega} |V_T(\omega)|^2 |H(\omega)|^2 \, d\omega \right] \\ &\simeq \frac{1}{2\pi T} [|V_T(-\omega_0)|^2 + |V_T(\omega_0)|^2] \Delta\omega \end{aligned} \tag{6.25}$$

For a real-valued signal, the negative frequency components contribute to half of the power and the positive frequency components contribute to the other half of the power. Using Eq. (6.6), the normalized output power density thus becomes

$$P_0 = \frac{1}{\pi T} |V_T(\omega_0)|^2 \Delta\omega \tag{6.26}$$

It is often difficult to evaluate the spectral density function accurately, but this problem is often avoided by relating the spectral density function to the corresponding autocorrelation function. The **autocorrelation function** $S(\tau)$ is defined as the time-average of the product of $v(t)$ and its delayed form $v(t+\tau)$:

$$\begin{aligned} S(\tau) &= \langle v(t)v(t+\tau)\rangle \\ &= \text{Lim}_{T\to\infty} \frac{1}{T} \int_{-T/2}^{T/2} v(t)v(t+\tau)\, dt \end{aligned} \tag{6.27}$$

Note that $S(0)$, in particular, is equal to the mean square voltage signal. Using Eq. (6.2), Eq. (6.27) can be written as

$$S(\tau) = \text{Lim}_{T\to\infty} \frac{1}{4\pi^2 T} \iint_{-\infty}^{\infty} \int_{-T/2}^{T/2} V_T(\omega) V_T(\omega') e^{j(\omega+\omega')t} e^{j\omega t}\, dt\, d\omega\, d\omega' \tag{6.28}$$

But as T approaches infinity,

$$\int_{-T/2}^{T/2} e^{j(\omega+\omega')t}\, dt \equiv 2\pi\delta(\omega+\omega') \tag{6.29}$$

Thus

$$\begin{aligned} S(\tau) &= \text{Lim}_{T\to\infty} \frac{1}{2\pi T} \iint_{-\infty}^{\infty} V_T(\omega) V_T(\omega')\delta(\omega+\omega') e^{j\omega t}\, d\omega\, d\omega' \\ &= \text{Lim}_{T\to\infty} \frac{1}{2\pi T} \int_{-\infty}^{\infty} |V_T(\omega)|^2 e^{j\omega t}\, d\omega \end{aligned} \tag{6.30}$$

Now, incorporating Eq. (6.10) into Eq. (6.30), we obtain

$$S(\tau) = \frac{1}{2} \int_{-\infty}^{\infty} S_T(\omega) e^{j\omega t}\, d\omega \tag{6.31}$$

Equation (6.31) is also known as the **Wiener-Khintchine theorem**. Since $S_T(\omega)$ is an even function of ω, Eq. (6.31) can be rewritten as

$$S(\tau) = \int_0^{\infty} S_T(\omega) \cos(\omega t)\, dt \tag{6.32}$$

showing that $S(\tau)$ is also an even function of τ (see Problem 6.16(a)). Equation (6.32) proves that the spectral density function $S_T(\omega)$ and the autocorrelation function $S(\tau)$ form a Fourier transform pair. Thus, if any one of these two functions is determined experimentally, the other is evaluated using Eq. (6.31) or Eq. (6.32). It is possible to determine the autocorrelation function first and then take an inverse Fourier transform of it to obtain the spectral density function.

This method is applicable to both random and nonrandom signals. Note that for the real-valued signals, the computation of the autocorrelation function turns out to be equivalent to taking the convolution of $v(-t)$ and $v(t)$.

The autocorrelation function is particularly useful in the detection or recognition of signals that are corrupted with additive noise. For example, consider a periodic uncorrupted signal such as that shown in Fig. 6.4(a). Its autocorrelation function is shown in Fig. 6.4(b). Figures 6.4(c) and 6.4(d), respectively, show a band-limited random noise waveform and its autocorrelation function. Finally, Figs. 6.4(e) and 6.4(f), respectively, show the corrupted signal and its autocorrelation function. In spite of the fact that even though the signal is barely identifiable in Fig. 6.4(e), the corresponding autocorrelation function of Fig. 6.4(f) clearly resembles the autocorrelation of the uncorrupted signal shown in Fig. 6.4(b). However, the autocorrelations of noise and signal are both simultaneously present in the output. This, therefore, poses a problem during the detection of aperiodic signals. This apparent problem is solved by making use of a cross-correlation operation.

The **cross-correlation** of $v_s(t)$ and $v_n(t)$ is defined as

$$S_{sn}(\tau) = \text{Lim}_{T\to\infty} \frac{1}{T} \int_{-T/2}^{T/2} v_s^*(t) v_n(t+\tau)\, dt \tag{6.33}$$

To illustrate the advantage of using cross-correlation, consider a random waveform $v_s(t)$ as shown in Fig. 6.5(a). Note that such a waveform is similar in nature to the noise of Fig. 6.4(c). Consequently, the autocorrelation of $v_s(t)$ is quite similar to that shown in Fig. 6.4(d). For the second function $v_n(t)$, we may choose the sum of a delayed waveform $v_s(t)$ (shifted in time by t_0) and a second random waveform $v_n'(t)$ so that $v_n(t) = v_s(t-t_0) + v_n'(t)$. Such a composite waveform is shown in Fig. 6.5(b). Provided that a detector has the waveform $v_s(t)$ stored in its memory, it can perform the cross-correlation operation between the waveforms of Figs. 6.5(a) and 6.5(b). The cross-correlation result is shown in Fig. 6.5(c). The value of the time delay t_0 is determined from the time span between the origin and the largest peak.

For simplicity, we restricted the mathematical treatment developed so far to only a single signal. However, in practice, the detector encounters a sequence of several such individual signals. The time-dependent random signal variable $i(t)$ can be assumed to be made up of a large number of individual signals $f(t-t_j)$ where t_j is the time of origin for the jth signal

$$i_T(t) = \sum_{j=1}^{N} f(t-t_j) \tag{6.34}$$

where N is the number of total events occuring in the period $-\frac{1}{2}T \leqslant t \leqslant \frac{1}{2}T$. An example of such an event can be the electron current caused by the photoemission from a source surface. The dc value of the detected current can be evaluated as

$$\langle i \rangle = T^{-1} \int_{-T/2}^{T/2} i_T(t)\, dt = \langle N \rangle \tag{6.35}$$

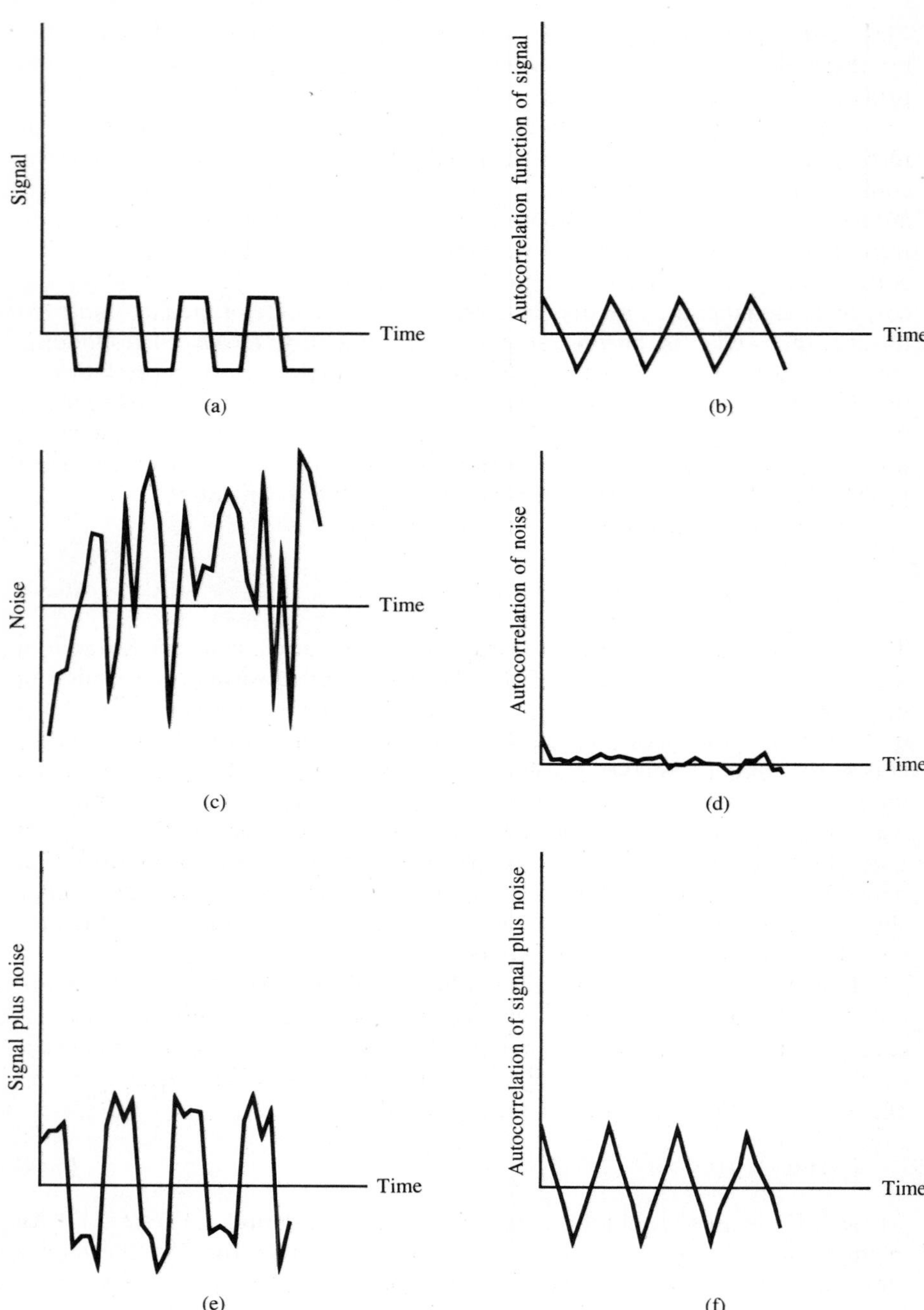

FIGURE 6.4

An autocorrelation example

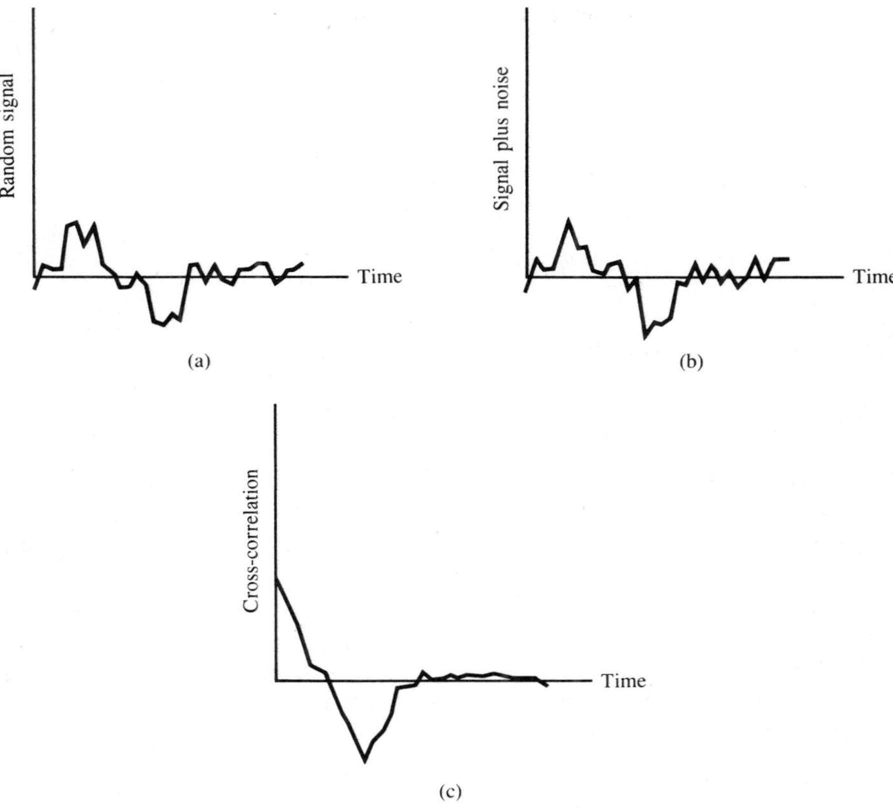

FIGURE 6.5

A cross-correlation example

where $\langle N \rangle$ is the average rate of occurrence of the signal and is given by

$$\langle N \rangle = \mathrm{Lim}_{T \to \infty} \frac{N}{T} \tag{6.36}$$

Equation (6.35) provides an average value for the current. To estimate the instantaneous value of the current, however, we obtain the spectrum of the current using Fourier transform relations as follows:

$$\begin{aligned} I_T(\omega) &= \sum_k^N \int_{-T/2}^{T/2} f(t - t_k) e^{-j\omega t}\, dt \\ &= \sum_k^N \int_{-T/2}^{T/2} f(t) e^{-j\omega t_k} e^{-j\omega t}\, dt \\ &= \sum_k^N F(\omega) e^{-j\omega t_k} \end{aligned} \tag{6.37}$$

where $F(\omega)$ represents the Fourier transform of an individual event. The duration of each individual event $f(t-t_k)$ is assumed to be very short when compared to T. As per Eq. (6.18), the calculation of the corresponding spectral distribution of power involves evaluating

$$\begin{aligned}|I_T(\omega)|^2 &= \sum_k^N F(\omega)e^{-j\omega t_k} \sum_l^N F(\omega)^* e^{j\omega t_l} \\ &= |F(\omega)|^2 \left[N + \sum_{k \neq l}^N \sum_l^N e^{j\omega(t_l - t_k)} \right] \end{aligned} \tag{6.38}$$

If we take the average of the right-hand side of Eq. (6.38) over an ensemble of a large number of identical systems, the contribution of the second random term approaches zero. Therefore, using Eq. (6.36), we can write

$$\langle |I_T(\omega)|^2 \rangle = |F(\omega)|^2 N = |F(\omega)|^2 \langle N \rangle T \tag{6.39}$$

The spectral density function can now be determined using Eq. (6.11) as

$$S_T(v) = \frac{2}{T} |I_T(2\pi v)|^2 = 2\langle N \rangle |F(2\pi v)|^2 = 2\langle i \rangle |F(2\pi v)|^2 \tag{6.40}$$

In deriving Eq. (6.40), we assumed the individual events to be identical except for the fact that they were displaced in time. In practice, individual events may have depended on one or more additional parameters. For example, for a subclass of events $f_\alpha(t-t_j)$ having α as its common parameter, the spectral density is given by

$$S_{T,\alpha}(v) = 2\langle N(\alpha) \rangle |F_\alpha(2\pi v)|^2 \,\Delta\alpha \tag{6.41}$$

where $F_\alpha(2\pi v)$ is the Fourier transform of $f_\alpha(t)$ and $\langle N(\alpha) \rangle \,\Delta\alpha$ is the average number of such events per unit time in the interval α and $\alpha + \Delta\alpha$ such that

$$\int_{-\infty}^{\infty} \langle N(\alpha) \rangle \,d\alpha = \langle N \rangle \tag{6.42}$$

Accordingly, the overall spectral density can be determined by adding the subclass spectral densities (each having a common parameter). Since the probability distribution of α, denoted by $p(\alpha)$, is equivalent to $\langle N(\alpha) \rangle / \langle N \rangle$, the overall spectral density is given by

$$\begin{aligned} S_T(v) &= \sum_\alpha S_{T,\alpha}(v) = 2 \sum_\alpha \langle N(\alpha) \rangle |F_\alpha(2\pi v)|^2 \,\Delta\alpha \\ &= 2\langle N \rangle \sum_\alpha p(\alpha) |F_\alpha(2\pi v)|^2 \,\Delta\alpha \\ &= 2\langle N \rangle \int_{-\infty}^{\infty} p(\alpha) |F_\alpha(2\pi v)|^2 \,d\alpha \\ &= 2\langle N \rangle \langle |F(2\pi v)|^2 \rangle = 2\langle i \rangle \langle |F(2\pi v)|^2 \rangle \end{aligned} \tag{6.43}$$

where

$$\int_{-\infty}^{\infty} p(\alpha)\, d\alpha = 1 \tag{6.44}$$

Whenever the individual events are not identical, Eq. (6.43) is used for evaluating the overall spectral density. Equation (6.40) is simpler but is used only in the case of identical events.

The autocorrelation function of $i(t)$ when Z is unity, for simplicity, can be determined using Eq. (6.27) as

$$\begin{aligned} S(\tau) &= \lim_{T\to\infty} \frac{1}{T} \int_{-T/2}^{T/2} i(t)i(t+\tau)\, dt \\ &= \lim_{T\to\infty} \frac{1}{T} \int_{-\infty}^{\infty} \sum_{k=1}^{N} f(t-t_k) \sum_{l=1}^{N} f(t+\tau-t_l)\, dt \\ &= \lim_{T\to\infty} \frac{1}{T} \int_{-\infty}^{\infty} \left[\sum_{k=1}^{N} f(t-t_k) f(t+\tau-t_k) \right. \\ &\qquad \left. + \sum_{k\neq l}^{N} \sum_{l=1}^{N} f(t-t_k) f(t+\tau-t_l) \right] dt \end{aligned} \tag{6.45}$$

The first term of Eq. (6.45) is the same as the correlation of $f(t)$ with itself. However, the second term involves no correlation and consequently it may be described by $\langle i \rangle^2$. Thus Eq. (6.45) reduces to

$$S(\tau) = \langle N \rangle \int_{-\infty}^{\infty} f(t)f(t+\tau)\, dt + \langle i \rangle^2 \tag{6.46}$$

Note that $S(\tau)$ is the same as the mean square signal when $\tau = 0$ (per Eq. (6.27)). But

$$S(0) = \langle N \rangle \int_{-\infty}^{\infty} [f(t)]^2\, dt + \langle i \rangle^2 \tag{6.47}$$

Consequently, the variance, defined as the mean square deviation of the signal, is given by

$$\langle (\Delta i)^2 \rangle \equiv \langle i^2 \rangle - \langle i \rangle^2 = \langle N \rangle \int_{-\infty}^{\infty} [f(t)]^2\, dt \tag{6.48}$$

Both $\langle i \rangle$ and $\langle (\Delta i)^2 \rangle$ are important parameters needed to characterize the quality of detection.

Detector noise refers to unwanted responses, random or deterministic, which interfere with the detection of a signal in a system. These noises arise from a variety of sources and can be classified as either artificial or naturally occurring. Usually a description of random noise in terms of time is meaningless. However, in bandpass systems where the bandwidth of noise is somewhat limited, it is possible to characterize noise in terms of time.

Noise directly affects our capacity to measure and quantify a signal. To understand the influence of noise in signal detection, consider the phasor representation of noise as shown in Fig. 6.6. In the figure, $V_{nc}(t)$ represents the in-phase component of the noise and $V_{ns}(t)$ represents the quadrature component. Both $V_{nc}(t)$ and $V_{ns}(t)$ are random and thus their phasor sum has a random amplitude V_n and a random phase θ_n. In phasor form, the instantaneous noise field can be written as

$$V_n(t) = [V_{nc}(t) + jV_{ns}(t)]e^{j\omega_0 t} \tag{6.49}$$

where ω_0 is the central frequency of the band. The constraint of having a narrow-band noise allows us to treat both $V_{nc}(t)$ and $V_{ns}(t)$ as slowly varying components when they are compared to $\exp(j\omega_0 t)$. Now, consider that this noise is accompanying a to-be-detected optical signal:

$$v_s(t) = V_s \sin \omega t \tag{6.50}$$

Taking the real part of Eqs. (6.49) and (6.50), we can express the total of the signal and noise fields at the detector as

$$\begin{aligned} v(t) &= \mathrm{Re}\{[V_{nc}(t) + j(V_{ns}(t) - V_s)]e^{j\omega t}\} \\ &\equiv \mathrm{Re}[V(t)e^{j\omega t}] \end{aligned} \tag{6.51}$$

where $V(t)$ corresponds to the resultant phasor as shown in Fig. 6.6.

Since the noise component of the total field is random, its probability of occurrence is dictated by a Gaussian probability distribution function. The average (or expected) value of a measurable noise parameter x is thus given by

$$\langle x \rangle = \int_{-\infty}^{\infty} x p(x)\, dx \tag{6.52}$$

where $p(x)$, describing the **Gaussian probability distribution** function, is defined as

$$p(x) = \frac{1}{2\pi\sigma^2} e^{-x^2/2\sigma^2} \tag{6.53}$$

where $\sigma > 0$ is the standard deviation of x from $\langle x \rangle$. Since both $V_{nc}(t)$ and $V_{ns}(t)$ are random, their root-mean-square average values are obtained using Eq. (6.52), yielding

$$\langle V_{nc}^2(t) \rangle^{1/2} = \langle V_{ns}^2(t) \rangle^{1/2} = \sigma \tag{6.54}$$

But the ensemble average of the random noise terms are given by

$$\langle V_{nc}(t) \rangle = \langle V_{ns}(t) \rangle = 0 \tag{6.55}$$

The total power delivered to a load of a unit impedance can now be calculated using Eq. (6.51) as

$$\begin{aligned} P(t) &= v(t)v^*(t) \\ &= [V_{nc}(t)]^2 + [V_{ns}(t)]^2 + V_s^2 - 2V_{ns}(t)V_s \end{aligned} \tag{6.56}$$

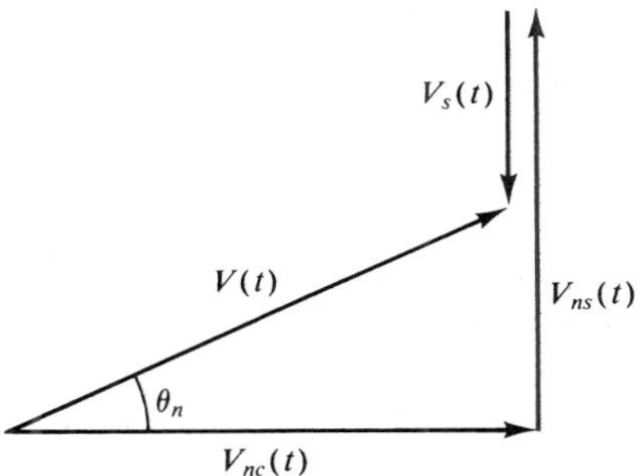

FIGURE 6.6

The phasor representation of a narrow-band noise

The average value of the power can now be determined from Eq. (6.56) by making use of Eqs. (6.54) and (6.55):

$$\begin{aligned}\langle P \rangle \equiv \langle P(t) \rangle &= V_s^2 + \langle V_{nc}^2(t) \rangle + \langle V_{ns}^2(t) \rangle \\ &= V_s^2 + 2\sigma^2 \\ &\equiv P_s^2 + 2\sigma^2\end{aligned} \tag{6.57}$$

where P_s is the signal power that would be detected when the noise components are absent. It is obvious that the measurement of $P(t)$ is limited by the randomness of noise components. The deviation of the time-averaged power from the actual power is often measured in terms of the root-mean-square (rms) power deviation. It is used to estimate the uncertainty involved in the measurement of power. The rms power deviation is evaluated using Eqs. (6.54) and (6.55) as

$$\begin{aligned}\sigma_p &= \langle [P(t) - \langle P \rangle]^2 \rangle^{1/2} \\ &= \langle [V_{nc}^2(t) + V_{ns}^2(t) - 2V_{ns}(t)V_s - \langle V_{nc}^2(t) \rangle - \langle V_{ns}^2(t) \rangle]^2 \rangle^{1/2} \\ &= [2\langle V_{ns}^4(t) \rangle + 4P_s \langle V_{ns}^2(t) \rangle - 2\sigma^4]^{1/2}\end{aligned} \tag{6.58}$$

However, Eq. (6.52) provides that

$$\langle V_{ns}^4(t) \rangle = 3\sigma^4 \tag{6.59}$$

and, therefore, Eq. (6.58) reduces to

$$\sigma_p = 2\sigma(\sigma^2 + P_s)^{1/2} \tag{6.60}$$

Equation (6.60) provides a remarkable result as it identifies the minimum possible signal power that is detectable in the presence of noise. It is obvious that the minimum detectable signal power $P_{\min}$ should be no less than the rms power deviation. Thus

$$P_{\min} = 2\sigma(\sigma^2 + P_{\min})^{1/2} \tag{6.61}$$

which, in turn, yields

$$P_{\min} = P_n(1 \pm 2^{1/2}) \tag{6.62}$$

where P_n is the noise power given by

$$P_n = \langle V_{nc}^2 \rangle + \langle V_{ns}^2 \rangle \tag{6.63}$$

Consequently, the minimum detectable signal power turns out to be equal to $(1 + 2^{1/2}) = 2.4142$ times the value of the average noise power. Thus the lower limit of the detectable signal power is determined by minimizing the value of the average noise power.

6.3 Noise Sources

The mathematical basis for the studies involving noise that was discussed in the last section can now be utilized to characterize noise sources. In most electro-optic applications, we seek to detect the weakest optical signals. But often it becomes difficult to realize such a system performance. Improving the signal-to-noise ratio of an electro-optic system may involve optimizing both the photodetector and the amplification circuitry that follows it. However, optimization of the system characteristics cannot be begun without an understanding of the various noise sources that affect the photodetection process. In practice, the detectors can involve any one or more of several different types of noises. Five of the most important types of noise follow:

1. *Shot noise* involves noises that can be considered as a series of randomly occurring independent events. The emission of electrons from the photocathode of a photoemissive detector is a manifestation of such noise.
2. *Johnson noise* (or *thermal noise*) is generally caused by the random motion of carriers in a conductor or a resistor. Almost all detectors are affected by Johnson noise.
3. *Generation-recombination noise* occurs in photoconductors and is caused by the generation and recombination of free charge carriers. It is introduced by the carrier concentration fluctuations Δn and Δp of the carrier densities n and p, respectively, that result in a fluctuation of device resistance. Often it is misunderstood and falsely taken to be a shot noise.
4. *Temperature fluctuation noise* is caused by the irregular exchange of heat between the detector and its immediate environment. Its effect becomes most pronounced in the smaller thermal detectors.
5. *Flicker noise* (or *contact noise*) is caused by the fluctuation of the work function. Very often the effect of this noise is reduced by improving the process used in manufacturing the detector surfaces. Contact noise is normally characterized by a noise intensity spectrum that has a constant $(1/f^{\alpha})$-form where $0.8 \leqslant \alpha \leqslant 2$ and f is the frequency. Flicker noise, like temperature fluctuation noise, affects only the thermal detectors.

For simplicity, we limit our detailed analysis to the three most important types of noise: shot noise, thermal noise, and generation-recombination noise. Note, however, that to achieve a high signal-to-noise ratio, the chosen photodetector should have a high quantum efficiency while, at the same time, the photodetector and amplifier noises should be maintained at a minimum. A thorough knowledge of the source characteristics, as well as the interrelationships of the various photodetector noises, is essential in coming up with a superior electro-optic system.

6.3.1 Shot Noise

Shot noise, also referred to as either photon noise or quantum noise, is generated in a detector whenever random independent events, such as the emission of photoelectrons, occur. This phenomenon is quite different from that of the conduction of electrons because the latter is basically described as a large-scale drift phenomenon. Noise introduced by the emitted electrons of a phototube cathode is a classic example of shot noise. There are two important aspects of shot noise in photoelectric detection:

1. The radiation of photons is random because the rate of arrival of photons follows a statistical process.
2. The kinetic energy of the electrons acquired in the device is a process that has a limited probability.

For most thermal and semiconductor detectors, however, it is proper to combine both of these aspects into a single source of shot noise.

Photon noise also refers to the noise generated by the discreteness of the radiation field. It can be classed into two types: (a) that due to the signal radiation alone and (b) that due to the background radiation only. To quantify photon noise, therefore, we are assumed to have an ideal photon detector that is detecting monochromatic radiation of a certain power. Such an ideal detector produces no current in the absence of any incident power, and it involves no noise except that which accompanies the incident radiation.

Since the emission of photons is a statistical process, it is appropriate to describe the photoemission in terms of its probability distribution function. A discrete process where photons may be either emitted or not emitted is generally a binomial distribution of events. The probability that exactly n photons will be emitted in m observations is thus given by

$$p_B(n) = \frac{m!}{n!(m-n)!}\left(\frac{\langle n\rangle}{m}\right)^n\left(1-\frac{\langle n\rangle}{m}\right)^{m-n} \tag{6.64}$$

where $\langle n\rangle = \eta Pt/h\nu$ is equal to the average number of photons emitted during those observations where η is the quantum efficiency, P is the constant power incident at the detector, t is the total time of observation, and ν is the frequency of the nearly monochromatic radiation. Note that $\langle n\rangle/m$ provides the probability of having one photoemission during a single observation. However, as the number of observations approaches infinity, Eq. (6.64) reduces to what is commonly referred to as the **Poisson's distribution** as follows:

$$p_p(n) = \mathrm{Lim}_{m\to\infty} \frac{\left(1-\frac{1}{m}\right)\left(1-\frac{2}{m}\right)\cdots\left(1-\frac{n-1}{m}\right)}{n!} \langle n\rangle^n \left(1-\frac{\langle n\rangle}{m}\right)^{m-n}$$
$$= \frac{\langle n\rangle^n}{n!} \mathrm{Lim}_{m\to\infty} \left(1-\frac{\langle n\rangle}{m}\right)^{m-n}$$
$$= \frac{\langle n\rangle^n}{n!e^{\langle n\rangle}} \tag{6.65}$$

Using the Poisson distribution function of Eq. (6.65), we can determine (see Problem 6.11) the expected value of n^2, represented as $\langle n^2\rangle$, to be $\langle n\rangle + \langle n\rangle^2$. Consequently, the root-mean-square noise level is given by

$$\sigma_{i,p} = [\langle n^2\rangle - \langle n\rangle^2]^{1/2} = \langle n\rangle^{1/2} \tag{6.66}$$

Accordingly, the associated **signal-to-noise ratio** present in the incident radiation is obtained as

$$\mathrm{SNR}_i = \frac{\langle n\rangle}{\sigma_i} = \langle n\rangle^{1/2} \tag{6.67}$$

In the event that the ideal detector has a less than perfect quantum efficiency, the signal-to-noise ratio is obtained by multiplying the right-hand side of Eq. (6.67) with the square root of the quantum efficiency.

It is important to realize that radiation from incoherent sources—e.g., a blackbody radiator—does not necessarily follow Poisson distribution because incoherent lights have time-dependent characteristics. All sources other than the lasers emit radiation whose characteristics obey the Bose-Einstein distribution (introduced in Chapter 2). In accordance with the Bose-Einstein distribution, the probability for a photon to have a total energy u at a frequency ν is given by

$$p_{\mathrm{BE}}(u) = \frac{e^{-u/kT}}{\int_0^\infty e^{-u/kT}\,du} \tag{6.68}$$

Therefore, the number of photons expected to be emitted by an incoherent source is

$$\langle n\rangle = \frac{\langle u\rangle}{h\nu}$$
$$= \frac{\int_0^\infty ue^{-u/kT}\,du}{h\nu\int_0^\infty e^{-u/kT}\,du} = \frac{e^{-h\nu/kT}}{1-e^{-h\nu/kT}} \tag{6.69}$$

Similarly, the expected value of n^2 can also be found for this distribution function. And finally the root-mean-square noise level can be found to be

$$\sigma_{i,\mathrm{BE}} = \left[\langle n \rangle \frac{e^{h\nu/kT}}{e^{h\nu/kT} - 1} \right]^{1/2} \tag{6.70}$$

Note, however, that whenever the incoherent source is operated either at shorter wavelength, or at lower temperature, or both, Eq. (6.70) reduces to Eq. (6.66). Consequently, under these conditions, the signal-to-noise ratios for both coherent and incoherent sources become identical.

Variations in the emission of photons are responsible for noise in the detector photocurrent. The average value of the detector photocurrent observed during a time interval τ is given by

$$\langle i \rangle = \frac{e\langle n \rangle}{\tau} \tag{6.71}$$

The mean square current deviation, averaged over many such independent observation periods, each of duration τ, is thus obtained as

$$\begin{aligned} \langle i_N^2 \rangle &= \langle (\Delta i)^2 \rangle = \langle (i - \langle i \rangle)^2 \rangle \\ &= \left(\frac{e}{\tau}\right)^2 \langle (n - \langle n \rangle)^2 \rangle \\ &= \left(\frac{e}{\tau}\right)^2 \langle n \rangle = \frac{e\langle i \rangle}{\tau} \end{aligned} \tag{6.72}$$

The width of each of the resulting current pulses is, however, much smaller than the sampling time τ. Again, the effective bandwidth of the detection system is normally given by

$$\mathrm{BW} = \int_0^\infty \int_{-\tau/2}^{\tau/2} \frac{1}{\tau} e^{-j\omega t}\, d\tau\, d\nu = \frac{1}{2\tau} \tag{6.73}$$

Thus the mean square noise current can be expressed as

$$\langle i_N^2 \rangle = 2e\langle i \rangle(\mathrm{BW}) \tag{6.74}$$

This result is often referred to as **Schottky's formula**.

Consider the spectral density of current in a vacuum photodiode. The photodiode current results from the random flow of carriers as illustrated in Fig. 6.7. In a vacuum photodiode, the electrons are released from one of the electrodes and are collected at the other electrode which is maintained at a relatively higher potential. Many such photodiodes can be cascaded to form a high gain detector known as a photomultiplier.

Provided the initial velocity of an electron is negligible, the current resulting

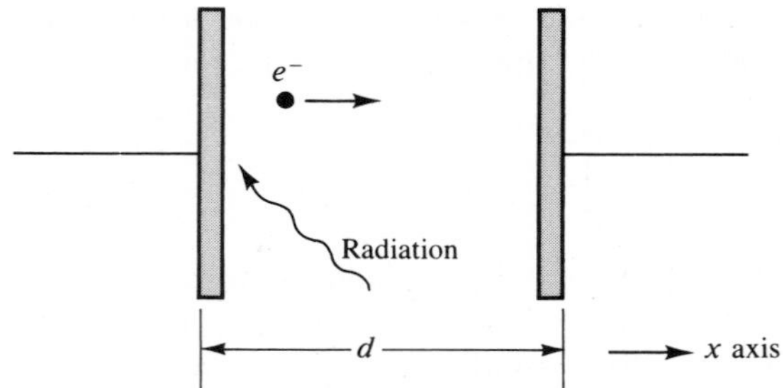

FIGURE 6.7
Moving electrons in a photodiode

from each emission of an electron is given by

$$i_e(t) = \begin{cases} \dfrac{2et}{\tau^2} = \dfrac{e}{d}\dfrac{dx}{dt} & 0 \leqslant t \leqslant \tau \\ 0 & \text{otherwise} \end{cases} \tag{6.75}$$

where τ is the average time taken by the electrons to reach the anode. The Fourier transform of this current pulse is obtained as follows:

$$\begin{aligned} I(\omega) &= \frac{2e}{\tau^2}\int_0^{\tau} te^{-j\omega t}\,dt \\ &= \frac{2e}{\omega^2\tau^2}[(1 + j\omega t)e^{-j\omega t} - 1] \end{aligned} \tag{6.76}$$

whose modulus square is

$$|I(\omega)|^2 = \left(\frac{2e}{\omega^2\tau^2}\right)^2\left[4\sin^2\left(\frac{1}{2}\omega\tau\right) + (\omega\tau)^2 - 2\omega\tau\sin(\omega\tau)\right] \tag{6.77}$$

Equation (6.77) is important because it provides the normalized detector power response at a given frequency. It is customary to characterize frequency response in terms of a cut-off frequency that is defined as the frequency value at which the power response drops to one-half of its maximum value. This half-power point occurs at a frequency approximately equal to $(2\tau)^{-1}$. For extremely small values of τ, however, Eq. (6.76) reduces to

$$I(\omega) \simeq e \tag{6.78}$$

The corresponding spectral power density function could be determined using Eq. (6.40) as

$$S_T(\nu) = 2e\langle i\rangle \tag{6.79}$$

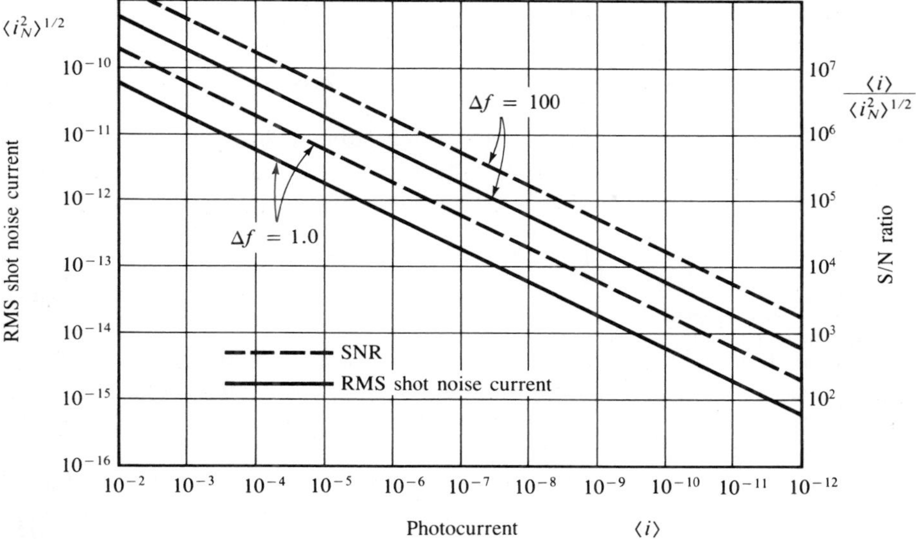

FIGURE 6.8

Shot noise characteristics. (Adapted from W. Budde, *Optical Radiation Measurement.* Vol. 4: *Physical Detectors of Optical Radiation.* Orlando, Fl.: Academic Press, 1983, p. 61.)

Accordingly, the power of the equivalent noise generator at ν is given by

$$\langle i_N^2 \rangle \equiv S_T(\nu)\,\Delta\nu = 2e\langle i \rangle\,\Delta\nu \tag{6.80}$$

which is identical to that given in Eq. (6.74). It is important to realize that this noise current, as well as the thermal noise current, is an alternating current with frequencies near ν and is a function of the bandwidth of the measuring device. The square root of the right-hand side of Eq. (6.79) is shown plotted in Fig. 6.8 for two constant values of bandwidth (1 Hz and 100 Hz). The graph also shows the ratio of photocurrent and noise current.

Consider now that an ideal detector of bandwidth BW and quantum efficiency η is exposed to an irradiation of signal power P_s. The corresponding signal current generated in the photodetector is

$$i_s = \frac{e\eta P_s}{h\nu} \tag{6.81}$$

while the root-mean-square noise current associated with the signal current is given using Eqs. (6.74) and (6.81):

$$\langle i_N^2 \rangle^{1/2} = e\left[\frac{2\eta \mathrm{P}_s(\mathrm{BW})}{h\nu}\right]^{1/2} \tag{6.82}$$

The associated signal-to-noise ratio is thus found to be

$$\mathrm{SNR}_{SL} = \frac{i_s}{\langle i_N^2 \rangle^{1/2}} = \left[\frac{\eta P_s}{2h\nu(\mathrm{BW})}\right]^{1/2} \tag{6.83}$$

When SNR <1, the signal remains completely submerged within the noise, but with SNR >1, the signal becomes identifiable. We often define the **noise equivalent power** (NEP), $P_{\mathrm{NEP},SL}$, as the value of incident radiation power that provides an SNR of unity.

$$P_{\mathrm{NEP},SL} = \frac{2h\nu(\mathrm{BW})}{\eta} = \frac{h\nu}{\eta\tau} \tag{6.84}$$

Note that the bandwidth of a filter is given by $(2\tau)^{-1}$ where τ is the sampling time. Interestingly, Eq. (6.84) implies that when quantum efficiency is unity, on the average one photon is detected by the detector during each of the measurement periods.

In practice, detectors are also exposed to a background power P_B in addition to the signal power P_s. For example, in the case of visible light, this background power can be due to solar scatter from the clouds, whereas in the case of infrared detection the background power can be due to surrounding thermal radiation. In the background-limited detectors, the rms noise current is given by

$$\langle i_N^2 \rangle^{1/2} = e\left[\frac{2\eta(P_s + P_B)(\mathrm{BW})}{h\nu}\right]^{1/2} \tag{6.85}$$

Accordingly, using Eqs. (6.81) and (6.85), the signal-to-noise ratio in such detection of photons is given by

$$\mathrm{SNR}_{BL} = P_s\left[\frac{\eta}{2h\nu(P_s + P_B)(\mathrm{BW})}\right]^{1/2} \tag{6.86}$$

The noise-equivalent power, for background-limited detection ($P_B \gg P_s$) can be thus determined as

$$P_{\mathrm{NEP},BL} = \left[\frac{2h\nu P_B(\mathrm{BW})}{\eta}\right]^{1/2} \tag{6.87}$$

Note that for most of the background-limited optical and infrared detectors, the minimum detectable power varies as the square root of the bandwidth rather than as the bandwidth itself. By comparison, the minimum detectable power in the source-limited case, such as in radio-frequency and microwave systems, varies as the bandwidth. It is no wonder, therefore, that the output power of an incoherent photon detector varies as the square of input power.

It is now appropriate to introduce a figure of merit, referred to as detectivity, which is used to characterize photodetectors. **Detectivity** is defined as the reciprocal of noise equivalent power. A large value for detectivity implies that the detector is highly sensitive. However, in most of the extended objects, the value of the background power P_B is dependent on the surface area A and, therefore, detectivity turns out to be a function of the detector surface area. Consequently, it is more common to use an area-independent figure of merit known as the **specific detectivity** D^* which is defined as

$$D^* = \frac{\sqrt{A(\text{BW})}}{P_{\text{NEP},BL}} = \sqrt{\frac{\eta}{2h\nu E_{e,B}}} \tag{6.88}$$

where $E_{e,B}$ is the background irradiance incident at the detector. Note that D^* is not only independent of the detector surface area but it is also independent of bandwidth.

Earlier, the vacuum photodiode was treated as an ideal detector having a limited quantum efficiency and a finite frequency cut off. It is probable, however, that thermionic emission from the cathode may also contribute to a dark current (defined as the current that exists even when illumination is absent). The dark current associated with a photocathode is described by the **Richardson-Dushmann equation**, which relates the dark current with the work function ϕ.

$$i_d = \frac{4\pi m_e e k^2 T^2 A}{h^3} e^{-\phi/kT} \tag{6.89}$$

where the photocathode area is denoted by A. In actuality, the photocathode dark current is a manifestation of a sequence of random events. The resulting shot noise is characterized by a background power given by

$$P_{B,\text{dark}} = \frac{i_d h\nu}{e\eta} \tag{6.90}$$

Accordingly, the dark current limited noise equivalent power for the device can be found from Eq. (6.87) as

$$P_{\text{NEP},DL} = \frac{h\nu}{\eta}\left[\frac{2i_d(\text{BW})}{e}\right]^{1/2} \tag{6.91}$$

The equivalent circuit of the detector showing both the signal and noise current generators is shown in Fig. 6.9.

Let us now attempt to derive the specific detectivity of an ideal photon detector, as shown in Fig. 6.10, when the photon detector is exposed to a background thermal radiator at temperature T. Let $\eta = 1$, for simplicity. Consider that the photodetector has a detector surface of area A and that it is collecting thermal energy through a cone of apex angle $2\theta_0$. Note that Eq. (2.41) provides the energy density $\rho(\nu)\,d\nu$ of the thermal radiator as

$$\rho(\nu)\,d\nu = \frac{8\pi\nu^2\,d\nu}{c^3}\frac{h\nu}{e^{h\nu/kT} - 1} \tag{6.92}$$

Thus the amount of thermal energy having a frequency bandwidth $d\nu$ which

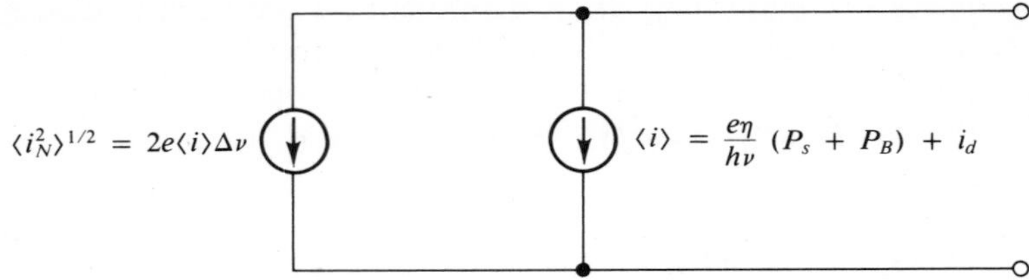

FIGURE 6.9

Equivalent circuit of the vacuum photodiode detector

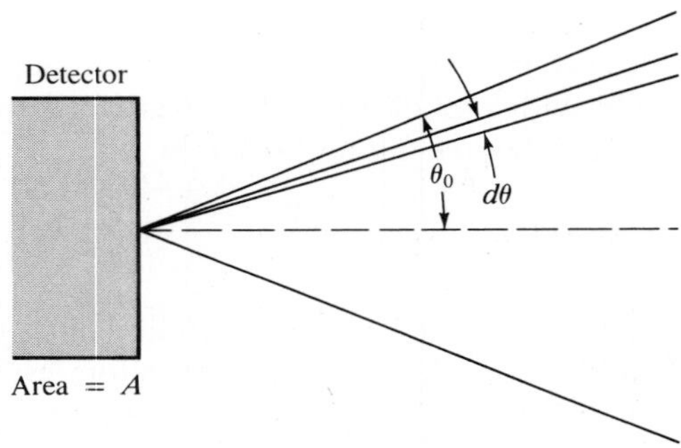

FIGURE 6.10

The geometric parameters involving a photodetector

passes through the detector surface is given by

$$\begin{aligned} dP_B &= \frac{c}{4\pi}\int_0^{\theta_0} (A\cos\theta)(2\pi\sin\theta\, d\theta)\rho(v)\, dv \\ &= \frac{8\pi^2 hv^3 A\, dv \sin^2\theta_0}{c^2[e^{hv/kT} - 1]} \end{aligned} \tag{6.93}$$

where $A\cos\theta$ is equivalent to the effective surface area and $2\pi\sin\theta\, d\theta$ represents the solid angle pertaining to the cone of our interest. The specific detectivity, for the set up of Fig. 6.10, can be expressed as

$$D^* = \sqrt{\frac{A}{2hvP_B}} = \sqrt{\frac{A}{2(hv)^2\langle R\rangle}} \tag{6.94}$$

where $\langle R\rangle$ is the average rate at which the photoevents occur. The photodetector collects all radiation as long as the frequency is larger than a cut-off frequency value ν_c. Accordingly, the average rate of photoevents is found to be

$$\langle R\rangle = \int_{\nu_c}^{\infty} \frac{1}{h\nu}\, dP_B$$

$$= \frac{8\pi^2 A}{c^2} \sin^2\theta_0 \int_{\nu_c}^{\infty} \frac{\nu^2}{e^{h\nu/kT}-1}\, d\nu$$

$$\simeq \frac{8\pi^2 A}{c^2} \sin^2\theta_0 \int_{\nu_c}^{\infty} \nu^2 e^{-h\nu/kT}\, d\nu \tag{6.95}$$

whenever $h\nu/kT \gg 1$. Defining $x = (h/kT)\nu$ and $x_c = (h/kT)\nu_c$, we get

$$\langle R\rangle = \frac{8\pi^2 A}{c^2} \sin^2\theta_0 \left(\frac{kT}{h}\right)^3 \int_{x_c}^{\infty} x^2 e^{-x}\, dx$$

$$= \frac{8\pi^2 A}{c^2} \sin^2\theta_0 \left(\frac{kT}{h}\right)^3 [x_c^2 + 2x_c + 2]e^{-x_c} \tag{6.96}$$

The value for $\langle R\rangle$ can then be used in Eq. (6.94) to yield the specific detectivity (see Problem 6.8). Figure 6.11 shows the specific detectivity for several of the familiar photodetectors. For comparison, D^* of an ideal detector is also shown. Due to the presence of G-R noise, D^* for the photoconductor is $2^{1/2}$ times lower than that for an ideal photon detector. For any given wavelength, the detector having the highest D^* value is usually the best choice.

Consider now a photomultiplier tube that has n dynodes, as shown in Fig. 4.17. The incoming current i_0 arriving at the first dynode has an accompanying noise current that is no different from what was derived earlier for the vacuum photodiode. Successive PMT stages amplify signal as well as noise currents and contribute to an overall power gain of δ^{2n} where n represents the number of dynodes and δ represents the average number of electrons emitted by each of the dynodes for every incident electron. Unfortunately, δ is not necessarily a constant and thus the SNR of a PMT is different from that of a vacuum photodiode. For δ up to the order of about 5, the probability distribution function $p(k)$ describing the probability of having k secondaries from each electron follows that of a Poisson distribution. Consequently, $\langle k\rangle = \langle\delta\rangle$ and $\langle k^2\rangle = \langle\delta\rangle^2 + \langle\delta\rangle$.

The current arriving at the mth dynode ($m \leqslant n$) of a PMT is equivalent to $\langle\delta\rangle^{m-1} i_0$. Since each of the random pulses is caused by a charge in the amount of ke, the mean square noise current is found using Eq. (6.74) as

$$\langle i_{N,m}^2\rangle \simeq 2(ke)[k\langle\delta\rangle^{m-1} i_0](\mathrm{BW})$$

$$= 2ei_0 k^2 \langle\delta\rangle^{m-1}(\mathrm{BW}) = 2ei_0\langle k^2\rangle\langle\delta\rangle^{m-1}(\mathrm{BW}) \tag{6.97}$$

where $\langle k^2\rangle$ has been used in place of k^2 because k is not a constant. By using $(\langle k^2\rangle - \langle\delta\rangle^2)$ instead of $\langle k^2\rangle$ in Eq. (6.97), we can determine the excess noise that is introduced by the current at the mth stage. The sum total of all noise arriving at the nth dynode is thus obtained from the product of excess noise and

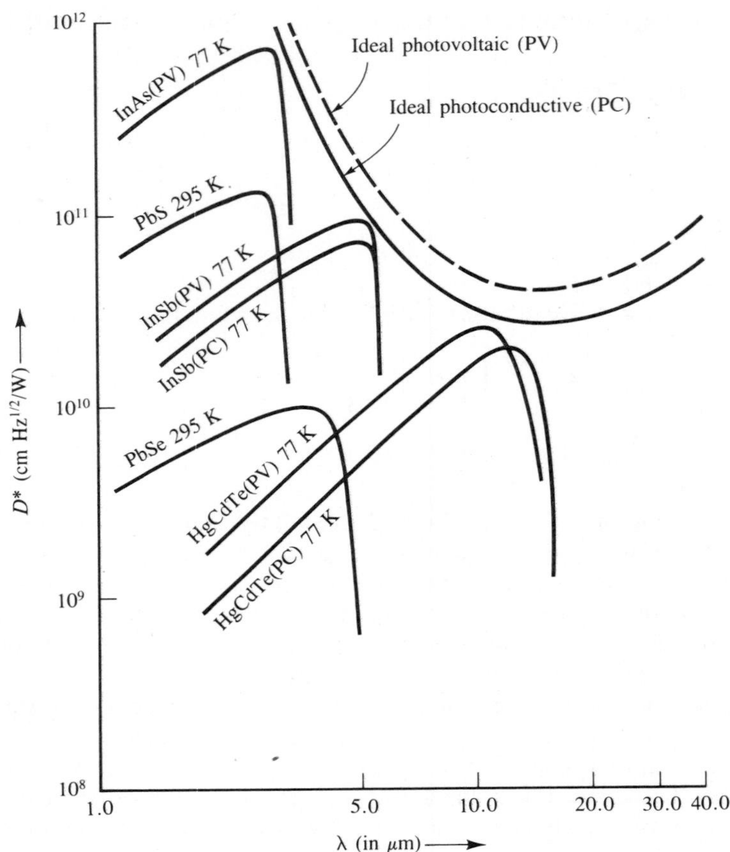

FIGURE 6.11

Specific detectivity for some of the detectors

gain of the remaining stages $\langle\delta\rangle^{2(n-m)}$. Accordingly, the effective excess mean square noise current contributed by the mth dynode is given by

$$\begin{aligned}\langle i_{N,m}^2\rangle_{\text{out}} &= \langle\delta\rangle^{2(n-m)}2ei_0\langle\delta\rangle^{m-1}(\text{BW})[\langle k^2\rangle - \langle\delta\rangle^2] \\ &= 2ei_0\langle\delta\rangle^{2n-m-1}(\text{BW})[\langle k^2\rangle - \langle\delta\rangle^2]\end{aligned} \tag{6.98}$$

The mean square noise of the PMT is determined by adding the sum of excess contributions of all dynodes and the mean square noise for an ideal photomultiplier as follows:

$$\begin{aligned}\langle i_N^2\rangle_{\text{PMT}} &= 2ei_0(\text{BW})[\langle\delta\rangle^{2n} + \{\langle k^2\rangle - \langle\delta\rangle^2\}\sum_{m=1}^{n}\langle\delta\rangle^{2n-m-1}] \\ &= 2ei_0(\text{BW})\langle\delta\rangle^{2n}[1 + \{\langle k^2\rangle - \langle\delta\rangle^2\}\sum_{m=1}^{n}\langle\delta\rangle^{-(m+1)}]\end{aligned}$$

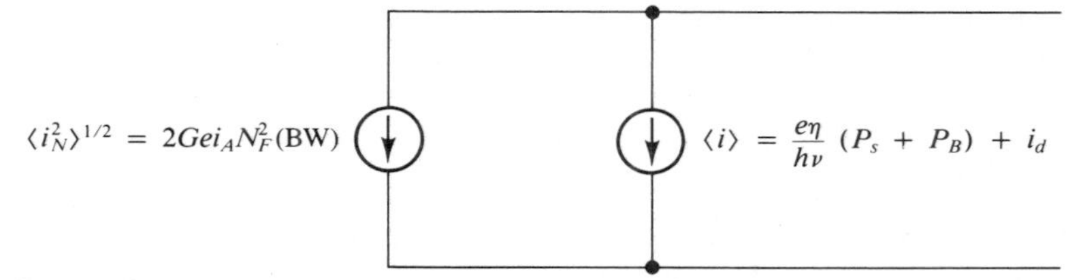

$\langle i_N^2 \rangle^{1/2} = 2Gei_A N_F^2(\mathrm{BW})$

$\langle i \rangle = \dfrac{e\eta}{h\nu}(P_s + P_B) + i_d$

FIGURE 6.12

Equivalent circuit of a PMT

$$= 2ei_0(\mathrm{BW})\langle\delta\rangle^{2n}\left[1 + \frac{\langle k^2\rangle - \langle\delta\rangle^2}{\langle\delta\rangle(\langle\delta\rangle - 1)}\right]$$

$$= 2ei_0(\mathrm{BW})\langle\delta\rangle^{2n}\left[\frac{\langle k^2\rangle - \langle\delta\rangle}{\langle\delta\rangle(\langle\delta\rangle - 1)}\right]$$

$$= 2ei_0(GN_F)^2(\mathrm{BW}) \tag{6.99}$$

where the noise factor N_F is defined as

$$N_F = \sqrt{\frac{\langle k^2\rangle - \langle\delta\rangle}{\langle\delta\rangle(\langle\delta\rangle - 1)}} \tag{6.100}$$

and G is the PMT current gain. Since $\langle k^2\rangle = \langle\delta\rangle^2 + \langle\delta\rangle$, N_F reduces to $\langle\delta\rangle/(\langle\delta\rangle - 1)$. Again, since the photomultiplier anode current is

$$i_A = Gi_0 \tag{6.101}$$

the mean square noise current is sometimes expressed as

$$\langle i_N^2\rangle_{\mathrm{PMT}} = 2Gei_A N_F^2(\mathrm{BW}) \tag{6.102}$$

Recall also that the total photocathode current consists of dark current, signal current, as well as background current, and is thus given by

$$i_0 = i_d + \frac{\eta e}{h\nu}(P_s + P_B) \tag{6.103}$$

Equation (6.102) can be used now to obtain the SNR of a PMT as

$$\mathrm{SNR}_{BL,\mathrm{PMT}} = \frac{i_s}{\sqrt{\langle i_N^2\rangle_{\mathrm{PMT}}}}$$

$$= \frac{\sqrt{\eta}P_s}{2h\nu(N_F)^2(\mathrm{BW})\left[P_s + P_B + \dfrac{h\nu i_d}{\eta e}\right]^{1/2}} \tag{6.104}$$

The equivalent circuit for a PMT is shown in Fig. 6.12. Note that the noise factor accounts for the fact that δ is not a constant in all of the dynodes. For the most typical photomultiplier, N_F approaches 1.22 whereas it approaches unity only in the case of an ideal PMT.

6.3.2 Thermal Noise

Thermal noise (also referred to as Johnson, or Nyquist, noise) is an integral part of the detectors that is caused by the random motion of charged carriers within the electrical elements of the detector. Even though the charge neutrality is maintained in the electrical elements, the random thermal motion of charge carriers gives rise to fluctuations in the carrier concentration gradient. The thermal agitation of electrons in the electrical elements, as well as the bandwidth effects of the processing filter to which these elements feed, results in voltage fluctuations. The thermal noise is not an inherent aspect of the detection process; however, it determines the sensitivity of the detector. If a second electrical element is now connected across the first element, the thermally induced voltage of the first gives rise to a current and causes a power transfer to the second element. Such a transfer of power between the electrical elements manifests itself as the thermal noise in a detector.

The thermal noise becomes more prominent in the low-frequency applications of photodetectors. However, an understanding of classical physics is not enough for the comprehension of thermal noise. Ideas from both field quantization and statistical mechanics are often applied simultaneously to the understanding of thermal noise. The probability of finding a physical system at temperature T in a given quantum state n, which is in thermal equilibrium with its environment, is dictated by the Boltzmann distribution formula

$$p(n) = \frac{e^{-nh\nu/kT}}{\sum_{n=0}^{\infty} e^{-nh\nu/kT}} \tag{6.105}$$

It is well known that the energy of a mode of frequency ν is given by $nh\nu$. Thus the average energy of the mode is determined as

$$\begin{aligned}\langle E \rangle &= h\nu \sum_{n=0}^{\infty} np(n) = h\nu \sum_{n=0}^{\infty} ne^{-nh\nu/kT}\{1 - e^{-h\nu/kT}\} \\ &= h\nu[1 - e^{-h\nu/kT}] \sum_{n=0}^{\infty} ne^{-nh\nu/kT} \\ &= h\nu[1 - e^{-h\nu/kT}]\left[-\frac{d}{d\alpha}\left(\sum_{n=0}^{\infty} e^{-\alpha n}\right)\right]\end{aligned} \tag{6.106}$$

where

$$\alpha = h\nu/kT \tag{6.107}$$

Equation (6.106) can be simplified further as

$$\langle E\rangle = h\nu[1 - e^{-h\nu/kT}]\left[-\frac{d}{d\alpha}\left(\frac{1}{1-e^{-\alpha}}\right)\right]$$

$$= \frac{h\nu}{e^{h\nu/kT}-1} \tag{6.108}$$

Consequently, the power ΔP associated with the mode in the frequency interval ν and $\nu + \Delta\nu$ is given by

$$\Delta P = \langle E\rangle\,\Delta\nu = \frac{h\nu\,\Delta\nu}{e^{h\nu/kT}-1} \tag{6.109}$$

$$\simeq kT\,\Delta\nu \tag{6.110}$$

when $kT \gg h\nu$. Equation (6.109) is referred to as the quantum generalization of the resistance-independent Nyquist formula. Note that the approximate expression of Eq. (6.110) is valid only at normal temperature and at frequencies up to the microwave region. At higher frequencies (10^{14} Hz or higher), Eq. (6.109) provides a more correct answer.

Under a matched condition, the average power $\langle P\rangle$ dissipated in a resistive element is given by

$$\langle P_{\text{th}}\rangle = \frac{\langle V_{N,\text{th}}^2\rangle}{4R} = kT\,\Delta\nu \tag{6.111}$$

where $\langle V_{N,\text{th}}^2\rangle$ is the mean square thermal noise voltage appearing across a load having resistance R. The rms thermal noise voltage can thus be determined as

$$\langle V_{N,\text{th}}^2\rangle^{1/2} = [4RkT\,\Delta\nu]^{1/2} \tag{6.112}$$

Note that the rms thermal noise voltage is independent of the operating frequency but dependent on the frequency bandwidth. The equivalent rms thermal noise current can be obtained by dividing the right-hand side of Eq. (6.112) by R:

$$\langle i_{N,\text{th}}^2\rangle^{1/2} = \sqrt{\frac{4kT\,\Delta\nu}{R}} \tag{6.113}$$

Figure 6.13 shows the equivalent noise sources of a resistive element while Fig. 6.14 illustrates the characteristics of the noise voltage and noise current for different values of resistance and bandwidths. In general, the spectral power density of thermal noise is given by

$$P(\omega) = \frac{h\nu}{2(e^{h\nu/kT}-1)} \tag{6.114}$$

which is a constant when $kT \gg h\nu$.

Another aspect that needs to be pointed out is that each of the resistive elements of the actual circuits has associated with it a small amount of

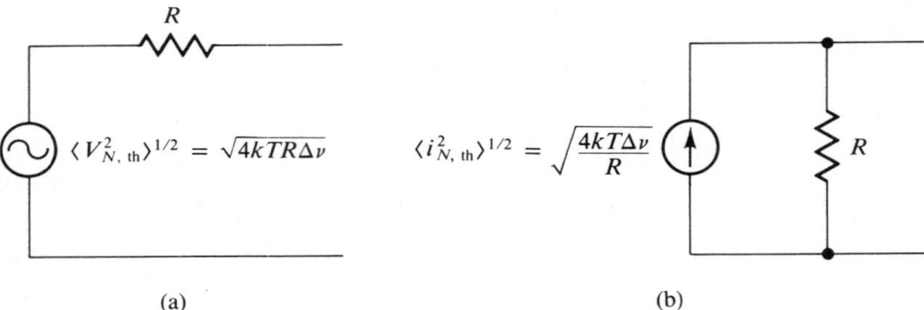

FIGURE 6.13

Equivalent thermal noise circuits

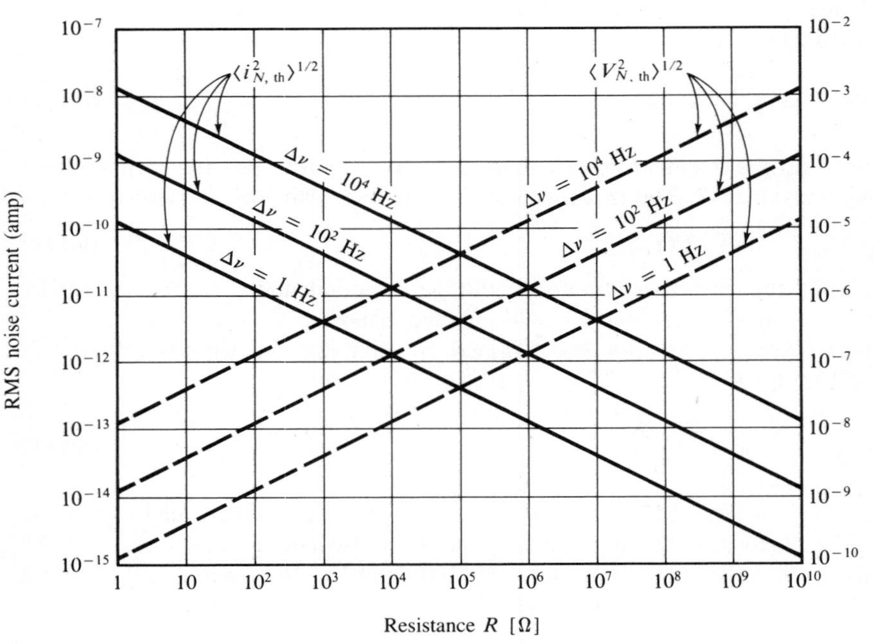

FIGURE 6.14

Root-mean-square noise current and voltage curves. (Adapted from W. Budde, *Optical Radiation Measurements*. Vol. 4: *Physical Detectors of Optical Radiation*. Orlando, Fl.: Academic Press, 1983, p. 63.)

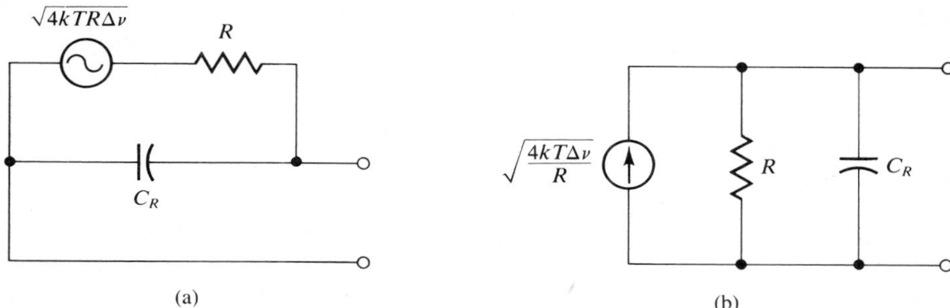

FIGURE 6.15

Equivalent thermal noise circuits including stray capacitances

capacitance. The presence of such capacitance generally reduces the noise voltage that appears across the resistor terminal because

$$\langle V_{N,\mathrm{th}}^2 \rangle = \frac{4kRT\,\Delta\nu}{1+(\omega RC_R)^2} \tag{6.115}$$

where C_R represents the stray capacitance of the resistive element. Figure 6.15 shows the thermal noise equivalent circuits that include a stray capacitance. Thermal noise occurs in all of the dissipative elements, regardless of whether or not they carry current. If the element carries a current, the noise is superimposed on the signal, which in turn affects the SNR.

The total mean square noise voltage associated with a network of m resistors of resistance R_i, connected in series, is given by

$$\langle V_{N,\mathrm{th}}^2 \rangle = 4k\,\Delta\nu \sum_{i=1}^{m} R_i T_i \tag{6.116}$$

where T_i is the operating temperature of the ith resistor. The equivalent noise temperature T_{eq} defined as

$$T_{\mathrm{eq}} = \frac{\sum R_i T_i}{\sum_i R_i} \tag{6.117}$$

is the temperature at which the series equivalent resistor will yield the same mean square noise voltage. Equation (6.116) can, therefore, be rewritten in terms of the series equivalent resistance R_{eq} as follows:

$$\langle V_{N,\mathrm{th}}^2 \rangle = 4k\,\Delta\nu T_{\mathrm{eq}} \sum_{i=1}^{m} R_i = 4k\,\Delta\nu T_{\mathrm{eq}} R_{\mathrm{eq}} \tag{6.118}$$

Similarly, the equivalent mean square noise current is given by

$$\langle i_{N,\mathrm{th}}^2 \rangle = \frac{4kT_{\mathrm{eq}}\,\Delta\nu}{R_{\mathrm{eq}}} \tag{6.119}$$

provided the equivalent noise temperature is given by

$$T_{\text{eq}} = \frac{\sum_i \frac{T_i}{R_i}}{\sum_i \frac{1}{R_i}} \tag{6.120}$$

Both shot noise and thermal noise are also referred to as **white noises**. If white noise alone is present in a photodetection, the SNR can be improved by extending the duration of measurement—i.e., reducing the bandwidth. If two or more sources of noise are present in a circuit, and if they are statistically independent, then the total noise is equal to the square root of the sum of squares of the individual noises. For example, when both shot-noise current and thermal-noise current are being generated simultaneously in a detector circuit, the total noise current $I_{N,\text{tot}}$ is given by

$$\begin{aligned} I_{N,\text{tot}} &= [i^2_{N,\text{shot}} + i^2_{N,\text{thermal}}]^{1/2} \\ &= \left[2e\langle i \rangle \, \Delta v + \frac{4kT\,\Delta v}{R} \right]^{1/2} \end{aligned} \tag{6.121}$$

using Eqs. (6.80) and (6.113). Provided the two noise currents are correlated to each other, the total noise is given by

$$(I_{N,\text{tot}})^2 = (I_{N,\text{shot}})^2 + (I_{N,\text{thermal}})^2 + 2C(I_{N,\text{shot}})(I_{N,\text{thermal}}) \tag{6.122}$$

where $-1 \leqslant C \leqslant 1$ is a correlation coefficient. When $C = 0$, Eq. (6.122) reduces to Eq. (6.121). However, when $C = \pm 1$, the two sources are said to be totally correlated yielding

$$I_{N,\text{tot}} = I_{N,\text{shot}} \pm I_{N,\text{thermal}} \tag{6.123}$$

6.3.3 Generation-Recombination Noise

Statistical fluctuations present in the rates of generation and recombination of the charge carriers in the sensing element—e.g., in a photoconductor—result in a noise that is generally referred to as the **generation-recombination** (G-R) **noise**. Consequently, the corresponding noise current produced in the detector consists of a set of random pulses whose widths fluctuate in accordance with the probability distribution of generation-recombination. The pulses are generally found to have a Poisson distribution.

The behavior of charge carriers is characterized by their average recombination lifetime τ. Thus, if $q(t)$ is the probability that a charge carrier has not yet recombined within time t, then we have

$$\frac{dq}{dt} = -\frac{q}{\tau} \tag{6.124}$$

which yields

$$q(t) = e^{-t/\tau} \tag{6.125}$$

Defining $p(t)\,dt$ as the probability that a recombination will occur during the time interval between t and $t + dt$, we find that

$$q(t) = 1 - \int_0^t p(t)\,dt \tag{6.126}$$

Accordingly, we can evaluate the recombination probability distribution function to be

$$p(t) = -\frac{dq}{dt} = \frac{e^{-t/\tau}}{\tau} \tag{6.127}$$

It can be seen using Eq. (6.127) that the average recombination time $\langle t \rangle$ is indeed equal to τ.

The contribution of the elemental current pulses to the mean square noise current is found to be

$$d\langle i_N^2 \rangle = 2(Ge)\frac{t}{\tau}\,d\langle i \rangle(\mathrm{BW}) \tag{6.128}$$

where G is the average device gain, Ge is the average charge in each of the pulses, and $d\langle i \rangle$ is the elemental current associated with each pulse of width t. However, $d\langle i \rangle$ can be determined from Eq. (6.127) as

$$d\langle i \rangle = \frac{t}{\tau}\,p(t)\langle i \rangle\,dt \tag{6.129}$$

which can be then incorporated in Eq. (6.128) to evaluate the total mean square noise current:

$$\langle i_N^2 \rangle = \frac{2Ge\langle i \rangle(\mathrm{BW})}{\tau^3}\int_0^\infty t^2 e^{-t/\tau}\,dt = 4Ge\langle i \rangle(\mathrm{BW}) \tag{6.130}$$

When Eq. (6.130) is compared with Schottky's formula (Eq. (6.74)), we note that this mean square noise current is two times larger than that due to a photodiode. This is a consequence of the fact that the fluctuating effect charge is now Ge rather than e.

The root mean square signal current in a photoconductor is given by

$$\langle i_s^2 \rangle^{1/2} = \frac{\eta G e P_s}{h\nu} \tag{6.131}$$

Again, the overall mean square noise current is determined by summing up the contributions due to shot noise, Johnson noise, and R-G noise present in both photocurrent and dark current. Thus the mean square noise current is given by

$$\langle i_N^2 \rangle_{\mathrm{photoconductor}} = \frac{4\eta G^2 e^2(\mathrm{BW})}{h\nu}\left[P_s + P_B + \frac{h\nu i_d}{\eta G e} + \frac{h\nu kT}{\eta G^2 e^2 R}\right] \tag{6.132}$$

Accordingly, the photoconductor SNR is found to be

$$\mathrm{SNR}_{\text{photoconductor}} = \frac{P_s}{2}\left[\frac{\eta}{h\nu(\mathrm{BW})\left[P_S + P_B + \dfrac{h\nu i_d}{\eta G e} + \dfrac{h\nu kT}{\eta G^2 e^2 R}\right]}\right]^{1/2} \tag{6.133}$$

which can then be used to determine the noise equivalent power.

In practice, it is desirable for the detection system to be limited by only photon noise. Cooling the photoconductor minimizes the contribution of dark current, while the effect of the thermal noise can be minimized by increasing both detector and load resistances. Thus the NEP of a detector limited primarily by G-R noise in the photocurrent is found to be

$$P_{\mathrm{NEP,GRL}} = \sqrt{\frac{4h\nu P_B(\mathrm{BW})}{\eta}} \tag{6.134}$$

Equation (6.134), when compared to Eq. (6.87), indicates that the NEP of a photoconductor limited by G-R noise is $2^{1/2}$ times larger than the NEP of a detector limited by only photon noise.

6.4 Heterodyne Detection

All of the detection schemes considered thus far are classified as direct detection type. In all of these schemes, the generated output signal (voltage or current) varies linearly as the incoming signal power. However, none of these detectors is suitable for modulation purposes because no information regarding the phase or frequency of the optical signal is retained by the output.

Consider the electric field of an optical carrier wave that can be described by

$$E_c(t) = E_{co}\cos(\omega_c t + \phi_c) \tag{6.135}$$

where E_{co}, ω_c, and ϕ_c, respectively, represent the amplitude, angular frequency, and phase of the field. When this wave impinges directly on a detector, the response of a detector may be given by

$$I_d = RE_{co}^2\langle\cos(\omega_c t + \phi_c)\rangle = \tfrac{1}{2}RE_{co}^2 \tag{6.136}$$

where R denotes the responsivity of the detector. Apparently, therefore, direct detection schemes are not viable for the purposes of the modulation of signals. A heterodyne detection scheme that retains the phase information at the output can be employed to overcome this limitation. It will be shown in this section that the heterodyne output signal (current or voltage) is proportional to the input optical signal. The scheme turns out to be very noisy; however, it is used frequently whenever a better frequency resolution is desirable.

The heterodyne detector is the optical analog of the radio-frequency receivers. Whereas the radio-frequency systems use a diode (also referred to as a square-law detector) whose voltage is reasonably proportional to the square of

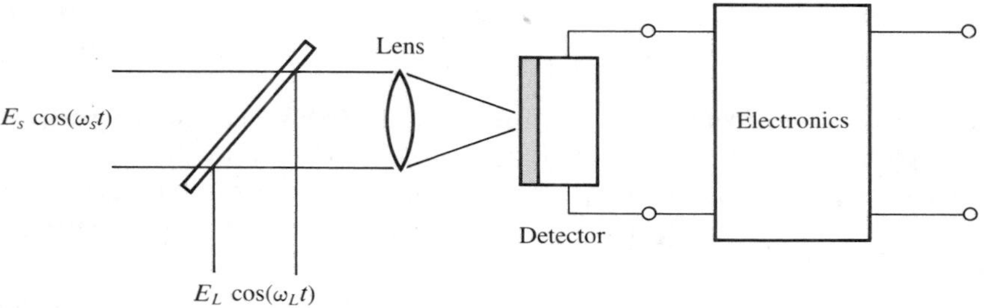

FIGURE 6.16

An optical heterodyne detector

the current, the optical heterodyne detectors employ photon detectors where signal current is proportional to the square of the electric field. Figure 6.16 shows the schematic of an optical heterodyne detector. The to-be-detected signal of frequency ω_s is passed through a transmitting mirror (which reflects very little) and is combined with an intense optical field, referred to as the local oscillator field, with a frequency of $\omega_L \geqslant \omega_s$ by means of a reflecting mirror. The total field is then converged to a photodetector and thereafter processed electronically. The total field at the detector is thus given by

$$E_t(t) = E_s e^{j\omega_s t} + E_L e^{j\omega_L t} \tag{6.137}$$

such that the two fields are linearly polarized in the same direction after being combined. Using Eq. (1.29), the cycle-averaged power corresponding to this field can be evaluated to give

$$P_t(t) = \frac{\alpha}{2}\sqrt{\frac{\varepsilon_0}{\mu_0}}\,[|E_s|^2 + |E_L|^2 + 2|E_s||E_L| \cos[(\omega_L - \omega_s)t + \phi]] \tag{6.138}$$

where α is a constant and ϕ is the relative phase between E_s and E_L which is given by

$$\phi = \tan^{-1}\left(\frac{\text{Im}(E_L)}{\text{Re}(E_L)}\right) - \tan^{-1}\left(\frac{\text{Im}(E_s)}{\text{Re}(E_s)}\right) \tag{6.139}$$

defining the signal and local oscillator currents, respectively, as

$$i_s = \frac{\eta\alpha e}{2hv}\sqrt{\frac{\varepsilon_0}{\mu_0}}\,|E_s|^2 \tag{6.140a}$$

and

$$i_L = \frac{\eta\alpha e}{2hv}\sqrt{\frac{\varepsilon_0}{\mu_0}}\,|E_L|^2 \tag{6.140b}$$

Both Eq. (6.138) and Eq. (6.81) can be used to provide the total current $i_t(t)$ as

$$i_t(t) = i_s + i_L + 2(i_s i_L)^{1/2} \cos[(\omega_L - \omega_s)t + \phi] \qquad \textbf{(6.141)}$$

The first two terms of Eq. (6.141) are both dc terms and, of the two, i_L is overwhelmingly large. The intermediate frequency (IF) term can be separated from the two dc terms by inserting an electrical bandpass filter (centered at $\omega_L - \omega_s$) at the output of the detector. Accordingly, the video signal current is thus

$$i_{vs} = 2(i_s i_L)^{1/2} \cos[(\omega_L - \omega_s)t + \phi] \qquad \textbf{(6.142)}$$

The video signal current is dependent on amplitude, angular frequency, and phase and, therefore, the heterodyne detector is more suitable for the purposes of modulation. Note that in the event the local oscillator is set to a frequency which is the same as that of the signal, the detector output can still be used for both amplitude and phase modulation. Such a technique is referred to as the **homodyne detection**. The strength of the heterodyne signal is often characterized by its mean square value given by

$$\langle i_{vs}^2 \rangle = 2 i_s i_L \qquad \textbf{(6.143)}$$

One can make the video signal current large by increasing the local oscillator power P_L. But an increasing local oscillator power increases the dc current, which in turn increases the amount of shot noise. The total heterodyne SNR is given by

$$\text{SNR}_{\text{heterodyne}} = \sqrt{\frac{\langle i_{vs}^2 \rangle}{2 e i_L (\text{BW})}} = \sqrt{\frac{i_s}{e(\text{BW})}} = \sqrt{\frac{\eta P_s}{h\nu(\text{BW})}} \qquad \textbf{(6.144)}$$

where P_s is the signal power. The video signal power is directly proportional to the input optical power. The SNR is inversely proportional to the bandwidth very much like that in the radio-frequency systems. Note also that when the local oscillator contributes to a single-frequency wave, the output video spectrum for a broadband signal becomes an exact duplicate of the input spectrum. The heterodyne detectors provide noise-free gain just as in a photomultiplier. Analysis shows that the use of a photoconductor in the heterodyne scheme can increase the minimum detectable signal by a factor of $2^{1/2}$.

6.5 · Detector Performance Parameters

The design and success of an optical sensor depends to a great extent on the performance of the detector. Two of these performance parameters, NEP and normalized detectivity, were discussed in previous sections. The NEP was defined as the power incident on the detector that produces an SNR of unity for a specified bandwidth. A third important parameter in common usage is defined in terms of the NEP and is referred to as the **noise equivalent input**, NEI. It is defined as the *irradiance incident on the optical detector that produces an SNR of unity*. In terms of NEP, it is expressed as

$$\mathrm{NEI} = \frac{\mathrm{NEP}}{\sqrt{A}} \tag{6.145}$$

Responsivity is another important detector performance parameter that is defined as the signal produced at the detector output caused by a unit radiant power incident upon the detector. In particular, the spectral responsivity is defined in terms of the differential wavelength $d\lambda$ centered at about λ by

$$R_\lambda = \frac{S_\lambda}{\Phi_{e\lambda}\, d\lambda} \tag{6.146}$$

where S_λ is the detector output either in amps or in volts and $\Phi_{e\lambda}$ is the spectral radiant flux that was introduced in Section 2.2.1. The total responsivity of the detector is obtained from the ratio of the total output signal and the total incident radiant flux as

$$R = \frac{\int_0^\infty S_\lambda\, d\lambda}{\int_0^\infty \Phi_{e\lambda}\, d\lambda} = \frac{\int_0^\infty R_\lambda \Phi_{e\lambda}\, d\lambda}{\int_0^\infty \Phi_{e\lambda}\, d\lambda} \tag{6.147}$$

Consequently, the total responsivity depends on both spectral properties of the source as well as on the response properties of the detector.

Most detectors respond to an impulse input function with an exponential rise or decay having a time constant τ. Consider the current response of the optical detector that is of the form

$$i(t) = i_0 e^{-t/\tau} \tag{6.148}$$

The corresponding current frequency response is thus determined as follows

$$i(\omega) = \int_0^\infty i(t) e^{-j\omega t}\, dt = \frac{i_0 \tau}{1 + j\omega\tau} \tag{6.149}$$

By taking the modulus of this expression, we can obtain the frequency response of the detector as

$$R_\lambda(\omega) = \frac{R_\lambda(0)}{\sqrt{1 + \omega^2\tau^2}} \tag{6.150}$$

Responsivity has, therefore, low-pass characteristics. Consequently, the detector can reproduce the input signal pulse only if its response time is short compared to the pulse width duration. Note also that Eq. (6.150) is identical in behavior to Eq. (4.11).

The normalized detectivity D^* was defined in Eq. (6.88) and is basically a measurable quantity independent of both detector area and bandwidth. Therefore, D^* may be used to compare directly the merits of detectors of different sizes and bandwidths. Alternatively, D^* can be expressed in terms of the SNR as

$$D^* = \frac{\sqrt{A(\mathrm{BW})}}{\Phi_e}(\mathrm{SNR}) \qquad \textbf{(6.151)}$$

D^* is often interpreted as the SNR of the detector when 1 W of the radiant power is incident on a detector of area 1 cm^2 with a bandwidth of 1 Hz. This interpretation should not be taken seriously because most of the detectors reach their limiting output well below 1 W and, therefore, may never have an SNR approaching the value of D^*. Since the detectivity is proportional to the responsivity, the spectral detectivity is expected to show a dependence on temporal frequency as

$$D^*_\lambda(\omega) = \frac{D^*_\lambda(0)}{(1+\omega^2\tau^2)^{1/2}} \qquad \textbf{(6.152)}$$

In many of the semiconductor detectors noise is found to be proportional to the fluctuation in the number of background photons. Such detectors are often referred to as the **background-limited photodetectors** (or BLIP for short). For example, the spectral detectivity of the BLIP photoconductive detector is given by

$$D^*_{\lambda,\mathrm{BLIP}} = \frac{1}{2hv}\left[\frac{\eta_\lambda}{E_{eB}}\right]^{1/2} \qquad \textbf{(6.153)}$$

whereas that for a photovoltaic detector is

$$D^*_{\lambda,\mathrm{BLIP}} = \frac{1}{hv}\sqrt{\frac{\eta_\lambda}{2E_{eB}}} \qquad \textbf{(6.154)}$$

where η_λ is the spectral quantum efficiency and E_{eB} is the total background irradiance at the detector.

Problems

6.1 A voltage signal described by $v(t) = e^{at}u(t)$ V is applied as a low-pass filter input where $u(t)$ is a unit step function. Calculate the energies of both input and output provided that the filter has a low frequency gain of unity.

6.2 A signal $v(t) = e^{-at}u(t)$ V is applied as a low-pass filter input where $u(t)$ is a unit step function. The magnitude of the transfer function is equal to $b/(\omega^2 + b^2)^{1/2}$. Determine the ratio a/b when exactly 50% of the input signal energy is transferred to the output.

6.3 Find the power spectral density of the periodic signal $v(t) = 0.5\cos(\omega_0 t + \theta)$. Determine the mean power across a 5-Ω load.

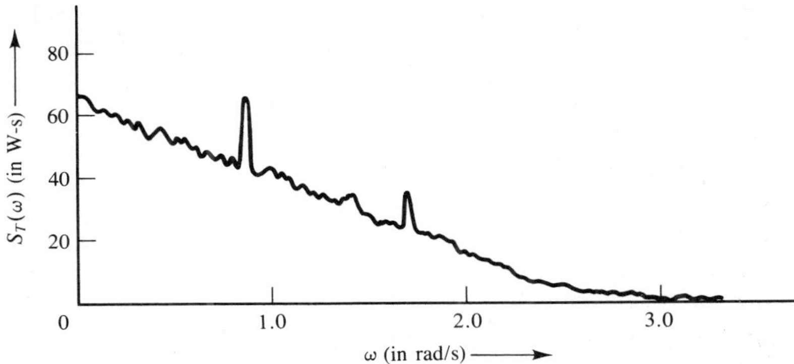

FIGURE 6.P1

6.4 Determine what periodic components are present in the signal whose power spectrum is provided in Fig. 6.P1. Estimate the length of the observation time (in seconds).

6.5 Find the autocorrelation function of the periodic signal given in Problem 6.3.

6.6 Determine and plot the autocorrelation function of a periodic triangular wave with a peak-to-peak amplitude V_m, period T, and a mean value of $\frac{1}{2}V_m$.

6.7 Determine the power spectral density of $F_1 e^{j\omega_0 t}$ by (a) first finding its autocorrelation function and then (b) by taking the Fourier transform of the autocorrelation function.

6.8 Calculate the detectivity for an ideal detector where $\theta_0 = \pi/2$ and $T =$ 300 K. Plot D^* as a function of the frequency of radiation.

6.9 Determine the equivalent noise temperature (for both noise current and noise voltage equivalent circuits) when $kT/h\nu$ is not larger than unity.

6.10 Obtain the rms thermal noise current and the rms thermal noise voltage curves for the case when $kT < h\nu$.

6.11 (a) Show that the Poisson distribution function is normalized. (b) Show that for n number of emissions of photons, $\langle n^2 \rangle = \langle n \rangle + \langle n \rangle^2$ as long as the emission of photons obeys Poisson distribution.

6.12 (a) Show that for Bose-Einstein statistics, the expected value of n^2 is given by $\langle n^2 \rangle = [e^{-h\nu/kT} + e^{-2h\nu/kT}]/[1 + e^{-h\nu/kT}]^2$ and (b) verify Eq. (6.70).

6.13 Determine the SNR for an ideal detector when the source is incoherent in nature and is operated either at very high temperature or at long wavelength or both.

6.14 For a photomultiplier having 5 dynodes, show that the noise factor approaches 1.12. Also show that it approaches unity in the case of an ideal PMT.

6.15 For a photoconductor current $i(t)$ that is equal to e/τ_d when $0 \leqslant t \leqslant \tau$ and 0 otherwise, show that the mean square noise current in the frequency range ν and $\nu + d\nu$ is given by

$$\langle i_N^2 \rangle = \frac{4e\langle I \rangle \left(\dfrac{\tau}{\tau_d} \right) \Delta\nu}{1 + \omega^2 \tau^2}$$

where τ is the lifetime of the photoexcited carriers, τ_d is the drift time, $\langle I \rangle$ is the average current equal to $\langle N \rangle e(\tau/\tau_d)$, and $\langle N \rangle$ is the average number of carriers generated per second.

6.16 Show that (a) $S(\tau) = S(-\tau)$ and (b) $S(0) \geqslant |S(\tau)|$.

6.17 Using Eq. (6.46), determine the spectral density function of randomly occurring identical events in terms of $F(\omega)$.

6.18 Compare the limiting SNRs of the heterodyne, direct, and homodyne detection, provided that the corresponding bandwidths are equivalent.

References and Suggested Readings

Boyd, R. W. *Radiometry and the Detection of Optical Radiation.* New York: Wiley, 1983.

Budde, W. "Physical Detectors of Optical Radiation." (Vol. IV of *Optical Radiation Measurement*). New York: Academic Press, 1973.

Dean, A. B., C. T. Elliot and A. M. White. "Photoconductive Detectors Employing an Optical Induced Nonlinear Resistance." *Opt. Engn.* Vol. 24 (1985): 668.

Dereniak, E. L., and D. G. Crowe. *Optical Radiation Detectors.* New York: Wiley, 1984.

Jenkins, T. E. *Optical Sensing Techniques and Signal Processing.* United Kingdom: Prentice Hall International (UK) Ltd, 1987.

Kingston, R. H. *Detection of Optical and Infrared Radiation* (2nd printing). Berlin: Springer-Verlag, 1979.

MacDonald, R. I. "High Gain Optical Detection with GaAs Field Effect Transistors." *Appl. Opt.* Vol. 20 (1981): 591.

McCreight, C. R., and J. H. Goebal. "Infrared Charge-Injection-Device Array Performance at Low Background." *Appl. Opt.* Vol. 20 (1981): 3189.

Seyrafi, K. *Electro-Optical Systems Analysis.* Los Angeles: Electro-Optical Research Company, 1973.

Shapiro, J. H. "Heterodyne Mixing Efficiency for Detector Arrays." *Appl. Opt.* Vol. 26 (1987): 3600.

Stremler, F. G. *Introduction to Communication Systems.* Reading, MA: Addison-Wesley, 1977.

Verdeyen, J. T. *Laser Electronics.* Englewood Cliffs, NJ: Prentice-Hall, 1981.

Yariv, A. *Optical Electronics* (3rd ed.). New York: Holt, Rinehart & Winston, 1985.

CHAPTER SEVEN

Modulation of Light

7.1 Introduction

Chapters 4 and 5 introduced a number of important electro-optic devices. However, the treatment of these devices will remain somewhat incomplete if we do not discuss the techniques of altering beams of light in amplitude and phase. The modulating methods consist of altering a particular property or a parameter of a given optical signal in proportion to a second signal. In general, such techniques are referred to as **modulation**. Often electro-optical modulation schemes are used to produce Q-switched lasers. Another potential use of modulation is in the area of electro-optic communications. For example, in amplitude-modulation based communication, the amplitude of an optical signal is altered (while its frequency and phase are kept invariant) by means of a stimulating signal.

Modulation of light is usually accomplished by two means: internal and external. Both of these methods have strict requirements, and the method that is ultimately chosen depends to a great extent on the optical source as well as on the modulating rate. Internal modulation is accomplished by directly influencing the source—for example, either by varying the pump power or the internal driving power or by inserting an optical modulator inside the cavity of a laser. It is possible to pump semiconductor lasers by means of amplitude-modulated current sources. However, switching high currents at high enough rates is often limited by the external circuitry. On the other hand, external modulation is applied only after the radiation has emerged from the source. Internal modulation has the advantage of involving less source power than that of external modulation because a good portion of the generated power in the latter is curtailed by the modulation scheme itself.

The characteristics of optical irradiance—e.g., amplitude, frequency, phase, and polarization—may be altered by introducing a modulator in the beam. The modulators are of many types, but the most important of these are based on electro-optic, magneto-optic, or acousto-optic (or elasto-optic) effects. Basically, these effects involve the application of a particular field that causes variation in either the refractive index or the polarization characteristics of a medium. Accordingly, the next section of this chapter reviews optical polarization, based on an understanding of Maxwell's equation of Chapter 1. This is followed by a detailed treatment of various modulation schemes.

The chapter on the modulation of light is then extended to include a discussion of the capabilities of a photographic recording medium (or plate) that continues to be the most fundamental element for optical processing and filtering systems. There is no doubt that a photographic-film-based recording scheme involves a cumbersome chemical development process that is also time consuming. While the nearly real-time devices like Pockels read-out optical modulator and liquid crystal light valves are being developed, there is little evidence at this time that these will be able to totally replace photographic film. To demonstrate its importance, in this chapter we use photographic film to produce pulse-width modulation of continuous-tone, two-dimensional images. The modulation concepts involving a photographic medium are general enough to be considered equally applicable to those involving the current and future generation real-time recording devices. The resulting binarized images can be employed in optical signal-processing systems as we show in the next chapter.

7.2 Polarization

When an electromagnetic field vector in free space oscillates in a specific plane, the wave is referred to as **plane-polarized**. An actual beam of randomly polarized light usually consists of a combination of many such plane-polarized waves. As we show later, it is also possible to extract polarized beams out of a beam of unpolarized light.

From our earlier discussion of Maxwell's equation, it is obvious that an electromagnetic wave can be characterized by the behavior of either only the electric field or only the magnetic field. Consider a plane wave whose wave vector (represented by the unit vector **k**) is in the direction of the z axis. The electric field vector may point along a direction perpendicular to **k**, whereas the magnetic field vector will point along $\mathbf{k} \times \mathbf{E}$. The monochromatic electric field wave can thus be written in terms of its rectangular coordinates x and y as

$$\mathbf{E}(\mathbf{r}, t) = E_x\mathbf{i} + E_y\mathbf{j} \tag{7.1}$$

where

$$E_x = A_x \cos(\omega t - kz + \phi_x) \tag{7.2a}$$

$$E_y = A_y \cos(\omega t - kz + \phi_y) \tag{7.2b}$$

i and **j** are unit vectors, respectively, along x and y axes; A_x and A_y are amplitudes, and ϕ_x and ϕ_y are phase angles, respectively, of the field components. The classification of waves as to the nature of their polarization in fact depends on (a) the relative phase $\Delta\phi = \phi_x - \phi_y$ and (b) the ratio of the amplitudes A_x/A_y. Provided we eliminate time and spatial dependencies from Eqs. (7.2a) and (7.2b), it is possible to characterize the path of the resultant field vector.

From Eqs. (7.2a) and (7.2b), we can readily obtain

$$\frac{E_x}{A_x}\cos\phi_y - \frac{E_y}{A_y}\cos\phi_x = -\sin(\omega t - kz)\sin(\Delta\phi) \tag{7.3a}$$

and

$$\frac{E_x}{A_x}\sin\phi_y - \frac{E_y}{A_y}\sin\phi_x = -\cos(\omega t - kz)\sin(\Delta\phi) \tag{7.3b}$$

Now, by adding the squares of Eqs. (7.3a) and (7.3b), we get

$$\left(\frac{E_x}{A_x}\right)^2 + \left(\frac{E_y}{A_y}\right)^2 - 2\left(\frac{E_x}{A_x}\right)\left(\frac{E_y}{A_y}\right)\cos(\Delta\phi) = \sin^2(\Delta\phi) \tag{7.4}$$

which is the equation of an ellipse. Consequently, the electromagnetic wave is generally referred to as having elliptical polarization. It implies that, in general, the resultant electric field vector **E** will both rotate and change its magnitude in a way such that the tip of the vector follows an elliptical path. For arbitrary values of A_x, A_y, and $\Delta\phi$, the corresponding ellipse can be just housed within a rectangle of sides $2A_x$ and $2A_y$, as shown in Fig. 7.1. Note that the angle θ formed by its major axis and x axis is given by

$$\theta = \tfrac{1}{2}\,\text{Arc tan}\left[\cos(\Delta\phi)\tan\left\{2\,\text{Arc tan}\left(\frac{A_y}{A_x}\right)\right\}\right] \tag{7.5}$$

and that the sum of the squares of the lengths of the semi-major and semi-minor axes is given by

$$a^2 + b^2 = A_x^2 + A_y^2 \tag{7.6}$$

Further, it should be noted that this elliptical path can be traced by rotating the **E** vector either in a clockwise or counterclockwise direction. The exact direction of tracing is, however, determined by the sign of $\Delta\phi$ and whether or not we are looking along or against the beam. For example, we can base our analysis while looking against the beam, in which case the beam is assumed to approach us out of the plane as shown in Fig. 7.1. The electric field vector is thus referred as

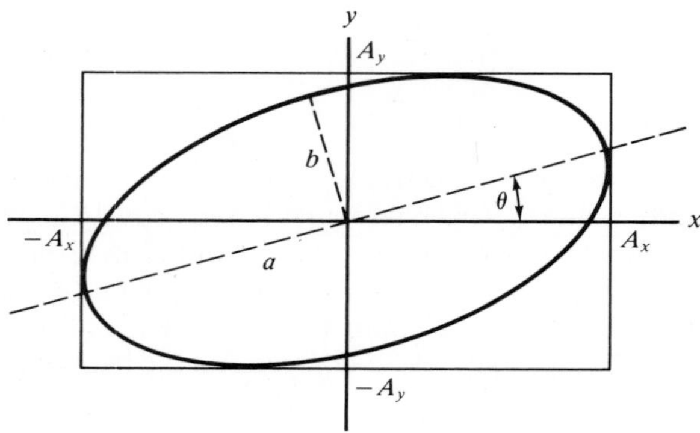

FIGURE 7.1

The characteristics of elliptically polarized light

having right-handed polarization (corresponding to clockwise rotation) when $\sin(\Delta\phi)$ is negative, and likewise it is referred to as having left-handed polarization (corresponding to counterclockwise rotation) when $\sin(\Delta\phi)$ is positive.

There are two special cases of Eq. (7.4) that deserve attention. A plane-polarized—sometimes referred to as linearly polarized—light is said to exist when the relative phase is an integral multiple of π, i.e., when $\Delta\phi = m\pi$ where m is any integer. Equation (7.4) then reduces to

$$E_y = (-1)^m \frac{A_y}{A_x} E_x \tag{7.7}$$

which is the equation of a straight line. A linearly polarized electric field at a fixed position thus undergoes a simple harmonic motion along a straight line. At a particular value of time, however, the field has a sinusoidal spatial dependence, which lies in a plane defined by the line $y = (-1)^m(A_y/A_x)x$ and the z axis. And, finally, if the input amplitudes are equal and if the relative phase is an integral multiple of $\pi/2$, Eq. (7.4) reduces to

$$E_x^2 + E_y^2 = A^2 \tag{7.8}$$

where $A \equiv A_x = A_y$. This special case of the electric field is referred to as the circularly polarized field.

An ordinary light source consists of numerous randomly oriented excited atomic emitters, each of which radiates a polarized wave train for about 10^{-8} s. New wave trains are generated randomly by these emitters and, consequently, the resultant light is referred to as unpolarized. There are a number of important techniques by which the polarization of an incoming light can be altered. These include (a) reflection and refraction, (b) scattering, (c) dichroism, and (d) birefringence.

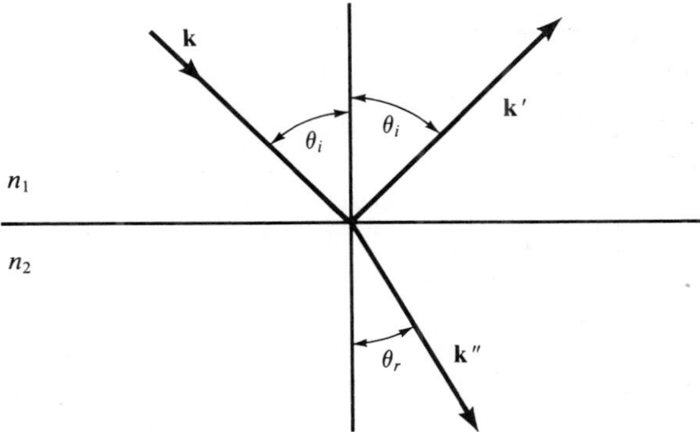

FIGURE 7.2
Refraction and reflection at a plane boundary

To study the impact of reflection and refraction on polarization of a light beam, consider the set up of Fig. 7.2, where **k**, **k**′, and **k**″ are used to denote coplanar wave vectors along the direction of the incident, reflected, and refracted beam, respectively. The respective direction of electric and magnetic vectors in all three beams are governed by Eq. (1.18):

$$\mathbf{k} \times \mathbf{E} = \mu_0 \omega \mathbf{H} \tag{7.9a}$$

$$\mathbf{k}' \times \mathbf{E}' = \mu_0 \omega \mathbf{H}' \tag{7.9b}$$

$$\mathbf{k}'' \times \mathbf{E}'' = \mu_0 \omega \mathbf{H}'' \tag{7.9c}$$

where the absence of prime, presence of single prime, and presence of double prime refer to incident, reflected, and refracted beams, respectively.

For the geometry of Fig. 7.2, we already said that reflection of light can be classified into two categories, TE (transverse electric) and TM (transverse magnetic). Accordingly, the coefficients of reflections, r_{TE} and r_{TM}, are found to be

$$r_{\mathrm{TE}} = \left.\frac{E'}{E}\right|_{\mathrm{TE}} = \frac{\cos\theta_i - \sqrt{n^2 - \sin^2\theta_i}}{\cos\theta_i + \sqrt{n^2 - \sin^2\theta_i}} \tag{7.10}$$

and

$$r_{\mathrm{TM}} = \left.\frac{E'}{E}\right|_{\mathrm{TM}} = \frac{\sqrt{n^2 - \sin^2\theta_i} - n^2\cos\theta_i}{\sqrt{n^2 - \sin^2\theta_i} + n^2\cos\theta_i} \tag{7.11}$$

where $n = n_2/n_1$. The square of these coefficients gives the reflectance, defined as the corresponding fraction of the incident optical energy that is reflected. From Eq. (7.11), we can see that the TM amplitude ratio for reflection becomes zero at a particular angle of incidence θ_i when

$$\theta_i = \text{Arc tan}\,(n) \tag{7.12}$$

This angle, also called **Brewster's angle**, is a particular incidence angle, corresponding to reflected light that consists only of an electric field vibrating perpendicular to the plane of incidence. Note also that for this particular orientation of incident beam, the angle between reflected and refracted beams is exactly $\pi/2$. Consequently, when the incident beam of light is linearly polarized in the TM mode, and when it arrives at the interface at Brewster's angle, there is no reflection at all. The light in that case is thus transmitted totally. Such a perfect window, often referred to as a Brewster's window, is used extensively in the design of lasers as indicated in Section 5.6. The active laser cavity is bounded by two Brewster's windows such that the light can make repeated passes through the windows. After repeated to-and-fro oscillations, the laser beam becomes completely polarized in the TM mode.

If the light incident at Brewster's angle is unpolarized, the reflected light is found to be polarized, whereas the refracted light is generally found to be partially polarized with more TM components than TE components present in it. Consequently, a stack of plates (placed parallel to each other) can be used to successfully produce a purely polarized refracted beam having only TM components.

If the incoming light is unpolarized, it can be represented by two orthogonal, incoherent, and equal components. And, consequently, the reflectance of the natural light is given by

$$R = \tfrac{1}{2}[|r_{\text{TE}}|^2 + |r_{\text{TM}}|^2] \tag{7.13}$$

Figure 7.3 shows the plot of Eq. (7.13) along with the squares of Eqs. (7.10) and (7.11). The middle plot corresponds to the reflectance of natural light.

Scattered radiation can often be polarized whenever there is an irregular distribution of minute scattering centers. Examples of such scattering centers can be found in the dust speckles suspended in air and even in the gas molecules of air. The oscillating electric field of the incident light induces these scattering centers to oscillate. Finally, the oscillation results in the reradiation of light in almost all directions. However, the reradiated light scattered at a direction perpendicular to that of the incident light is generally found to be linearly polarized. Consequently, when the sun is not directly overhead, the scattered light that descends vertically downward is found to be partially polarized. In any event, the introduction of polarization by scattering is an inefficient process and cannot be relied upon to produce pure polarized light.

Dichroism is a property of a particular class of anisotropic optical media that can be used to produce polarized light. A number of minerals and organic compounds exhibit the dichroic property, whereby light propagating through the media is partially absorbed, depending on field orientation. Light oscillating along only one particular direction is transmitted completely through the media. A tourmaline crystal—or better yet a Polaroid (film of polyvinyl alcohol

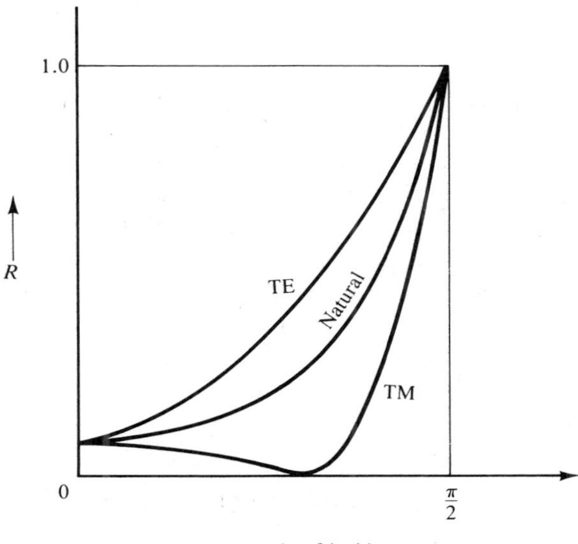

FIGURE 7.3

Reflectance versus incident light for TE, TM, and natural light

impregnated with iodine)—is a typical example of such a material that has a preferred direction, referred to as the transmission axis, along which the transmitted light becomes linearly polarized. A sheet of Polaroid film is often referred to as a polarizer because it can extract linearly polarized light from incoming unpolarized light. A second Polaroid sheet (with its transmission axis at 90° to that of a polarizer), referred to as an analyzer, will result in no transmission, when placed next to a polarizer, as shown in Fig. 7.4. As the analyzer is rotated while the polarizer is kept fixed, the light transmitted by the polarizer-analyzer combination increases, reaching a maximum when the two transmission axes are perfectly aligned. With I_p representing the maximum output intensity, the transmitted intensity can be expressed by **Malus's law** in terms of the angle between the transmission axes:

$$I_t = I_p \cos^2 \theta \tag{7.14}$$

Accordingly, the analyzer affords a means of determining whether the incoming light is polarized or not. This is accomplished by rotating the analyzer with respect to the polarizer. For an incoming unpolarized or circularly polarized light, no fluctuation in intensity is observed at the output. However, for an incoming elliptically polarized light, the transmitted light is seen to fluctuate between a maximum and minimum. Those two extremes correspond to the two cases when the analyzer is parallel to the major and minor axes of the ellipse.

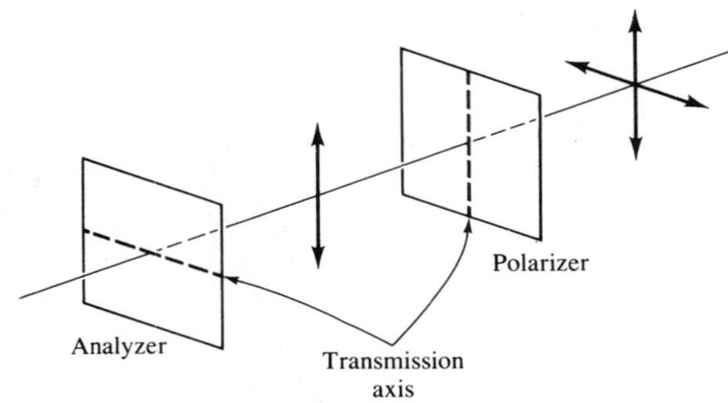

FIGURE 7.4

A polarizer-analyzer pair

Appropriately, we can use the concept of degree of polarization V defined as

$$V = \frac{I_{\max} - I_{\min}}{I_{\max} + I_{\min}} \tag{7.15}$$

to characterize the quality of a light beam. It should be noted that since intensity absorption varies exponentially with the thickness of the medium, the irradiance at a depth of z in the absorber is given by

$$I(z) = I_i e^{-\alpha z} \tag{7.16}$$

where I_i represents the incident irradiance and α is the absorptivity of the medium. For the perfect dichroic material, α is expected to be independent of wavelength. However, this is not so in a Polaroid H-sheet, especially at the blue region of the spectrum. Consequently, a combination of polarizer and analyzer, such as that shown in Fig. 7.4 using a Polaroid H-sheet, will contribute to a blue transmitted light.

It is now appropriate to describe a particular scheme whereby logical operations between polarization-encoded binary inputs are performed by means of a lensless, shadow-casting system. This system, referred to as the **polarization-encoded optical shadow-casting** (POSC) **system** and shown in Fig. 7.5, has a remarkably high speed of operation. Since all of the operations are linear, the computation speed is limited only by the speed with which light travels between the input and the output planes. The system uses spatially encoded two-

FIGURE 7.5 (opposite)

(a) An optical shadow-casting system. [Adapted from Y. Ichioka, and J. Tanida, "Optical Parallel Logic Gates Using a Shadow-Casting System for Optical Digital Computing," *Proc. IEEE.*, Vol. 72 (1984), p. 787. © 1984 IEEE]; (b) Shadow-casting on an individual pixel

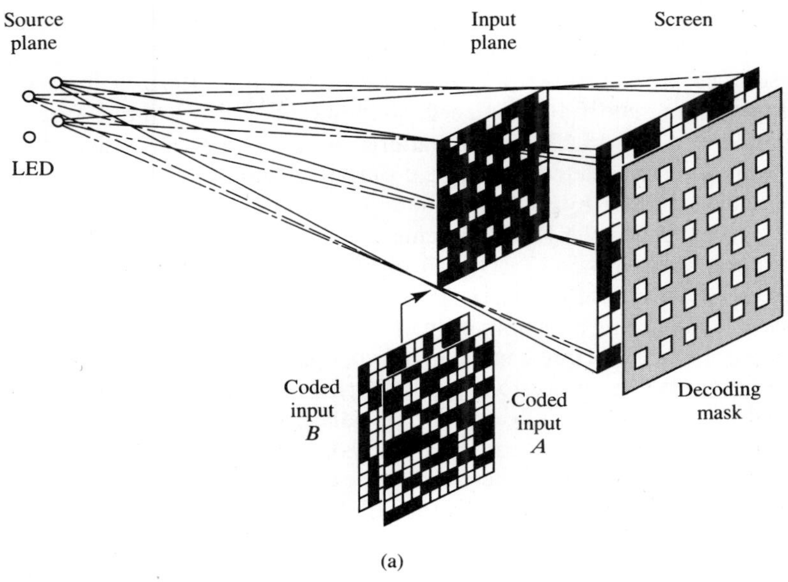
Source
plane
LED
Input
plane
Screen
Coded
input
B
Coded
input
A
Decoding
mask

(a)

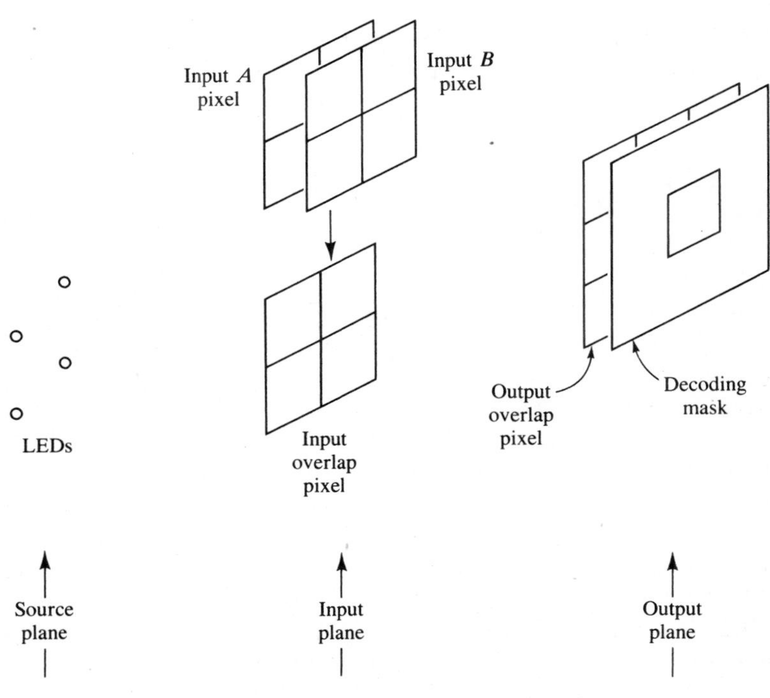
Input A
pixel
Input B
pixel
LEDs
Input
overlap
pixel
Output
overlap
pixel
Decoding
mask
Source
plane
Input
plane
Output
plane

(b)

dimensional binary patterns as inputs that are placed in contact at the input plane. The overlapped coded pattern is illuminated by a set of $N \times N$ LED sources. For example, a 3×3 output overlap pattern is formed from overlapping the relatively displaced shadows of a 2×2 input pixel when illuminated by a 2×2 LED source pattern. A decoding mask is thereafter used to spatially filter and detect the logical output. A variety of logical operations can be realized, either by changing the input LED pattern or by switching the output mask type. In the POSC scheme, 2×2 LED sources are allowed to be either unpolarized and on (TRUE, T) or off (FALSE, F), or to be horizontally polarized (H) or vertically polarized (V). Likewise each of the 2×2 pixels of the inputs is encoded using a combination of transparent (T code), opaque (F code), vertically polarized (V code), and horizontally polarized (H code) codes. Again, the output mask can be of four types: horizontally polarized (H), vertically polarized (V), transparent (T), and mixed polarized (VH or HV). Accordingly, up to four different outputs can be obtained. Note that the output obtained using mixed polarized mask is detected at the central output micro-pixel using a detector whose threshold exceeds 50% of the maximum detectable light intensity.

POSC systems are based on the following principles: (a) the overlap of a transparent code and a given input code leaves the input code unaltered, whereas the overlap of any given input code and an opaque code nullifies the effect of input code; (b) the overlap of any one of the codes with itself reproduces the same code; and (c) the overlap of a pair of orthogonal codes (e.g., either H and V or T and F) results in an opaque output. For example, consider three inputs, A, B, and C, whose true (binary 1) values are coded in a 2×2 pixel format respectively as

$$A = \begin{pmatrix} \mathrm{T} & \mathrm{F} \\ \mathrm{H} & \mathrm{T} \end{pmatrix} \quad B = \begin{pmatrix} \mathrm{T} & \mathrm{T} \\ \mathrm{H} & \mathrm{F} \end{pmatrix} \quad C = \begin{pmatrix} \mathrm{V} & \mathrm{H} \\ \mathrm{T} & \mathrm{H} \end{pmatrix} \tag{7.17}$$

and are overlapped at the POSC input plane. Assume further that the four LEDs (unpolarized) are turned on and that two different masks (one vertically polarized and the other horizontally polarized) are used for obtaining two different outputs. A horizontally polarized mask will detect only T and H codes if they appear at the central micro-pixel of the output overlap pattern. Similarly, the vertically polarized mask extracts only T and V codes.

Equation (7.17) provides the encodings for the three inputs (having binary 1 values), and it is also possible to generate the encodings for the complements of the inputs (having binary 0 values) from Eq. (7.17), simply by replacing T, F, V, and H codes with F, T, H, and V codes, respectively. When any one of the eight binary input combinations of A, B, and C is introduced at the POSC input, we can see that the corresponding detected output follows the truth table shown in Table 7.1. The presence of four source LEDs causes the four relatively displaced shadows of the input overlap pixel to overlap again at the output plane. The combined effect of all four LEDs shows up only at the central micro-pixel of

TABLE 7.1

A Full Adder Example

Inputs			*Outputs*	
A	*B*	*C*	*Vertically Polarized Mask*	*Horizontally Polarized Mask*
0	0	0	0	0
0	0	1	1	0
0	1	0	1	0
0	1	1	0	1
1	0	0	1	0
1	0	1	0	1
1	1	0	0	1
1	1	1	1	1

the output overlap pixel. The output mask can then decode the output from the central micro-pixel position. Comparing each of the input combinations and their corresponding outputs (as listed in Table 7.1) convinces us that this particular POSC system functions like a digital full adder, where the outputs generated using vertically and horizontally polarized output masks correspond to sum and carry out, respectively.

POSC outputs are determined by a number of factors—the source pattern, mask type, and encoding pattern of the inputs. The selection of source, mask, and encoding pattern is thus critical in determining what kind of a function the POSC system will perform. By careful analysis, we can determine the input encoding patterns necessary for realizing other arithmetic and or logic functions.

The last but not the least important means of yielding plane-polarized light involves the phenomenon of birefringence, also referred to as double refraction, that occurs primarily in mediums having directional optical properties. Such a medium is referred to as **anisotropic**. Depending on the direction of polarization, the beam transmitted through anisotropic crystals—e.g., calcite and quartz—may be subjected to a birefringence that produces two separate beams (referred to as ordinary and extraordinary beams) in the same direction, but with two different velocities. A detailed discussion of these phenomena appear in the next section.

7.3 Light Propagation in Crystals

Crystals are generally anisotropic. This implies that the polarization produced by the crystal varies in a manner depending on the direction of the applied field. Consequently, the speed of propagation of light in a crystal is a strong function of propagation direction as well as polarization of light. Not all crystals exhibit birefringence. The class of crystals that do not show this property is known as the cubic class.

For a nonmagnetic medium ($\mu = 1$), we can use Maxwell's equations (Eqs. 1.1–1.4) to obtain

$$\nabla \times (\nabla \times \mathbf{E}) = \omega^2 \mathbf{D} \tag{7.18}$$

where the **D**, **E**, and **H** vectors have sinusoidal dependence of the form $\exp[j(\omega t - \mathbf{k} \cdot \mathbf{r})]$ such that $\mathbf{k} = k\mathbf{k}'$ where $\mathbf{k}'$ is a unit vector in the direction of $\mathbf{k}$ and $k = |\mathbf{k}|$. However, we can see that

$$\nabla \times (\nabla \times \mathbf{E}) = -k^2[\mathbf{k}' \times (\mathbf{k}' \times \mathbf{E})] = -k^2[\mathbf{k}'(\mathbf{k}' \cdot \mathbf{E}) - \mathbf{E}] \tag{7.19}$$

Accordingly, using Eqs. (7.18) and (7.19), we get

$$\mathbf{D} = \frac{n^2}{c^2}[\mathbf{E} - \mathbf{k}'(\mathbf{k}' \cdot \mathbf{E})] \tag{7.20}$$

The electric displacement vector **D** is orthogonal to the direction of propagation. But for all nonabsorbing dielectric medium, **D** is also given in terms of the dielectric tensor $\boldsymbol{\varepsilon}$ by

$$\mathbf{D} = \boldsymbol{\varepsilon}\mathbf{E} \tag{7.21}$$

where

$$\boldsymbol{\varepsilon} = \begin{bmatrix} \varepsilon_x & 0 & 0 \\ 0 & \varepsilon_y & 0 \\ 0 & 0 & \varepsilon_z \end{bmatrix} \tag{7.22}$$

and ε_x, ε_y, and ε_z are nonnegative principal dielectric constants of $\boldsymbol{\varepsilon}$. In the case of isotropic (cubic) crystals, these three values are identical, whereas in uniaxial (trigonal, tetragonal, or hexagonal) crystals, only two of these three principal values are equal and in biaxial (triclinic, monoclinic, orthorhombic) crystals, all three values are different. Naturally occurring crystals are generally biaxial; however, almost all crystals that are used in electro-optic devices are uniaxial.

In a uniaxial crystal, we can choose the axes so that two of the principal dielectric constants become identical. For simplicity, let $\varepsilon_x = \varepsilon_y \equiv (n_0/c)^2$ and $\varepsilon_z \equiv (n_E/c)^2$. We may further assume, without any loss of generality, that the unit vector $\mathbf{k}'$ has no y component and that it forms an angle α_0 with the z axis. Often in such a set up, the z axis is referred to as the **optic axis**. Equation (7.20) can now be rewritten for the uniaxial crystal to give

$$n^2 \sin\alpha_0(E_x \sin\alpha_0 + E_z \cos\alpha_0) - (n^2 - n_0^2)E_x = 0 \tag{7.23a}$$

$$(n^2 - n_0^2)E_y = 0 \tag{7.23b}$$

$$n^2 \cos\alpha_0(E_x \sin\alpha_0 + E_z \cos\alpha_0) - (n^2 - n_E^2)E_z = 0 \tag{7.23c}$$

A quick analysis will convince us that there are two valid solutions of Eqs. (7.23a) through (7.23c). The one corresponding to nonzero E_y yields what is known as the **ordinary wave solution**, and the one corresponding to $E_y = 0$ provides the **extraordinary wave solution**. The ordinary wave electric field vector, when orthogonal to both $\mathbf{k}'$ and the optic axis, corresponds to both $E_x = E_z = 0$ and $n^2 = n_O^2$. The corresponding electric displacement vector is given by

$$\mathbf{D}_O = \left(\frac{n_O}{c}\right)^2 \mathbf{E} \tag{7.24}$$

where $c/n_O \equiv v_O$ is the velocity of the ordinary wave.

The extraordinary wave solution, which corresponds to $E_y = 0$, nonzero E_x, and nonzero E_z, is obtained by solving Eqs. (7.23a) and (7.23c) simultaneously. To obtain a valid solution, we set the determinant formed by the coefficients of E_x and E_z equal to zero. Thus

$$(n^2 \sin^2 \alpha_O - n^2 + n_O^2)(n^2 \cos^2 \alpha_O - n^2 + n_E^2) - (n^2 \sin \alpha_O \cos \alpha_O)^2 = 0 \tag{7.25}$$

which gives

$$\frac{\sin^2 \alpha_O}{n_E^2} + \frac{\cos^2 \alpha_O}{n_O^2} = \frac{1}{n^2(\theta)} \tag{7.26}$$

Equation (7.26) describes an ellipsoidal contour for the extraordinary wave index surface.

As long as the propagation is along the optic axis, the index of refraction remains n_O, but for the propagation along a plane normal to the optic axis (x–y plane), the index of refraction becomes n_E. For all other directions, at any wavelength, the index of refraction for the ordinary wave is a constant, whereas the index of refraction for the extraordinary wave falls somewhere between n_E and n_O. The exact value of the extraordinary wave index of refraction depends on the angle formed by the incident ray with the direction of the optic axis. We can conclude, therefore, that for the cases of uniaxial crystals, the index surfaces consist of a sphere of radius n_O and an ellipsoid of revolution defined by Eq. (7.26), and that the two are tangent to each other only at points along the direction of optic axis. In fact, uniaxial crystals are those crystals that have only one optic axis with $\Delta n = |n_E - n_O|$ providing a measure of birefringence.

A detailed analysis reveals that birefringence is only a special case of the phenomenon whereby biaxial crystals have two optic axes. In any event, the uniaxial crystals can be of two types: positive ($n_E > n_O$) and negative ($n_E < n_O$). Figure 7.6 shows the corresponding index surfaces where vector **S** is used to represent the Poynting vector. Note that in the positive uniaxial crystal, $v_O \geqslant v_E$ and, in the negative uniaxial crystal, $v_E \geqslant v_O$ where $v_E \equiv c/n_E$ represents the extraordinary wave velocity. Table 7.2 lists some of the typical uniaxial crystals and their indices of refraction. From among the other important birefringent crystals, lithium niobate ($LiNbO_3$) and lithium tantalate ($LiTaO_3$) are positive crystals, whereas potassium dihydrogen phosphate (KH_2PO_4)—KDP—and

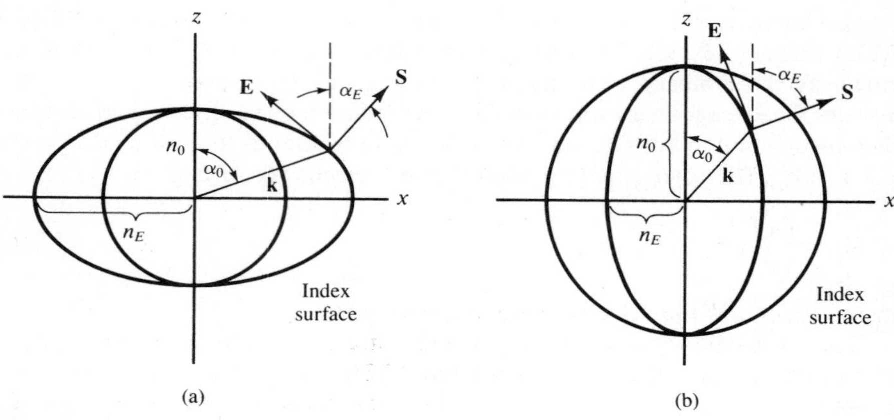

FIGURE 7.6

Index surfaces for (a) a positive uniaxial crystal and (b) a negative uniaxial crystal

ammonium dihydrogen phosphate ($NH_4H_2PO_4$)—ADP—are negative birefringent crystals.

A uniaxial crystal like calcite can be cut and polished to act as a polarizing material such that its optic axis can form an arbitrary angle with the incident light. The plane defined by the optic axis and the direction of propagation is ordinarily known as the **principal section**. We can identify four possible cases for the orientation of the crystal depicted in Fig. 7.7, where the lines or dots within the crystal specify the direction of the optic axis.

In Fig. 7.7(a), the incident propagation direction is along the optical axis

TABLE 7.2

The Characteristics of Selected Uniaxial Crystals

Type	*Name*	n_O	n_E
Positive	Ice	1.309	1.313
	Rutile	2.620	2.900
	Quartz	1.544	1.553
	Zircon	1.923	2.900
Negative	Apatite	1.6461	1.6417
	Beryl	1.598	1.590
	Calcite	1.660	1.490
	Sodium nitrate	1.5854	1.3369
	Tourmaline	1.670	1.640

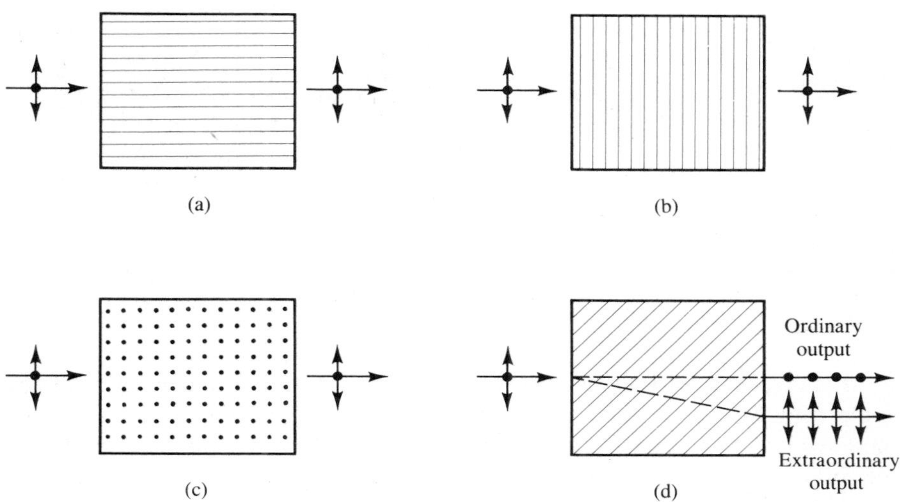

FIGURE 7.7

Orientation of the optic axis in birefringent crystal: (a) along the input, (b–c) normal to the input, and (d) at an arbitrary angle other than those described

and, therefore, there is no effect on the polarization of the input. Both ordinary and extraordinary waves propagate through the crystal with the same speed. The effective refractive index of the crystal, in this case, is n_O. In parts (b) and (c), the direction of propagation is normal to the optic axis such that in one case, the incoming polarization direction is along the optic axis, whereas in the other case it is normal to the optic axis. Consequently, the ordinary and extraordinary waves travel straight through the crystal, but with different velocities, because there are two refractive indices involved in the process. The resulting output beam is associated with a relative phase difference. If the crystal has a thickness of d, the corresponding phase difference is given by

$$\Delta\phi = \frac{2\pi d}{\lambda}|n_E - n_O| \tag{7.27}$$

where λ is the wavelength of light in vacuum. The phase difference can therefore be adjusted by using crystals of different thicknesses. When $\Delta\phi = \pi/2$, the crystal is called a **quarter-wave plate** and, similarly, when $\Delta\phi = \pi$, it is referred to as the **half-wave plate**. These may also be referred to as the **zero-order plates** because thicker plates can also be fabricated so that the relative phase between the two waves increases by an additional amount $2\pi m$, where m is an integer. A crystal that allows a continuous change in thickness, therefore, can serve as an adjustable retardation plate. Such especially designed crystals are called **compensators**.

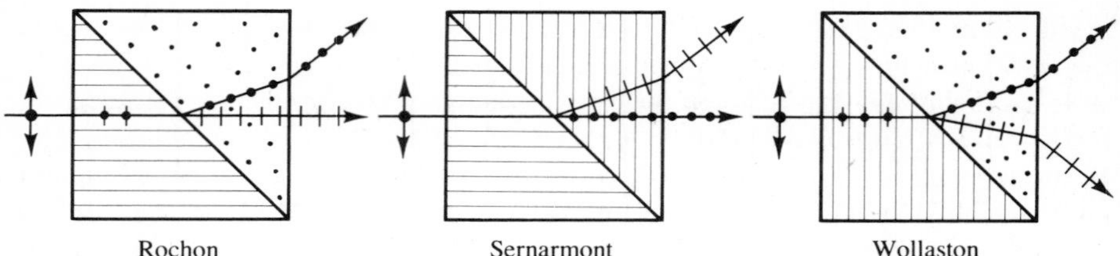

FIGURE 7.8
Polarizing prisms

Finally, Fig. 7.7(d) shows the case where the optic axis makes an arbitrary angle with the beam direction. In such a case, the incident beam is doubly refracted into two beams so that the polarization of one is orthogonal to that of the other. The beam that appears undeviated at the output, having its polarization perpendicular to the optic axis, is the ordinary beam. The other beam, with its polarization in a direction perpendicular to that of the ordinary wave, does not follow Snell's law and is referred to as the extraordinary beam. Often it is necessary to physically separate these linearly polarized beams. Two quartz prisms can be combined in many varied ways, as shown in Fig. 7.8, to accomplish such a feat. Note that it is also possible to separate the beams with a prism combination that has both of the optic axes directed along the same direction. However, in such a set up, either air or a suitable cement material must be used to separate the two prisms. Such a combination—e.g., a Glan prism and a Nicol prism—separates the beams by means of internal reflection at the prism junction.

There is a class of material that can rotate the plane of polarization of light passing through them. This unique property is referred to as **optical activity**. This phenomenon occurs because of a certain twisting of the orbits of electrons in the presence of an oscillatory field. The amount of rotation per unit length of travel is generally referred to as the specific rotatory power. Substances like sodium chloride, cinnabar, quartz, benzil, and certain sugars are examples of such optically active substances. Depending on the point of view of an observer, the substances are called **dextrorotatory** and **levorotatory**, depending on whether the rotation of the plane of polarization is clockwise or counterclockwise. Note that amount of optical activity in most substances is a function of the wavelength of light.

7.4 Electro-Optic Modulation

Electro-optic modulators consist of a dielectric medium and a means for applying an electric field to it. Application of an electric field causes the refractive index of the dielectric medium to be modified. The mechanism involved in this phenomenon is referred to as the electro-optic effect of the dielectric medium. The mechanism is often used for realizing both amplitude as well as the phase modulation of optical signals. The application of such a modulation scheme exists in optical communications, Q-switching of lasers, and beam deflections. In principle, electro-optic modulators cause a change in phase shift, which in turn is either a linear or a quadratic function of the applied electric field. This change in phase shift implies a change in optical length or index of refraction.

In an electro-optic crystal, such as those mentioned in the last section, the change in the index of refraction n along a crystal axis may be expressed in series form as

$$\Delta\left(\frac{1}{n^2}\right) = pE + kE^2 + \cdots \tag{7.28}$$

where E is the electric field, p is the linear electro-optic coefficient, and k is the quadratic electro-optic coefficient. In useful crystals, either the linear electro-optic effect (referred to as the Pockels effect) or the quadratic electro-optic effect (referred to as the Kerr effect) is predominant. In either case, the index of refraction will change at the modulation rate of the electric field. The effect allows a means of controlling the intensity or phase of the propagating beam of light. A Pockels cell uses the linear effect in crystals, whereas a Kerr cell uses the second-order electro-optic effect in various liquids and ferroelectrics; however, the former requires far less power than the latter to achieve the same amount of rotation and thus is used more widely. The Pockels effect, in particular, depends on the polarity of the applied electric field.

In the last section, we considered the propagation of light in crystals. Accordingly, in the absence of an external electric field, the indices of refraction along the rectangular coordinate axes of a crystal are related by the index ellipsoid:

$$\left(\frac{x}{n_x}\right)^2 + \left(\frac{y}{n_y}\right)^2 + \left(\frac{z}{n_z}\right)^2 = 1 \tag{7.29}$$

A careful examination will convince us that Eq. (7.26) is only a special case of Eq. (7.29). In the presence of an arbitrary electric field, however, the linear change in the coefficients of the index ellipsoid can be represented by

$$\Delta\left(\frac{1}{n^2}\right)_i = \sum_j p_{ij} E_j \tag{7.30}$$

where p_{ij}s are referred to as Pockels constants, $i = 1, 2, 3, \ldots, 6$; and $j = x, y, z$. The 6×3 electro-optic matrix having p_{ij} as its elements is often called the **electro-optic tensor**. In centrosymmetric crystals, all eighteen elements of the tensor are zero, whereas in the triclinic crystals, all elements are nonzero. But in the great majority of crystals, while some of the elements are zero, many of the elements have identical values. Table 7.3 lists the characteristics of the electro-optic tensors for some of the more important noncentrosymmetric crystals.

Determing the electro-optic effect in a particular modulator thus involves using the characteristics of the crystal in question and finding the allowed polarization directions for a given direction of propagation. A knowledge of refractive indices along the allowed directions can be used to decompose the incident optical wave along those allowable polarization directions. Thereafter, we can determine the characteristics of the emergent wave. Consider, for example, the case of KDP, whose nonzero electro-optic tensor components are p_{41}, p_{52}, and p_{63}. Using Eqs. (7.29) and (7.30), we can write the equation of the effective index ellipsoid for KDP as

$$\frac{x^2 + y^2}{n_O^2} + \left(\frac{z}{n_E}\right)^2 + 2p_{41}yzE_x + 2p_{41}xzE_y + 2p_{63}xyE_z = 1 \tag{7.31}$$

with $n_x = n_y \equiv n_O$ and $n_z = n_E$ for this uniaxial crystal where $p_{41} = p_{52}$. To be specific, let us restrict ourselves to the case in which the external field is directed along only the z direction. Equation (7.31), for such a case, reduces to

$$\frac{x^2 + y^2}{n_O^2} + \left(\frac{z}{n_E}\right)^2 + 2p_{63}xyE_z = 1 \tag{7.32}$$

TABLE 7.3

Some of the Important Electro-Optic Constants

Crystal	*Nonzero Elements (10^{-12} m/V)*	*Typical Refractive Index*
$BaTiO_3$	$p_{13} = p_{23} = 8.0$ $p_{33} = 23.0$ $p_{42} = p_{51} = 820.0$	$n_x = n_y \equiv n_O = 2.437$ $n_E = n_z = 2.365$
KDP	$p_{41} = p_{52} = 8.6$ $p_{63} = 10.6$	$n_x = n_y \equiv n_O = 1.51$ $n_E = n_z = 1.47$
ADP	$p_{41} = p_{52} = 28.0$ $p_{63} = 8.5$	$n_x = n_y \equiv n_O = 1.52$ $n_E = n_z = 1.48$
GaAs	$p_{41} = p_{52} = p_{63} = 1.6$	$n_x = n_y \equiv n_O = 3.34$
CdTe	$p_{41} = p_{52} = p_{63} = 6.8$	$n_x = n_y \equiv n_O = 2.6$
Quartz	$p_{11} = 0.29$ $p_{21} = p_{62} = -0.29$ $p_{41} = 0.2$ $p_{52} = -0.2$	$n_x = n_y \equiv n_O = 1.546$ $n_z \equiv n_E = 1.555$
$LiNbO_3$	$p_{13} = p_{23} = 8.6$ $p_{33} = 30.8$ $p_{22} = 3.4$ $p_{12} = p_{61} = -3.4$ $p_{42} = p_{51} = 28$	$n_x = n_y \equiv n_O = 2.286$ $n_z \equiv n_E = 2.200$
CdS	$p_{13} = p_{23} = 1.1$ $p_{33} = 2.4$ $p_{42} = p_{51} = 3.7$	$n_x = n_y \equiv n_O = 2.46$ $n_z \equiv n_E = 2.48$

Equation (7.32) can be transformed to have a form of the type

$$\left(\frac{x'}{n_{x'}}\right)^2 + \left(\frac{y'}{n_{y'}}\right)^2 + \left(\frac{z'}{n_{z'}}\right)^2 = 1 \tag{7.33}$$

which has no mixed terms. The parameters x', y', and z' of Eq. (7.33) denote the directions of the major axes of the index ellipsoid in the presence of the external field. The lengths of the major axes of the index ellipsoid are given by $2n_{x'}$, $2n_{y'}$, and $2n_{z'}$, respectively.

By comparing Eqs. (7.32) and (7.33), it is obvious that z and z' are parallel to each other. Again the symmetry of x and y in Eq. (7.32) suggests that x' and y' are related to x and y by a rotation of 45°. The transformation between the coordinates (x, y) and (x', y') is given by

$$\begin{bmatrix} x \\ y \end{bmatrix} = \begin{bmatrix} \cos\left(\frac{1}{4}\pi\right) & -\sin\left(\frac{1}{4}\pi\right) \\ \sin\left(\frac{1}{4}\pi\right) & \cos\left(\frac{1}{4}\pi\right) \end{bmatrix} \begin{bmatrix} x' \\ y' \end{bmatrix} \tag{7.34}$$

which, when substituted in Eq. (7.32), results in

$$\left[\frac{1}{n_0^2} + \mathrm{p}_{63} E_z\right] x'^2 + \left[\frac{1}{n_0^2} - \mathrm{p}_{63} E_z\right] y'^2 + \left(\frac{z'}{n_E}\right)^2 = 1 \tag{7.35}$$

Comparing Eqs. (7.33) and (7.35) and using the differential relation $dn = \frac{1}{2}n^3 d(1/n^2)$, we find that

$$n_{x'} = n_0 - \tfrac{1}{2} n_0^3 \mathrm{p}_{63} E_z \tag{7.36a}$$

$$n_{y'} = n_0 + \tfrac{1}{2} n_0^3 \mathrm{p}_{63} E_z \tag{7.36b}$$

$$n_{z'} = n_E \tag{7.36c}$$

when $(1/n_0^2) \gg \mathrm{p}_{63} E_z$. The velocity of propagation of an emerging wave polarized along the x' axis differs from that of an emerging wave polarized along the y' axis. The corresponding phase shift difference between the two waves (referred to as electro-optic retardation) after having traversed a thickness W of the crystal is given by

$$\Delta\phi = \frac{2\pi W}{\lambda} |n_{x'} - n_{y'}| = \frac{2\pi W}{\lambda} n_0^3 \mathrm{p}_{63} E_z \tag{7.37}$$

Provided that V is the voltage applied across the crystal, the retardation is then given by

$$\Delta\phi = \frac{2\pi}{\lambda} n_0^3 \mathrm{p}_{63} V \tag{7.38}$$

The emergent light is in general elliptically polarized. It becomes circularly polarized only when $\Delta\phi = \frac{1}{2}\pi$ and linearly polarized when $\Delta\phi = \pi$. Often the retardation is also given by $\pi V/V_\pi$ where $V_\pi (\equiv \lambda/2n_0^3 \mathrm{p}_{63})$ is the voltage necessary to produce a retardation of π.

7.4.1 Electro-Optic (E-O) Modulators

In the last section, we showed that it is possible to control optical retardation and, thus, the flow of optical energy in noncentrosymmetric crystals by means of voltage. This capability of retardation serves as the basis of modulation of light. Figure 7.9 shows a schematic of a system showing how a KDP modulator can be used to achieve amplitude modulation. This particular set up is also known as Pockels electro-optic amplitude modulator. The crossed polarizer-analyzer combination of the set up is necessary for the conversion of phase-modulated light into amplitude-modulated light. Further, in this set up, the induced electro-optic axes of the crystal make an angle of 45° with the analyzer-polarizer axes.

Elliptically polarized light emerges from the E-O crystal since, upon modulation, the two mutually orthogonal components of the polarized beams travel inside the crystal with different velocities. To feed an electric field into the system, the end faces of the E-O crystal are coated with a thin conducting layer that is also transparent to optical radiation. As the modulating voltage is changed, the eccentricity of the ellipse changes. The analyzer allows a varying amount of outgoing light in accordance with the modulating voltage applied across the E-O crystal.

The electric field components associated with the optical wave, immediately upon emerging from the electro-optic crystal, are respectively

$$E_{x'} = A \tag{7.39a}$$

$$E_{y'} = A\, e^{-j\Delta\phi} \tag{7.39b}$$

Thus the total field that emerges out of the analyzer is evaluated by summing the $E_{x'}$ and $E_{y'}$ components (along the analyzer axis), which gives us

$$\begin{aligned} E_0 &= [A\, e^{-j\Delta\phi} - A] \cos(\tfrac{1}{4}\pi) \\ &= \frac{A}{\sqrt{2}}[e^{-j\Delta\phi} - 1] \end{aligned} \tag{7.40}$$

The resulting irradiance of the transmitted beam is therefore given by

$$I_0 = \text{const}\,(E_0)(E_0^*) = \text{const}\, 2A^2 \sin^2\left(\frac{\Delta\phi}{2}\right) \equiv I_i \sin^2\left(\frac{\Delta\phi}{2}\right) \tag{7.41}$$

where I_i is the irradiance of the light incident on the input side of the E-O crystal. One can rewrite Eq. (7.41) as

$$\frac{I_0}{I_i} = \sin^2\left[\tfrac{1}{2}\pi \frac{V}{V_\pi}\right] \tag{7.42}$$

where $V_\pi \equiv |\lambda/(2p_{63} n_0^3)|$ is the voltage required for having the maximum transmission. Often V_π is also referred to as the **half-wave voltage** because it corresponds to a relative spatial displacement of $\lambda/2$ or to an equivalent phase difference of π.

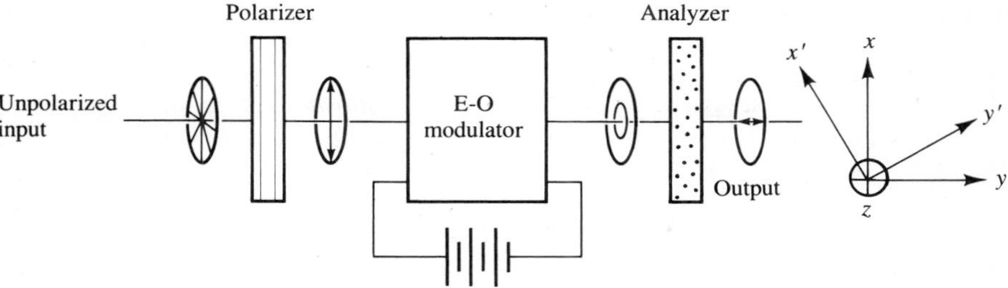

FIGURE 7.9

Pockels electro-optic amplitude modulator: (a) set up and (b) orientation of the crystal axes

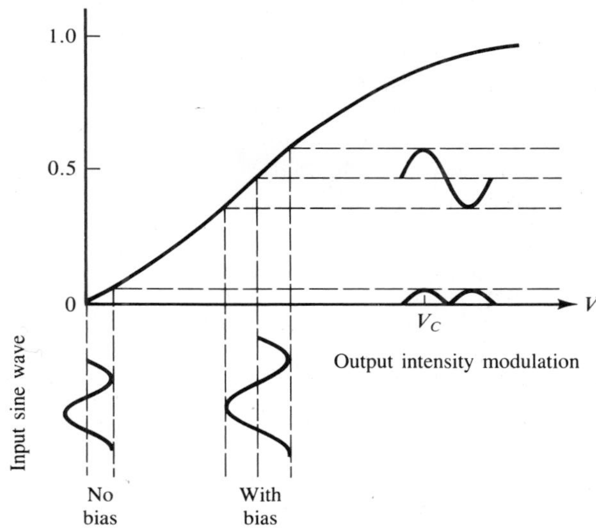

FIGURE 7.10

Transmission versus crystal voltage in a Pockels E-O amplitude modulator

Figure 7.10 shows the transmission characteristics of the cross-polarized E-O modulator as a function of the applied voltage. It can be seen that the modulation is nonlinear. In fact, for small voltages, the transmission is proportional to V^2. The effectiveness of an E-O modulator is often enhanced by biasing it with a fixed retardation of $\pi/2$. A small sinusoid voltage will then result in a nearly sinusoidal modulation of the transmitted intensity. This is achieved by introducing a quarter-wave plate between the polarizer and the E-O crystal. The quarter-wave plate shifts the E-O characteristics to the 50% transmission point. With the quarter-wave plate in place, the net phase difference between the two emerging waves becomes $\Delta\phi' = \Delta\phi + (\pi/2)$. Thus the output transmission is then given by

$$\frac{I_0}{I_i} = \sin^2\left(\frac{\pi}{4} + \frac{1}{2}\pi\frac{V}{V_\pi}\right) \simeq \frac{1}{2}\left(1 + \sin\left(\pi\frac{V}{V_\pi}\right)\right) \tag{7.43}$$

Note that with no modulating voltage, the modulator output intensity transmission reduces to 0.5. Again, for small V, the transmission factor varies linearly as the crystal voltage.

For an input sine wave modulating voltage, the transmission can be expressed as

$$\frac{I_0}{I_i} = \tfrac{1}{2}[1 + \sin\{mV_p \sin(\omega_m t)\}] \tag{7.44}$$

where m is a constant of proportionality, V_p is the peak modulating voltage, and ω_m is the modulation angular frequency. When $mV_p \ll 1$, the intensity modulation becomes a replica of the modulating voltage. The irradiances of the transmitted beam begin to vary with the same frequency of the sinusoid voltage. If, however, the condition $mV_p \ll 1$ is not satisfied, the intensity variation becomes distorted. Note that Eq. (7.44) can be expanded in terms of Bessel functions of the first kind to give

$$\frac{I_0}{I_i} = \tfrac{1}{2} + J_1(mV_p)\sin(\omega_m t) + J_3(mV_p)\sin(3\omega_m t) + J_5(mV_p)\sin(5\omega_m t) + \cdots \tag{7.45}$$

since

$$\sin(x \sin y) = 2[J_1(x)\sin(y) + J_3(x)\sin(3y) + J_5(x)\sin(5y) + \cdots] \tag{7.46}$$

when $J_0(0) = 1$ and, for nonzero n, $J_n(x) = 0$. The amount of distortion in the modulation process is often characterized by the ratio between the square root of sum of harmonic amplitude squares and the fundamental term amplitude. Therefore,

$$\text{distortion}\,(\%) = \frac{\{[J_3(mV_p)]^2 + [J_5(mV_p)]^2 + \cdots\}^{1/2}}{J_1(mV_p)} \times 100 \tag{7.47}$$

Consider the set up of Fig. 7.11 where the electro-optic crystal is oriented in such a way that the incident beam is polarized along one of the birefringent axes, say x'. In this specific case, the state of polarization is not changed by the applied electric field. However, the applied field changes the output phase by an amount

$$\Delta\phi_{x'} = \frac{\omega W}{c}|\Delta n_{x'}| \tag{7.48}$$

where W is the length of the E-O crystal. For a sinusoidal bias field $E_z =$

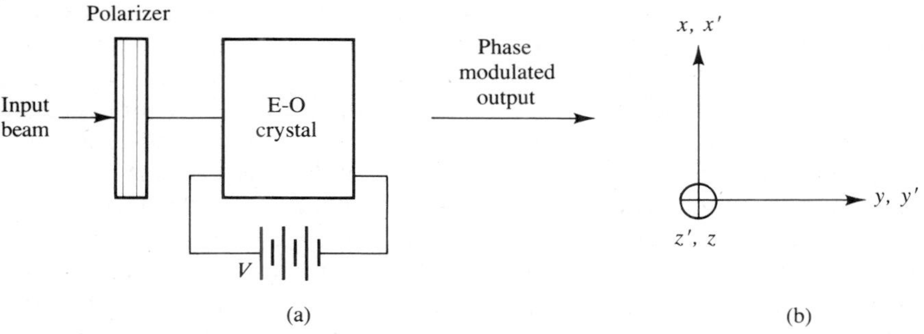

FIGURE 7.11

A phase modulator using an E-O crystal: (a) the set up and (b) the orientation of crystal axes

$E_{z,p} \sin(\omega_m t)$ and an incident field $E_{in} = E_{in,p} \cos(\omega t)$, the transmitted field is given by

$$\begin{aligned} E_{out} &= E_{in,p} \cos\left[\omega t - \frac{\omega W}{c}\{n_0 - \Delta\phi_{x'}\}\right] \\ &= E_{in,p} \cos\left[\omega t - \frac{\omega W}{c}\{n_0 - \tfrac{1}{2} n_0^3 \text{p}_{63} E_{z,p} \sin(\omega_m t)\}\right] \end{aligned} \quad \textbf{(7.49)}$$

If the constant phase factor $\omega W n_0/c$ is neglected, the transmitted electric field can be rewritten as

$$E_{out} = E_{in,p} \cos[\omega t + \delta \sin(\omega_m \text{t})] \quad \textbf{(7.50)}$$

where $\delta \equiv \frac{1}{2}(\omega W n_0^3 \text{p}_{63} E_{z,p}/c)$ is the phase modulation index. Note that this phase modulation index is one-half of the retardation $\Delta\phi$ (as given by Eq. (7.38)). Using Eq. (7.46) and the relationship

$$\cos(x \sin y) = J_0(x) + 2[J_2(x)\cos(2y) + J_4(x)\cos(4y) + \cdots] \quad \textbf{(7.51)}$$

we can rewrite Eq. (7.50) as

$$\begin{aligned} E_{out} = E_{in,p}[&J_0(\delta)\cos(\omega t) + J_1(\delta)[\cos\{(\omega + \omega_m)t\} - \cos\{(\omega - \omega_m)t\}] \\ &+ J_2(\delta)[\cos\{(\omega + 2\omega_m)t\} - \cos\{(\omega - 2\omega_m)t\}] \\ &+ J_3(\delta)[\cos\{(\omega + 3\omega_m)t\} - \cos\{(\omega - 3\omega_m)t\}] + \cdots] \end{aligned} \quad \textbf{(7.52)}$$

Accordingly, in this case, the optical field is seen to be phase-modulated with energy distribution in the sidebands varying as a function of the modulation index δ. We observe that while the E-O crystal orientation of Fig. 7.10 provides an amplitude modulation of light, the set up of Fig. 7.11 can provide a phase modulation of light. Both of these modulators are called **longitudinal effect devices** because in both cases the electric field is applied in the same direction as that of propagation. In both of these cases, the electric field is applied either by electrodes with small apertures in them or by making use of semi-transparent conducting films on either side of the crystal. This arrangement, however, is not too reliable because the field electrodes tend to interfere with the optical beam.

Alternatively, transverse electro-optic modulators can be used as shown by the system shown in Fig. 7.12. The polarization of the light lies in the x'-z plane at a 45° angle to the x' axis while light propagates along the y' axis and the field is applied along z. With such an arrangement, the electrodes do not block the incident optical beam and, moreover, the retardation can be increased by introducing longer crystals. In the longitudinal case, the amount of retardation is proportional to V and is independent of the crystal length W according to Eq. (7.38). Using Eqs. (7.36a) and (7.36c), the retardation caused by the transverse E-O amplitude modulator is given by

$$\Delta\phi_t = \phi_{x'} - \phi_{z'} = \frac{2\pi W}{\lambda}\left[(n_0 - n_E) - n_0^3 \rho_{63}\frac{V}{d}\right] \tag{7.53}$$

where n_0 and n_E are the refractive indices, respectively, along ordinary and extraordinary axes. Note the $\Delta\phi_t$ has a voltage-independent term that can be used to bias the irradiance transmission curve. The half-wave voltage can be reduced by a long but thin E-O crystal. Such an E-O crystal allows the transverse E-O modulators to have a better frequency response but at the cost of having small apertures.

There are occasions when we might be interested in driving the modulating signals to have large bandwidths at high frequencies. This can happen when we decide to use the wide frequency spectrum of a laser. To meet the demand of such a scenario, the modulator capacitance that is caused by the parallel-plate electrodes and the finite optical transit time of the modulator limits both bandwidth and the maximum modulation frequency. Consider the equivalent circuit of a high-frequency, electro-optic modulator as shown in Fig. 7.13. Let R_{in} be the total internal resistance of the modulating source V, while C represents the parallel-plate capacitance of the E-O crystal. When R_{in} is greater than the capacitive impedance, a significant portion of the modulating voltage is engaged across the internal resistance, thus making the generation of electro-optic retardation relatively insignificant. In order to increase the proportion of the modulating voltage that is engaged across the E-O crystal, it is necessary to connect a parallel resistance-inductance circuit in parallel with the modulator. The load R_L is chosen to be very large when compared to R_{in}. The choice guarantees that most of the modulating voltage is employed across the E-O crystal. At the resonant frequency $\nu_0 = [1/\{2\pi(LC)^{1/2}\}]$, the circuit impedance is equivalent to load resistance. However, this system imposes a restriction on the bandwidth and makes it finite. The bandwidth is given by $[1/\{2\pi R_L C\}]$ and is centered at the resonant frequency ν_0. Beyond this bandwidth, the modulating voltage is generally wasted across R_{in}. Consequently, for the modulated signal to be an exact replica of the modulating signal, the maximum modulation

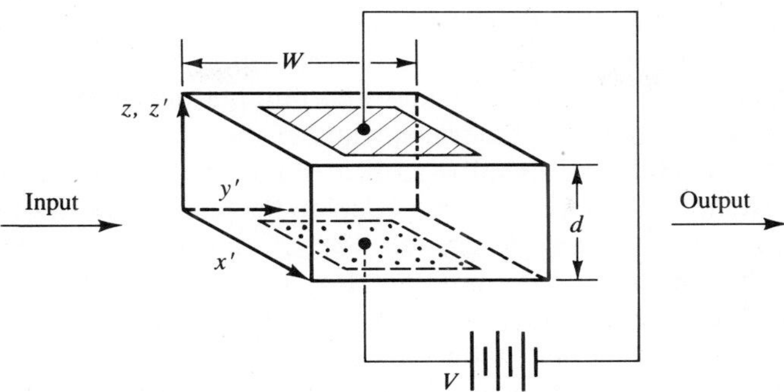

FIGURE 7.12

A transverse E-O amplitude modulator

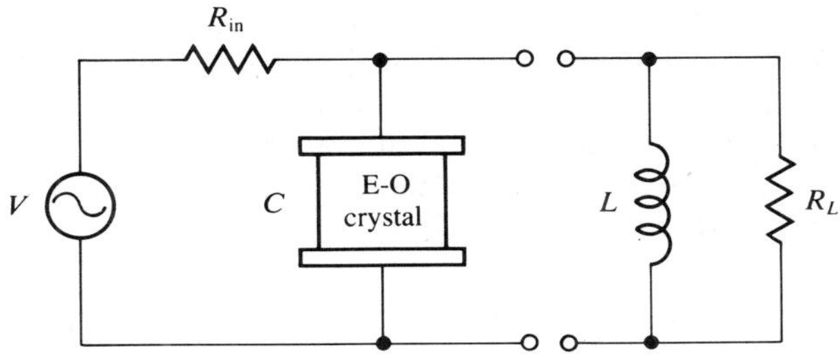

FIGURE 7.13

A circuit equivalent to an E-O modulator

bandwidth must not be allowed to exceed $\Delta\nu$ where $\Delta\nu$ is usually dictated by the specific application.

The power P needed to drive the E-O crystal is given by $\frac{1}{2}(V_{max}^2/R_L)$ where $V_{max}(\equiv E_{z,max} W)$ is the peak modulating voltage that produces the peak retardation $\Delta\phi_{max}$. Using Eq. (7.38) we can thus relate the driving power and the modulation bandwidth by

$$P = \frac{(\Delta\phi_{max})^2 \lambda^2 C\,\Delta\nu}{4\pi \mathrm{p}_{63}^2 n_0^6}$$

$$= \frac{(\Delta\phi_{max})^2 \lambda^2 A K_s \varepsilon_0\,\Delta\nu}{4\pi \mathrm{p}_{63}^2 n_0^6 W} \tag{7.54}$$

since at the modulation frequency ν_0, the parallel-plate capacitance C is given by $AK_s\varepsilon_0/W$ where A is the cross-sectional area of the crystal, K_s is the relative dielectric constant for the material, and W is the plate separation between the two electrodes.

As long as the modulating frequency is relatively low, the modulating voltage remains appreciably constant across the crystal. If the above condition is not fulfilled, however, the maximum allowable modulation frequency is found to be restricted substantially by the transit time of the crystal. To overcome this restriction, the modulating signal is applied transversely to the crystal in the form of a traveling wave with a velocity equal to the phase velocity of the optical signal propagating through the crystal. The transmission line electrodes are provided with a matched impedance at the termination point as shown in Fig. 7.14(a). The optical field is then subjected to a constant refractive index as it passes through the modulator, thus making it possible to have higher modulation frequencies. The traveling modulation field at a time t will have the form

$$E(t, z(t)) = E_p \exp\left[j\omega_m \left\{ t - \frac{z}{v_p} \right\} \right]$$

$$= E_p \exp\left[j\omega_m \left\{ t - \frac{c}{nv_p}(t - t_0) \right\} \right] \quad \textbf{(7.55)}$$

where v_p is the phase velocity of the modulation field, ω_m is the modulating angular frequency, and t_0 is defined as a reference to account for the time when the optical wavefront enters the E-O modulator. The electro-optic retardation due to the field can be written in accordance with Eq. (7.37) as

$$\Delta\phi = \Phi \frac{c}{n} \int_{t_0}^{t_0 + t_t} E(t, z(t))\, dt \quad \textbf{(7.56)}$$

where $\Phi = 2\pi n_0^3 p_{63}/\lambda$ and $t_t = nW/c$ is the total transit time (i.e., time taken by light to travel through the crystal). Eq. (7.56) can be evaluated to give the traveling wave retardation as

$$\Delta\phi_{\text{travel}} = (\Delta\phi)_0 \frac{e^{j\omega_m t_0}[e^{j\omega_m t_t[1 - (c/nv_p)]} - 1]}{j\omega_m t_t \left(1 - \dfrac{c}{nv_p}\right)} \quad \textbf{(7.57)}$$

where $(\Delta\phi)_0 = (\Phi c t_t E_p/n)$ is the peak retardation. The reduction factor

$$F_{\text{travel}} = \frac{e^{j\omega_m t_t[1 - (c/nv_p)]} - 1}{j\omega_m t_t \left(1 - \dfrac{c}{nv_p}\right)} \quad \textbf{(7.58)}$$

provides the amount of reduction in the maximum retardation owing to transit-time limitation. If instead we had begun to calculate the retardation for a sinusoidal modulation field that has the same value throughout the modulator, the retardation would have been

$$\Delta\phi_{\text{nontravel}} = (\Delta\phi)_0 \frac{e^{j\omega_m t_0}[e^{j\omega_m t_t} - 1]}{j\omega_m t_t} \quad \textbf{(7.59)}$$

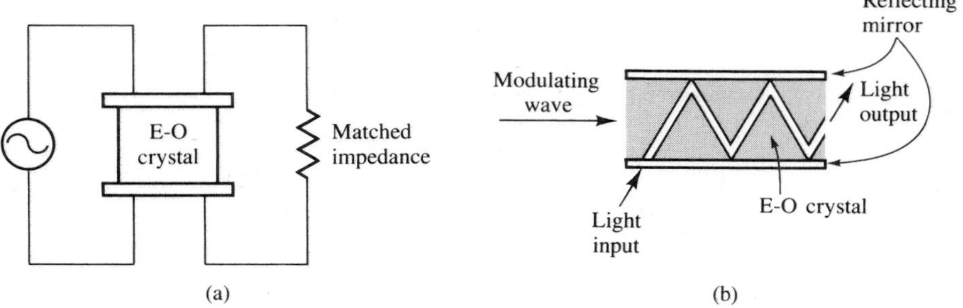

FIGURE 7.14

(a) Ideal traveling wave E-O modulator and (b) zig-zag traveling wave E-O modulator

By comparing Eqs. (7.57) and (7.59), we find that the two expressions are identical except that in the traveling wave, the E-O modulator t_t is replaced by $t_t\{1-(c/nv_p)\}$. The reduction factor in this latter case is given by

$$F_{\text{nontravel}} = \frac{e^{j\omega_m t_t} - 1}{j\omega_m t_t} \tag{7.60}$$

In the case of a nontraveling system, the reduction factor is unity only when $\omega_m t_t \ll 1$—that is, when the transit time is smaller than the smallest modulation period. But in the case of the traveling system, F approaches unity whenever the two phase velocities are equal—that is, when $c = nv_p$. Thus, in spite of the limitation of transit time, the maximum retardation is realized using the traveling wave modulator.

In practice, it might become very difficult to synchronize both electrical and optical waves. For a perfect synchronization, we expect to use an E-O crystal for which $n = K_s^{1/2}$. But for most naturally occurring materials, n is less than $K_s^{1/2}$. Thus synchronization is achieved either by including air gaps in the electrical waveguide cross section or by slowing down the optical wave by means of a zig-zag modulator as shown in Fig. 7.14(b).

A useful amplitude modulator involving waveguides, known as the Mach-Zehnder modulator, consists of neither polarizer nor analyzer but only E-O material. The set up, as shown in Fig. 7.15, splits the incoming optical beam into a "waveguide Y" and then recombines them. If the phase shift present in both of the arms is identical, all of the input power minus the waveguide loss reappears at the output. One arm of the Mach-Zehnder set up is provided with an electric field placed across it such that the amplitude of the field can be varied. The output power is modulated by changing the voltage applied across the waveguides. When a sufficiently high voltage is applied, the net phase shift difference between the arms can become 180°, thus canceling the power output altogether. Because of the small dimensions involved in electrode separations, relatively small switching voltages are used in such a modulation scheme.

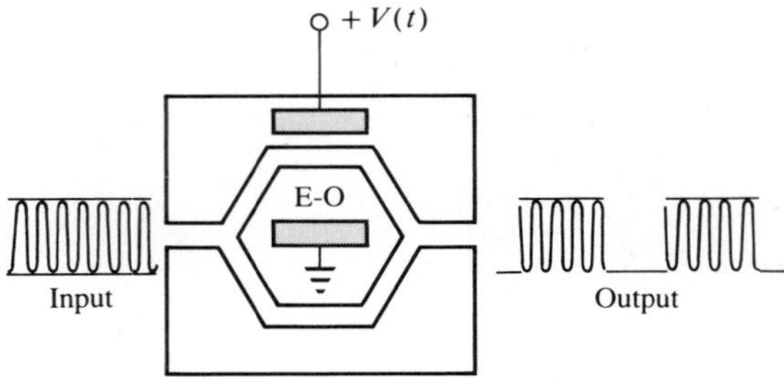

FIGURE 7.15
Mach-Zehnder waveguide E-O modulator

Figure 7.16(a) shows a prism deflector (see Problem 7.12) that uses bulk electro-optic effect so that we can deflect an optical beam by means of externally applied electric voltage. The electric voltage induces regions having different refractive indices, as a result of which the prism steers the refracted beam of light. Stacks of prisms can be used to provide a larger angle of deflection. Figure 7.16(b) shows an interesting application where birefringent crystals are combined with a Pockels cell modulator to form a digital electro-optic deflector. By applying an electric field, we can rotate the direction of polarization by $\pi/2$. Accordingly, by manipulating voltages (V_1 and V_0), we can shift the input optical beam to any one of the four spatial locations at the output. Similarly, by using n Pockels cell modulators and n birefringent crystals, we can shift light to a total of 2^n spatial locations. A set up such as that of Fig. 7.16(b) can be considered for various optical computing applications.

A device by the name of **Pockels read-out optical modulator** (PROM) can be fabricated by having an E-O crystal, such as $Bi_{12}SiO_{20}$ or ZnSe, sandwiched between two transparent electrodes as shown in Fig. 7.17(a). There is also an insulating layer between the E-O crystal and the electrode. A dc voltage can create mobile carriers, which in turn cause the voltage in the active crystal to decay. The device is normally exposed with the illumination pattern of a blue light. The voltage in the active area that corresponds to the brightest zones of the input pattern decays, while that corresponding to the comparatively darker area either does not change or changes very little. The input intensity pattern is thus transformed into a spatial voltage pattern. On the other hand, the read-out beam usually uses a red light. Note that the sensitivity of an E-O crystal is much higher in the blue region than in the red region. Such a choice of the read-out beam ensures that the read-out beam may not cause a change in the stored voltage pattern. In the read-out mode, the regions having the least amount of voltage decay act like a half-wave retarder. The reflected light, whose polarization is thus a function of the voltage pattern, is then passed through a polarizer to reproduce the output. For an eraser light E, the amplitude of the output is

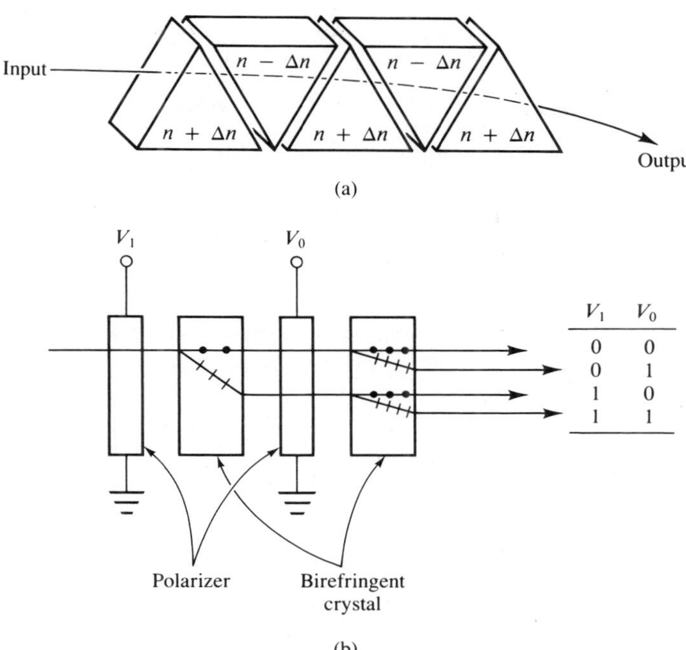

FIGURE 7.16

(a) Prism deflector and (b) digital E-O deflector

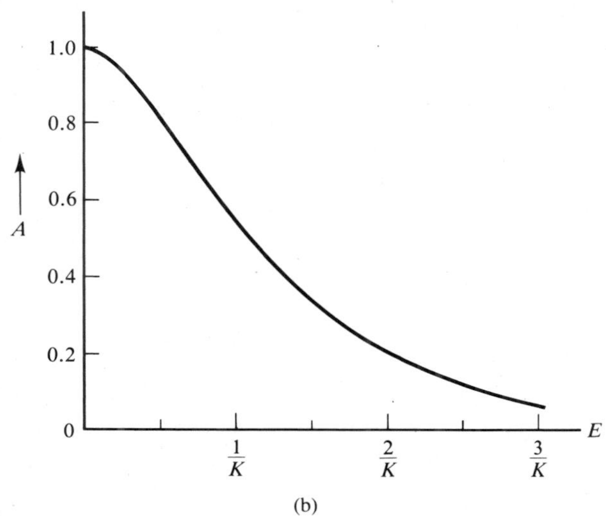

FIGURE 7.17

(a) PROM device and (b) its characteristics

found to be

$$A = A_0 \sin\left(\frac{\pi V_0}{V_{1/2}}\right) e^{-KE} \tag{7.61}$$

where A_0 is the amplitude of input read-out beam, V_0 is the voltage applied across the E-O crystal, $V_{1/2}$ is the half-wave voltage, and K is a positive constant. In the reflection read-out mode, $V_{1/2} \simeq 2V_0$. The amplitude of the reflectance of the PROM when plotted against the exposure is surprisingly found to be similar to that of a photographic film with a nearly linear region between $E = 0$ and $E = 2/K$.

So far this section has dealt primarily with the Pockels effect of electro-optic crystals. However, there are many isotropic media available that behave like uniaxial crystals when subjected to an electric field E. The change in refractive index Δn of those isotropic media varies as the square of the electric field. The resulting optic axis points parallel to the applied field. A Kerr modulator is produced by placing one of these media between the crossed polarizers. Modulation at frequencies up to 10^{10} Hz has been realized using a Kerr modulator. The difference between the two refractive indices that correspond respectively to light polarized parallel to the induced optic axis and light polarized perpendicular to the induced optic axis is provided by

$$\Delta n = |n_{\parallel} - n_{\perp}| = \kappa \lambda E^2 \tag{7.62}$$

where κ is the Kerr constant of the material. At room temperature, the Kerr constant has a typical value of 0.7×10^{-12}, 3.5×10^{-12}, and 4.5×10^{-10} cm/V^2, respectively, for benzene, carbon disulphide, and nitrobenzene.

The applied electric field induces an electric moment, which in turn reorients the molecules in a manner so that the medium becomes anisotropic. The delay between the application of the field and the appearance of the effect is, though not negligible, on the order of 10^{-12} s. A liquid Kerr cell containing nitrobenzene, as shown in Fig. 7.18, has been used for many years, but it has the disadvantage of requiring a large driving power. This problem is often overcome by using instead mixed ferroelectric crystals at a temperature near the Curie point, where ferroelectric materials start exhibiting optoelectric properties. Potassium tantalate niobate (KTN) is an example of such a mixture of two crystals, where one has a high Curie point and the other has a low Curie point, but the Curie point of the compound lies very close to room temperature. The transmittance characteristics of a Pockels cell and a Kerr cell are shown, for comparison, in Fig. 7.19.

In Chapter 5, we discussed two particular schemes for realizing a Q-switched laser. The first involves rotating end mirrors, and the second involves using a saturable absorber such as bleachable dye in the resonator. However, it is also possible to realize Q-switching nonmechanically—by subjecting either an E-O crystal or a liquid Kerr cell to an electric field. Nonmechanical schemes are relatively more precise in controlling the losses. Such a system is shown in Fig. 7.20. When there is no field, there is no rotation. But in the presence of the

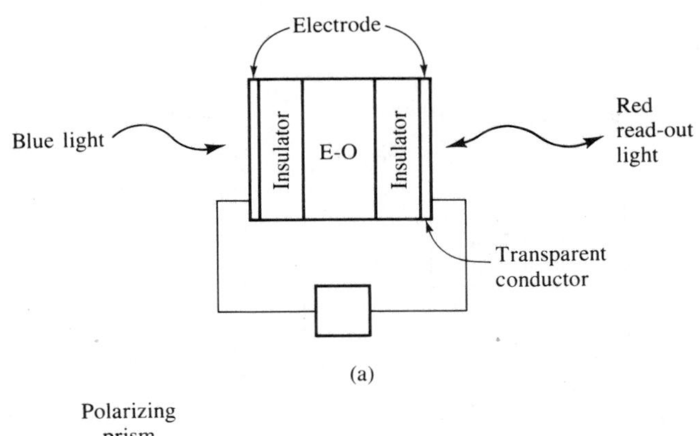

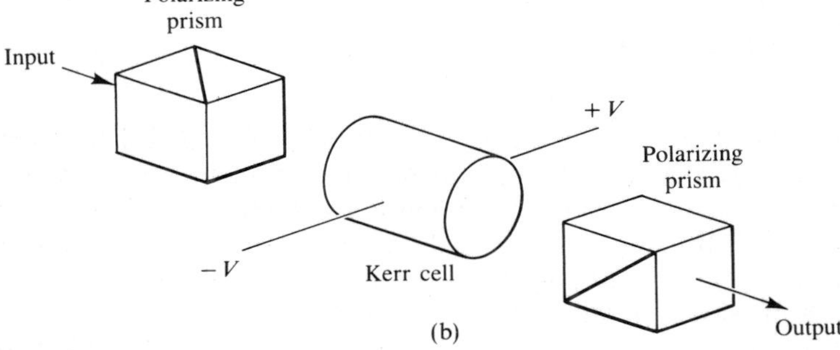

FIGURE 7.18

A Kerr cell light modulator

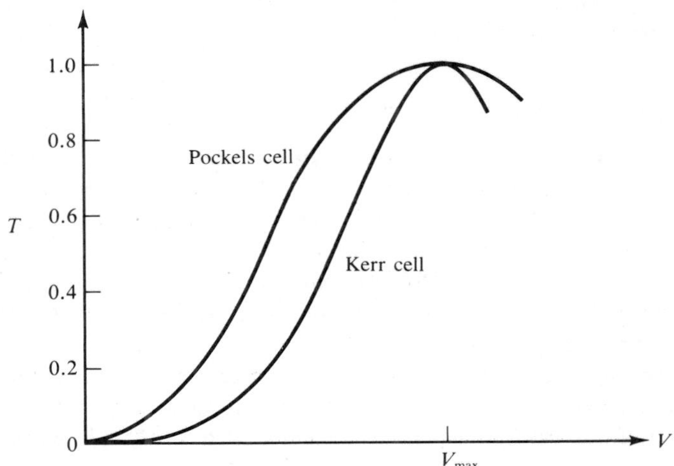

FIGURE 7.19

Comparative transmitters versus voltage characteristics for the E-O modulators

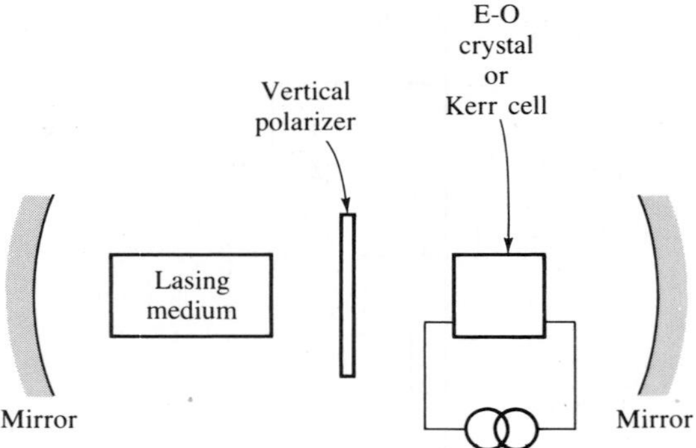

FIGURE 7.20

A Q-switched laser system

field, an E-O device can introduce a rotation of $\pi/2$. Thus, when the E-O device is assembled along with a polarizer, the combination works as a shutter. Because of the vertical polarizer, the light coming out of the lasing medium is plane-polarized. Because the polarized light has to traverse through the E-O device twice before coming back to the polarizer, only half of the voltage required to produce a rotation of $\pi/2$ is applied to the system. Accordingly, the polarizer of the E-O system blocks the light from coming to the main chamber of the laser. This is equivalent to causing a loss in the laser resonator. Such a loss, when suitably timed, can be made to produce a pulsed laser output with each pulse having an extremely high intensity. The voltages required to introduce appropriate fields are usually in the order of kilovolts, and thus it becomes possible to Q-switch at a rate of only nanoseconds or less.

7.4.2 Liquid Crystal Light Valve

The liquid crystal light valve (LCLV) is a specific spatial light modulator (SLM) with which one can imprint a pattern on a beam of light in nearly real-time. The two aspects of this device that are particularly important for discussion here are the modulation of light and the mechanism for addressing the device. Besides LCLVs, there are many more SLMs that are currently being considered for electro-optic applications. However, most of these devices follow only variations of the same physical principle.

The term **liquid crystal** (LC) refers to a particular class of materials whose rheological behavior is similar to that of liquids, but whose optical behavior is similar to that of crystalline solids over a given range of temperature. In particular, a type of LC, referred to as the nematic LC, is commonly used in LCLV as well as in most other LC devices. Lately, ferroelectric LCs are also being used in real-time display devices. In comparison, ferroelectric LC-based devices respond at a faster rate than the nematic LC-based devices. Nematic

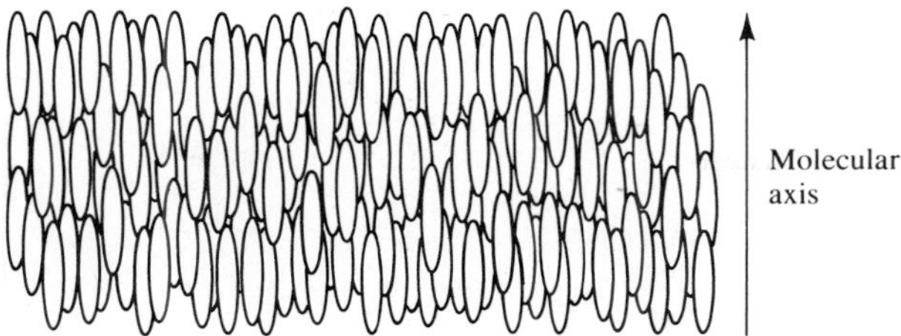

FIGURE 7.21

An aggregate of nematic liquid crystals

LCs generally consist of thin, long molecules, all of which tend to be aligned in the same direction, as shown in Fig. 7.21. When an electric field is applied to an LC layer, the molecules are reoriented in space due to both field and ionic conduction effects. The facts that LCs have a rodlike cylindrical molecular shape and that they tend to be aligned are prime reasons for yielding two E-O effects: electric field effect and birefringence. In the "off" state, an LCLV utilizes the properties of a nematic cell, while in the "on" state, it utilizes the birefringence properties.

The shape of the LC molecule introduces a polarization-dependent variation of the index of refraction that contributes to its birefringence characteristics. The difference between the two indices of refraction given by $\Delta n \equiv |n_{\parallel} - n_{\perp}|$ is a measure of the anisotropy of the material, where $n_{\parallel}$ represents the index of refraction for the component of light parallel to the molecular axis, and $n_{\perp}$ represents the index of refraction for the light component having an orthogonal polarization. Since LC molecules tend to be aligned, a bulk LC sample exhibits the same anisotropy as that exhibited by an individual LC molecule. In fact, a birefringent LC cell is normally formed by stacking LC layers parallel to the cell wall. Accordingly, all of the LC molecules are aligned along only one direction. When linearly polarized (at 45° to the alignment axis) light enters an LC cell of thickness D, the parallel polarization component lags behind the orthogonal component (due to positive anisotropy) by $\Delta\phi = [2\pi D(\Delta n)]/\lambda$. In general, the transmitted light turns out to be elliptically polarized. With a suitable choice of D, we can force the LC cell to behave like a half-wave or a quarter-wave plate. When compared to other traditional materials, LCs are preferable for making such retardation plates. In a typical LC material, Δn is a nonnegligible quantity. For example, for a typical LC, Δn is about 0.2 in the infrared, whereas Δn is only about 0.0135 in CdS in the infrared; in the visible, say, for quartz, Δn can be as low as 0.009. Thinner LC cells produce a comparatively large value of $\Delta\phi$. Note that a thinner LC cell allows for a larger acceptance angle for the incoming light, whereas a thicker, solid crystal like quartz forms a cell that is extremely sensitive to the angle of incidence. Further, LC cells can be grown to have reasonably large aperture sizes, suitable for handling higher laser power; however, the size will be limited by the tolerance for optical flatness.

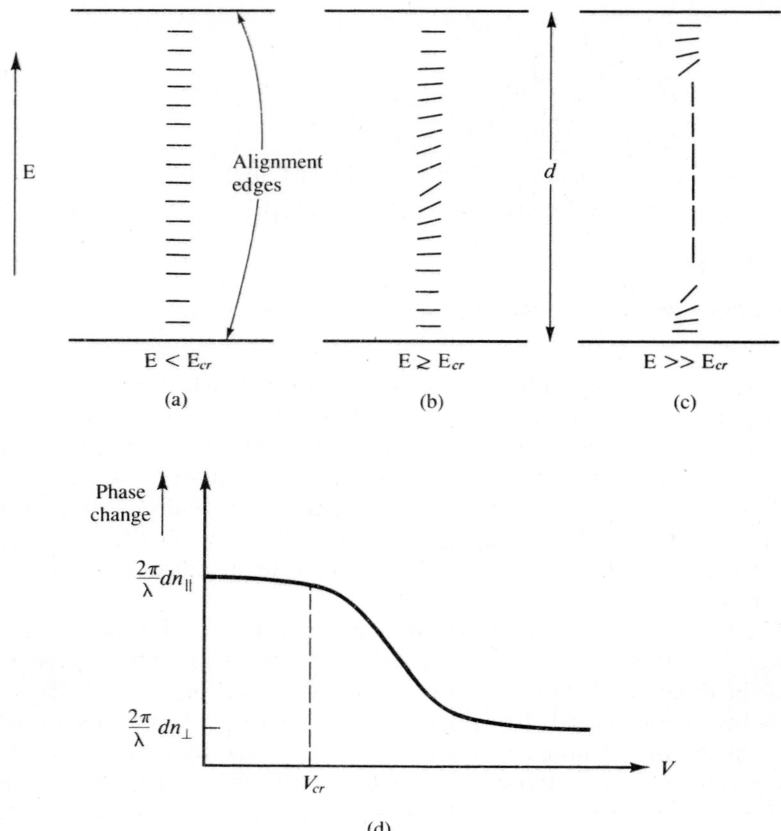

FIGURE 7.22

Field effects in liquid crystals

For positive anisotropy, there is a region of applied voltage over which the LC molecules may gradually rotate, introducing a variable phase delay in the output light. This feature, as illustrated in Fig. 7.22, can be used to create a voltage-controlled phase shifter. As stated earlier, a typical LC device combines the characteristics of both birefringence and field effects. A typical SLM structure is shown in Fig. 7.23, where the cell is organized in the form of a quarter-wave plate. Consider an incoming light that is linearly polarized at an angle of 45° to the direction of alignment. The transmitted light is then found to be circularly polarized, but the reflected light that passes back through the quarter-wave cell becomes polarized in a direction perpendicular to that of the incident light. The first polarizer acts as an analyzer and thus blocks the light. In the presence of an external voltage, however, the birefringence can be reduced. The polarization characteristics of the resultant reflected light are changed so

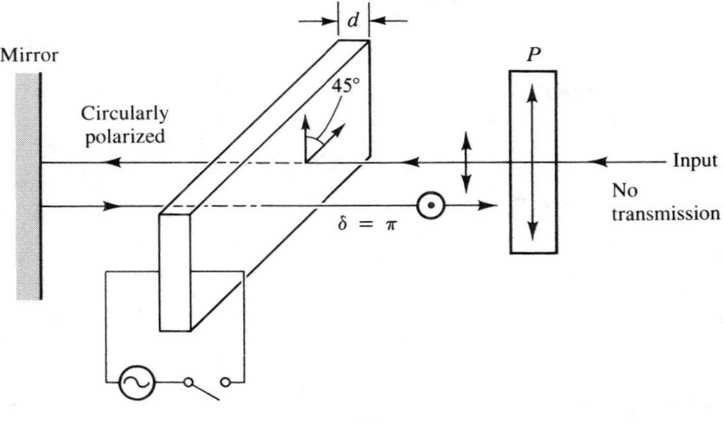

(a) Off state

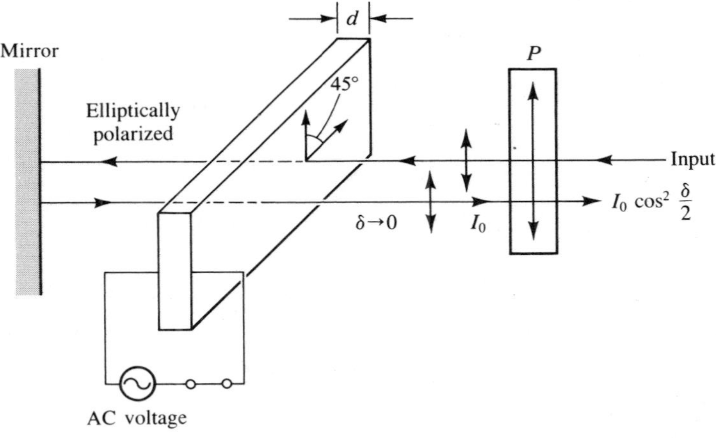

(b) On state

FIGURE 7.23

A spatial light modulator configuration

that the analyzer cannot block all of the reflected light. The external voltage can thus be used to control the transmission of light. In particular, transmission is zero when voltage is zero, but transmission reaches a maximum at a high enough voltage. It is possible for the transmission to have a nonbinary value when voltage is set between the two extremes. Devices made using this SLM configuration are generally very sensitive to variations in cell thickness and light wavelength.

Interestingly, the twisted nematic LCs are also used in the common liquid crystal display (LCD) as shown in Fig. 7.24. Instead of having parallel layers of

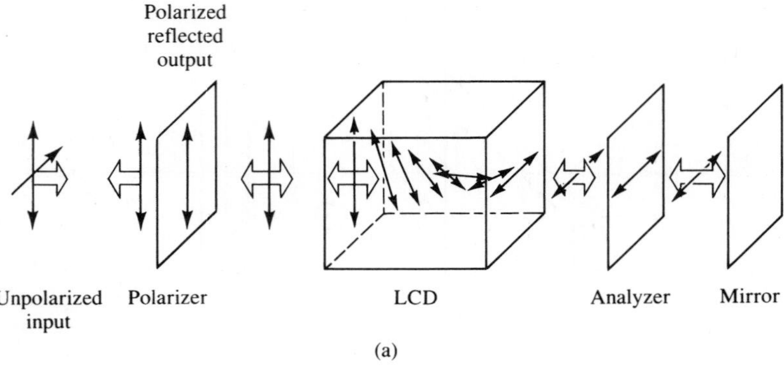

(a)

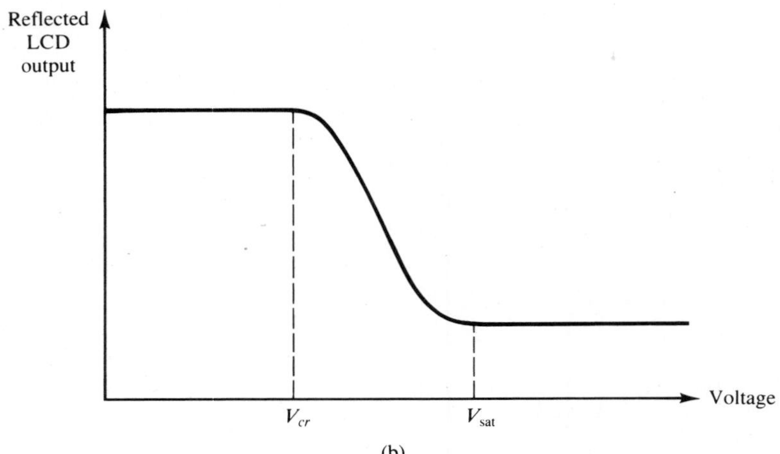

(b)

FIGURE 7.24

The LCD display: (a) the system and (b) its characteristics

LC stacks, the alignment layers at the opposite faces of the cell are maintained at 90° to each other. The remaining layers, depending on their positions, are oriented in a manner such that there is a gradual change in orientation from one end of the cell to the other. The molecules of a stack tend to line up in the same common direction, but they also tend to align themselves with those of the neighboring stacks. The tilt of the molecular axes changes gradually between the two edges. An externally applied voltage can generally overcome the effects of alignment force and, with sufficient voltage, the molecules can fall back to their isotropic states. When the voltage is withdrawn, the light going through the cell undergoes a rotation of 90° and thus passes through the analyzer. But under the influence of voltage, light falls short of a 90° rotation and, as a result, the analyzer can block most of the light.

When the voltage is low or completely withdrawn, LC molecules remain gradually twisted across the cell from the alignment direction of one wall to that of the other wall. Such smooth transition is referred to as **adiabatic**. The light polarization is able to follow this slowly varying twist primarily because of the facts that the cell width is larger than the light wavelength and that the width of each molecule is smaller than the light wavelength. As voltage is increased, LC molecules tend to be reorganized along the direction of the applied field. But in order for this to happen, the LC molecules have to overcome the alignment forces of the cell walls. The molecules located at the very center are farthest away from the walls and are, therefore, more likely to be reorganized.

As the tilt angle approaches $\pi/2$, molecules fail to align themselves with their immediate neighbors. In fact, in the extreme case of a $\pi/2$ tilt, the cell splits into two distinct halves. While the molecules in one-half are aligned along with one wall, the molecules in the other half are aligned along with the other wall. Such a nonadiabatic set up is modeled simply as a birefringent cell with an elliptically polarized transmission. Consequently, this state corresponds to maximum transmission. Other than the two extremes, slowly varying and abrupt, there exists transmitted light with an intermediate degree of elliptical polarization. The amount of polarization in the transmission is thus controllable by an externally applied voltage that allows the LCLV to operate with gray levels. Note that the maximum birefringence occurs when the incident light forms an angle of $\pi/4$ with the molecular axis. Thus nematic LCs having a 45° twist provide the maximum LCLV transmission.

The LCLV functions quite similarly to an LCD with only slight differences. The transmission characteristics of the analyzer in the case of an LCLV are the inverse of those in an LCD. This is indicated in Fig. 7.25. The LCLV requires only a little energy to produce an output, but it requires a large energy to yield its maximum output. This allows the device to produce a positive image when it is addressed optically. Again, while an LCD deals with binary transmission, an LCLV utilizes gray levels (corresponding to intermediate levels of transmission) to accurately represent an image. Thus an LCLV is often operated in the transition region that exists between the minimum and the maximum transmission. To produce such an operating characteristic, therefore, an LCLV is organized differently from an LCD.

The light incident on a cell normally strikes its surface along the normal. The incident beam of light becomes linearly polarized in the direction of the molecular alignment of the first layer. In the absence of an applied voltage, the polarization direction rotates through an angle of $\pi/4$ along the helical twist of the LCs. The returning reflected light from the mirror undergoes a further twist of $\pi/4$, amounting to a total of $\pi/2$ rotation, and is thus blocked by the analyzer. With the application of external voltage, however, light transmission increases because light can no longer follow the twist. Thus, by a combination of both birefringence and field effects, the twisted nematic cell can produce E-O modulation.

Typical LCDs are addressed via electrode leads, each of which is connected

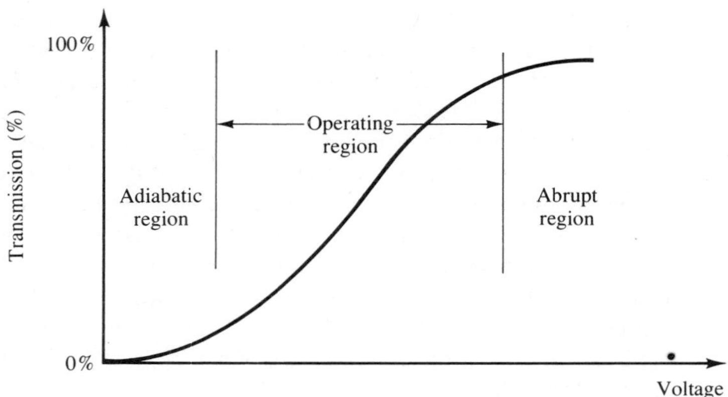

FIGURE 7.25

The transmission characteristics of an LCLV

to only one display segment. Such an addressing technique poses a serious problem when the number of leads begins to increase. LCLVs are, however, addressed differently—by means of optics. Optical addressing allows the image information to be fed into an LCLV in parallel. Thus, in the case of an LCLV, the frame time is same as the response time of only one pixel. By comparison, in a scanning display, a frame time may equal the response time of one pixel multiplied by the total number of pixels. For example, a typical 20″ × 20″ flat-panel display may consist of up to 1000 × 1000 pixels, with two to three thousand pixels per dimension possible. The simplicity of optical addressing is thus obvious. Optical addressing is also preferable because it provides better resolution.

A typical optically addressed LCLV is shown in Fig. 7.26. A photoconductive material is used to transfer the optical input to an equivalent voltage on the LC layer. The photoconductor is highly resistive, and it thus utilizes most of the voltage when there is no incident light. Very little or no voltage can be applied across the LC layer to limit the transmission of the read-out light. In the presence of incident light, the resistivity of the photoconductor decreases, as a result of which more voltage appears across the LC layer. The intensity of the incident light thus engages a proportional amount of voltage across the LC layer. Accordingly, an input intensity variation will manifest itself as a voltage variation across the LC layer. Again, the coherent read-out beam illuminating the back of the LCLV is reflected back but modulated by the birefringent LC layer. Thus the input optical image is transferred as a spatial modulation of the read-out beam. The dielectric mirror present in the device provides optical isolation between the input incoherent beam and the coherent read-out beam. In practice, however, the LCLV is driven by an ac voltage. For frequencies with periods less than the molecular response time, the LCLV responds to the rms

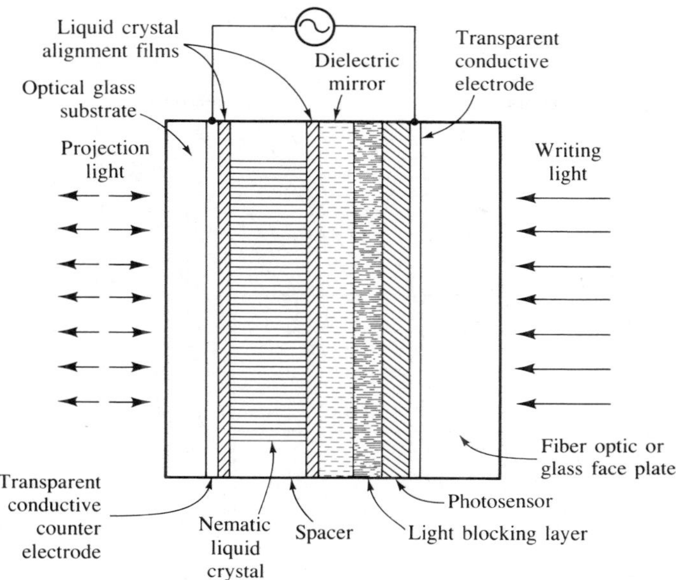

FIGURE 7.26

A schematic of the LCLV. (Adapted from F. T. S. Yu, *Optical Information Processing.* New York: Wiley, 1983, p. 140. © 1983 by John Wiley & Sons.)

value of the voltage. An ac-driven LCLV provides flexibility in choosing both the type of photoconductor as well as the arrangement of the intermediate layers.

For the ac-driven LCLV, CdS is generally chosen as the photoconductor, while CdTe is used as the light-blocking material. The CdTe layer isolates the photoconductor from any read-out light that gets through the mirror. It is possible to feed the optical input data by means of a CRT, fixed masks, or even an actual real-time imagery. A typical LCLV has a 25-mm diameter aperture, a 3 μm LC layer, a 15 ms response time, 60 lines/mm resolution at 50% modulation transfer function (MTF), a 100:1 average contrast ratio, and a lifetime of several thousand operating hours.

A more recent innovation is the CCD-addressed LCLV as shown in Fig. 7.27. The CCD structure is introduced at the input surface of the semiconductor wafer in the LCLV. The first CCD register is fed with charge information until it is full. The content of this CCD register is next clocked into the next CCD register. The serial input register is again filled and then emptied in parallel. This process is repeated until the total CCD array is loaded with a complete frame of charge information. An applied voltage can then cause the charge to migrate across the silicon to the mirror surfaces. The CCD-addressed LCLV generally

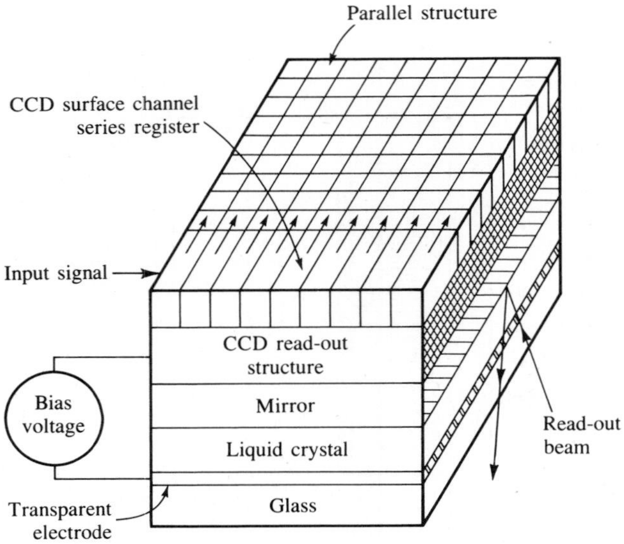

FIGURE 7.27
CCD-addressed LCLV

requires a positive pulse to transfer the charge. This particular display device is very attractive because of its high speed, high sensitivity, high resolution, and low noise distortion.

The LCLV is able to provide image intensification because the read-out beam may be as much as five orders of magnitude brighter than the write beam. The efficient isolation provided by both the mirror and the light-blocking layer is responsible for such intensification as well as for the wavelength conversion between the two beams. An LCLV is thus ideally suited to process infrared imagery. Infrared images typically consist of weak signals that require amplification before being processed further. In addition, it is easier to perform optical processing in the visible domain than in the infrared. LCLVs have already been applied to radar-signal processing, digital-optical computation, optical correlation, and optical-image processing; in fact, they have many more applications. However, the response time of an LCLV (determined by the finite time it takes to rotate the LC molecules) is questionable for many of the operations. New LC materials that may improve the response time are currently being developed. The LCLV can be operated with either all coherent, all incoherent, or mixed read-and-write beams. But the real limitation of this device happens to be its inflated cost.

Often in an adaptive system it becomes necessary to measure the difference between a signal and an estimate of that signal. This measured difference is often used to improve future estimates of the signal. Figure 7.28 shows such a feedback system, which uses a special spatial light modulator, known as a microchannel

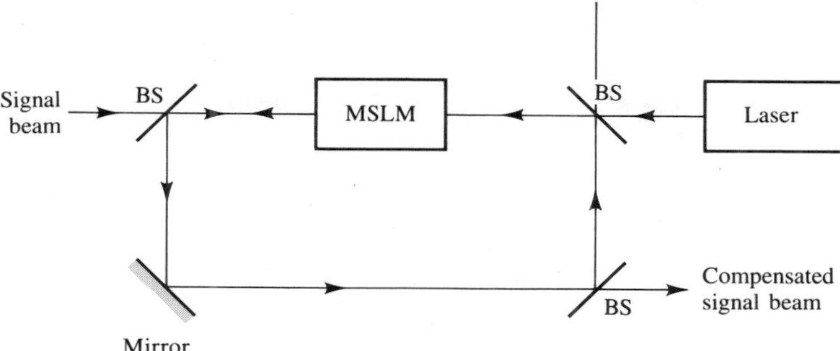

FIGURE 7.28

An adaptive compensating system

spatial light modulator, MSLM. The MSLM consists of an E-O crystal, a dielectric mirror, a microchannel plate, and a photocathode. The signal wave front passes through the beam splitter (BS) and is then reflected by an optically controlled MSLM that serves as a phase shifter. After further reflections, a part of the beam passes through a second beam splitter to generate a compensated signal beam. The remaining portion of the beam is mixed with a local oscillator to produce an error signal, which in turn is used to control further phase shifting as well as to maintain phase compensation.

7.5 Acousto-Optic Modulation

The terms **acousto-optic** (A-O) or **elasto-optic** effects are used interchangeably to indicate that the refractive index of a medium is being changed either by a mechanically applied strain or by ultrasonic waves. Accordingly, an acousto-optic modulator consists of a medium whose refractive index undergoes a sinusoidal variation in the presence of an externally applied ultrasonic signal, as shown in Fig. 7.29. The solid lines indicate the regions of maximum stress, and the dashed lines indicate the regions of minimum stress. There are many materials, such as water, lithium niobate, lucite, cadmium sulphide, quartz, and rutile, that exhibit changes in the refractive index once they are subjected to strain.

As the light enters an A-O medium, it experiences a higher value of refractive index at the region of maximum stress, and thus advances with a relatively lower velocity than those wave fronts that encounter the regions of minimum stress. The resultant light wave front thus inherits a sinusoidal form. The variation in the acoustic wave velocity is generally negligible, and so we may safely assume that the variation of refractive index in the medium is stationary as far as the optical wave front is concerned. A narrow collimated beam of light

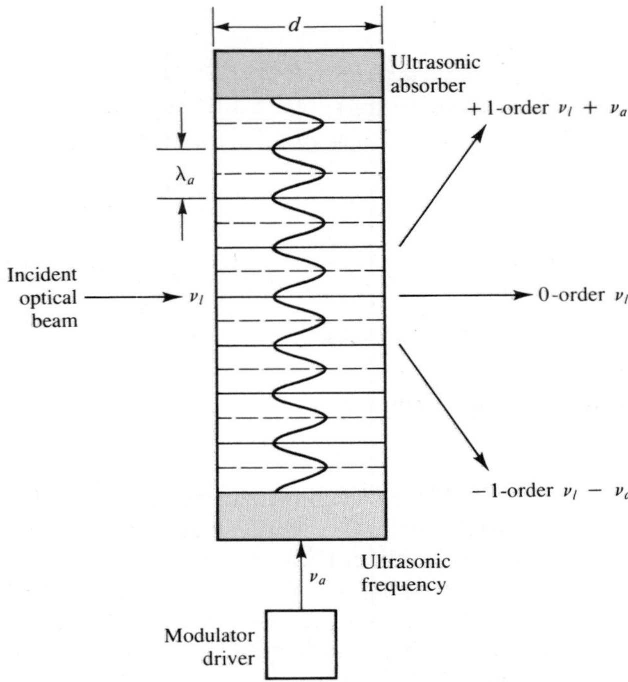

FIGURE 7.29

An acousto-optic modulator

incident upon such a medium is thus scattered into primary diffraction orders. In most practical cases, higher diffraction orders have negligible intensities associated with them. The zero-order beam generally has the same frequency as that of the incident beam, while the frequencies of the ± 1 orders undergo a frequency modulation.

In order to appreciate the basics of acousto-optic effects, we can consider the collisions of photons and phonons. Note that light consists of photons while an acoustic wave consists of phonons characterized by their respective momentums, $\hbar\mathbf{k}_l$ and $\hbar\mathbf{k}_a$, where $\mathbf{k}_l$ and $\mathbf{k}_a$ are the respective wave vectors. Likewise, the energies of a photon and a phonon are given respectively by $h\nu_l$ and $h\nu_a$, where ν_l and ν_a are the respective frequencies. Consider the scenario where the photon-phonon collision results in a scattered photon whose wave vector is given by $\mathbf{k}_l'$, as illustrated in Fig. 7.30. The condition for the conservation of momentum, when applied to this collision, yields

$$k_l \cos\theta = k_l' \cos\theta' \qquad \textbf{(7.63a)}$$

and

$$k_a = k_l \sin\theta + k_l' \sin\theta' \qquad \textbf{(7.63b)}$$

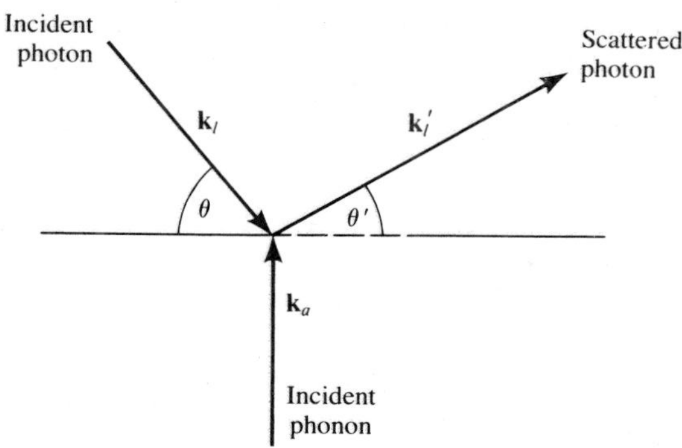

FIGURE 7.30

Photon-phonon collision with the annihilation of phonon

where θ and θ' are the angles formed by the incident and the scattered photons respectively. Consequently, the angle of the scattered photon is evaluated to give

$$\theta' = \text{Arc}\tan\left[\frac{k_a}{k_l}\sec\theta - \tan\theta\right] \tag{7.64}$$

It is reasonable to assume that $k_a \ll k_l$ and thus, for small values of θ, Eq. (7.64) reduces to

$$\theta' \simeq \left[\frac{c}{v_a v_l}\right] \nu_a - \theta \tag{7.65}$$

where v_a is the acoustic velocity. Equation (7.65) explicitly shows that the angle formed by the scattered photon is proportional to the acoustic frequency. By measuring the deflection angle, we can estimate the acoustic frequency. Further, Eq. (7.64) reveals that for an incident angle $\theta \equiv \theta_B = \sin^{-1}(k_a/2k_l)$ in an isotropic medium, $\theta = \theta'$ and $k_l = k_l'$. At this particular angle of incidence, referred to as the Bragg angle, photon momentum is conserved and the diffraction efficiency reaches a maximum. Note that the power in the scattered beam varies with θ and reaches a maximum when θ is exactly equal to the Bragg angle.

It is important to realize that the acousto-optical effect is produced by multiple collisions of photons and phonons. In any event, the scope of Eq. (7.65) is somewhat valid in most practical devices. The condition for conservation of energy is only approximately valid in photon-phonon collision. However, in practice, the frequency of the scattered photon $\nu_l \simeq \nu_l'$ since $\nu_a \ll \nu_l$. In anisotropic materials, k_l' approaches rk_l where r is the ratio of the refractive indices corresponding to the diffracted and incident waves, respectively. Equations

(7.63a) and (7.63b) can be modified to give

$$\theta = \sin^{-1}\left[\frac{k_a}{2k_l}\left\{1+\left(\frac{k_l}{k_a}\right)^2(1-r^2)\right\}\right] \quad \textbf{(7.66a)}$$

and

$$\theta' = \sin^{-1}\left[\frac{k_a}{2rk_l}\left\{1-\left(\frac{k_l}{k_a}\right)^2(1-r^2)\right\}\right] \quad \textbf{(7.66b)}$$

But to have valid solutions, the condition

$$1-\frac{k_a}{k_l} \leqslant r \leqslant 1+\frac{k_a}{k_l} \quad \textbf{(7.67)}$$

must be satisfied. It is obvious that θ and θ' are equal only when $r = 1$ because it is not possible to have $r = (k_a/k_l) - 1$ when $k_a \ll k_l$. Thus the phenomenon $\theta = \theta'$ is associated only with the Bragg angle of incidence and the condition $r = 1$. In general, for an incident wave vector, there are two values of k_a (and thus k_l') that satisfy the condition of conservation of momentum. Note that in anisotropic media, the conservation of momentum is satisfied over a wider range of acoustic frequencies or incident light beam directions than is normally realizable in isotropic materials. Consequently, in acousto-optical devices, birefringent diffraction plays a dominant role in determining modulation.

The diffraction of the light beam in A-O modulators is justifiably associated with a diffraction grating set up by the acoustic waves. The exact characteristics of this diffraction are indicated by the parameter $Q = k_a^2 d/k_l$ where d is the width of the acousto-optic device. When $Q < 1$, the diffraction is said to operate in the **Raman-Nath regime**, and when $Q \gg 1$, the diffraction is said to operate in the **Bragg regime**. In the region where $1 \leqslant Q \leqslant 10$, the diffraction has a characteristic that is a mixture of the two extremes. Since Q is directly proportional to d, a higher Q requires lesser drive power for any given interaction efficiency. In the Raman-Nath regime, the acousto-optic grating can be treated as a simple grating such that

$$m\lambda = \lambda_a \sin\theta_m \quad \textbf{(7.68)}$$

where λ_a is the acoustic wavelength, m is an integer, and θ_m is the corresponding angle of diffraction. By comparison, in the Bragg regime, the acoustic field acts very much like a "thick" diffraction grating, requiring that

$$\theta = \theta' = \sin^{-1}\left(\frac{m\lambda}{2\lambda_a}\right) \quad \textbf{(7.69)}$$

Bragg diffraction is identical to that of a plane grating only when the angle of incidence equals the diffracting angle. All of the reflected waves, except those for which $\theta = \theta'$, interfere constructively, producing a very strong first-order component.

The fraction of the light diffracted is often characterized by the diffraction efficiency η, defined as $(I_0 - I)/I_0$ where I is the output irradiance in the diffraction orders and I_0 is the output irradiance in absence of the acoustic waves. While the diffraction efficiency of the Raman-Nath grating is only about 0.35, it approaches 1.00 for the Bragg case. At the Bragg angle, the diffraction efficiency is given by $\sin^2[(\pi\,\Delta n\,d)/(\lambda\cos\theta_B)]$, where Δn is the amplitude of the refractive index fluctuation.

A Bragg cell can be used to switch light beam directions by turning on and off the acoustic source. The intensity of the diffracted light, however, depends on the amplitude of the acoustic wave. An amplitude modulation of the acoustic wave will, therefore, produce amplitude-modulated light beams. But again, the movement of the acoustic waves produces a moving diffraction grating, as a result of which the frequencies of the diffracted beams are Doppler-shifted by an amount $\pm m v_a$. This frequency shifting can be effectively manipulated to design frequency modulators. A-O modulator transfer function is sinusoidally dependent on the input voltage; however, this presents no difficulty in on-and-off modulation. For analog modulation, it is necessary to bias only the modulator at a carrier frequency such that the operating point is in an approximately linear region of operation. When compared with an E-O modulator that consumes voltage on the order of $\simeq 10^3$ V, an acousto-optic modulator requires only a couple of volts. But since the acoustic wave propagation is slow, the A-O devices are far slower than the E-O devices. The characteristics of an A-O device are often limited by the frequency response of the acoustic source, figure of merit, and acoustic attenuation. Most of the A-O materials are lossy. Materials with high figures of merit normally have a higher attenuation. The most commonly used A-O materials are quartz, tellurium dioxide, lithium niobate, and gallium phosphide. The materials with lower figures of merit are also used, but they operate with a higher drive power. A practical limit for small devices is a drive power density of 100 to 500 W/mm^2, provided there is a proper heat sink. Bandwidths of up to 800 MHz are common in most commercial A-O modulators.

A-O modulators are used widely in a large number of applications, such as laser ranging, signal-processing systems, optical computing, medium-resolution high-speed optical-beam deflectors, acoustic traveling-wave lens devices, and mode-locking. Figure 7.31 shows a system where A-O modulators are used to support beam-scanning of a laser printer. In laser printers, a rotating drum with an electrostatically charged photosensitive surface (a film of cadmium sulfide or selenium on an aluminium substrate) is used so that a modulated laser beam can repeatedly scan across the rotating surface to produce an image. The most commonly used beam-scanning system utilizes an He-Ne laser, a modulator, and a rotating polygonal prism. The He-Ne laser is preferred over other lasers because the photosensitive layer is sensitive to its output. But since it is difficult to modulate an He-Ne laser internally, the modulation is done externally. We can also use an E-O cell for the modulator, but an A-O modulator is preferred because of its ability to operate with unpolarized light, and also because it requires a low-voltage power supply.

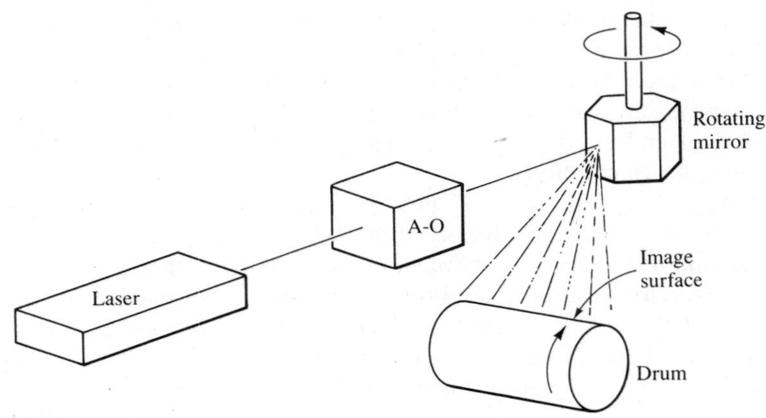

FIGURE 7.31

The beam-scanning system for the laser printer

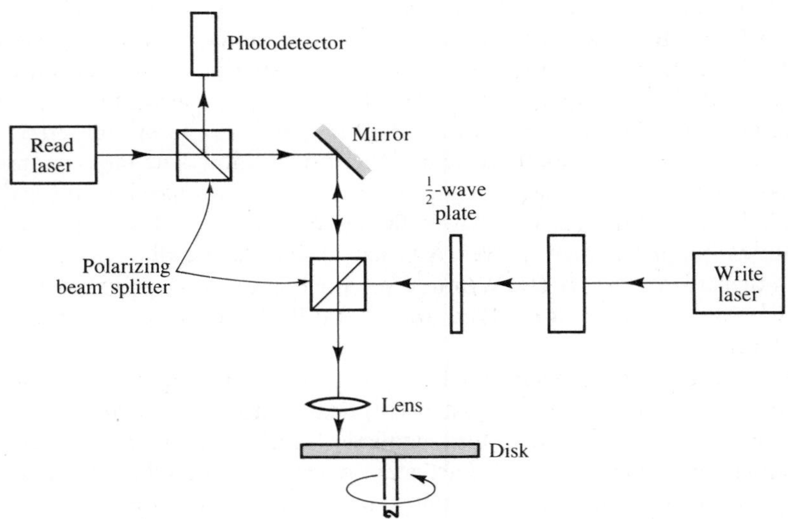

FIGURE 7.32

A schematic of a direct-read-after-write optical disk system

A-O modulators are also used in systems involving optical disks. Quite like their audio counterparts, optical disks store information in optical tracks. But in the case of optical disks, there is neither a groove nor a continuous line present, but rather "pits" that are small areas providing an optical contrast with respect to their surroundings. The varied reflectances along the track represent the information stored. These disks are versatile in the sense that they can be used for both direct read-out and recording. Figure 7.32 shows one such direct-read-after-write optical disk system. The write laser usually has more power than the read laser. The more sophisticated systems are arranged to have angular and polarization separation of beams to ensure that the read beam does not interfere with the reflections of the write beam.

7.6 Magneto-Optic Devices

The Faraday effect is the simplest of all the magneto-optic effects, whereby the rotation of polarization takes place only when an external magnetic field is applied to the medium. Materials that are candidates for displaying this effect include common salt (NaCl), zinc sulphide, quartz, and yttrium iron garnate. When a plane-polarized light beam passes through any one of these materials in the presence of a magnetic field, its plane of polarization is found to rotate by an amount proportional to the magnetic field component along the direction of propagation. The phenomenon is very similar to what we described earlier as optical activity, except for the fact that the sense of rotation caused by the Faraday effect is independent of the direction of propagation of the light beam. The sense of rotation is determined by the direction of the magnetic field. Reversing the direction of field reverses the sense of rotation. Thus it is possible to double the amount of rotation by reflecting the incoming light back through the Faraday effect device.

The rotation of the plane of polarization Φ in a Faraday effect device is given in terms of the magnetic flux B by

$$\Phi = VBd \cos \alpha \qquad \textbf{(7.70)}$$

where V is the Verdet constant of the material, d is the device length, and α is the angle between the direction of the magnetic flux and the direction of propagation of the optical beam. In general, the Faraday effect is small and is dependent on wavelength since the Verdet constant is dependent on the wavelength of light.

A Faraday rotator, when used along with a polarizer and an analyzer pair, can act as an optical isolator, as shown in Fig. 7.33. The analyzer is rotated relative to the polarizer by an amount $\pi/4$. The light passing from left to right is polarized along the horizontal plane of the polarizer. The Faraday rotator length and the fields can be adjusted for a clockwise rotation of $\pi/4$. The analyzer allows the Faraday rotator output to be transmitted with minimal attenuation. The reflected light wave will retrace the path but will undergo a further rotation of $\pi/4$ before reaching the polarizer. But by this time the light beam polarization is devoid of components along the polarizing direction of the polarizer. The device thus prevents the reflected light from reaching the source. A similar arrangement can also be used for modulating the light intensity by

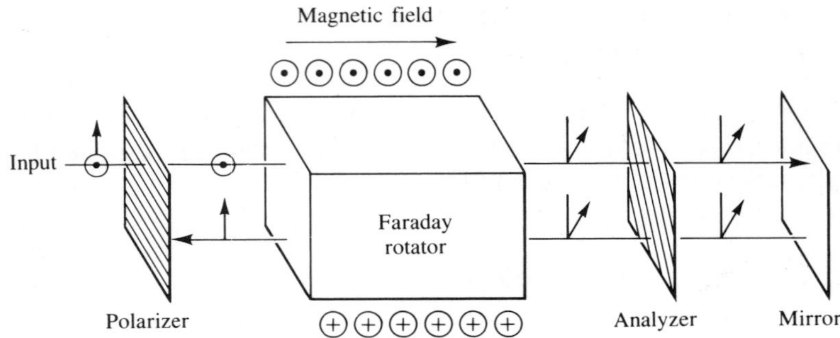

FIGURE 7.33

An optical isolator using a Faraday rotator

modulating the magnetic field. Note, however, that the maximum frequency of such modulators is determined by the inductance of the coil that is used in generating the magnetic flux.

There are two more but less significant magneto-optic effects: the Cotton-Mouton effect and the Kerr magneto-optic effect. The former is the magnetic equivalent of the E-O Kerr effect. When liquids such as nitrobenzene or carbon disulfide are exposed to a transverse magnetic field, they behave similarly to a uniaxial crystal whose optical axis lies along the magnetic field and contributes to birefringence. The Kerr magneto-optic effect, on the other hand, is observed when linearly polarized light is reflected off the polished pole of an electromagnet. The reflected light becomes elliptically polarized with respect to the oscillation of the incident light. The contribution of this effect is very minimal. To make a meaningful device, it is necessary to introduce multilayer coatings to the pole surface so that the light can be subjected to multiple reflections.

In general, it is easier to generate electric fields than magnetic fields. For this reason, E-O devices are preferred over magneto-optic devices. One possible use of magneto-optic devices is in large-capacity computer memories. Such magneto-optic memories can be read via the Faraday effect or the Kerr magneto-optic effect. In the presence of a magnetic field, the ferromagnetic material can rotate the plane of polarization of the incident light. The writing of memory is achieved by raising the temperature of the ferromagnetic elements above the Curie point by means of laser heating, and then cooling in the presence of the magnetic field. Accordingly, a particular magnetization direction is imprinted in the ferromagnetic elements. The magnetization along a particular direction may be used to represent a binary 1, while that along the opposite direction may represent a binary 0.

7.7 Image Binarization Using the Photographic Process

In an earlier section, we introduced a real-time E-O light modulator LCLV as a recording device. The LCLV is capable of converting an incoherent optical image into a coherent image which can then be used in almost all types of coherent optical processors. But it is worthwhile to note that the maximum resolution of an LCLV is only on the order of 60 lines/mm. Accordingly, LCLV cannot accommodate for holographic displays. There are, of course, other real-time recording materials, such as Pockels read-out optical modulator (PROM) and photoplastic devices. PROM devices, however, have been observed to be ineffective when read over an extended period of time in the presence of strong illumination. Further, PROMs and LCLVs are both too expensive at the current time, and this limits their use as practical recording devices. Photoplastic devices are cheap, but they respond very slowly and the corresponding SNR is relatively poor due to the random thickness variations of the developed thermoplastics. On the other hand, the photographic medium is not only slow—its development process is quite messy. However, it is still doubtful whether the real-time devices can ever totally replace photographic films and plates. They can serve as input mediums to feed input data into optical processors, as mediums to record the processor output, and also as mediums for introducing filter characteristics into an optical processor (to be introduced in Chapter 8). Often these devices are used to play all three of these roles.

Appropriately, this section describes how to achieve spatial modulation of two-dimensional images with the help of a given recording medium and especially designed screens, referred to as halftone or contact screens. Such spatially modulated images are often used as inputs to coherent optical processors for filtering operations.

In the printing and graphic arts industries, the halftone technique has long been used in the reproduction of continuous-tone photographs. Basically, a screen is used to transform an original photograph into a high-contrast halftone negative or positive that contains arrays of opaque dots with sizes that vary according to the optical density of the original. Printing plates can then be photo-reproduced from the halftone photograph.

Screens are classified into contact and noncontact types according to the way they are used during the exposure to make the halftone. In the use of noncontact screens, a distance is maintained between the screen and the high-contrast film during exposure. The difficulty of determining the optimum value of this distance between the screen and the high-contrast film for halftone reproduction is avoided by using contact screens that are placed in close contact with the film during exposure.

The halftone contact screen technique has enough flexibility that it can be used for nonlinear optical-image processing. In this context, it may be noted that linear operations, such as convolution, correlation, and Fresnel and Fourier transformations, can be realized very easily with optical processors. On the other hand, digital processors can readily achieve nonlinear transformations. Thus any extension of the capabilities of optical systems to include nonlinear operations will enhance their suitability as viable alternatives to purely digital systems. It has been demonstrated that logarithmic transformation, level-slice, edge-tailored bandpass, exponential transformation, analog-to-digital conversion, equidensitometry, and pseudocolor operations can be realized using halftone screen processing. The importance of the application of this technique to image processing is thus clear.

Nonlinear optical-image processing using contact screens consists of either one or both of the following steps. The first step involves a mapping of the input intensity I_{in} through a mapping function f such that $I_{out} = f(I_{in})$ where I_{out} is the output intensity. It is important to note that both the input and output intensities are functions of spatial coordinates. The mapping function f can be made either linear or nonlinear through the use of contact screens. The output is generally in the form of a pulse-width or pulse-area modulated nonlinear image of the original. In the second step, when the modulated binary output in transparency form is placed at the input plane of a coherent optical-image processing system as shown in Fig. 7.34, many orders of diffraction patterns appear at the Fourier plane. A spatial filter can then be used at the Fourier plane to select or to modify any one or several diffraction orders of the spectrum. The corresponding output image can then be found at the focal plane of the second lens. The intensity at any point in this plane can be made to vary as a nonlinear function of the intensity at the corresponding point in the input image. Thus the overall nonlinear relationship between the input and the output depends on spatial filtering as well as on the contact screen process used in the first step.

An essential element used in the contact screen process is the contact screen itself. A contact screen normally consists of a periodic array of continuous-tone cell patterns on a glass or plastic substrate. The contact screen transforms the continuous-tone image into a binary image through a contact printing process. In the process, the original continuous-tone photographic transparency or intensity distribution is contact-printed through the screen onto a high-contrast recording medium (ideally having $\gamma = \infty$). When the exposed high-contrast film is developed, the original gray tones can be represented by the average transmittance of either area-modulated opaque dots if a two-dimensional screen is used or pulse-width-modulated opaque bars if a one-dimensional screen is used.

The contact screen process using a specific one-dimensional screen can be illustrated with the help of Fig. 7.35. In Fig. 7.35(a), the transmittance $T^s(x)$ of several unit cells of the screen are plotted as a function of x. We assume that the continuous-tone original has a maximum spatial frequency of less than the frequency of the screen. This assures that the input intensity transmittance T^P remains constant over the region of at least one unit cell, provided the edge effects are not considered. During the contact-printing process, the screen is sandwiched between the continuous-tone original and the recording film,

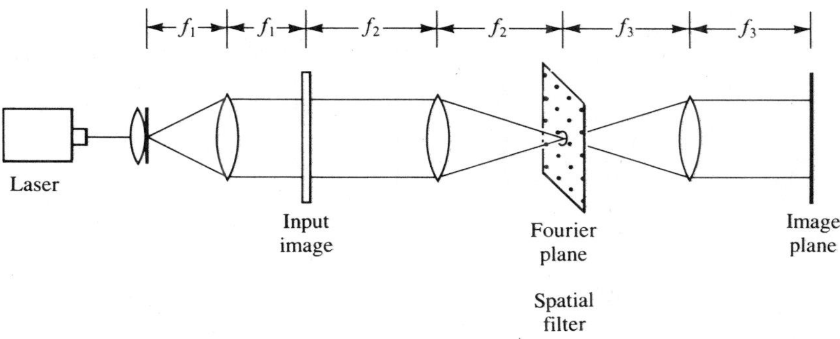

FIGURE 7.34

A coherent optical-image processor

preferably while the film is resting on a vacuum system. When the screen is in contact with the original, and when the combined transmittance $T^s(x)\,T^P$ is equal to or greater than the threshold transmittance T^{th}, the developed film has a transmittance of nearly 0; otherwise, its transmittance is approximately 1, as shown in Fig. 7.35(d). For convenience, we may also define a threshold transmittance as

$$T^{\text{th}} = \frac{I_{\text{th}}}{I_{\text{in}}} \tag{7.71}$$

where I_{th} and I_{in} denote the threshold energy and the energy of the input illumination, respectively. The resulting photograph with binary levels of transmittance is called a halftone photograph, which is nothing but a spatially sampled version of the original.

For simplicity, the derivations of screen transmittance corresponding to a particular nonlinear mapping of the input are limited to screens of nonlinear, one-dimensional (line) cell patterns only. As shown in Fig. 7.35(d), each group of equal-width halftone transmittance pulses corresponds to a constant value of input photographic transmittance T^P. The screen transmittance is generally assumed to be periodic such that

$$T^s(x) = T^s(x + X) \tag{7.72}$$

where X is the screen period.

The overall transmittance of a unit cell T^H can be defined as the ratio of the width of the transparent region to the total width of the unit cell. When the original input is sufficiently sampled, T^H represents the regional average transmittance of the halftone photograph.

For $T^P \geqslant T^{\text{th}}/T^s$, the width of the transparent bar within each unit cell of the halftone photograph is equal to $(X - \chi)$ where χ is the particular value of x

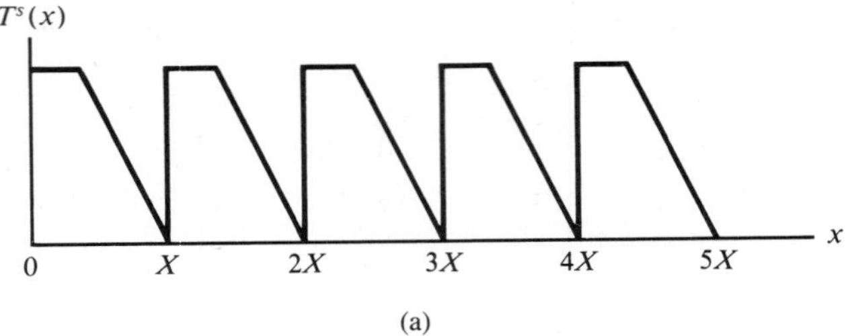

(a)

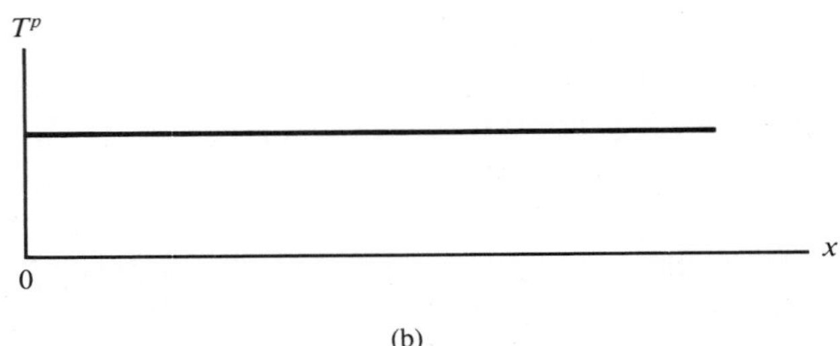

(b)

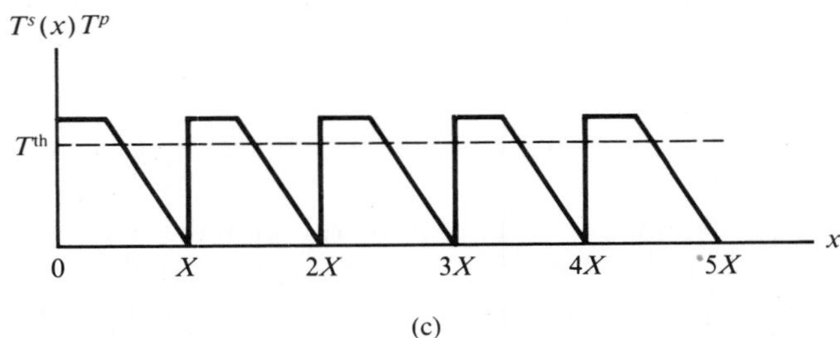

(c)

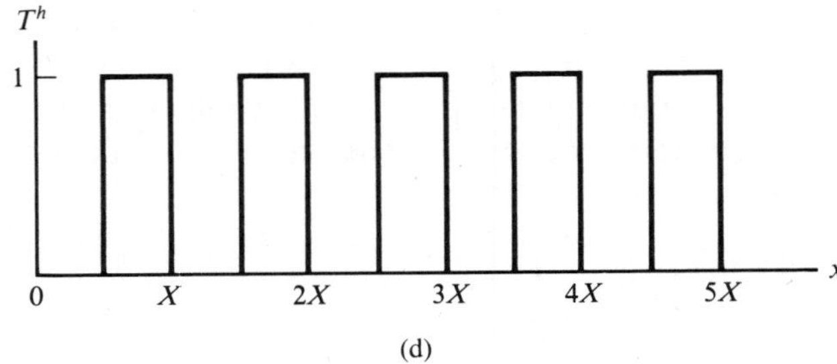

(d)

FIGURE 7.35

Contact screen process: (a) transmittance of a few neighboring screen cells, (b) input transmittance, (c) combined transmittance, and (d) transmittance of the resultant halftone photograph

for which

$$T^s(x)\,T^P = T^{\text{th}} \tag{7.73}$$

When $T^P < T^{\text{th}}/T^s$, the film is not sufficiently exposed and, therefore, the corresponding region of the halftone photograph has a transmittance of nearly 1 (almost clear). For simplicity, we assume that the transmittance is unity.

Based on the preceding discussion, T^H may be written as

$$T^H = \begin{cases} 1 - \dfrac{x}{X} & T^s T^P \geqslant T^{\text{th}} \\ 1 & T^s T^P < T^{\text{th}} \end{cases} \tag{7.74}$$

On the other hand, screen transmittance is given by

$$T^s(x) = \begin{cases} \dfrac{T^{\text{th}}}{T^P} & T^P \geqslant T^{\text{th}} \\ 1 & T^P < T^{\text{th}} \end{cases} \tag{7.75}$$

In general, the functional dependence of T^H on T^P may be expressed as

$$T^H = f(T^P) \tag{7.76}$$

where the function f determines the actual nonlinear relationship. Equation (7.73) can then be used to determine x in terms of T^P as

$$x = [1 - f(T^P)]X \qquad T^P \geqslant T^{\text{th}} \tag{7.77}$$

Consider two specific cases for the purpose of illustration: (a) a logarithmic case $f(T^P) = \alpha - \beta T^P$ and (b) the exponential case $f(T^P) = \alpha \exp(\beta T^P)$ where α and β are two constants such that $1 \geqslant f(T^P) \geqslant 0$. When function (a) is substituted into Eq. (7.77), we have

$$x = \left[1 - \alpha + \beta \ln\left(\frac{T^s(x)\,T^P}{T^s(x)}\right)\right]X \tag{7.78}$$

and when function (b) is used, Eq. (7.77) becomes

$$x = [1 - \alpha e^{\beta[T^s(x)T^P/T^s(x)]}]X \tag{7.79}$$

Using Eqs. (7.78), (7.75), and (7.73), we obtain

$$T^s(x) = \begin{cases} \dfrac{T^{\text{th}}}{e^{(\alpha - 1 + (x/X))/\beta}} & 1 \geqslant e^{(\alpha - 1 + (x/X))/\beta} \geqslant T^{\text{th}} \\ 1 & \text{otherwise} \end{cases} \tag{7.80}$$

for the logarithmic case, and similarly using Eqs. (7.79), (7.75), and (7.73), we get

$$T^s(x) = \begin{cases} \dfrac{\beta T^{\text{th}}}{\ln\left[\left(1 - \dfrac{x}{X}\right)\dfrac{1}{\alpha}\right]} & 1 \geqslant \dfrac{1}{\beta}\ln\left[\left(1 - \dfrac{x}{X}\right)\dfrac{1}{\alpha}\right] \geqslant T^{\text{th}} \\ 1 & \text{otherwise} \end{cases} \tag{7.81}$$

for the exponential case.

The screen design process developed thus far was formulated with the unrealistic assumption that a hard-clipping recording medium is used for making the halftone image. But this assumption is an oversimplification of the actual behavior of the recording medium, as is evident from our discussion in Chapter 4. An uncompensated screen such as the one developed here will, when used, produce gray areas in addition to the usual binary black and white. It is noteworthy that in the case of real-time replacement for photographic film, the assumption is more disastrous because nearly all current devices are intended to operate only linearly with moderate contrast (γ of 3) over a broad dynamic range (transmittances of 0.001 to 1.0). The precompensation design analysis that follows takes into consideration the actual response of the recording medium. Both density and spatial modulations are utilized to realize more reliable contact screens.

For simplicity, the design is limited to only nonsymmetric, one-dimensional screens. Assume further that $T^s(0) > T^s(X)$. The optical energy reaching the film at position x is given by

$$E(x) = eT^P\, T^s(x) \tag{7.82}$$

where e is the light intensity incident on the continuous-tone input transparency. Let T^h be the transmittance of the resulting halftone photograph, which is related to the incident energy by

$$T^h(E) = \begin{cases} 1 & \text{if } 0 \leqslant E \leqslant E_1 \\ 0 & \text{if } E_2 \leqslant E \leqslant \infty \end{cases} \tag{7.83}$$

The general nature of $T^h(E)$ is illustrated in Fig. 7.36. On the other hand, the overall transmittance of a unit cell of the halftone image T^H may be defined as the average of the displayed transmittance over a single screen cell. T^H can thus be expressed in terms of T^h as

$$T^H = \frac{1}{X}\int_0^X T^h(eT^P\, T^s(x))\, dx \tag{7.84}$$

The design problem therefore reduces to determining $T^s(x)$, provided the functional relationship $T^H = f(T^P)$ and the exact form of $T^h(E)$ are specified. Furthermore, the precompensated screen profile is expected to compensate for the nonbinary nature of the recording medium. The halftone image generated

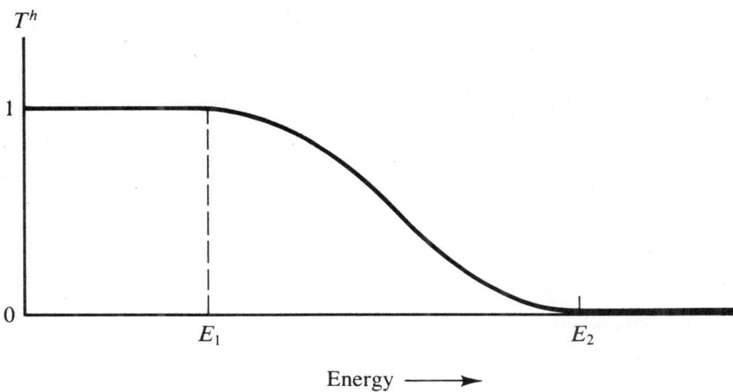

FIGURE 7.36

The general characteristics of the halftone transmittance versus the exposing energy

thereby is expected to be binary in nature. It will thus be possible to identify a specific transmittance value $T_{\text{eq}}^{\text{th}}$, referred to as the equivalent threshold transmittance, such that

$$T^h(x) = \begin{cases} 1 & \text{if } T^P T^s(x) < T_{\text{eq}}^{\text{th}} \\ 0 & \text{if } T^P T^s(x) \geqslant T_{\text{eq}}^{\text{th}} \end{cases} \tag{7.85}$$

Accordingly, T^H may be given by

$$T^H = \begin{cases} 1 - \dfrac{x}{X} & \text{if } T^s T^P \geqslant T_{\text{eq}}^{\text{th}} \\ 1 & \text{if } T^s T^P < T_{\text{eq}}^{\text{th}} \end{cases} \tag{7.86}$$

Further, the screen transmittance is expressed as

$$T^s(x) = \begin{cases} \dfrac{T_{\text{eq}}^{\text{th}}}{T^P} & \text{if } T^P \geqslant T_{\text{eq}}^{\text{th}} \\ 1 & \text{if otherwise} \end{cases} \tag{7.87}$$

Equations (7.87) can thus be used to yield

$$\begin{aligned} x \equiv \tau(T^s(x)) &= [1 - f(T^P)]X \\ &= \left[1 - f\left(\frac{T_{\text{eq}}^{\text{th}}}{T^s(x)}\right)\right]X \qquad T^P \geqslant T_{\text{eq}}^{\text{th}} \end{aligned} \tag{7.88}$$

where $T^H = f(T^P)$ and τ denotes the inverse function of $T^s(x)$. We can solve Eq. (7.88) to determine the screen transmittance after having identified $T_{\text{eq}}^{\text{th}}$ and the form of f.

We can rewrite Eq. (7.84) as

$$T^H = \frac{1}{X}\left[\int_0^{x_2} + \int_{x_2}^{x_1} + \int_{x_1}^{X}\right] T^h(E)\, dx \tag{7.89}$$

where

$$eT^P T^s(x_1) = E_1 \tag{7.90}$$

$$eT^P T^s(x_2) = E_2 \tag{7.91}$$

Using Eq. (7.83), we find

$$\int_0^{x_2} T^h(E)\, dx = 0 \tag{7.92}$$

and

$$\int_{x_1}^{X} T^h(E)\, dx = X - x_1 \tag{7.93}$$

Equation (7.89), therefore, reduces to

$$T^H = \frac{1}{X}[X - \tau((E_1)/(eT^P))] + \frac{1}{X}\int_{x_2}^{x_1} T^h(E)\, dx \tag{7.94}$$

Substituting $\Upsilon = T^s(x)$, we can evaluate the last term of Eq. (7.94) by the method of parts as follows:

$$\begin{aligned}\frac{1}{X}\int_{x_2}^{x_1} T^h(E)\, dx &= \frac{1}{X}\int_{T^s(x_2)}^{T^s(x_1)} T^h(E)\tau'(\Upsilon)\, d\Upsilon \\ &= \frac{1}{X}\left[\tau\left(\frac{E_1}{eT^P}\right) - \int_{T^s(x_2)}^{T^s(x_1)} eT^P T^{h'}(E)\tau(\Upsilon)\, d\Upsilon\right]\end{aligned} \tag{7.95}$$

where $\tau'(\Upsilon) \equiv d\tau/d\Upsilon$ and $T^{h'}(E) \equiv dT^h/d\Upsilon$. Therefore, Eq. (7.94) simplifies to

$$T^H = \frac{1}{X}\left[X - \int_{T^s(x_2)}^{T^s(x_1)} eT^P T^{h'}(E)\tau(\Upsilon)\, d\Upsilon\right] \tag{7.96}$$

$T^{\text{th}}_{\text{eq}}$ can now be obtained by introducing Eq. (7.88) into Eq. (7.96). As a first example, let us consider the case of linear mapping $T^H = \alpha - \beta T^P$, where α and β are two constants such that for all T^P, $1 \geqslant T^H \geqslant 0$. Accordingly, Eq. (7.96) can be written as

$$\begin{aligned}T^H = \alpha - \beta T^P &= \frac{1}{X}\left[X - \int_{T^s(x_2)}^{T^s(x_1)} eT^P T^{h'}(E) X\left[1 - \alpha + \beta\frac{T^{\text{th}}}{\Upsilon}\, d\Upsilon\right]\right. \\ &= \alpha - \beta T^{\text{th}}_{\text{eq}}\, eT^P \left.\int_{T^s(x_2)}^{T^s(x_1)} \frac{T^{h'}(E)}{\Upsilon}\, d\Upsilon\right]\end{aligned} \tag{7.97}$$

For the linear mapping we can therefore conclude

$$T_{\mathrm{eq}}^{\mathrm{th}} = \left[-\int_{T^s(x_1)}^{T^s(x_2)} \frac{eT^{h'}(E)}{\Upsilon}\, d\Upsilon \right]^{-1}$$

$$= \left[-\int_{E_1}^{E_2} \frac{eT^{h'}(E)}{E}\, dE \right]^{-1} \tag{7.98}$$

and, consequently, the corresponding precompensated screen transmittance is obtained from Eq. (7.88) as

$$T^s(x) = \beta \left\{ -\int_{E_1}^{E_2} \left[eT^{h'}(E) \frac{dE}{E} \right] \left[\alpha - 1 + \frac{x}{X} \right] \right\}^{-1} \tag{7.99}$$

Similarly, for the logarithmic mapping $T^H = a - b \ln(T^P)$, the equivalent threshold transmittance and the precompensated screen transmittance are respectively given by

$$\ln T_{\mathrm{eq}}^{\mathrm{th}} = \int_{T^s(x_1)}^{T^s(x_2)} eT^P T^{h'}(E) \ln\left(\frac{E}{e}\right) d\Upsilon \tag{7.100}$$

and

$$T^s(x) = \exp\left[-\int_{E_1}^{E_2} T^{h'}(E) \ln\left(\frac{E}{e}\right) dE - b\left\{ a - 1 + \frac{x}{X} \right\} \right] \tag{7.101}$$

The general nature of Eqs. (7.99) and (7.101) can now be utilized to yield known special cases.

For a familiarly assumed hard-clipping recording medium, the halftone transmittance is given by

$$T^h(E) = \begin{cases} 1 & \text{if } E < E^{\mathrm{th}} \\ 0 & \text{if } E \geqslant E^{\mathrm{th}} \end{cases} \tag{7.102}$$

where E^{th} is the threshold energy of exposure such that the threshold transmittance $T^{\mathrm{th}} = E^{\mathrm{th}}/e$. The incorporation of Eq. (7.101) into Eq. (7.98) gives us

$$T_{\mathrm{eq}}^{\mathrm{th}} = T^{\mathrm{th}} \tag{7.103}$$

for the linear mapping and, therefore,

$$T^s(x) = \beta T^{\mathrm{th}} \left[\alpha - 1 + \frac{x}{X} \right]^{-1} \tag{7.104}$$

which, for the special case of $\alpha = 0$ and $\beta = -1$, reduces to

$$T^s(x) = \frac{T^{\mathrm{th}}}{1 - \dfrac{x}{X}} \tag{7.105}$$

For the finite-gamma recording medium,

$$T^h(E) = \begin{cases} \left(\dfrac{E}{E^{\text{th}}}\right)^{-\gamma} & \text{for } E^{\text{th}} \leqslant E \\ 1 & \text{for } E < E^{\text{th}} \end{cases} \tag{7.106}$$

Precompensated screen transmittance for a linear mapping, therefore, results in

$$T^s(x) = \frac{\gamma + 1}{\gamma} \frac{T^{\text{th}}}{1 - \dfrac{x}{X}} \tag{7.107}$$

The screen criterion that follows Eq. (7.99) is the most accommodating and is highly precompensating for the true nature of the recording medium, while that following Eq. (7.107) is good only for those recording mediums whose H-D curves can be approximated by straight lines. Again, Eq. (7.105), which is a special case of Eq. (7.99), represents the transmittance profile of only uncompensated screens. All of the equations describing the design of nonsymmetric screens are equally applicable for the design of symmetric screens, provided the term $[1-(x/X)]$ has been replaced with $[1-(2x/X)]$.

In Chapter 8, we use two specially designed contact screens (logarithmic and exponential) for logarithmic filtering and the subsequent exponential processing to effectively realize an optical homomorphic image processor. Such an optical information processing system is used for separating multiplicative noises from the signal content of a two-dimensional image.

Problems

7.1 Calculate the reflectance of light when the sine of the angle of incidence exceeds the refractive index. Discuss the physical significance of your result.

7.2 Obtain the ray velocity surfaces for uniaxial anisotropic crystals. Describe the physical significance of the velocity surfaces under all pertinent conditions.

7.3 Refer to Fig. 7.6 and show that $\Delta\alpha \equiv \alpha_E - \alpha_0$ for a uniaxial crystal is a maximum when $\alpha_0 = \tan^{-1}(n_E/n_0)$. Determine the maximum value of $\Delta\alpha$.

7.4 For a POSC system described in Section 7.2, determine the two outputs corresponding to (a) vertically polarized and (b) horizontally polarized masks when

$$A = \begin{pmatrix} \text{H} & \text{H} \\ \text{T} & \text{V} \end{pmatrix} \qquad B = \begin{pmatrix} \text{H} & \text{V} \\ \text{T} & \text{H} \end{pmatrix} \qquad C = \begin{pmatrix} \text{T} & \text{H} \\ \text{V} & \text{H} \end{pmatrix}$$

Comment on the overall characteristics of the device.

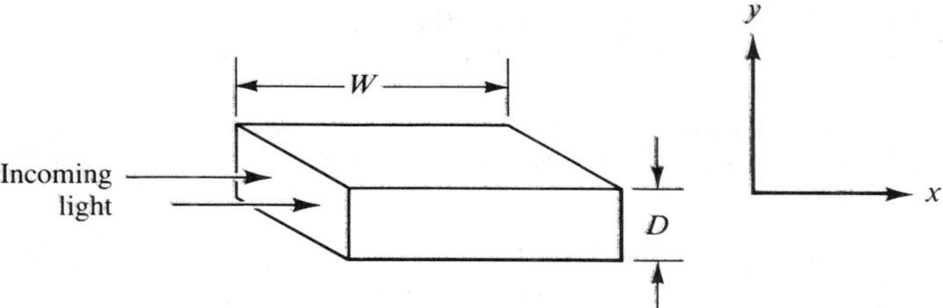

FIGURE 7.P1

7.5 Repeat Problem 7.4 when the inputs are

$$A=\begin{pmatrix}\mathrm{V} & \mathrm{F}\\ \mathrm{H} & \mathrm{V}\end{pmatrix}\qquad B=\begin{pmatrix}\mathrm{T} & \mathrm{V}\\ \mathrm{H} & \mathrm{H}\end{pmatrix}\qquad C=\begin{pmatrix}\mathrm{V} & \mathrm{H}\\ \mathrm{T} & \mathrm{H}\end{pmatrix}$$

7.6 For a two-input POSC system, determine the corresponding outputs for all of the four possible masks. The inputs are

$$A=\begin{pmatrix}\mathrm{V} & \mathrm{V}\\ \mathrm{H} & \mathrm{H}\end{pmatrix}\qquad B=\begin{pmatrix}\mathrm{H} & \mathrm{F}\\ \mathrm{H} & \mathrm{F}\end{pmatrix}$$

7.7 Find the allowable directions of polarizations and their refractive indices when $LiNbO_3$ is used as an E-O modulator and the external field is applied along its optical axis. Under what condition is the emergent light circularly polarized?

7.8 Calculate the retardation for a transverse E-O modulator using a GaAs cubic crystal in terms of pertinent system parameters.

7.9 At what value of the applied voltage is the rate of change of intensity transmission in a Pockels KDP modulator independent of the presence of a quarter-wave plate? The wavelength of the given light is 550 nm.

7.10 Derive Eq. (7.59).

7.11 Show that a traveling wave modulator increases the frequency limit by a factor $[1-\{c/(nv_p)\}]^{-1}$ where v_P is the phase velocity of the modulation field.

7.12 For the crystal shown in Fig. 7.P1, the refractive index is given by $n(y)=n+(dn/dy)$. Show that the angle of deflection is given by $|W(dn/dy)|$.

7.13 Show that the contact screen characteristics given by

$$T^{s}(x)=T^{\text{th}}\left\{\beta\left[\frac{\gamma}{\gamma+n}\left\{\alpha-1+\frac{x}{X}\right\}\right]^{-1}\right\}^{1/n}$$

satisfy the nonlinear mapping $T^{H}=\alpha-\beta(T^{P})^{n}$ where α and β are two constants such that for all T^{P}, $1\geqslant T^{H}\geqslant 0$, X is the cell period of the line screen, and T^{th} is the threshold transmittance.

References and Suggested Readings

Awwal, A. A. S., M. A. Karim, and A. K. Cherri. "Polarization-Encoded Optical Shadow-Casting Scheme: Design of Trinary Combinational Logic Units." *Appl. Opt.* Vol. 26 (1987): 4814.

Awwal, A. A. S., and M. A. Karim. "Polarization-Encoded Optical Shadow-Casting Programmable Logic Array: Simultaneous Generation of Multiple Outputs." *Appl. Opt.* Vol. 27 (1988): 932.

Awwal, A. A. S., and M. A. Karim. "Polarization-Encoded Optical Shadow-Casting: Design of a *J-K* Flip-Flop." *Appl. Opt.* Vol. 27 (1988): 3719.

Awwal, A. A. S., and M. A. Karim. "Polarization-Encoded Optical Shadow-Casting: Direct Implementation of a Carry-Free Adder." *Appl. Opt.* Vol. 28 (1989): 785.

Awwal, A. A. S., and M. A. Karim. "Associative Polarization-Encoded Optical Shadow-Casting: Gray Image Encoding for Serial and Parallel Operations." *Appl. Opt.* Vol. 28 (1989): 284.

Bademian, L. "Parallel-Channel Acousto-Optic Modulation." *Opt. Engn.* Vol. 25 (1986): 303.

Berg, N. J., and J. N. Lee. *Acousto-Optic Signal Processing.* New York: Marcel Dekker, 1983.

Bleha, W. P., L. T. Lipton, E. Wiener-Avner, J. Grinberg, P. G. Reif, D. Casasent, H. B. Brown, and B. V. Markevitch. "Application of the Liquid Crystal Light Valve to Real-Time Optical Data Processing." *Opt. Engn.* Vol. 17 (1978): 371.

Casasent, D. "Acoustooptic Linear Algebra Processors: Architectures, Algorithms, and Applications." *Proc. IEEE.* Vol. 72 (1984): 831.

Casasent, D. "Acoustooptic Transducers in Iterative Optical Vector-Matrix Processors." *Appl. Opt.* Vol. 21 (1982): 1859.

Chang, I. C., and D. L. Hecht, "Characteristics of Acousto-Optic Devices for Signal Processors." *Opt. Engn.* Vol. 21 (1982): 76.

Davis, J. A., T. Day, R. A. Lilly, and H. K. Liu. "Multichannel Optical Correlator/Convolver Utilizing the Magnetooptical Spatial Light Modulator." *Appl. Opt.* Vol. 26 (1987): 2479.

Dewey, A. G. "Laser-Addressed Liquid Crystal Displays." *Opt. Engn.* Vol. 23 (1984): 30.

Dziedzic, J. M., R. H. Stolen, and A. Ashkin. "Optical Kerr Effect in Long Fibers." *Appl. Opt.* Vol. 20 (1981): 1403.

Efron, U., J. Grinberg, P. O. Braatz, M. J. Little, P. G. Reif, and R. N. Schwartz. "The Silicon Liquid Crystal Light Valve." *J. Appl. Phys.* Vol. 57 (1985): 1356.

Efron, U., P. O. Braatz, M. J. Little, R. N. Schwartz, and J. Grinberg. "Silicon Liquid Crystal Light Valves: Status and Issues." *Opt. Engn.* Vol. 22 (1983): 682.

Fatehi, M. T., K. C. Wasmundt, and S. A. Collins, Jr. "Optical Logic Gates Using Liquid Crystal Light Valve: Implementation and Application Example." *Appl. Opt.* Vol. 20 (1981): 2250.

Fatehi, M. T., K. C. Wasmundt, and S. A. Collins, Jr. "Optical Flip-Flops and Sequential Logic Circuits Using a Liquid Crystal Light Valve." *Appl. Opt.* Vol. 23 (1984): 2163.

Fisher, A. D., and C. Warde. "Technique for Real-Time High-Resolution Adaptive Phase Compensation." *Opt. Lett.* Vol. 8 (1983): 353.

Fisher, A. D., L. C. Ling, J. N. Lee, and R. C. Fukuda. "Photoemitter Membrane Light Modulator." *Opt. Engn.* Vol. 25 (1986): 261.

Gerritsen, H. J., and M. L. Jepsen. "Holographic Recordings of Nematic LC Displays." *Appl. Opt.* Vol. 26 (1987): 3717.

Goodman, J. W. *Introduction to Fourier Optics.* New York: McGraw-Hill, 1968.

Gottlieb, M., and F. Schaff. "Acoustooptic Refractive IR Chopper." *Appl. Opt.* Vol. 20 (1981): 3403.

Gregory, D. A., and L. L. Huckabee. "Acoustooptically Addressed Fourier Transform Matched Filtering." *Appl. Opt.* Vol. 24 (1985): 859.

Horner, H. L. (Ed). *Optical Signal Processing.* San Diego, CA: Academic Press, 1987.

Johnson, E. L., and M. A. Karim. *Digital Design: A Pragmatic Approach.* Boston: PWS Publishers, 1987.

Karim, M. A., A. A. S. Awwal, and A. K. Cherri. "Polarization-Encoded Optical Shadow-Casting Logic Units: Design." *Appl. Opt.* Vol. 26 (1987): 2720.

Karim, M. A. "Realization of Precompensated Monotonic Contact Screens." *Opt. Lett.* Vol. 9 (1984): 527.

Karim, M. A. "Fabrication of Precompensated Contact Screens for Finite-Gamma Recording Media." *J. Appl. Photogr. Engn.* Vol. 9 (1983): 100.

Karim, M. A. "Analysis of Fidelity Image Reproduction by Halftone Screens of Specific Designs." *Curr. Sc.* Vol. 49 (1980): 921.

Karim, M. A., and H. K. Liu. "Analysis of Tone Reproduction with Square, Circular, and Concentric-Ring Patterned Screens of Linear Cell Transmittance." *Opt. Lett.* Vol. 5 (1980): 132.

Karim, M. A., D. Cook, and P. F. McManamon. "Digital Beam Steering System Using Modules of Ferroelectric Liquid Crystals and Wollaston Prism." *Technical Digest of 1988 Annual Meeting of Optical Society of America*, 30 October–4 November, Santa Clara, CA: (1988): 118.

Khoo, I. C., and Y. R. Shen. "Liquid Crystals: Nonlinear Optical Properties and Processes." *Opt. Engn.* Vol. 24 (1985): 579.

Klein, M. V., and T. E. Furtak. *Optics* (2nd ed.). New York: Wiley, 1986.

Laycock, L. C., M. W. McCall, and C. R. Petts. "A Compact Real-Time Optical Processing System." *GEC J. Res.* Vol. 2, No. 2 (1984): 82.

Lee, H. "Optical Isolator Using Acoustooptic and Faraday Effects." *Appl. Opt.* Vol. 26 (1987): 969.

Liu, H. K., and M. A. Karim. "Graphic Analysis of Tone Reproduction by Contact Screens with Known Nonlinear Spatial Transmittances." *Opt. Lett.* Vol. 4 (1979): 408.

Lohmann, A. W., and J. Weigelt. "Spatial Filtering Logic Based on Polarization." *Appl. Opt.* Vol. 26 (1987): 131.

Marom, E. "Real-Time Image Subtraction Using a Liquid Crystal Light Valve." *Opt. Engn.* Vol. 25 (1984): 274.

Minemoto, T., S. Numata, and K. Miyamoto. "Optical Parallel Logic Gate Using Spatial Light Modulators with Pockels Effect: Implementation Using Three PROM Devices." *Appl. Opt.* Vol. 25 (1986): 948.

Naor, D., A. Flusberg, and I. Itzkan. "IR Laser Photography with Silver-Halide Emulsion." *Appl. Opt.* Vol. 20 (1981): 2474.

Pedrotti, F. L., and L. S. Pedrotti. *Introduction to Optics.* Englewood Cliffs, NJ: Prentice Hall, 1987.

Psaltis, D. "Optical Image Correlation Using Acoustooptic and Charge-Coupled Devices." *Appl. Opt.* Vol. 21 (1982): 491.

Reinish, R., P. Vincent, M. Neviere, and E. Pic. "Fast Pockels Light Modulator Using Guided Wave Resonance." *Appl. Opt.* Vol. 24 (1985): 2001.

Rhodes, W. T., and P. S. Guilfoyle. "Acoustooptic Algebraic Processing Architectures." *Proc. IEEE.* Vol. 72 (1984): 820.

Shimomura, T., and S. Kobayashi. "Color Contrast Criteria in a Guest-Host Mode Liquid Crystal Display." *Appl. Opt.* Vol. 20 (1981): 819.

Shirasaki, M., H. Takamatsu, and T. Obokata. "Bistable Magnetooptic Switch for Multimode Optical Fibers." *Appl. Opt.* Vol. 21 (1982): 1943.

Sirat, G., and N. Ben-Yosef. "A Monolithic Piezo-Electric Elasto-Optic Image Modulator." *Proc. 10th Int. Optical Comp. Conf.*, MIT, Cambridge, MA: (April 1983): 42.

Soref, R. A. "Electrooptic 4 × 4 Matrix Switch for Multimode Fiber-Optic Systems." *Appl. Opt.* Vol. 21 (1982): 1386.

Tai, A. M. "Low-Cost LCD Spatial Light Modulator with High Optical Quality." *Appl. Opt.* Vol. 25 (1986): 1380.

Upatneik, J. "Portable Real-Time Coherent Optical Correlator." *Appl. Opt.* Vol. 22 (1983): 2798.

VanderLugt, A. "Bragg Cell Diffraction Patterns." *Appl. Opt.* Vol. 21 (1982): 1092.

Verber, C. M., R. P. Kenan, and J. R. Busch. "Correlator Based on an Integrated Spatial Light Modulator." *Appl. Opt.* Vol. 20 (1981): 1626.

Warde, C., and J. Thackara. "Operating Modes of the Microchannel Spatial Light Modulator." *Opt. Engn.* Vol. 22 (1983): 695.

Warde, C., A. M. Weiss, A. D. Fisher, and J. I. Thackara. "Optical Information Processing Characteristics of the Microchannel Spatial Light Modulator." *Appl. Opt.* Vol. 20 (1981): 2066.

Yeh, P. "Dispersive Magnetooptic Filters." *Appl. Opt.* Vol. 21 (1982): 2069.

Yu, F. T. S. *Optical Information Processing.* New York: Wiley, 1983.

Yu, F. T. S., X. Lu, and M. Cao. "Application of a Magnetooptic Spatial Light Modulator to White-Light Processing." *Appl. Opt.* Vol. 23 (1984): 4100.

Yu, F. T. S., S. Jutamulia, and D. A. Gregory. "Real-Time Liquid Crystal TV XOR- and XNOR-Gate Binary Image Subtraction Technique." *Appl. Opt.* Vol. 26 (1987): 2738.

CHAPTER EIGHT

Fourier-Optics-Based Devices and Systems

8.1 Introduction

Most of the earlier chapters dealt with developing an understanding of electro-optic devices—physical principles, their designs and characteristics. In this chapter, many of those devices are configured to form meaningful **systems** and, in particular, a Fourier-optics-based system. Fourier-optics-based devices and systems are used primarily for processing two-dimensional optical data or images. In particular, the principal components of such a system are often a lens and a monochromatic light source which, in combination, function as an optical image processor that can perform a two-dimensional Fourier transform. The reason a system can perform a complex operation like a two-dimensional Fourier transform is that under suitable conditions, the output of a thin biconvex lens is equivalent to a spatial Fourier transform of the input.

Optical signal processors can be either coherent or incoherent, depending on the type of source used. As we show in this chapter, in such a system a thin lens is able to transform the two-dimensional optical information into its two-dimensional frequency format in a manner quite like that realized in an electronic data processor, but without any scanning. This unique property of parallelism and the possibility of achieving massive interconnections have encouraged present-day optical scientists and engineers to pursue the processing, not only of analog signals, but also of digitized binary signals. Their work is expected to yield schemes as well as system components for futuristic optical computing machines that can be used at least in special-purpose computing applications.

An understanding of Fourier-optics-based optical signal processors begins with a description of diffraction. Diffraction refers to the phenomenon that causes the propagation of a light beam to deviate from that predicted by geometrical optics. An ordinary treatment of geometric optics is unable to describe the "smearing" of the shadow's edges that occurs after light has passed through a small aperture, such as a pinhole or a narrow slit. There are two distinct types of diffraction, Fraunhofer and Fresnel. Fraunhofer diffraction prevails when both the incident and diffracted waves can be approximated by plane waves; Fresnel diffraction occurs when either incident or diffracted beams are nonplanar.

Huygens's principle is used often to explain diffraction effects by assuming that each point of the optical wave front behaves like a secondary source of light. Consequently, the wave fronts resulting from each of these secondary point sources spread out radially in all possible directions. We shall use the principles of superposition of plane waves to describe the secondary wave fronts in terms of the paraxial Fresnel diffractions and, finally, in terms of the far-field Fraunhofer diffraction limit. The diffraction effect of a thin lens on a plane beam of light passing through it is discussed next. We show that diffraction as well as phase transformations contribute to the spatial Fourier transforming property of a lens. Finally, a number of important Fourier-optics-based systems are described in terms of several specific optical signal-processing applications.

8.2 The Scalar Theory of Diffraction

To know the effective scalar field amplitude at a point due to a number of monochromatic sources, all having the same wavelength, we need to have an idea about the position, amplitude, and phases of each of these sources. From Huygens's principle of superposition, we can then determine the scalar amplitude of the net field at the point of interest. However, it can be shown that an exact knowledge of the sources is not always necessary in characterizing the net field amplitude at the point of observation. Instead, it is enough to know the amplitude and the spatial derivative of the amplitude along the normal, on an arbitrary surface enclosing the point of observation but not enclosing the source.

Consider a closed surface S that encloses a volume V as shown in Fig. 8.1. If g and h are two scalar-point functions such that their first and second derivatives are continuous inside and on the surface, then according to Green's identity

$$\iint_S \left[h\frac{dg}{dn} - g\frac{dh}{dn} \right] dS = \iiint_V (h\nabla^2 g - g\nabla^2 h)\, dV \qquad \textbf{(8.1)}$$

where d/dn represents a differentiation operator directed outward along the normal to the surface. In particular, if both g and h are wave functions satisfying an equation like Eq. (1.10) with a harmonic time-dependence of the form $\exp(\pm j\omega t)$, then the volume integral of Eq. (8.1) vanishes. Accordingly, for such

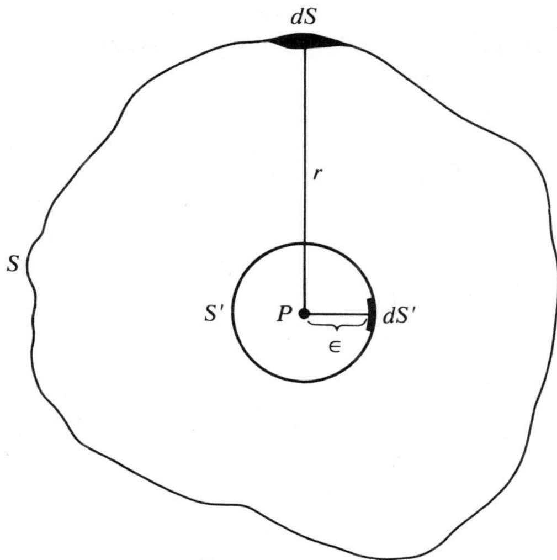

FIGURE 8.1
Surface of integration for determining Kirchhoff's integral

wave functions, we obtain

$$\iint_S \left[h\frac{dg}{dn} - g\frac{dh}{dn} \right] dS = 0 \tag{8.2}$$

Consider further that the scalar function h has the form

$$h = \frac{h_0}{r} e^{jkr} \tag{8.3}$$

where the time-dependent factor has been neglected for convenience. Note that this particular function represents a spherical wave that converges to a point, say P, enclosed within the surface S. Interestingly, however, h becomes infinite at P and, therefore, during the integration of Eq. (8.2), the point P is excluded. Accordingly, to evaluate Eq. (8.2), we subtract the contribution corresponding to the integral over a small sphere (of radius $\varepsilon > 0$ and surface area S' surrounding P). Over this small sphere, the operator d/dn can be replaced by the partial derivative $-\partial/\partial r$. Equation (8.2) thus reduces to

$$\iint_S \left[\frac{e^{jkr}}{r}\frac{dg}{dn} - g\frac{d}{dn}\left(\frac{e^{jkr}}{r}\right) \right] dS + \iint_{S'} \left[g\frac{\partial}{\partial r}\left(\frac{e^{jkr}}{r}\right) - \frac{e^{jkr}}{r}\frac{\partial g}{\partial r} \right] dS' = 0 \tag{8.4}$$

The second term of Eq. (8.4) can be simplified further to give

$$\iint_S \left[\frac{e^{jkr}}{r}\frac{dg}{dn} - g\frac{d}{dn}\left(\frac{e^{jkr}}{r}\right) \right] dS + \iint \left[-\frac{g}{r^2} + \frac{jkg}{r} + \frac{1}{r}\frac{\partial g}{\partial r} \right] e^{jkr} \varepsilon^2 \, d\Omega = 0 \tag{8.5}$$

where $d\Omega$ represents an element of a solid angle taken at the sphere of radius ε. Note that the second and third components of the second integral approach zero as r approaches zero. The second term of Eq. (8.5) thus approaches $-4\pi g(P)$ as r approaches zero. Equation (8.5) can thus be reorganized to give

$$g(P) = \frac{1}{4\pi}\iint_S \left[\frac{e^{jkr}}{r}\frac{dg}{dn} - g\frac{d}{dn}\left(\frac{e^{jkr}}{r}\right)\right] dS \tag{8.6}$$

This equation, known as the **Kirchhoff's integral**, shows that the scalar field at point P (i.e., when $r = 0$) can be determined from a knowledge of both the field and its outward derivative over a surface that surrounds the observation point P.

Kirchhoff's integral is extremely useful because it can be used to analyze the general problem of diffraction. Consider an aperture of arbitrary shape in an otherwise opaque partition, as shown in Fig. 8.2. Our goal is to calculate the field at point P due to spherical waves emanating from the point source P'. To apply Kirchhoff's integral to this problem, we choose a closed surface of integration to include the aperture as well as the point of observation. However, to evaluate the integral of Eq. (8.6), we need to make two basic assumptions. First, the wave function g and its derivative dg/dn are both assumed to be approximately zero at all points except at the aperture. Secondly, the values of the wave function g and its derivative dg/dn, anywhere at the aperture in presence of the screen, remain the same even when the screen is removed. These assumptions are reasonable enough, and the result obtained thereby agrees well with the experimental findings.

The spherical wave function g at the aperture is given by

$$g|_{\text{aperture}} = \frac{g_0}{r'} e^{jkr'} \tag{8.7}$$

where r' is the radial position of a point on the aperture relative to the source P'. Equation (8.6) can thus be evaluated over the aperture area dA to give g at P as

$$\begin{aligned} g(x, y) &= \frac{g_0}{4\pi}\iint \left[\frac{e^{jkr}}{r}\frac{d}{dn}\left(\frac{e^{jkr'}}{r'}\right) - \frac{e^{jkr'}}{r'}\frac{d}{dn}\left(\frac{e^{jkr}}{r}\right)\right] dA \\ &= \frac{g_0}{4\pi}\iint \left[\frac{e^{jkr}}{r}\cos(\mathbf{n}\cdot\mathbf{r}')\frac{\partial}{\partial r'}\left(\frac{e^{jkr'}}{r'}\right) - \left(\frac{e^{jkr'}}{r'}\right)\cos(\mathbf{n}\cdot\mathbf{r})\frac{\partial}{\partial r}\left(\frac{e^{jkr}}{r}\right)\right] dA \\ &= \frac{jkg_0}{4\pi}\iint \frac{e^{jk(r+r')}}{rr'}[\cos(\mathbf{n},\mathbf{r}') - \cos(\mathbf{n},\mathbf{r})]\, dA \end{aligned} \tag{8.8}$$

where $(\mathbf{n}, \mathbf{r}')$ and $(\mathbf{n}, \mathbf{r})$ denote the angles between the vectors (respectively, $\mathbf{r}'$ and $\mathbf{r}$) and the normal $\mathbf{n}$ to the surface of integration. It was assumed that both $\mathbf{r}$ and $\mathbf{r}'$ are large when compared to the radiation wavelength involved. Equation (8.8) is often referred to as the **Fresnel-Kirchhoff diffraction formula**, which in essence is the mathematical statement of Huygens's principle. It is interesting to note

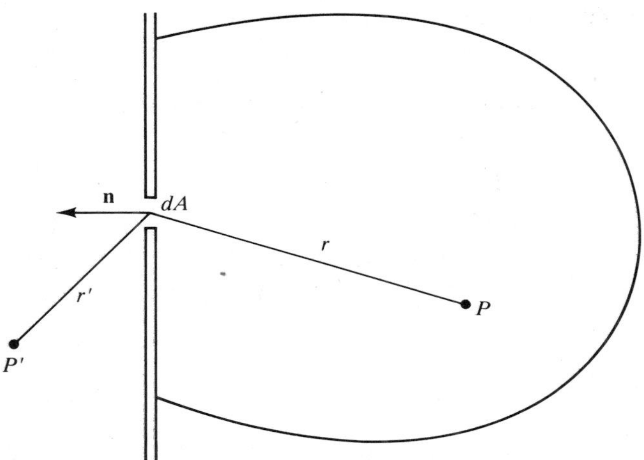

FIGURE 8.2

Diffraction geometry for evaluating a Fresnel-Kirchhoff formula

that Eq. (8.8) cannot distinguish between the source and the observer. The same result is obtained if the source is located at P and the observer is placed at P'. The so-called obliquity factor given by $[\cos(\mathbf{n}, \mathbf{r}) - \cos(\mathbf{n}, \mathbf{r}')]$ in effect reduces the incident amplitude. Its value is never greater than 2 nor less than 0. The presence of the factor j at the exponent implies that the diffracted waves are shifted in phase from the incident wave by $\pi/2$.

For simplicity, let us assume that the aperture lies in the plane (x_0, y_0) while the observation is being made at the plane (x, y), which is parallel to the plane of the aperture but at a distance d away. Therefore, the radial distance r is given by

$$\begin{aligned} r &= [d^2 + (x - x_0)^2 + (y - y_0)^2]^{1/2} \\ &= d\left[1 + \left(\frac{x - x_0}{d}\right)^2 + \left(\frac{y - y_0}{d}\right)^2\right]^{1/2} \\ &\simeq d\left[1 + \frac{1}{2}\left(\frac{x - x_0}{d}\right)^2 + \frac{1}{2}\left(\frac{y - y_0}{d}\right)^2\right] \end{aligned} \tag{8.9}$$

when $|(x - x_0)/d| < 1$ and $|(y - y_0)/d| < 1$. This particular assumption concerning the location of the observation point is referred to as the **Fresnel approximation**. Thus, when d is sufficiently large, the observer is said to be in the Fresnel diffraction region.

When the aperture dimensions are reasonably small, the value of the obliquity factor approaches 2. In the Fresnel approximation, Eq. (8.8) thus reduces to

$$g(x, y) = \iint W(x_0, y_0; x, y) g(x_0, y_0)\, dx_0\, dy_0 \tag{8.10}$$

where the weighting function W is given by

$$W(x_0, y_0; x, y) = \frac{1}{j\lambda}\frac{e^{jkr}}{r}$$
$$\simeq \frac{e^{jkd}}{j\lambda d} e^{[(j\pi/\lambda d)\{(x-x_0)^2+(y-y_0)^2\}]} \quad \textbf{(8.11)}$$

since $k = 2\pi/\lambda$. Physically, Eq. (8.10) implies that the spherical Huygens's wavelets can in effect be approximated by quadratic surfaces. Expanding these quadratic terms, the scalar field at P can thus be expressed as

$$g(x, y) = \frac{e^{jkd}}{j\lambda d} e^{(j\pi/\lambda d)(x^2+y^2)} \iint_{-\infty}^{\infty} g(x_0, y_0) e^{(j\pi/\lambda d)(x_0^2+y_0^2)} e^{-(j2\pi/\lambda d)(x_0 x + y_0 y)} dx_0\, dy_0 \quad \textbf{(8.12)}$$

Note that the superposition integral is subject to infinite limits because, according to the boundary conditions, $g(x_0, y_0)$ is nonzero only within the aperture.

It may be appropriate now to remember the two-dimensional Fourier transform pair. In Chapter 6, we made use of Fourier transform operations, but that use was limited to only one-dimensional cases. Furthermore, the Fourier transforms involved time as a variable in terms of which the functions were evaluated. In optics, images and data are usually available in two-dimensional format. Accordingly, optical processors involve a two-dimensional Fourier transform of the function $f(x, y)$, $F(k_x, k_y)$, which is defined by

$$F(k_x, k_y) = \iint_{-\infty}^{\infty} f(x, y) e^{j(k_x x + k_y y)}\, dx\, dy \quad \textbf{(8.13)}$$

where k_x and k_y are the angular spatial frequencies along x and y, respectively. Like its one-dimensional counterpart, $f(x, y)$ can be derived from $F(k_x, k_y)$ as well by

$$f(x, y) = \left(\frac{1}{2\pi}\right)^2 \iint_{-\infty}^{\infty} F(k_x, k_y) e^{-j(k_x x + k_y y)}\, dk_x\, dk_y \quad \textbf{(8.14)}$$

By comparing Eqs. (8.12) and (8.13), we note, however, that apart from the multiplicative amplitude and phase factors that are independent of (x, y), the scalar field at P can be found by taking the Fourier transform of $g(x_0, y_0) \exp[j\pi(x_0^2 + y_0^2)/(\lambda d)]$, provided the spatial frequencies are given by $v_x = x/\lambda d$ and $v_y = y/\lambda d$, respectively. The implication of such a relationship is astounding. The output and the input are related by a simple Fourier transform operation. Clearly, the accuracy of this startling conclusion rests on the attainment of the Fresnel approximation. Under such an approximation, either the source or the observation point is close enough to the diffracting aperture so that the wave-front curvature is not negligible. There is, however, a second type of recognizable diffraction referred to as Fraunhofer diffraction. It is realized when the distance between the source and the diffracting aperture and that between the aperture and the observation point are both very large. Thus, in the Fraunhofer domain, the curvature of both the incident and diffracted waves is negligible. The Fraunhofer approximation given by

$$d \gg \text{Max}\left[\frac{\pi(x_0^2 + y_0^2)}{\lambda}\right] \tag{8.15}$$

reduces the quadratic phase factor of Eq. (8.12) to unity. Thus Eq. (8.12) reduces to

$$g(x, y) = \frac{e^{jkd}}{j\lambda d} e^{(j\pi/\lambda d)(x^2+y^2)} \iint_{-\infty}^{\infty} g(x_0, y_0) e^{-j2\pi(\nu_x x_0 + \nu_y y_0)}\, dx_0\, dy_0 \tag{8.16}$$

Consequently, under the Fraunhofer approximation, aside from the multiplicative factors, the field at P is obtained simply by taking the Fourier transform of $g(x_0, y_0)$, evaluated at spatial frequencies $\nu_x = x/\lambda d$ and $\nu_y = y/\lambda d$. It must, however, be noted that no transfer function can be uniquely associated with Fraunhofer diffraction because the related approximation destroys the space invariance of the diffraction equation. Fraunhofer diffraction happens to be a special case of Fresnel diffraction and is mathematically more simple to comprehend. The analytical expressions derived thus far can be used to predict the field amplitude distribution across the diffraction pattern for any given aperture. The square of the amplitude distribution can be used in turn to yield the corresponding intensity distribution. Often intensity distribution is more comprehendable since it corresponds to what would be detected by any detector, including our eyes.

Consider, for example, a circular aperture of radius a, as shown in Fig. 8.3(a), which is being illuminated by a plane wave. For convenience, let us choose the origin of the coordinate system to be at the center of the aperture. Consequently for any point P_0 on the aperture, the rectangular coordinates are given by

$$x_0 = \rho \cos \phi \tag{8.17a}$$

$$y_0 = \rho \sin \phi \tag{8.17b}$$

where $\rho = (x_0^2 + y_0^2)^{1/2}$ is the radial distance between the center of the circle and point P_0 and ϕ is the angle made by the vector $\boldsymbol{\rho}$ with the x_0 axis. The spatial frequencies are given, respectively, by

$$\nu_x = (\sin\theta \cos\alpha)/\lambda \tag{8.18a}$$

$$\nu_y = (\sin\theta \sin\alpha)/\lambda \tag{8.18b}$$

where θ is the angle between the line C_0P and the normal to the planes, and α is the angle between the line PC and the x axis. For the case of an input plane wave, we can substitute $g(x_0, y_0)$ with a constant, say g_0. Equation (8.16) may

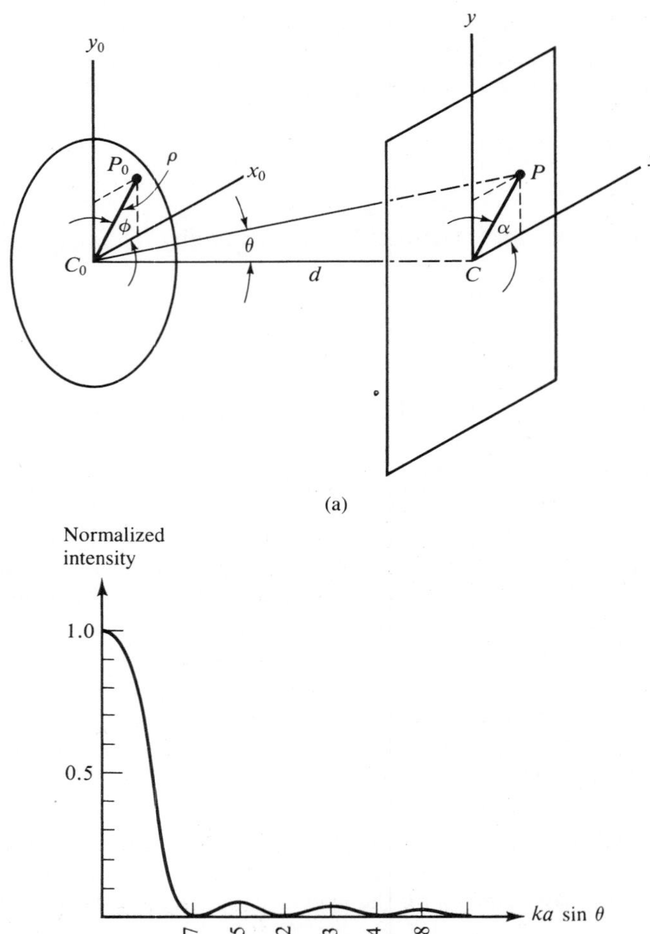

FIGURE 8.3

(a) Fraunhofer diffraction geometry for a circular aperture and (b) Airy pattern

thus be evaluated using Eqs. (8.17a), (8.17b), (8.18a) and (8.18b) as follows:

$$g(x, y) = g_0 \frac{e^{jkd}}{j\lambda d} e^{(j\pi/\lambda d)(x^2+y^2)} \int_0^a \rho \, d\rho \int_0^{2\pi} e^{[-jk\rho \sin\theta(\cos\phi\cos\alpha + \sin\phi\sin\alpha)]} \, d\phi$$

$$= g_0 \frac{e^{jkd}}{j\lambda d} e^{(j\pi/\lambda d)(x^2+y^2)} \int_0^a \rho \, d\rho \int_0^{2\pi} e^{[-jk\rho \sin\theta \cos(\phi-\alpha)]} \, d\phi \qquad \textbf{(8.19)}$$

But the requirement of having a periodic function of periodicity 2π provides that

$$\int_0^{2\pi} \cos(\phi - \alpha)\, d\phi = \int_0^{2\pi} \cos\phi\, d\phi \tag{8.20}$$

Thus Eq. (8.19) can be reduced to

$$\begin{aligned} g(x, y) &= 2\pi g_0 \frac{e^{jkd}}{j\lambda d} e^{(j\pi/\lambda d)(x^2+y^2)} \int_0^a J_0(k\rho \sin\theta)\rho\, d\rho \\ &= 2\pi g_0 \frac{e^{jkd}}{j\lambda d(k\sin\theta)^2} e^{(j\pi/\lambda d)(x^2+y^2)} \int_0^{ka\sin\theta} J_0(x)x\, dx \end{aligned} \tag{8.21}$$

where the Bessel function of the first kind of order zero is defined as

$$J_0 = \frac{1}{2\pi} \int_0^{2\pi} e^{-jx\cos y}\, dy \tag{8.22}$$

Interestingly, the general interrelationship rule of Bessel functions of the first kind is given by

$$x^{n+1}J_n(x) = \frac{d}{dx}[x^{n+1}J_{n+1}(x)] \tag{8.23}$$

Thus Eq. (8.21) reduces to

$$\begin{aligned} g(x, y) &= \frac{2\pi g_0}{(k\sin\theta)^2} \frac{e^{jkd}}{j\lambda d} e^{(j\pi/\lambda d)(x^2+y^2)} \int_0^{ka\sin\theta} \frac{d}{dx}(xJ_1(x))\, dx \\ &= \frac{2\pi g_0 a\, e^{jkd}}{k\sin\theta\, j\lambda d} e^{(j\pi/\lambda d)(x^2+y^2)} J_1(ka\sin\theta) \end{aligned} \tag{8.24}$$

Consequently, the intensity distribution of the diffracted output is given by

$$I(x, y) = g(x, y)g^*(x, y) = I_0\left[\frac{2J_1(ka\sin\theta)}{ka\sin\theta}\right]^2 \tag{8.25}$$

where all constant factors have been lumped into one constant I_0. This constant value is equivalent to the intensity measured at the central spot when $\theta = 0$. The intensity distribution given by Eq. (8.25) is also referred to as the Airy pattern. A plot of this pattern are shown in Fig. 8.3(b). The maxima of this pattern is obtained when $(ka\sin\theta) = 0, 5.1365, 8.4163, 11.6208, \ldots$, whereas the minima of the intensity variation occur at $(ka\sin\theta) = 3.8327, 7.0152, 10.1724, \ldots$. A more exhaustive treatment can lead to the general case of Fresnel diffraction if we start the analysis with Eq. (8.12) instead of beginning it with Eq. (8.16).

8.3 Fourier Transform Properties of Lenses

It was stated earlier that even an ordinary lens functions as an optical data processor. In this section, we are concerned, in particular, with the effect of a thin lens on an input optical beam. A lens is generally referred to as thin when a ray incident at any point of its surface emerges from the other side at approximately the same height. If we can neglect the effects of reflection and absorption, we can regard the thin lens as contributing to an (x, y)-dependent phase delay only. The phase delay is caused by the thickness variation of the lens.

Consider an aberrationless thin lens, as shown in Fig. 8.4, with an input object placed at a distance u in front of the lens. It has already been shown (see Eqs. (1.42) and (1.43)) that under paraxial approximation of geometric optics, the image will be located at a distance v from the lens such that

$$\frac{1}{u} + \frac{1}{v} = \frac{1}{f} \tag{8.26}$$

where f is the focal length of the thin lens. The phase factor at a transverse plane placed immediately in front of the thin lens is given by

$$e^{jk\sqrt{x_0^2 + y_0^2 + u^2}} \simeq e^{jk\{u + [(x_0^2 + y_0^2)/2u]\}} \tag{8.27}$$

provided that the lens axis is coincident with the z axis and that $x, y \ll u$ (this condition is referred to as the **paraxial approximation**). The wave front that converges to the image point can be characterized by its radius of curvature v. Thus it is (also under paraxial approximation) given by

$$e^{-jk\{v + (x_0^2 + y_0^2)/2v\}} \tag{8.28}$$

Note that while the positive sign of Eq. (8.27) corresponds to a diverging wave front, the negative sign of Eq. (8.28) corresponds to a converging wave front. We can now determine the effective phase factor introduced by a thin lens during its optical transformation using Eq. (8.26) as

$$\begin{aligned}\text{phase factor} &= \frac{e^{-jk\{v + [(x_0^2 + y_0^2)/2v]\}}}{e^{jk\{u + [(x_0^2 + y_0^2)/2u]\}}}\\ &= e^{-jk(u+v)}e^{-(j\pi/\lambda)[(1/u)+(1/v)](x_0^2+y_0^2)}\\ &= e^{-jk(u+v)}e^{-(j\pi/\lambda f)(x_0^2+y_0^2)}\end{aligned} \tag{8.29}$$

The first factor of Eq. (8.29) can be neglected because it is independent of both x_0 and y_0. Consequently, the effect of using a thin lens is to multiply the incident phase distribution by a

$$\text{phase factor} = p(x_0, y_0)e^{-(j\pi/\lambda f)(x_0^2+y_0^2)} \tag{8.30}$$

where

$$p(x_0, y_0) = \begin{cases} 1 & \text{for } x_0^2 + y_0^2 \leq a^2 \\ 0 & \text{for } x_0^2 + y_0^2 > a^2 \end{cases} \tag{8.31}$$

is referred to as the pupil function and a is the radius of the thin lens. Note that f is negative in the event the lens is biconcave.

Consider now an object, with amplitude transmittance $t_a(x_0, y_0)$, which is placed immediately in front of a thin lens of focal length f. Consider further that

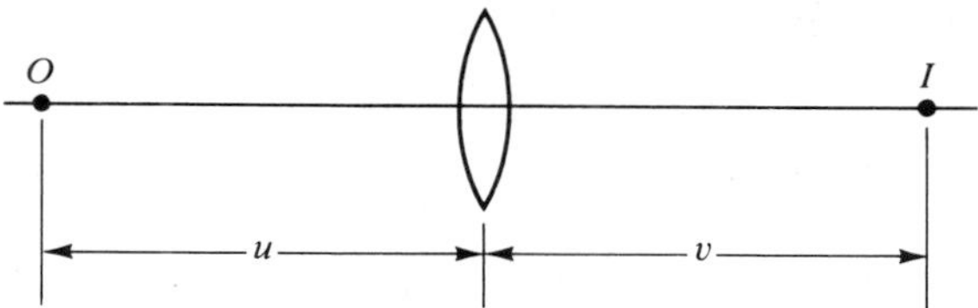

FIGURE 8.4

The object and its image due to a thin lens

the object is illuminated by a monochromatic plane wave of amplitude A. The amplitude distribution immediately behind the thin lens can be thus obtained using Eq. (8.30) as

$$g(x_0, y_0) = At_a(x_0, y_0)p(x_0, y_0)e^{-(j\pi/\lambda f)(x_0^2+y_0^2)} \tag{8.32}$$

Equation (8.16) can be then used to obtain the Fresnel diffraction distribution at the back focal plane of the thin lens. Thus, setting $d=f$, we obtain

$$g(x, y) = \frac{e^{jkf}}{j\lambda f} e^{(j\pi/\lambda f)(x^2+y^2)} \iint_{-\infty}^{\infty} At_a(x_0, y_0)p(x_0, y_0)e^{-(j2\pi/\lambda f)(x_0 x + y_0 y)}\, dx_0\, dy_0 \tag{8.33}$$

When the physical extent of the object is smaller than that of the lens, the pupil function can be ignored for all practical purposes. Accordingly, for such a case,

$$g(x, y) = A\frac{e^{jkf}}{j\lambda f} e^{(j\pi/\lambda f)(x^2+y^2)} \iint_{-\infty}^{\infty} t_a(x_0, y_0)e^{-(j2\pi/\lambda f)(x_0 x + y_0 y)}\, dx_0\, dy_0 \tag{8.34}$$

The significance of Eq. (8.34) becomes evident when we compare it with Eq. (8.13). Equation (8.34) implies that a thin lens performs a Fourier transform operation of a two-dimensional, complex amplitude distribution in parallel. At the focal plane (also referred to as the **Fourier plane**), the amplitude and phase are both influenced by the amplitude and phase of the object Fourier components evaluated at spatial frequencies $v_x = x/\lambda f$ and $v_y = y/\lambda f$. Because of the quadratic phase factors, the Fourier transform relationship between the object and the focal plane image is not an exact one. In many cases, however, the intensity profile at the Fourier plane is of more interest because it provides the power spectrum of the object image. The intensity distribution, resulting from an object of amplitude transmittance $t_a(x_0, y_0)$, is given by

$$I(x, y) = \frac{A^2}{\lambda^2 f^2} |F\{t_a(x_0, y_0)\}|^2 \tag{8.35}$$

where F represents the Fourier transform operation.

Let us again consider the set up of Fig. 8.4. The object having an amplitude transmission distribution $t_a(x, y)$ is now placed at a distance d_1 in front of the thin lens whose focal length is f. Consider further that a plane wave of amplitude A is illuminating the object. The Fresnel diffraction of the input field $At_a(x, y)$

produces an output field distribution immediately beyond the lens. Using Eqs. (8.10) and (8.30), we find that the output field distribution is

$$g(x_0, y_0) = A\frac{e^{jkd_1}}{j\lambda d_1} t_a(x, y) \otimes e^{(j\pi/\lambda d_1)(x^2+y^2)} e^{-(j\pi/\lambda f)(x^2+y^2)} \tag{8.36}$$

where the pupil function has been set to unity for simplicity, and $\otimes$ denotes the spatial convolution operation defined as

$$f_1(x, y) \otimes f_2(x, y) = \iint_{-\infty}^{\infty} f_1(\xi, \eta) f_2(x-\xi, y-\eta)\, d\xi\, d\eta \tag{8.37}$$

such that

$$F\{f_1(x, y) \otimes f_2(x, y)\} = F\{f_1(x, y)\} F\{f_2(x, y)\} \tag{8.38}$$

Finally, Eq. (8.10) can be used again to obtain the field distribution at a plane d_2 behind the lens as

$$\begin{aligned} g(x, y) &= g(x_0, y_0) \otimes \frac{e^{jkd_2}}{j\lambda d_2} e^{(j\pi/\lambda d_2)(x^2+y^2)} \\ &= -A\frac{e^{jk(d_1+d_2)}}{\lambda^2 d_1 d_2} \iiiint_{-\infty}^{\infty} t_a(\xi, \eta) e^{-(j\pi/\lambda f)(\xi'^2+\eta'^2)} \\ &\quad \times e^{(j\pi/\lambda d_1)[(\xi'-\xi)^2+(\eta'-\eta)^2]} e^{(j\pi/\lambda d_2)[(x-\xi')^2+(y-\eta')^2]} d\xi\, d\eta\, d\xi'\, d\eta' \end{aligned} \tag{8.39}$$

For the special case of $d_1 = d_2 = f$, Eq. (8.39) simplifies to

$$\begin{aligned} g(x, y) &= -\frac{Ae^{j2kf}}{(\lambda f)^2} \iiiint_{-\infty}^{\infty} t_a(\xi, \eta) e^{(j\pi/\lambda f)(x^2+y^2+\xi^2+\eta^2)} \\ &\quad \times e^{(j\pi/\lambda f)[\xi'^2 - 2\xi'(x+\xi)]} e^{(j\pi/\lambda f)[\eta'^2 - 2\eta'(y+\eta)]} d\xi\, d\eta\, d\xi'\, d\eta' \\ &= -A\frac{e^{j2kf}}{(\lambda f)^2} \iint_{-\infty}^{\infty} t_a(\xi, \eta) e^{(j\pi/\lambda f)(x^2+y^2+\xi^2+\eta^2)} e^{-(j\pi/\lambda f)[(x+\xi)^2+(y+\eta)^2]} \\ &\quad \times d\xi\, d\eta \int_{-\infty}^{\infty} e^{-(j\pi/\lambda f)\rho^2} d\rho \int_{-\infty}^{\infty} e^{-(j\pi/\lambda f)\sigma^2} d\sigma \\ &= \frac{jAe^{j2kf}}{\lambda f} \iint t_a(\xi, \eta) e^{-(j2\pi/\lambda f)(x\xi+y\eta)} d\xi\, d\eta \end{aligned} \tag{8.40}$$

It is now obvious that the Fourier plane field distribution is the same as the Fourier transform of the front focal plane field distribution, provided the transform is evaluated at spatial frequencies $v_\xi = x/\lambda f$ and $v_\eta = y/\lambda f$. Also note that when the object is placed at the front focal plane, no phase curvature appears in the transformation relation. We can also show that when the object plane is located behind the lens but within the focal length of the lens, the field distribution includes a quadratic phase factor. However, using this geometric set up, we can increase the scale of the transform by moving the object toward the lens, and we can decrease it by moving the object closer to the focal point. It is also interesting to note that the image as well as the Fourier transform of the

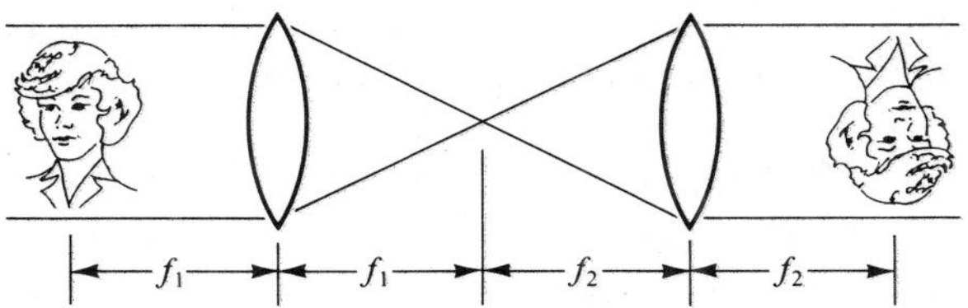

FIGURE 8.5

Image formation using two successive Fourier transforms

object are simultaneously realized using only one thin lens. The field distribution obtained thereby correctly represents the intensity of the object only—not of the phase of the object. Thus, instead of using a single lens, we use a two-lens system like the one shown in Fig. 8.5 to restore both the amplitude and phase information. The second Fourier transform operation that is obtained using the second lens recovers the input image at the output plane.

8.4 Optical Information-Processing Systems

The quality of a meaningful optical system is often characterized by the impulse response function of the system. Appropriately, the impulse response function of a system is defined as the output of the system to which an impulse—i.e., a delta functionlike—signal is applied. The response function can be used effectively to estimate the output characteristics of the system, even when the system is fed with an arbitrary input signal. Only the most ideal optical imaging system images a point source object as a point output image. Most practical systems—e.g., a lens system—result in an output that has a finite width instead of having a sharp width. In fact, a system that has a lower quality criterion will have a larger response width.

Let the impulse response of an optical system be described by $q(x, y)$ while the image plane is described by the x-y plane. Consider a distributed source u that is positioned in the object plane. A distributed source can always be considered as an ensemble of an infinite number of point sources. For simplicity, the amplitude of a point source at (x_0, y_0) is described by $u(x_0, y_0)$. The resultant output field amplitude due to all of these point sources is obtained by adding the amplitude contributions of all the constituent point sources. The output field amplitude is thus given by

$$\begin{aligned} g(x, y) &= \iint_{-\infty}^{\infty} u(x_0, y_0) q(x - x_0, y - y_0)\, dx_0\, dy_0 \\ &= u(x, y) \otimes q(x, y) \end{aligned} \tag{8.41}$$

Accordingly, the amplitude and, therefore, the intensity of the output image for an arbitrary input can be determined from the impulse response function $q(x, y)$. However, it should be noted that Eq. (8.41) is valid only as long as the system magnification is unity.

The Fourier transform $Q(v_x, v_y)$ of the impulse response function $q(x, y)$ is referred to as the **coherent transfer function** (CTF). Using the rules of Fourier transform integrals involving convolution, Eq. (8.41) can be thus rewritten as

$$G(v_x, v_y) = U(v_x, v_y)Q(v_x, v_y) \tag{8.42}$$

where $G(v_x, v_y)$ and $U(v_x, v_y)$ are the Fourier transforms of $g(x, y)$ and $u(x, y)$, respectively. The CTF is thus obtained simply as

$$Q(v_x, v_y) = \frac{G(v_x, v_y)}{U(v_x, v_y)} \tag{8.43}$$

It is appropriate to compare the effects of using either a coherent or an incoherent source in optical information-processing systems. When the source is coherent, like a monochromatic laser source, the image plane impulse responses vary in perfect unison. The impulse responses are, therefore, added in the complex-amplitude domain. By comparison, when the source is incoherent, the image plane impulse responses vary randomly. Therefore, image plane impulse responses are added only in the power or intensity domain. It follows, therefore, that while a coherent information-processing system is linear in complex amplitude, an incoherent information-processing system is linear in intensity.

In a coherent optical system, $u(x_0, y_0, t) = u(x_0, y_0)$ where t denotes time of origination of light at the source. The output intensity distribution of a coherent optical system is found using Eq. (8.41):

$$I(x, y) = |u(x, y) \otimes q(x, y)|^2 \tag{8.44}$$

But when the distributed light source is incoherent, the process of evaluating the intensity function becomes extremely complicated. Assuming that the propagation medium is not subjected to any change, we can obtain the incoherent output intensity as

$$\begin{aligned} I(x, y) &= \langle [\textstyle\iint u(x_0, y_0, t) q(x - x_0, y - y_0)\, dx_0\, dy_0] \\ &\qquad \times [\textstyle\iint u(x_0, y_0, t) q(x - x_0, y - y_0)\, dx_0\, dy_0]^* \rangle \\ &= \langle \textstyle\iiiint q(x - x_0, y - y_0) q^*(x - x'_0, y - y'_0) u(x_0, y_0, t) u^*(x'_0, y'_0, t) \\ &\qquad \times dx_0\, dy_0\, dx'_0\, dy'_0 \rangle \\ &= \textstyle\iiiint \langle u(x_0, y_0, t) u^*(x'_0, y'_0, t) \rangle q(x - x_0, y - y_0) q^*(x - x'_0, y - y'_0) \\ &\qquad \times dx_0\, dy_0\, dx'_0\, dy'_0 \end{aligned} \tag{8.45}$$

where $\langle\,\rangle$ denotes taking an average in time. Recall that in an incoherent system, the phase variation between two point source inputs is statistically independent. Consequently, the time-averaged term of Eq. (8.45) is zero except when $x_0 = x'_0$

and $y_0 = y'_0$. Accordingly, Eq. (8.45) reduces to

$$\begin{aligned} I(x, y) &= \iiiint |u(x_0, y_0)|^2 \delta(x_0 - x'_0)\delta(y_0 - y'_0) q(x - x_0, y - y_0) \\ &\quad \times q^*(x - x'_0, y - y'_0)\, dx_0\, dy_0\, dx'_0\, dy'_0 \\ &= \iint |u(x_0, y_0)|^2 |q(x - x_0, y - y_0)|^2\, dx_0\, dy_0 \\ &= |u(x, y)|^2 \otimes |q(x, y)|^2 \end{aligned} \tag{8.46}$$

The output intensity expressions for the coherent optical system (as given by Eq. (8.44)) and for the incoherent optical system (as given by Eq. (8.46)) are different from one another. The difference lies in the order in which the square modulus and the convolution operations occur in the expressions.

A transfer function, referred to as the **optical transfer function** (OTF), is often defined for incoherent optical information-processing systems for the purpose of system characterization. However, unlike the CTF, the OTF is defined in terms of the Fourier transform of the intensity rather than that of the amplitude. Using Eq. (8.46), we can write

$$\mathscr{I}(\nu_x, \nu_y) = \mathscr{U}(\nu_x, \nu_y)\mathscr{Q}(\nu_x, \nu_y) \tag{8.47}$$

where the normalized Fourier transforms of $I(x, y)$, $|u(x, y)|^2$, and $|q(x, y)|^2$, respectively, are defined according to

$$\mathscr{I}(\nu_x, \nu_y) = \frac{F\{I(x, y)\}}{F\{I(x, y)\}|_{\nu_x = \nu_y = 0}} \tag{8.48a}$$

$$\mathscr{U}(\nu_x, \nu_y) = \frac{F\{|u(x, y)|^2\}}{F\{|u(x, y)|^2\}|_{\nu_x = \nu_y = 0}} \tag{8.48b}$$

$$\mathscr{Q}(\nu_x, \nu_y) = \frac{F\{|q(x, y)|^2\}}{F\{|q(x, y)|^2\}|_{\nu_x = \nu_y = 0}} \tag{8.48c}$$

The normalization included in these definitions is for the convenience of mathematics and to assure the fact that the intensity is always a nonnegative quantity.

It is interesting to note that the OTF as given by Eq. (8.48c) can be expressed in terms of the CTF as follows:

$$\begin{aligned} \mathscr{Q}(\nu_x, \nu_y) &= \frac{F\{q(x, y)q^*(x, y)\}}{\iint_{-\infty}^{\infty} Q(\xi, \eta)Q^*(\zeta - \nu_x, \eta - \nu_y)\, d\zeta\, d\eta|_{\nu_x = \nu_y = 0}} \\ &= \frac{Q(\nu_x, \nu_y) \oplus Q(\nu_x, \nu_y)}{\iint_{-\infty}^{\infty} |Q(\zeta, \eta)|^2\, d\zeta\, d\eta} \end{aligned} \tag{8.49}$$

where $\oplus$ denotes a correlation operation defined as

$$f_1(x, y) \oplus f_2(x, y) = \iint_{-\infty}^{\infty} f_1^*(\zeta - x, \eta - y) f_2(\zeta, \eta)\, d\zeta\, d\eta \tag{8.50}$$

such that

$$F\{f_1(x, y) \oplus f_2(x, y)\} = F^*\{f_1(x, y)\}F\{f_2(x, y)\} \tag{8.51}$$

Therefore, the autocorrelation of the CTF is the same as the OTF. The significance of this obvious conclusion is that an OTF of a diffraction-limited system extends to a frequency value that is twice the cut-off frequency of a CTF. However, you should not necessarily conclude that the resolution of an incoherent system is better than that of a coherent system. While the coherent cut off refers to the maximum spatial frequency component of the output amplitude, the incoherent cut off refers to the maximum component of the output intensity. The modulus of an OTF, referred to as a **modulation transfer function** (MTF), is equivalent to the ratio of the contrast of the output image to that of the input image.

We may characterize an optical system by its ability to distinguish between two closely spaced point sources. The resolution of an optical system is therefore related to the width of the impulse response. A system is said to have a better resolution only when its impulse response is narrower. A suitable resolution criterion, referred to as the Rayleigh criterion, is used often for system characterization. According to it, two incoherent point sources are considered resolvable only when the center of the Airy pattern produced by one point source and the first minima of the Airy pattern produced by the second point source are coincident. From Eq. (8.25), the minimum resolvable separation of ideal point images is given by

$$\delta = 1.22\frac{\lambda d}{l} \tag{8.52}$$

where l is the diameter of the circular aperture stop and d is the distance between the exit pupil and the observation plane.

8.5 Spatial Filtering Using Binary Filters

Spatial frequency filtering is an information-processing operation in which certain desired spatial frequencies of an input image are either removed or altered by filters placed at the Fourier domain plane. Consider the coherent optical information-processing system of Fig. 8.6. A coherent beam diverging from a pinhole spatial filter placed at the focal plane of lens L_1 results in a plane wave front. This wave front is used to illuminate the input object. Lens L_2 transforms the input image (which is located at its front focal plane) and, as a result, the frequency spectrum of the input is displayed in the back focal plane of L_2. The back focal plane of L_2, referred to as the Fourier plane, also happens to be the front focal plane of lens L_3. Therefore, L_3 retransforms the Fourier plane spectrum and forms an image at the output plane of the system. The filters necessary for the particular spatial filtering operation are placed in the Fourier plane and thus a filtered image appears at the output plane.

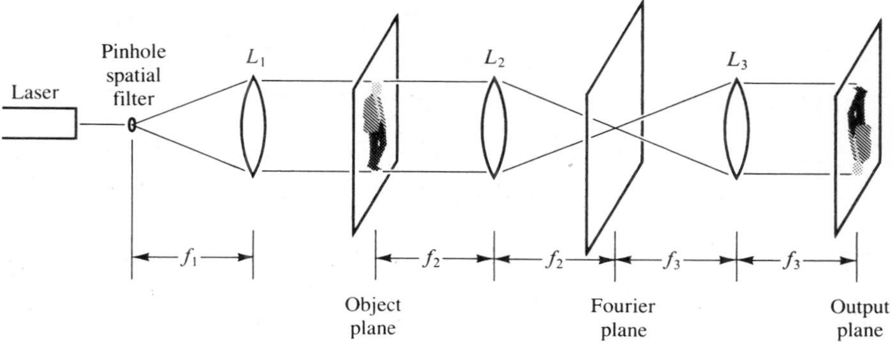

FIGURE 8.6

A coherent optical information-processing system

Several kinds of filters are used in spatial filtering operations: binary, amplitude, phase, and complex filters. **Binary filters** constitute the simplest type of filter. Stops and slits are typical examples of such a filter. The transmittance of stops and slits is either unity or zero. Those among the binary filters that pass only lower spatial frequency components are referred to as **low-pass filters**. Similarly, those that pass higher spatial frequency components are referred to as **high-pass filters**. Finally, certain binary filters, known as **bandpass filters**, are designed to pass only selected spatial frequency components.

Consider a line diffraction grating of finite extension b which consists of slits of width a and cell period X (see Fig. 8.7(a)). The amplitude transmission of such a line grating is given by

$$t_a(x, y) \equiv t_a(x) = \sum_{n=-\infty}^{\infty} \operatorname{rect}\left(\frac{x - nX}{a}\right) \operatorname{rect}\left(\frac{x}{b}\right) \tag{8.53}$$

where

$$\operatorname{rect}(\zeta) = \begin{cases} 1 & |\zeta| \leq \frac{1}{2} \\ 0 & \text{otherwise} \end{cases} \tag{8.54}$$

But

$$\begin{aligned} \operatorname{rect}\left(\frac{x}{a}\right) \otimes \operatorname{comb}\left(\frac{x}{X}\right) &= X \int \operatorname{rect}\left(\frac{\zeta}{a}\right) \sum_{n=-\infty}^{\infty} \delta(x - nX - \zeta)\, d\zeta \\ &= X \sum_{n=-\infty}^{\infty} \operatorname{rect}\left[\frac{x - nX}{a}\right] \end{aligned} \tag{8.55}$$

since

$$\operatorname{comb}\left(\frac{x}{X}\right) = \sum_{n=-\infty}^{\infty} \delta\left(\frac{x}{X} - n\right) = X \sum_{n=-\infty}^{\infty} \delta(x - nX) \tag{8.56}$$

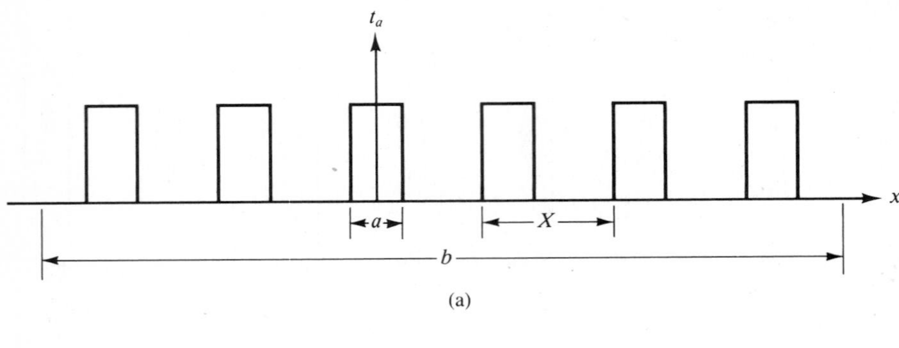

(a)

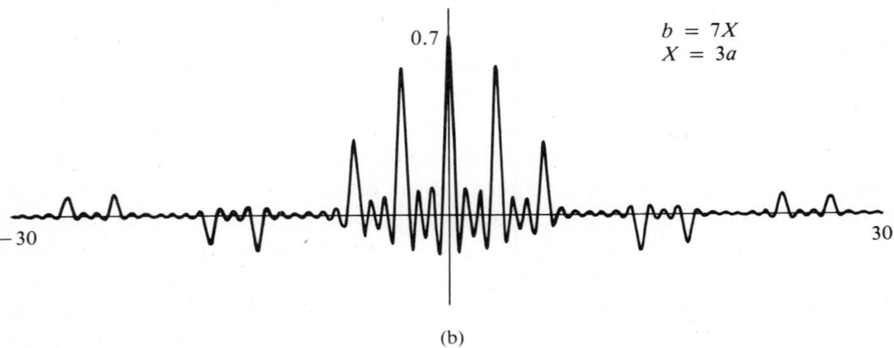

(b)

FIGURE 8.7

(a) A line diffraction grating of finite dimension and (b) its diffraction pattern at the Fourier plane

When this line grating, as described in Eq. (8.53), is introduced in the object plane of the coherent optical information-processing system, the corresponding field distribution at the Fourier plane is found to be proportional to

$$\begin{aligned} F\{t_a(x, y)\} &= \frac{1}{X} F\left\{\left[\operatorname{rect}\left(\frac{x}{a}\right) \otimes \operatorname{comb}\left(\frac{x}{X}\right)\right] \operatorname{rect}\left(\frac{x}{b}\right)\right\} \\ &= \frac{1}{X} F\left\{\operatorname{rect}\left(\frac{x}{a}\right) \otimes \operatorname{comb}\left(\frac{x}{X}\right)\right\} \otimes F\left\{\operatorname{rect}\left(\frac{x}{b}\right)\right\} \\ &= \frac{1}{X}\left[F\left\{\operatorname{rect}\left(\frac{x}{a}\right)\right\} F\left\{\operatorname{comb}\left(\frac{x}{X}\right)\right\}\right] \otimes F\left\{\operatorname{rect}\left(\frac{x}{b}\right)\right\} \end{aligned} \quad \textbf{(8.57)}$$

using Eq. (8.38) and the fact that

$$F\{f_1(x, y) f_2(x, y)\} = F\{f_1(x, y)\} \otimes F\{f_2(x, y)\} \quad \textbf{(8.58)}$$

In this context, it is noteworthy that

$$F\left\{\operatorname{rect}\left(\frac{x}{a}\right)\right\}=\frac{\sin(\pi v_x a)}{\pi v_x}\equiv a\operatorname{sinc}(av_x) \tag{8.59a}$$

and

$$F\left\{\operatorname{comb}\left(\frac{x}{X}\right)\right\}=X\sum_{n=-\infty}^{\infty}\delta(v_x X-n)=X\operatorname{comb}(v_x X) \tag{8.59b}$$

where $v_x = x/\lambda f$. Equation (8.57) can now be evaluated using Eqs. (8.59a) and (8.59b) as follows

$$\begin{aligned} F\{t_a(x,y)\} &= \frac{ab}{X}\left\{\operatorname{sinc}(av_x)\sum_{n=-\infty}^{\infty}\delta\left(v_x-\frac{n}{X}\right)\right\}\otimes\operatorname{sinc}(bv_x) \\ &= \frac{ab}{X}\sum_{n=-\infty}^{\infty}\operatorname{sinc}\left(\frac{an}{X}\right)\delta\left(v_x-\frac{n}{X}\right)\otimes\operatorname{sinc}(bv_x) \\ &= \frac{ab}{X}\sum_{n=-\infty}^{\infty}\operatorname{sinc}\left(\frac{an}{X}\right)\operatorname{sinc}\left[b\left\{v_x-\frac{n}{X}\right\}\right] \\ &= \frac{ab}{X}\left[\operatorname{sinc}(bv_x)+\sum_{n\neq 0}^{\infty}\operatorname{sinc}\left(\frac{an}{X}\right)\operatorname{sinc}\left[b\left\{v_x-\frac{n}{X}\right\}\right]\right] \end{aligned} \tag{8.60}$$

The frequency spectra of the input image, as will be seen in the Fourier plane, thus consist of a bright central spot and many higher order diffraction spots whose intensity diminishes more or less continually at higher diffraction orders. Figure 8.7(b) shows the diffraction pattern corresponding to the grating of Fig. 8.7(a). The third lens L_3 performs a Fourier transform of this diffraction pattern and thus reproduces an output image that is proportional to the input image with its coordinate system reflected. The reflected system of the coordinate justifies the fact that an inversion has been caused by double Fourier transformations. The finite size of the aperture at the Fourier plane may limit the transmitted spatial frequencies and, in addition, may cause the system to modify the transmitted image.

Now consider introducing a particular binary filter at the Fourier plane to cause what has been referred to as spatial filtering. Assume further that $b \gg X$ such that there is no significant overlapping of diffraction orders. First, let us introduce a rather restrictive low-pass filter that passes only the zeroth diffraction order. According to Eq. (8.60), therefore, only the Fourier plane information corresponding to $[ab\operatorname{sinc}(bv_x)]/X$ centered at $v_x = 0$ is transmitted by this binary filter. The amplitude field distribution at the output plane is thus given by

$$g(x,y)|_{n=0}=F\left\{\frac{ab}{X}\operatorname{sinc}(bv_x)\right\}=\frac{a}{X}\operatorname{rect}\left(\frac{x}{b}\right) \tag{8.61}$$

The output consists of a broad luminating line of width b and contains no information at all about the input periodicity. Next, consider a low-pass filter that allows the two first-order diffraction spots ($n = \pm 1$) as well as the central brightest spot to pass through. The amplitude transmittance of the resulting output is accordingly given by

$$g(x, y)|_{n=0,\pm 1} = F\left\{\frac{ab}{X}\left[\operatorname{sinc}(bv_x) + \operatorname{sinc}\left(\frac{a}{X}\right)\left\{\operatorname{sinc}\left[b\left(v_x + \frac{1}{X}\right)\right]\right.\right.\right.$$
$$\left.\left.\left. + \operatorname{sinc}\left[b\left(v_x - \frac{1}{X}\right)\right]\right\}\right]\right\}$$
$$= \frac{a}{X}\operatorname{rect}\left(\frac{x}{b}\right)\left[1 + 2\cos\left(\frac{2\pi x}{X}\right)\operatorname{sinc}\left(\frac{a}{X}\right)\right] \qquad \textbf{(8.62)}$$

This output amplitude distribution, unlike that described by Eq. (8.61), has the same periodicity as that of the input. A pinhole spatial filter like the one used in the optical information processor of Fig. 8.6 is an example of this latter binary filter. Accordingly, the pinhole spatial filter removes spurious fringe patterns that are otherwise present in the laser output. A lens is often used to bring a laser beam to sharp focus. A fine pinhole placed at the focal plane acts like a low-pass filter and is able to remove higher order spatial frequencies, thus improving the quality of the beam. With a low-pass filter that allows more diffraction orders to pass through, the output begins to appear more like the input and the corners appear rounded. And when all of the diffraction orders are allowed to pass through the filter, the output becomes identical to the input. Interestingly, if a bandpass filter capable of passing only the ± 1 order is introduced at the Fourier plane, the resulting output image appears to have a periodicity that is half of what it is in the input grating.

Finally, consider the effect of having a high-pass binary filter that blocks only the central brightest diffraction spot. The output amplitude distribution of this spatially filtered image is found to be given by

$$g(x, y)|_{n\neq 0} = F\left\{\frac{ab}{X}\sum_{n\neq 0}^{\infty}\operatorname{sinc}\left(\frac{an}{X}\right)\operatorname{sinc}\left[b\left(v_x - \frac{n}{X}\right)\right]\right\}$$
$$= \frac{1}{X}\left\{\operatorname{rect}\left(\frac{x}{b}\right)\left[\operatorname{rect}\left(\frac{x}{a}\right)\otimes\operatorname{comb}\left(\frac{x}{X}\right)\right] - a\operatorname{rect}\left(\frac{x}{b}\right)\right\} \qquad \textbf{(8.63)}$$

The first term of Eq. (8.63) has been normalized to unity. Consequently, the coefficient of the second term determines the actual characteristics of the output amplitude. When $a/X = \frac{1}{2}$—i.e., when the line diffraction has a 50-50 duty cycle—the field amplitude at the output plane is either $\frac{1}{2}$ or $-\frac{1}{2}$ at all x such that $g(x, y)$ is in phase with $t_a(x, y)$. The resulting output is characterized by a uniform intensity and not by periodicity.

Another interesting case is obtained when $a/X > \frac{1}{2}$. The output field amplitude is then either $1 - (a/X)$ or $-a/X$ such that $g(x, y)$ and $t_a(x, y)$ are in phase. The corresponding output is characterized by an intensity distribution such that the darker portions of the input appear brighter at the output and the brighter portions of the input appear darker at the output. Thus a contrast reversal is obtained using such selective high-pass spatial filtering. However, when $a/X < \frac{1}{2}$, the output intensity distribution and the input intensity distribution are both found to be in phase. High-pass spatial filters block the lower diffraction orders. Consequently, the output image obtained using high-pass filtering consists of only the edges of the input. As more central diffraction orders are blocked, the detail of the edge of the object is continually enhanced.

8.6 Nonlinear Optical Signal Processing Using Contact Screens

In the last section, we explored how filters—binary filters in particular—can be used at the Fourier plane to alter the input image. Specially designed binary filters can be employed in applications such as edge enhancement and image reversal. It should be borne in mind that up to this point, nothing was done directly to the input to affect it *before* it was introduced to the optical system of Fig. 8.6. Nonlinear optical signal processing that is accomplished with the use of contact screens is comparatively different in that the inputs are modulated before being introduced to the optical system. In Chapter 7, we explored how a halftone contact screen can be used to realize pulse-width or pulse-area modulation of the input. In addition, we determined the transmittance characteristics of the contact screens according to the type of nonlinear transformation desired. The intensity at any point of the output plane of the optical information-processing system depends on spatial filtering as well as on contact screen preprocessing of the input.

While the halftone technique can achieve a multitude of nonlinear transformations, in this section we shall focus our attention on a particular optical processing system used to separate noise that is multiplicatively combined with the signal. In many imaging cases, such as x-ray radiograms and images with speckle noises, the images are multiplied by noise. The removal of noises from these images is often desirable or necessary. Logarithmic filtering has been used for the separation of optical images that are multiplied by or convolved with noise in linear homomorphic digital electronic signal-processing systems in which the optical image and the noise are assumed to be separated in the spatial frequency domain.

Linear filtering systems are often sufficient to separate a signal and noise that are added to one another. But when the signal and noise are nonadditively combined, a linear system cannot be used to separate them. If the signal is multiplied to noise, the spectra of the signal and noise remain entwined through a convolution operation. This makes signal extraction in a linear system very difficult if not impossible. This dilemma is resolved in a digital processor by generalizing the concept of a linear system to include a broader class of operations. The generalized system is called a **homomorphic system**. A system characterized by the operator H is said to be homomorphic if

$$H[f_1(x) \square f_2(x)] = H[f_1(x)] \square H[f_2(x)] \tag{8.64}$$

and

$$H[c \cap \{f_1(x) \square f_2(x)\}] = c \cap H[f_1(x) \square f_2(x)] \quad \textbf{(8.65)}$$

for any two arbitrary signals $f_1(x)$ and $f_2(x)$ where $\square$ denotes a to-be-defined rule of combination and $\cap$ denotes a scalar multiplication. Specifically when $\square$ represents an addition and $\cap$ represents simple multiplication, the homomorphic system reduces to a conventional linear system. Furthermore, a homomorphic system also includes a system that is invertible—i.e., a system for which output is caused by only one input. An invertible system with transformation Γ and an inverse transformation Γ^{-1} has the following properties:

$$\Gamma\{\Gamma^{-1}(f_1) \square \Gamma^{-1}(f_2)\} = f_1 \square f_2 \quad \textbf{(8.66)}$$

and

$$\Gamma\{c \cap \Gamma^{-1}(f)\} = c \cap f \quad \textbf{(8.67)}$$

With a suitable choice of these operations, the system satisfies the generalized principle of superposition.

A homomorphic system can be used to treat the problem of multiplicative noise when the signal transmittance T^P is multiplied by noise transmittance T^N. In particular, the homomorphic system can be represented by a cascade of three operations as illustrated in Fig. 8.8. The first phase involves transforming a corrupted signal into additive components $\ln(T^P)$ and $\ln(T^N)$. The second phase involves separating the signal and noise by means of spatial filtering at the Fourier domain of a linear system. Thereafter, in the third phase, an exponential transformation is performed to restore the signal.

The input image $T^P T^N$ can be transformed into a logarithmic halftone image by means of a logarithmic contact screen. The halftone transparency then consists of additive components $\ln(T^P)$ and $\ln(T^N)$. This halftone image is placed at the input plane of the optical information-processing system while the spatial filtering is performed at the Fourier plane to separate noise from the signal. The image recovered at the output plane is then exponentially transformed by means of an exponential contact screen to restore the original signal.

In general, we can assume that the input image transmittance $T(x, y)$ can be expressed as a product of n constituent transmittances $T_i(x, y)$,

$$T(x, y) = \prod_{i=1}^{n} T_i(x, y) \quad \textbf{(8.68)}$$

The various $T_i(x, y)$ components are separated into additive terms by a logarithmic transformation through contact screen printing. The input image is logarithmically transformed by contact printing it with a logarithmic contact screen onto a recording medium. The resulting halftone negative is then contact-printed onto another high-contrast negative to produce a halftone positive. This particular halftone positive is then introduced at the input plane of the system.

Following logarithmic transformation, the input image transmittance

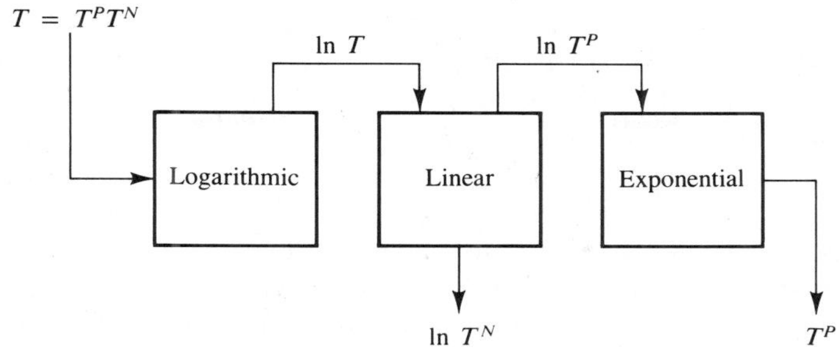

FIGURE 8.8

A homomorphic optical processing scheme. (M. A. Karim and H. K. Liu. "An Optical Homomorphic Image-Processing System." *Opt. Lett.* Vol. 7, p. 371, 1982.)

$T(x, y)$ becomes

$$\ln T(x, y) = \sum_{i \neq j}^{n} \ln T_i(x, y) + \ln T_j(x, y) \tag{8.69}$$

For illustration, $T_j(x, y)$ can be used to represent multiplicative noise. By introducing the halftone positive at the input plane of the coherent optical processor, we obtain, at the Fourier plane,

$$F\{\ln T(x, y)\} = F\{\ln T_j(x, y)\} + F\left\{\sum_{i \neq j}^{n} \ln T_i(x, y)\right\} \tag{8.70}$$

Provided the spectral patterns are not overlapped, a spatial filter can be introduced at the Fourier plane to block out and hence remove the term $F\{\ln T_j(x, y)\}$. The signal related frequency plane components are given by

$$F\left\{\sum_{i \neq j}^{n} \ln T_i(x, y)\right\} = F\{\ln T(x, y)\} - F\{\ln T_j(x, y)\} \tag{8.71}$$

The use of a second Fourier transform lens will yield at the image plane the term

$$F^{-1}F\left\{\sum_{i \neq j}^{n} \ln T_i(x, y)\right\} = \sum_{i \neq j}^{n} \ln T_i(x, y) = \ln T(x, y) - \ln T_j(x, y) \tag{8.72}$$

However, in the case of linear filtering, the Fourier transform of Eq. (8.65) will lead to

$$\begin{aligned} F\{T(x, y)\} &= F\left\{\prod_{i \neq j}^{n} T_i(x, y)\right\} \\ &= F\{T_1(x, y)\} \otimes F\{T_2(x, y)\} \otimes \cdots \otimes F\{T_n(x, y)\} \end{aligned} \tag{8.73}$$

It is thus not possible to single out and remove any noise component in the case of linear filtering.

In the following, we discuss the difference between logarithmic and linear filtering in more detail. As we did previously, we assume the input transmittance $T(x, y)$ to be a product of the signal transmittance $T^P(x, y)$ and the noise transmittance T^N.

$$T(x, y) = T^P(x, y) T^N(x, y) \tag{8.74}$$

For mathematical simplicity, we consider only $T^N(x, y) = T^N(x)$. Obviously, this is not an oversimplification of realistic circumstances. For example, in biomedical x-ray imagery, a one-dimensional grid made of lead strips is often used in front of the film-screen cassette to absorb scattered radiation. Such absorbing structures introduce one-dimensional, gridlike multiplicative noise in the radiograph. Provided the noise transmittance function T^N of period X satisfies the conditions for Fourier series expansion, it can be represented by

$$T^N(x) = \frac{1}{2} T_0^{N1} + \sum_{i=1}^{\infty} [T_i^{N1} \cos(ix) + T_i^{N2} \sin(ix)] \tag{8.75}$$

where

$$T_i^{N1} = \frac{2}{X} \int_{-X/2}^{X/2} T^N(x) \cos\left(\frac{2\pi i x}{X}\right) dx \tag{8.76a}$$

$$T_i^{N2} = \frac{2}{X} \int_{-X/2}^{X/2} T^N(x) \sin\left(\frac{2\pi i x}{X}\right) dx \tag{8.76b}$$

A logarithmic contact screen of period W can be designed, in accordance with Eq. (7.80), such that the average transmittance of the logarithmic halftone image T^H is related to the input transmittance $T(x, y)$ by

$$T^H = \frac{\ln(T) - \ln(T_{\min})}{\ln(T_{\max}) - \ln(T_{\min})} \tag{8.77}$$

where $T_{\max}$ and $T_{\min}$ are the maximum and the minimum values of $T(x, y)$. For the purpose of generality, we consider that the parallel cell lines of the contact screen are directed along the w direction, which may not necessarily be along either the x or y direction.

The Fourier transform of the logarithmic halftone transmittance can be written using Eqs. (8.74) and (8.77) as

$$F(T^H) = F(T^{HP}) + 2 \sum_{m=1}^{\infty} \int_{-\infty}^{\infty} T^{HP}(\zeta) \operatorname{sinc}(m T^{HP}(\zeta)) \cos(2\pi m v_{ow} \zeta) \times e^{-j2v_w \xi} d\zeta + \frac{1}{2} a_o \delta(v_x, v_y) + \sum_{n=1}^{\infty} (a_n + j b_n) \delta(v_x - n v_{ox}, v_y) \tag{8.78}$$

where

$$a_n = \frac{2}{X} \int_{-X/2}^{X/2} \frac{\ln T^N(x)}{k} \cos\left(\frac{2\pi n x}{X}\right) dx \tag{8.79a}$$

$$b_n = \frac{2}{X} \int_{-X/2}^{X/2} \frac{\ln T^N(x)}{k} \sin\left(\frac{2\pi n x}{X}\right) dx \tag{8.79b}$$

$$k = \ln(T_{\max}) - \ln(T_{\min}) \tag{8.80}$$

$$\nu_{ow} = \frac{1}{W} \tag{8.81}$$

$$\nu_{ox} = \frac{1}{X} \tag{8.82}$$

$$\operatorname{sinc}(z) = \frac{\sin(\pi z)}{\pi z} \tag{8.83}$$

where ν_i is the spatial frequency corresponding to the i axis and T^{HP} is the average transmittance of the logarithmic halftone image obtained using an input of transmittance $T^P(x, y)$. For the case of linear filtering, however, the Fourier transform of the input transmittance $T(x, y)$ becomes

$$\begin{aligned} F(T^P T^N) &= F(T^P) \otimes F(T^N) \\ &= F(T^P) \otimes \left[\frac{1}{2} T_0^{N1} \delta(\nu_x, \nu_y) + \sum_{n=1}^{\infty} (T_n^{N1} + jT_n^{N2}) \delta(\nu_x - n\nu_{0x}, \nu_y)\right] \\ &= \frac{1}{2} T_0^{N1} \Upsilon^P(\nu_x, \nu_y) + \sum_{n=1}^{\infty} (T_n^{N1} + jT_n^{N2}) \Upsilon^P(\nu_x - n\nu_{ox}, \nu_y) \end{aligned} \tag{8.84}$$

where

$$F(T^P) \equiv \Upsilon^P \tag{8.85}$$

At the Fourier plane, if only the zero-order spectral dot is allowed to pass through a pinhole spatial filter—i.e., when $m = n = 0$—Eq. (8.78) reduces to

$$F(T^H)|_{m=n=0} = F(T^{HP}) + \tfrac{1}{2} a_0\, \delta(\nu_x, \nu_y) \tag{8.86}$$

for logarithmic spatial filtering. The corresponding output image has a transmittance proportional to

$$F^{-1}[F(T^H)|_{m=n=0}] = T^{HP}(-x - y) + \tfrac{1}{2} a_0 \tag{8.87}$$

It can be seen that the output transmittance is proportional to the sum of a constant bias term $\frac{1}{2}a_0$ and the logarithm of T^P. On the other hand, for linear filtering, the zero-order term in Eq. (8.84) is simply

$$F(T^P T^N)|_{n=0} = \tfrac{1}{2} T_0^{N1} \Upsilon^P(\nu_x, \nu_y) \tag{8.88}$$

The output image accordingly has a transmittance proportional to

$$F^{-1}[F(T^P T^N)|_{n=0}] = \tfrac{1}{2} T_0^{N1} T^P(-x, -y) \tag{8.89}$$

It can be seen that the output transmittance is still proportional to the input signal but is reduced by a factor of $\frac{1}{2}T_0^{N1}$. Clearly, the linearly filtered output is different from the logarithmically filtered output. However, the comparison of the two processes in terms of only zero-order is too idealistic. In practice, it is not easy to isolate the zero-order spectral dot by a pinhole spatial filter of finite aperture. In realistic cases, higher order terms (say, ± 1 order) cannot be totally eliminated. If, for simplicity, only ± 1 spectral orders are present along with the zero-order spectral dot, then the reconstructed transmittance of the output image is proportional to

$$F^{-1}[F(T^H)|_{m=0;n=0,\pm 1}] = T^{HP}(-x,-y) + \tfrac{1}{2}a_0 + a_1 + jb_1 \tag{8.90}$$

for logarithmic filtering, and that for linear filtering is given by

$$F^{-1}[F(T^P T^N)|_{n=0,1} = \tfrac{1}{2}T_0^{N1}T^P(-x,-y) + (T_1^{N1} + jT_1^{N2})T^P(-x-X,-y) \tag{8.91}$$

A comparison of Eqs. (8.90) and (8.91) reveals that the logarithmically filtered output contains the logarithm of the signal and a complicated bias term, but the linearly filtered output is proportional to the signal transmittance overlapped with a reduced and displaced duplicate of itself. A similar phenomenon exists if additional higher orders are included in the output. The bias present in the logarithmically filtered output tends to add a constant glare to the logarithm of the signal. On the other hand, the linearly filtered output corresponds to a blurred signal because the signal overlaps with its own displaced replicates.

If the noise is assumed to have a finite period equal to or larger than the spatial width of the input photograph along x, it is necessary to include a few more spectral orders other than the ± 1 orders in Eqs. (8.90) and (8.91). This is due to the fact that there is hardly any spacing between the lower order spectral dots. With additional spectral dots, the linearly filtered output image deteriorates further because of repeated blurring.

For the purpose of illustration, the noise transmittance $T^N(x)$ is assumed to be a periodic Ronchi-ruling type function like the one shown in Fig. 8.9 such that its period X is much smaller than the dimensions of the input image:

$$T^N(x) = \text{rect}\left(\frac{2x}{X}\right) \otimes \text{comb}\left(\frac{x}{X}\right) \tag{8.92}$$

where

$$\text{rect}(z) = \begin{cases} a & |z| \leqslant \frac{1}{2} \\ b & |z| > \frac{1}{2} \end{cases} \tag{8.93}$$

$$1 \geqslant a \geqslant b > 0 \tag{8.94}$$

By substituting T^N into Eqs. (8.76a), (8.76b), (8.79a), and (8.79b), we obtain

$$T_0^{N1} = a + b \tag{8.95a}$$

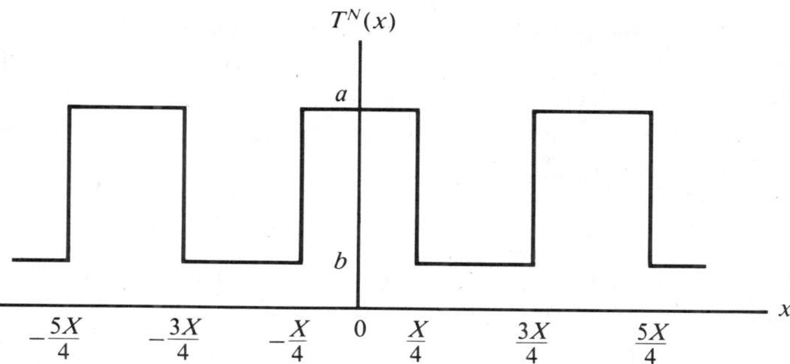

FIGURE 8.9

Ronchi-ruling type noise transmittance $T^N(x)$

$$T_1^{N1} = \frac{2(a-b)}{\pi} \tag{8.95b}$$

$$T_1^{N2} = 0 \tag{8.95c}$$

$$a_0 = \frac{\ln(ab)}{k} \tag{8.95d}$$

$$a_1 = \frac{2\ln\left(\frac{a}{b}\right)}{\pi k} \tag{8.95e}$$

and

$$b_1 = 0 \tag{8.95f}$$

Therefore, for the case of a Ronchi-ruling type of noise, Eqs. (8.87) and (8.89) through (8.91), characterizing output image transmittance, reduce to

$$F^{-1}[F(T^H)|_{m=n=0}] = T^{HP}(-x, -y) + \frac{\ln(ab)}{2k} \tag{8.96a}$$

$$F^{-1}[F(T^P T^N)|_{n=0}] = \tfrac{1}{2}(a+b)T^P(-x, -y) \tag{8.96b}$$

$$F^{-1}[F(T^H)|_{m=0;n=0,\pm 1}] = T^{HP}(-x, -y) + \frac{(\pi+4)\ln(a)+(\pi-4)\ln(b)}{2\pi k} \tag{8.97a}$$

and

$$F^{-1}[F(T^P T^N)|_{n=0,\pm 1}] = \tfrac{1}{2}(a+b)T^P(-x, -y) + \frac{2(a-b)}{\pi} T^P(-x-X, -y) \tag{8.97b}$$

respectively.

In order to complete the homomorphic system, we subject the logarithmically filtered output image to an exponential transformation by means of an exponential contact screen. The exponential contact screen can be designed using Eq. (7.81) in such a way that the average transmittance of the exponential halftone photograph is related to the input transmittance T by

$$T^H_{\exp} = e^{k(T-1)} \tag{8.98}$$

The image that is restored by exponentially transforming the zero-order logarithmically filtered output, therefore, has an average transmittance proportional to

$$T^H_r|_{m=n=0} = c_1 T^P \tag{8.99}$$

where

$$c_1 = \frac{\sqrt{ab}}{T_{\min} e^k} \tag{8.100}$$

Similarly, if the logarithmically filtered image were to include the first-order noise, the restored image would have an average transmittance proportional to

$$T^H_r|_{m=0;n=0,\pm 1} = c_2 T^P \tag{8.101}$$

where

$$c_1 = \frac{b}{T_{\min} e^k}\left(\frac{a}{b}\right)^{(1/2)[1+(4/\pi)]} \tag{8.102}$$

We can conclude from Eqs. (8.99) and (8.101) that the restored output is indeed proportional to the signal transmittance. Furthermore, unlike linear filtering, the output image of a homomorphic system is free from blurring.

The Fourier plane spectra corresponding to Eqs. (8.78) and (8.84) for the logarithmic and linear inputs are shown in Figs. 8.10(a) and 8.10(b), respectively. For the sake of illustration, we discuss the noise of frequencies of approximately 2.0, 1.0, and 0.5 lines/mm. The difference between the two can be readily observed. The former shows spectra of the photograph that are additive and well distributed in the $\nu_x - \nu_y$ frequency plane, while the latter shows the resultant intermodulation components of the spectrum restricted only along the ν_x axis. In the former, signal information shows up along the vertical axis, and the noise information is constrained along the horizontal direction. It can also be seen that for both cases, the noise frequency components along the horizontal axis are located closer to one another with decreasing noise frequency.

Figure 8.11 presents a comparison of the results of the logarithmically filtered and the linearly filtered output of an input photograph multiplied by noises of several spatial frequencies. Both logarithmic and linear filtering are very effective in removing high-frequency noises. But for noise of frequency less than 0.5 lines/mm, the linearly filtered output remains blurred as predicated by Eq. (8.97b). Finally, Fig. 8.12 shows the homomorphically processed output.

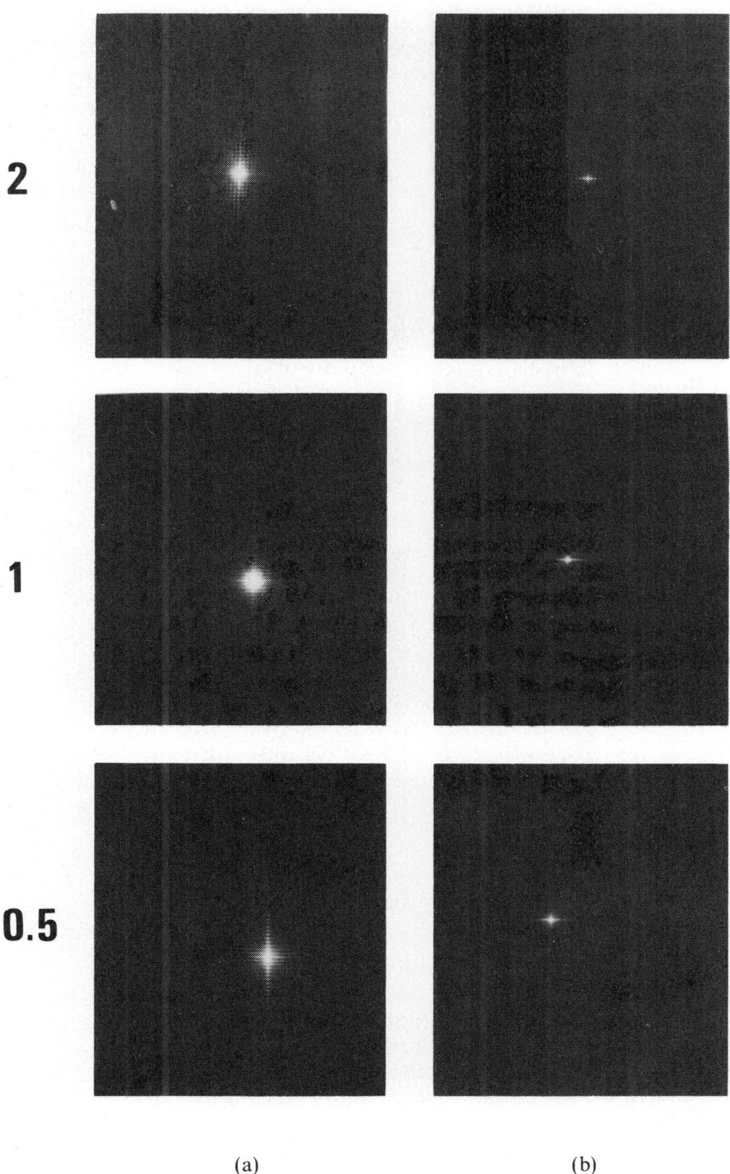

FIGURE 8.10

Fourier plane spectra of (a) logarithmic filtering and (b) linear filtering for noise frequencies of 0.5, 1.0, and 2.0 lines/mm. (M. A. Karim and H. K. Liu. "Linear Versus Logarithmic Spatial Filtering in the Removal of Multiplicative Noises." *Opt. Lett.*, Vol. 6, p. 207, 1981.)

FIGURE 8.11

(a) Input photograph with signal multiplied to the gratinglike noises, (b) corresponding logarithmically filtered output, and (c) corresponding linearly filtered output. (M. A. Karim and H. K. Liu. "Linear Versus Logarithmic Spatial Filtering in the Removal of Multiplicative Noises." ***Opt. Lett.*****, Vol. 6, p. 207, 1981.)**

It is appropriate to note that the set up of Fig. 8.6 uses photographic film. Consequently, the filtering process is not only time consuming but also difficult to control because of the messy development process involved. Alternatively, however, we can use a real-time recording device such as the LCLV to replace the photographic medium. Accordingly, we can use the optical system of

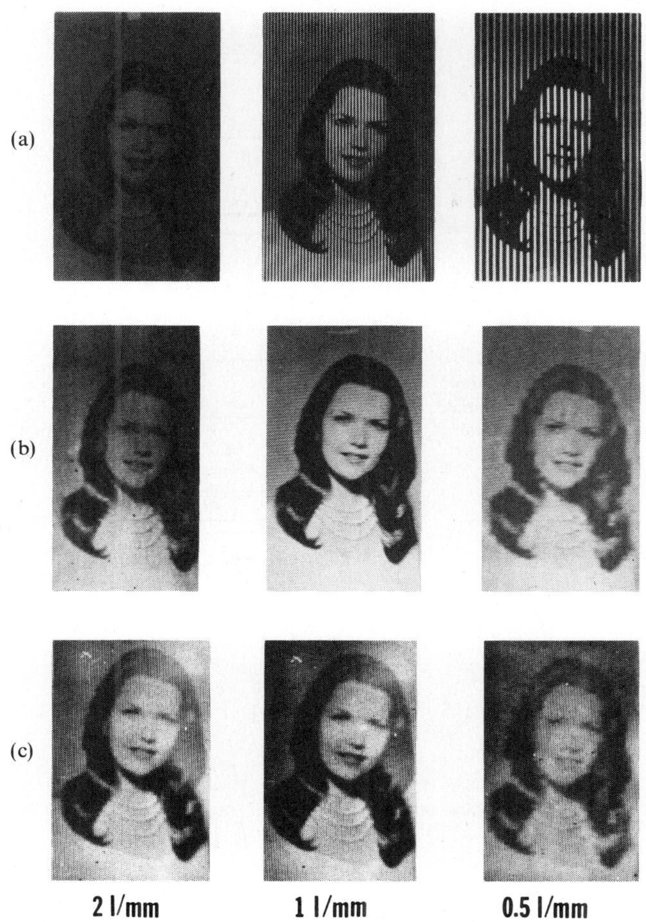

FIGURE 8.12

The output images of the purely optical homomorphic system. (M. A. Karim and H. K. Liu. "All-Optical Image-Processing System." *Opt. Lett.*, Vol. 7, p. 371, 1982.)

Fig. 8.13 to realize the same spatial filtering results but in (nearly) real-time. While the input photographic transparency and the screen are in perfect contact, they are introduced at the input plane. This particular arm of the system is exposed by an incoherent light source. The resultant halftone image is projected by the LCLV so that it is Fourier transformed in the coherent arm of the system. The Fourier plane pattern is then transformed again by a lens to yield the filtered output.

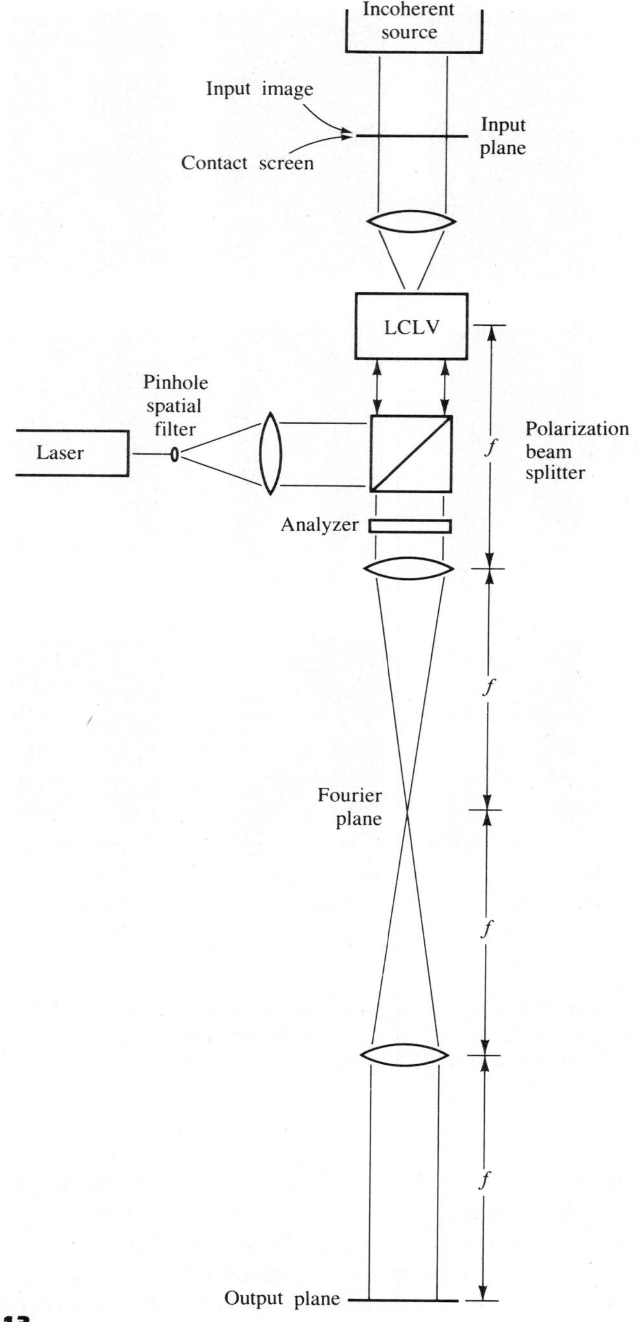

FIGURE 8.13

A real-time nonlinear signal processor using LCLV

8.7 Apodization

Apodization refers to a process by which an aperture function is altered to redistribute the energy in a diffraction pattern. In practice, apodization is used mostly for reducing the intensity concentration of secondary diffraction orders. Consider, for example, a single-slit aperture of width a, as shown in Fig. 8.14(a), whose diffraction pattern has the form of a sinc function. Now we can apodize the single-slit aperture by making use of a cosine function such as $\cos(\pi x/a)$, where $-a/2 \leqslant x \leqslant a/2$, as shown in Fig. 8.14(a). The diffraction pattern of the apodized slit is then given by

$$F\left\{\cos\left(\frac{\pi x}{a}\right)\right\} = \left[\frac{1}{v_x - \frac{\pi}{a}} - \frac{1}{v_x + \frac{\pi}{a}}\right] \cos\left(\frac{v_x a}{2}\right) \tag{8.103}$$

Comparing Eqs. (8.59a) and (8.103) we find that apodization reduces the amplitude substantially, and hence also the intensity, of the secondary maxima when compared to that of the central maxima. The two diffraction patterns are shown in Fig. 8.14(b). An apodization such as that of this example is found to suppress higher order diffraction spots. In fact, the resolution of astronomical telescopes is often enhanced by apodizing the circular apertures in such a way that the diffraction rings that otherwise appear around the images are diminished.

In a wide variety of applications it may be necessary to rectify Gaussian laser beams. It is often desirable to have an output irradiance that is a constant over a large cross-sectional area. The most popular means of doing this involves extracting the slowly varying central core of a diverging beam by means of a pinhole. No doubt the resulting output is no longer Gaussian, but this technique has a very low conversion efficiency because it wastes the greater portion of light energy. Interestingly enough, a binary area-modulated spatial filter can also be used to smooth the Gaussian energy distribution, producing a beam of uniform circular intensity. Consider a two-dimensional binary output, equivalent to a continuous inverse Gaussian filter, that can be used to attenuate the central core of the beam more than it attenuates the edges. Such a binary input can be introduced in a coherent optical system to yield a uniform circular beam of light at or beyond the output plane.

The transmission function of a two-dimensional binary input $T(r)$, as shown in Fig. 8.15, is given by

$$T(r) = \text{cyl}\left(\frac{r}{2w_0}\right) + \sum_{n=1}^{n} \text{ring}\left(\frac{r - nW}{w_n}\right) \tag{8.104}$$

where W is the period, w_n is the width of the nth concentric ring, r is the radial coordinate, the cylindrical cyl function is given by

$$\text{cyl}\left(\frac{r}{\rho}\right) = \begin{cases} 1 & \text{if } 0 < r < \rho/2 \\ 0 & \text{otherwise} \end{cases} \tag{8.105}$$

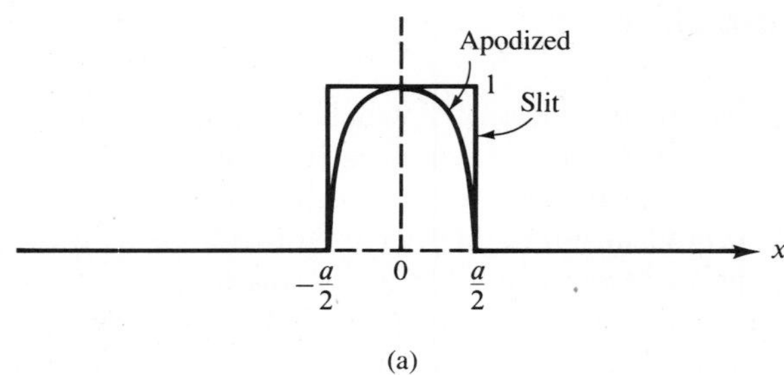

(a)

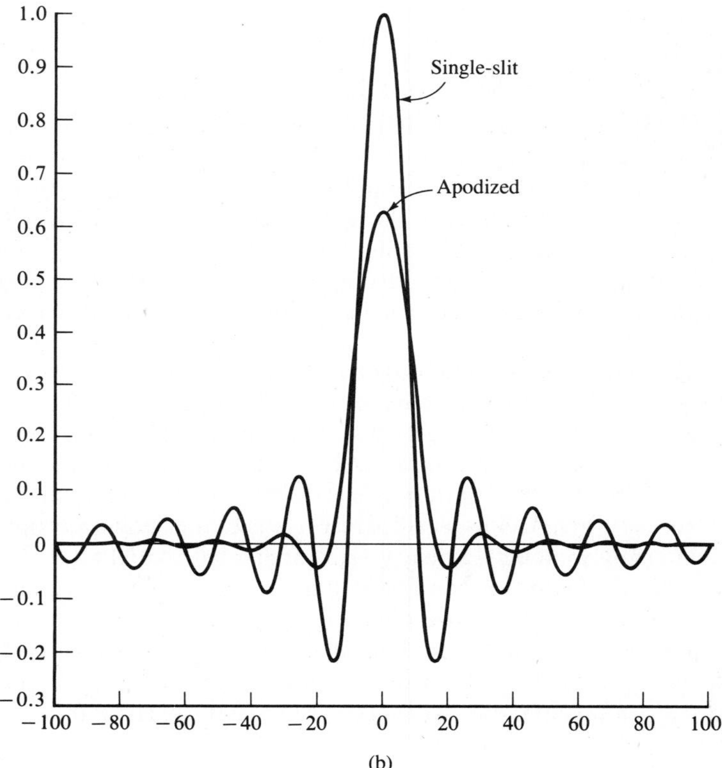

(b)

FIGURE 8.14

(a) Single-slit and apodized aperture and (b) the corresponding diffraction patterns

and the ring function is given by

$$\operatorname{ring}\left(\frac{r-\rho}{R}\right) = \begin{cases} 1 & \text{if } \rho - \frac{1}{2}R < r < \rho + \frac{1}{2}R \\ 0 & \text{otherwise} \end{cases} \tag{8.106}$$

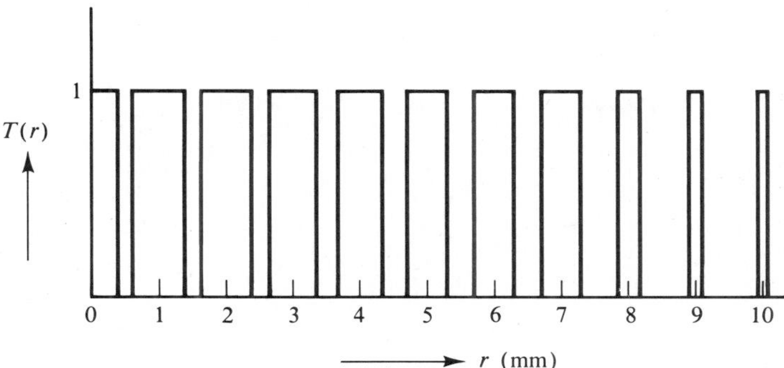

FIGURE 8.15

Characteristics of the inverse Gaussian binary input. (M. A. Karim, A. M. Hanafi, F. Hussain, S. Mustafa, Z. Samberid, and N. M. Zain. "Realization of a Uniform Circular Source Using a Two-Dimensional Binary Filter." ***Opt. Lett.*****, Vol. 10, p. 470, 1985.)**

Since only a symmetric, two-dimensional binary input is necessary for apodizing the circularly symmetric Gaussian laser beam, for simplicity Eq. (8.104) can be reduced to an equivalent one-dimensional format. The Gaussian intensity in the TEM_{00} mode appearing at the to-be-introduced binary input (placed at the object plane) is

$$I(r) = I_{00}e^{-(r/r_0)^2} \qquad \textbf{(8.107)}$$

where I_{00} is the peak intensity (at the beam axis) and r_0 is the Gaussian beam radius. The amplitude of light at a plane immediately behind the input is given by

$$a(r) = [g(r)]^{1/2}T(r) \qquad \textbf{(8.108)}$$

where $g(r) = [I(r)/I_{00}]$. The Fourier spectrum corresponding to Eq. (8.108) is spatially filtered when it is passed through a pinhole of width Δ. The resulting amplitude of output light at or beyond the output plane of the coherent optical system is thus found to be

$$t(r) = [g(r)]^{1/2} - \left\{[g(r)]^{1/2}\left[\text{comb}\left(\frac{r}{W}\right) \otimes \text{rect}\left(\frac{r}{w_n}\right)\right]\right\} \otimes \text{sinc}\left(\frac{r\Delta}{\lambda f}\right) \qquad \textbf{(8.109)}$$

where f is the focal length of the lenses and λ is the wavelength of light. The next step for designing the binary filter is to solve for w_n numerically in terms of $g(r)$, W, f, Δ, and λ, when $t(r)$ is assumed to be constant over the largest possible cross-sectional area.

The computed values for w_n for an 11-ring binary input are listed in Table 8.1 when $\lambda = 632.8$ nm, $f = 38.1$ cm, the diameter of the Fourier-plane pinhole is 15 μm, and $r_0 = 14.2$ mm. The binary input having the profile listed in Table 8.1 is then introduced at the object plane of the system. This binary filter

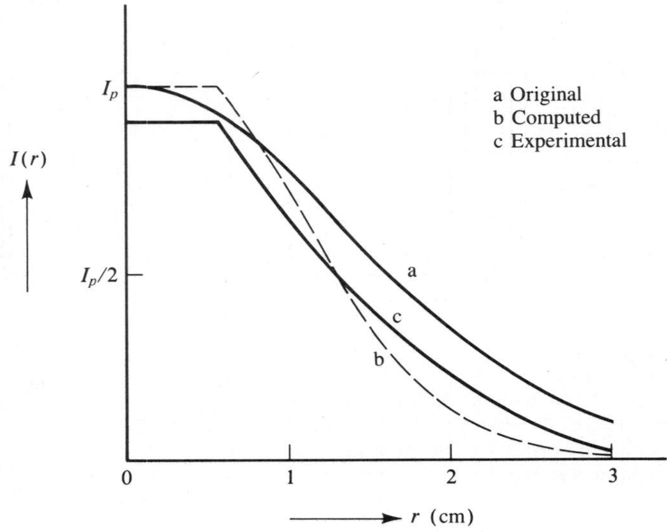

FIGURE 8.16

The output intensity versus radial distance. (M. A. Karim, A. M. Hanafi, F. Hussain, S. Mustafa, Z. Samberid, and N. M. Zain. "Realization of a Uniform Circular Source Using a Two-Dimensional Binary Filter." *Opt. Lett.*, Vol. 10, p. 470, 1985.)

TABLE 8.1

Computed Values of w_n for the Inverse Gaussian Binary Input

n	w_n(in mm)
0	0.260
1	0.676
2	0.669
3	0.649
4	0.623
5	0.592
6	0.557
7	0.551
8	0.330
9	0.190
10	0.120

M. A. Karim, A. M. Hanafi, F. Hussain, S. Mustafa, Z. Samberid, and N. M. Zain. "Realization of a Uniform Circular Source Using a Two-Dimensional Binary Filter." *Opt. Lett.*, Vol. 10, p. 470, 1985.

is expected to convert the Gaussian beam profile into a uniform one. The resulting values of the output intensities are measured and then plotted as a function of radial distances, as shown in Fig. 8.16. The measured output intensities indicate a variation of less than 3% over a diameter of 12 mm. In addition, the uniform intensity is equal to $0.75I_{00}$. A technique such as the one just described is valuable in transferring uniform energy to a target of large cross section.

8.8 Optical Processing Using Complex Filters

Complex filters are those spatial filters that act on both the amplitude and the phase of an input image. Often it is difficult to physically construct a filter having a desired complex transmittance. The photographic process is generally capable of recording only amplitude information and fails to store the phase information. Certainly the generation of a complex spatial filter by a computer is a possibility. However, in 1964 Vander Lugt conceived and demonstrated a highly practical method for constructing complex spatial filters. This method is based on interferometry and is closely related to what we refer to as holography.

The interferometric experimental set up of Fig. 8.17 can be used to synthesize a complex spatial filter. The point source, placed at the focal point S of lens L_1, produces a parallel beam of light which is then split into two portions. A portion passing through a prism P contributes to a parallel beam of light that strikes the photographic film placed in the filter plane at an angle θ. The other portion of light passes through a transparency of transmittance $h(x, y)$, which is positioned at the focal plane in front of lens L_2. Consequently, the information recorded at the filter plane is obtained from the interference between the prism-deflected plane beam of light and the Fourier transform of $h(x, y)$, $H(\nu_x, \nu_y)$. The amplitude distribution at the filter plane is given by

$$\Psi(\nu_x, \nu_y) = H_0 e^{-j\phi} + Ae^{-j2\pi\beta x} \qquad \textbf{(8.110)}$$

where $\beta = \sin\theta/\lambda$, and H_0 and ϕ represent the amplitude and phase of $H(\nu_x, \nu_y)$, respectively. The developed film negative will have a transmittance given by

$$\begin{aligned} T(x, y) &= 1 - \Psi(\nu_x, \nu_y)\Psi^*(\nu_x, \nu_y) \\ &= 1 - H_0^2 - A^2 - 2AH_0 \cos(2\pi\beta x - \phi) \end{aligned} \qquad \textbf{(8.111)}$$

Thus the recorded transparency (prepared using either a positive or a negative film) is associated with the phase of the filter. Once this frequency plane mask has been synthesized, it can then be introduced at the Fourier plane of the coherent optical processor. A number of examples of practical interest involving complex spatial filters are considered next.

Incidentally, the technique for recording complex spatial filters is known as **holography**. The fine-grain photographic film used to record the object information (for both amplitude and phase) acts basically as a diffraction screen, also

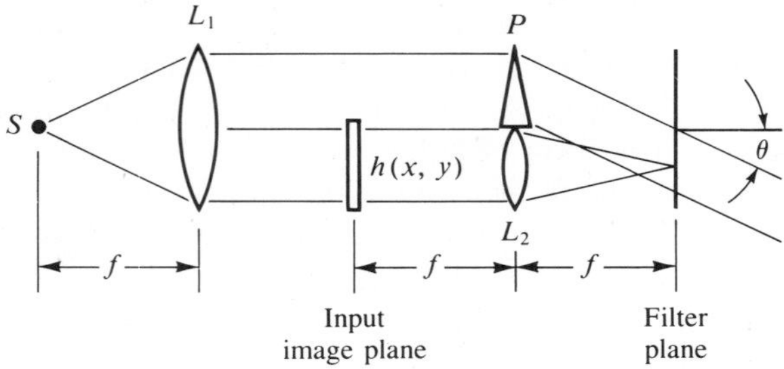

FIGURE 8.17

Set up for synthesizing complex spatial filter

referred to as a hologram. To make a hologram, the coherent laser output is divided into two separate beams, one of which has a complex amplitude given by $A_0 \exp(j\phi)$; it is the reflected wave front referred to as the object field. The other beam, referred to as the reference beam, is made to arrive on a holographic recording medium without striking the object. A hologram recording system is shown in Fig. 8.18(a). The recording medium is exposed simultaneously to both the reference beam, $A \exp[j2\pi(\zeta x + \eta y)]$, and the reflected beam. The resulting interference pattern recorded by the film constitutes a hologram. In fact, a hologram contains all the information needed to reproduce the wave field of the object. It can be used subsequently for reconstructing the object (in its three-dimensional form) by means of a coherent optical set up that is the same as the one shown in Fig. 8.18(b).

When the developed hologram is illuminated with a beam of light $A \exp[j2\pi(\zeta x + \eta y)]$ proportional to the reference beam, the transmitted wave field becomes

$$\begin{aligned} t(x, y) &= [A_0^2 + A^2 + 2AA_0 \cos(\phi - 2\pi\{\zeta x + \eta y\})]Ae^{j2\pi(\zeta x + \eta y)} \\ &= (A_0^2 + A^2)Ae^{j2\pi(\zeta x + \eta y)} + A^2 A_0 e^{j\phi} \\ &\quad + A^2 A_0 e^{j4\pi(\zeta x + \eta y)} e^{-j\phi} \end{aligned} \tag{8.112}$$

Hologram reconstruction thus involves three components—a direct beam and two first-order diffracted beams on either side of the direct beam. The first term of Eq. (8.112) corresponds to the direct beam, whereas the second term, which is proportional to the complex amplitude of object beam, represents one of the diffracted beams responsible for producing a virtual image. The last term of Eq. (8.112) is responsible for yielding a real image of the object.

There are a multitude of applications for holograms. One remarkable use involves making an optical memory bank. For example, Fig. 8.19 shows a system for recording a holographic disk. The digital information that is stored in a liquid crystal display is holographically recorded through a mask onto a rotating holographic disk. A series of holograms can be recorded as the disk is

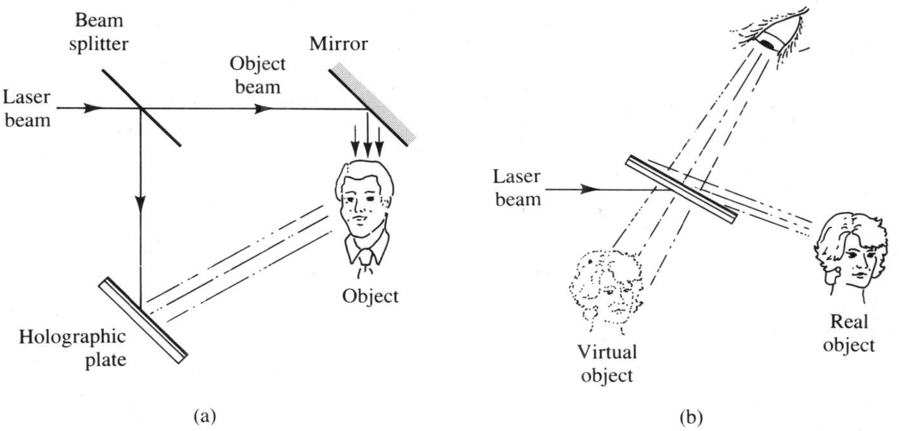

FIGURE 8.18

Holography: (a) recording and (b) reconstruction

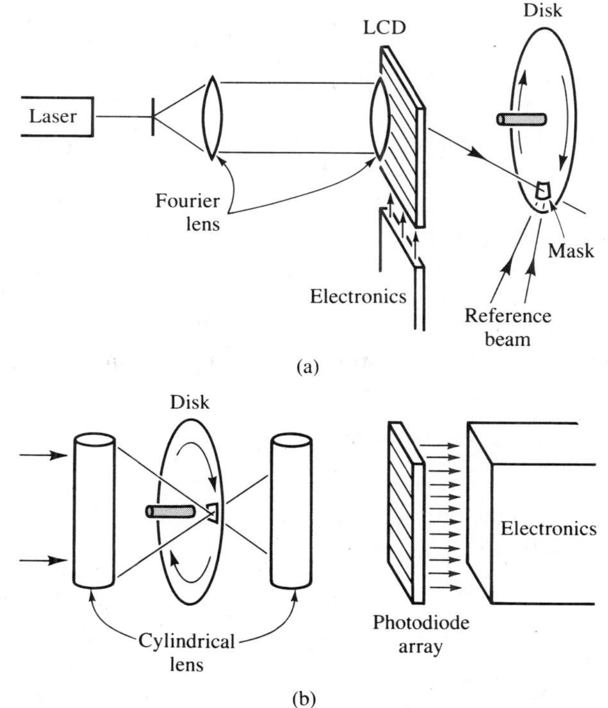

FIGURE 8.19

Holographic disk: (a) recording system and (b) playback system

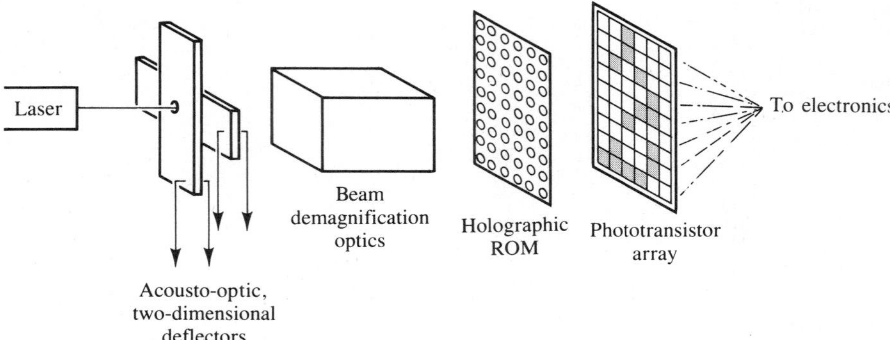

FIGURE 8.20

A hybrid electro-optic system using holographic ROM

rotated and also as the mask is moved radially. Both the reference beam and the object beam are introduced through the mask. The information of the hologram is retrieved by a read-out beam by scanning the rotating holographic disk. The holographic data that are reconstructed by a cylindrical lens are then picked up by the photodiode array. The system is somewhat analogous to that of the optical disk of Fig. 7.32.

Figure 8.20 shows an example of a system that uses a two-dimensional binary (each of the pixels is either a 0 or a 1) pattern as a read-only memory (ROM) of a hybrid electro-optic computer. Whereas the system of Fig. 8.19 involves a one-dimensional holographic recording, this hologram is recorded and retrieved in a two-dimensional format. The acousto-optic, two-dimensional deflector selects a particular hologram from the holographic ROM. The reconstructed real image is projected onto the receiving photosensitive element, which in turn passes the information to the number cruncher.

8.8.1 Phase Contrast Microscopy

It is extremely difficult to render visible a transparent object whose refractive index differs slightly from that of the surrounding transparent medium. A case in point is the determination of the structure of an object such as a biological cell using an ordinary microscope. In circumstances such as this, phase variations on the surface of the cell—due to either change in the refractive index or variation in thickness—become particularly useful. In fact, spatial filtering techniques sometimes use phase variations in the form of a complex spatial filter, known as a phase filter, which alters only the phase of the frequency spectrum.

To understand phase contrast microscopy, consider a phase object whose transmittance function is given by

$$t_a(x, y) = e^{j\phi(x,y)} \tag{8.113}$$

where $\phi(x, y)$ is the phase factor describing the object. Assuming that this phase difference is very small, we can express the transmittance function as

$$t_a(x, y) \simeq 1 + j\phi(x, y) \tag{8.114}$$

If a transparency such as this is introduced at the input plane of the coherent optical processor of Fig. 8.6, then the corresponding amplitude distribution at the Fourier plane is found to be

$$T_a(x, y) = F\{t_a(x, y)\} = \delta(v_x)\,\delta(v_y) + jF\{\phi(x, y)\} \tag{8.115}$$

Here the first term represents the diffraction pattern of the input, which is zero except for $v_x = v_y = 0$. On the other hand, the second term represents the diffraction pattern of the phase factor. Note also that the two terms are 90° out of phase by virtue of $+j$ factor. If a suitable phase plate is introduced now at the Fourier plane, it is possible to shift the phase of the second term by an additional 90°. Such a phase plate was described earlier as having a thickness equal to $\lambda/[4(n-1)]$, where n is the refractive index of the material of the plate and λ is the wavelength of light. The intensity pattern available at the output plane is consequently given by

$$\begin{aligned} I(x, y) &= |F[j\delta(x) + jF\{\phi(x, y)\}]|^2 \\ &= |j + j\phi(x, y)|^2 \simeq 1 + 2\phi(x, y) \end{aligned} \tag{8.116}$$

The first term of the output function is just a constant factor representing the image function of the input. Clearly, the second term of the output is related to the phase variation of the subject that is under study. Note also that if the phase delay introduced in the system were $3\pi/2$ instead of $\pi/2$, the output intensity would become $1 - 2\phi(x, y)$. This particular phenomenon is referred to as the **negative phase contrast**. While in the former phase contrast (referred to as the positive phase contrast) the thicker regions appear relatively brighter than the thinner regions, in the latter phase contrast microscopy the thicker regions appear relatively darker than the thinner regions.

It is interesting to note that the output contrast can be enhanced further by both attenuating the zero frequency term at the Fourier plane and by subjecting it to a phase shift of $\pi/2$. For an attenuating factor of α, therefore, the output intensity becomes

$$I(x, y) = |F[j\alpha\delta(x) + jF\{\phi(x, y)\}]|^2 = \alpha[\alpha + 2\phi(x, y)] \tag{8.117}$$

The image contrast is often given by

$$\text{contrast} = \frac{I(x, y) - \text{background intensity}}{\text{background intensity}} \tag{8.118}$$

Consequently, in the absence of attenuation, the contrast is only $2\phi(x, y)$, and in the presence of attenuation, the contrast becomes $2\phi(x, y)/\alpha$. A smaller value of α corresponds to an image having a higher contrast.

8.8.2 Pattern Recognition

Pattern, or character, recognition basically involves detecting the presence of a target pattern $h(x, y)$ in the input pattern of an image. It is very important in statistics, target detection, quality control, etc. To achieve pattern recognition, it is necessary to conduct a cross-correlation operation between the two functions $h(x, y)$ and $f(x, y)$.

Consider the function given by

$$\Psi = F\{f(x, y)\}[F\{h(x, y)\}]^* \tag{8.119}$$

The Fourier transform of this composite function is obtained as follows:

$$\begin{aligned} F(\Psi) &= \iint F\{f(x, y)\}[F\{h(x, y)\}]^* e^{-j2\pi(\nu_x x + \nu_y y)}\, d\nu_x\, d\nu_y \\ &= \iint F\{f(x, y)\} \iint h^*(\zeta, \eta) e^{j2\pi(\nu_x \zeta + \nu_y \eta)}\, d\zeta\, d\eta\, e^{-j2\pi(\nu_x x + \nu_y y)}\, d\nu_x\, d\nu_y \\ &= \iint h^*(\zeta, \eta) f(\zeta - x, \eta - y)\, d\zeta\, d\eta \\ &= \iint f(\zeta, \eta) h^*(x + \zeta, y + \eta)\, d\zeta\, d\eta \end{aligned} \tag{8.120}$$

Thus it is obvious that the Fourier transform of $F\{f(x, y)\}[F\{h(x, y)\}]^*$ is equivalent to a cross-correlation operation between $f(x, y)$ and $h(x, y)$. Consequently, the set up of Fig. 8.6 can be treated as an optical pattern recognition system when $f(x, y)$ is introduced at the input plane and a complex spatial filter characterized by $[F\{h(x, y)\}]^*$ is inserted at the Fourier plane. If $h(x, y)$ does indeed exist in $f(x, y)$, the output appearing beyond lens L_3 will show a strong autocorrelation peak at the location of matching. Obviously, therefore, the character recognition problem reduces to generating a complex spatial filter whose transmittance is proportional to $[F\{h(x, y)\}]^*$.

Equation (8.111) provides the intensity transmittance of a complex spatial filter $T(x,y)$ synthesized using a negative photographic transparency. As a result of having passed through the complex spatial filter placed at the Fourier plane, the field immediately behind the spatial filter is $F\{f(x, y)\}T(x, y)$. The output of the coherent optical processor is, therefore, given by

$$\begin{aligned} F[F\{f(x, y)\}T(x, y)] &= F[(F\{f(x, y)\})[1 - |H(\nu_x, \nu_y)|^2 - A^2]] \\ &\quad - F[F\{f(x, y)\}2AH_0 \cos(2\pi\beta x - \phi)] \\ &= F[(F\{f(x, y)\})(1 - |H(\nu_x, \nu_y)|^2 - A^2)] \\ &\quad - \iint h(\zeta, \eta) f(\beta - x - \zeta, -y - \eta)\, d\zeta\, d\eta \\ &\quad - \iint f(\zeta, \eta) h^*(x + \zeta + \beta, y + \eta)\, d\zeta\, d\eta \end{aligned} \tag{8.121}$$

The first term of Eq. (8.121), which manifests itself at the origin of the output plane, is not of much significance in the filtering operation. The second and the third terms are, however, of particular interest. The second term corresponds to a convolution of $f(x, y)$ and $h(x, y)$, and the output corresponding to it is centered at the coordinates $(\beta, 0)$ of the output plane. The third term corresponds to the cross-correlation of $f(x, y)$ and $h(x, y)$, and the output corresponding to it is centered at the coordinates $(-\beta, 0)$ of the output plane.

Obviously, the convolution and cross-correlation terms can be sufficiently separated by choosing a sufficiently high carrier frequency β—i.e., by introducing the reference wave at a steeper angle. If $h(x, y)$ is indeed present in $f(x, y)$, the corresponding output at the position of the cross-correlation field will show up as a bright identifiable peak.

It may be appropriate to note that an optical correlator like the one just described can be realized using either acousto-optic or spatial light modulators. The principal element of an acousto-optic signal processor is a signal input device that serves to convert electric signals to a spatially distributed optical signal. Such an acousto-optic device, referred to as a Bragg cell, was discussed in an earlier chapter. Acoustic-optic signal processors can be either space-integrating or time-integrating. Figure 8.21(a) shows a typical acousto-optic space-integrating correlator. The fixed spatial signal $f_2(x, y)$ is stored on a high-spatial-frequency carrier in a reference transparency that is then correlated with an incoming temporal signal $f_1(t)$ applied at the A–O cell. A single order diffracted by the acoustic grating is selected and imaged onto $f_2(x, y)$. The single order transmitted by the reference grating is next detected at the output. As the acoustic signal $f_1(t)$ propagates through the A–O cell, the effective time delay between the stored reference $f_2(x, y)$ and the incoming signal changes. The time-dependent output is determined by the cross-correlation of the two signals as a function of delay. Figure 8.21(b) shows the corresponding time-integrating A–O correlator. The signal $f_1(t)$ is likewise introduced at the A–O cell, whereas $f_2(t)$ is introduced via the intensity modulation of the light source. A spatial filter is used in this case to image the A–O output onto a detector array that is capable of time-integrating the signals incident upon it. Each of the detectors in the array accumulates the output incident upon it, but with a different constant relative delay between the two signals.

Figure 8.22 shows a real-time coherent correlator that uses a binary magneto-optic spatial light modulator (SLM) for the purpose of recording. The SLM device used in this set up consists of 48×48 elements. It should be noted that polarizers are not shown for simplicity. The particular SLM used for introducing the input is operated in the binary amplitude (on-off) modulation mode, whereas the second SLM for introducing the filter is operated as a binary phase ($0°$–$180°$) modulator. Both of these devices are controllable by a desk-top computer. Figure 8.23(a) shows the input pattern while Fig. 8.23(b) shows the expected correlation intensity pattern when a filter designed for detecting X is used in the set up. Figures 8.23(c) and 8.23(d) show the actual experimental results with two bright peaks at the corresponding positions of the two X characters.

It is often desirable to introduce both the object and the reference images at the same plane of the optical correlator. Such an optical pattern recognition system is referred to as a **joint transform image correlator**. Figure 8.24 shows an interesting set up for doing just that. The interference between the Fourier transforms of both object and reference images is obtained using Fourier lenses. A CCD placed at the Fourier plane can record the intensity of this interference

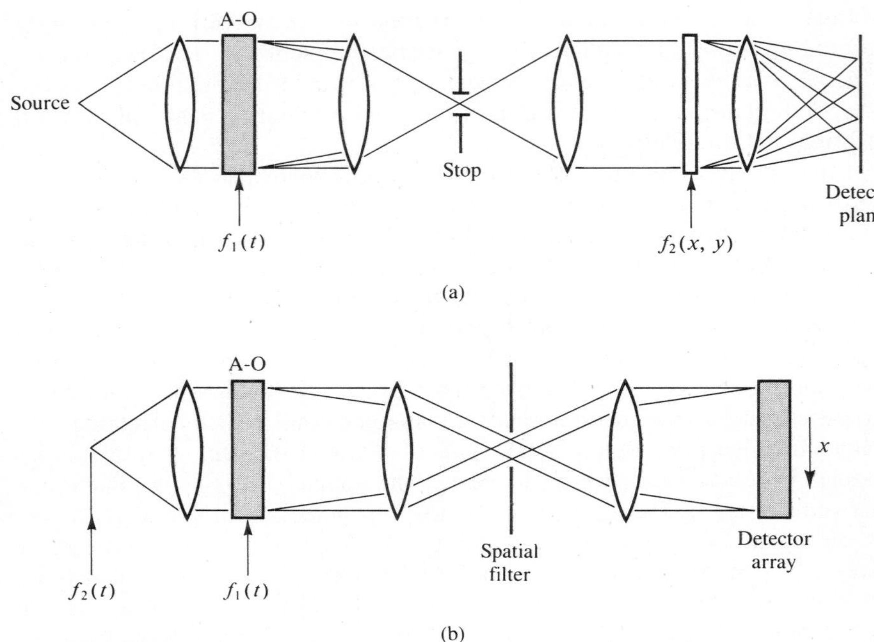

FIGURE 8.21

The A–O correlator: (a) space-integrating and (b) time-integrating

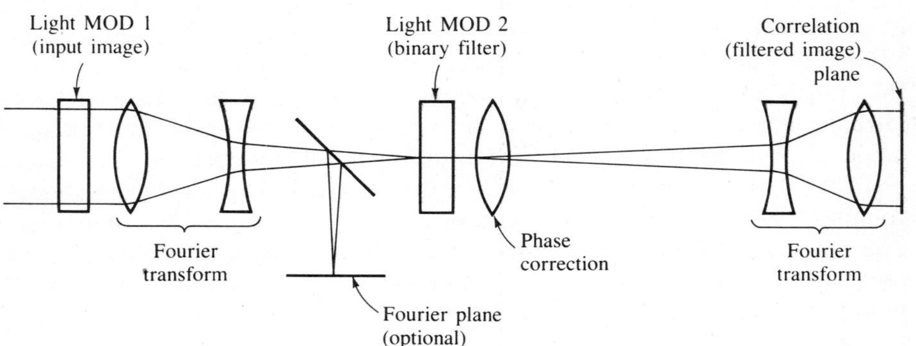

FIGURE 8.22

A light-mod correlator system. (D. L. Flannery, A. M. Biernacki, J. S. Loomis, and S. L. Cartwright. "Real-Time Coherent Correlator Using Binary Magneto-Optic Spatial Light Modulators at Input and Output Planes." *Appl. Opt.*, Vol. 25, p. 466, 1986.)

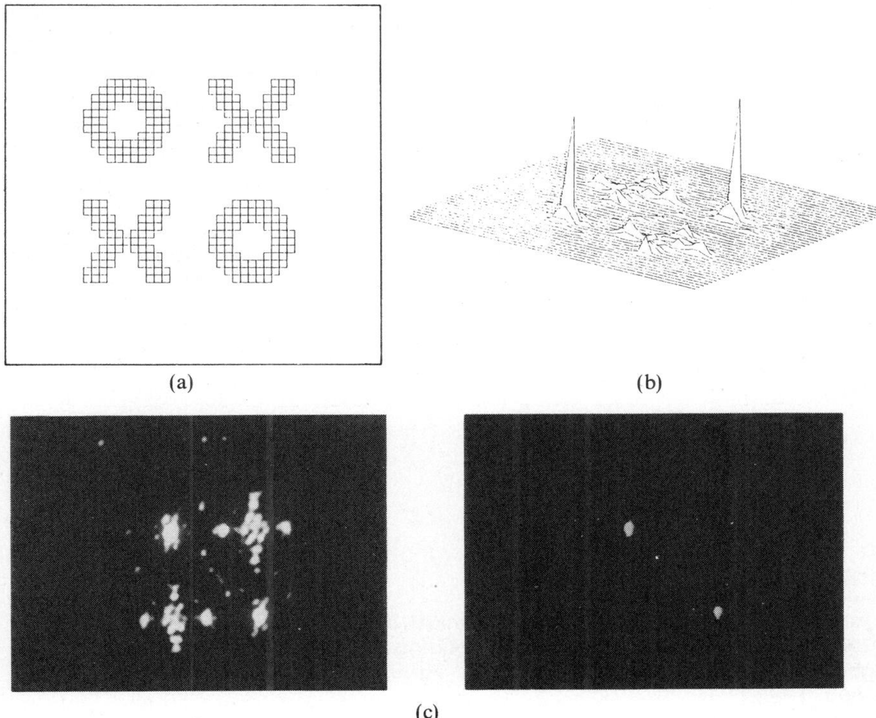

(a) (b) (c)

FIGURE 8.23

(a) The input having two Xs, (b) expected correlation intensity pattern, and (c) correlation intensity pattern for (i) higher brightness setting and (ii) lower brightness setting. (D. L. Flannery, A. M. Biernacki, J. S. Loomis, and S. L. Cartwright. "Real-Time Coherent Correlator Using Binary Magneto-Optic Spatial Light Modulators at Input and Output Planes." ***Appl. Opt.*****, Vol. 25, p. 466, 1986.)**

pattern. A binary SLM connected to the CCD array can then read out the binarized interference intensity pattern. The pattern is thereafter subjected to an inverse Fourier transform operation to produce the correlation results at the output. The intensity distribution at the CCD is given by

$$I = |F(o)|^2 + |F(r)|^2 + \{F(o)\}\{F(r)\}^* e^{-j\phi} + \{F(o)\}^*\{F(r)\} e^{j\phi} \tag{8.122}$$

where o and r represent the input object and the reference signals, respectively, and ϕ is a function of the spatial frequency coordinates, wavelength, and focal length. An inverse Fourier transform of this intensity distribution produces correlation results at the output. In this system, the intensity is thresholded at the CCD before the intensity distribution is retransformed by the second

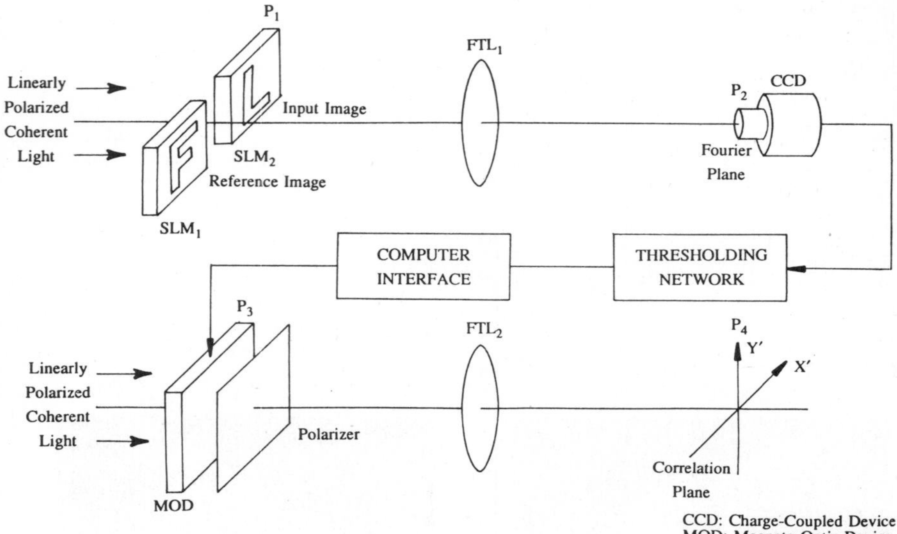

FIGURE 8.24

A joint transform image correlator system. (B. Javidi, and C. J. Kuo. "Joint Transform Image Correlator Using a Binary Spatial Light Modulator at the Fourier Plane." *Appl. Opt.*, Vol. 27, p. 663, 1988.)

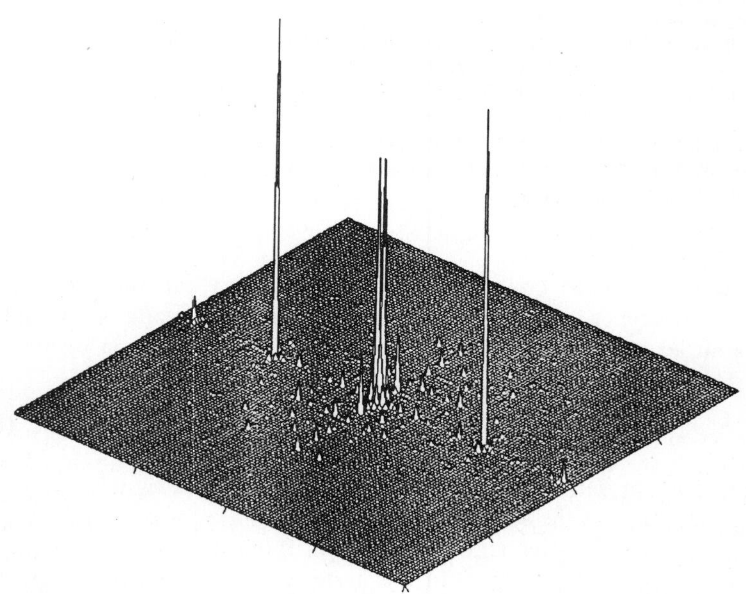

Figure 8.25(a)

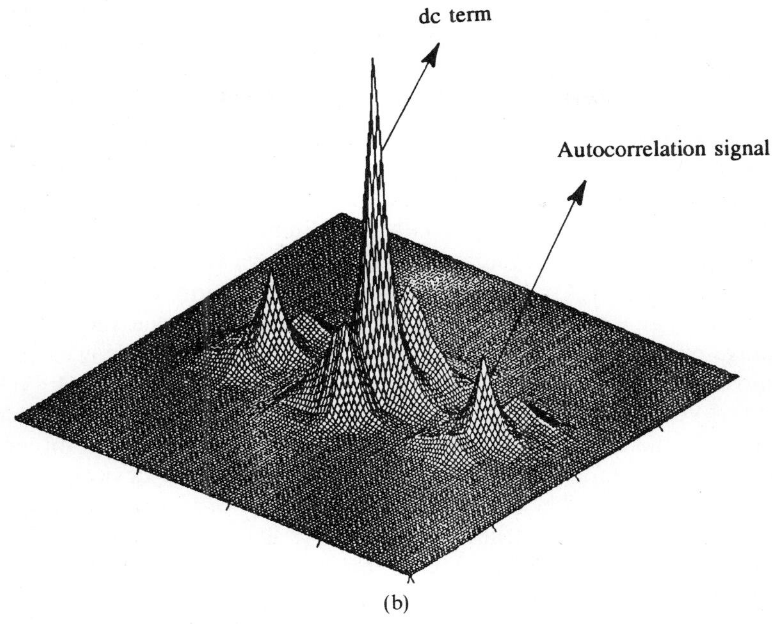

FIGURE 8.25

Autocorrelation results when characters F are used as input signals using (a) thresholded binarized intensity and (b) nonbinarized intensity. (B. Javidi, and C. J. Kuo. "Joint Transform Image Correlator Using a Binary Spatial Light Modulator at the Fourier Plane." *Appl. Opt.*, Vol. 27, p. 663, 1988.)

Fourier lens. Figure 8.25 shows the comparative results of using a nonbinarized and thresholded binary intensity pattern at the Fourier plane. The relative intensity values between the two justify the fact that binarizing the intensity not only does not degrade the performance of the correlator but also results in better discrimination.

8.9 Optical Computing Systems

A more recent area of activity involving electro-optics involves the development of devices and systems aimed at achieving computation capability. The main objective is to use photons, instead of electrons, in logical and arithmetic computations. In Chapter 7, we introduced a lensless shadow-casting scheme for realizing digital combinational logic operations, both simple and complex. In this section, we concentrate on Fourier-optics-based (with lens) systems that can be used for computation. It may seem odd that optical fields are being proposed for digital logic operations. Optical fields propagate mostly linearly, whereas the flow of electrons can be easily forced into discrete binary levels. However, there are enough advantages of an optical system to merit a flurry of recent activity in this field.

Optical sources have a large time bandwidth that approaches $\simeq 1$ GHz for some of the laser diodes. In addition, the involved optical systems have a large space bandwidth product. A multitude of resolvable elements can be made to operate in parallel. In the case of incoherent illumination, they are mutually independent; in the case of coherent illumination, they contribute toward realizing an extremely high degree of interconnectivity. In addition, two or more optical fields can propagate through each other without affecting one another. To take advantage of these benefits in manipulating data at high speeds, we can follow several routes: (a) devise gate levels units—e.g., AND, OR, NAND, NOR, X-OR, etc.; (b) devise set ups for executing entire algorithms; (c) devise systems for register level operations such as addition, subtraction, multiplication, etc.

The gate level operations may involve bistable optical devices or schemes like shadow-casting. The most mature level of operations involving the execution of an entire algorithm includes that of achieving a Fourier transform, for example, in a single pass through a thin lens. The register level schemes, on the other hand, can not only have multiple logic gates in combination but also combine with other similar subsystems to form robust systems.

Let us first explore a few of the schemes that can be used to achieve basic arithmetic operations between multielement, two-dimensional inputs. Consider two inputs, for example, described by $f_1(x, y)$ and $f_2(x, y)$, that are to be added or subtracted. An experimental set up such as the one illustrated in Fig. 8.26 can be used for such a purpose. The main beam is split into two portions by means of a beam splitter. These two beams are recombined by means of two suitably placed reflecting mirrors and a second beam splitter. The two two-dimensional inputs are inserted in these beams, and a phase shifter is introduced in one of the arms. The lens is able to combine the two input functions at the output plane. By adjusting the phase shifter, one can obtain either addition or subtraction. When the optical path difference caused by the phase shifter equals an even multiple of $\lambda/2$, the operation becomes additive, and if it equals an odd multiple of $\lambda/2$, the operation becomes subtractive. We can also use a system consisting of a holographic shear lens like the one shown in Fig. 8.27 for this purpose. The two image transparencies are inserted in the image plane of the coherent adder-subtracter system. The holographic lens shears and recombines the image at the output. To realize subtraction, we must tilt the shear lens at an angle so that the two optical paths differ by $\lambda/2$. This latter system is different from the former because here the input images are introduced side by side. Another variation of this system is shown in Fig. 8.28. The input images are illuminated by white light. A red filter is placed next to the upper input image, and a green filter is placed next to the lower input image. With proper alignment of the Wollaston prism, the input images can be combined to produce a pseudocolor difference image at the output. The location(s) where the two images have identical

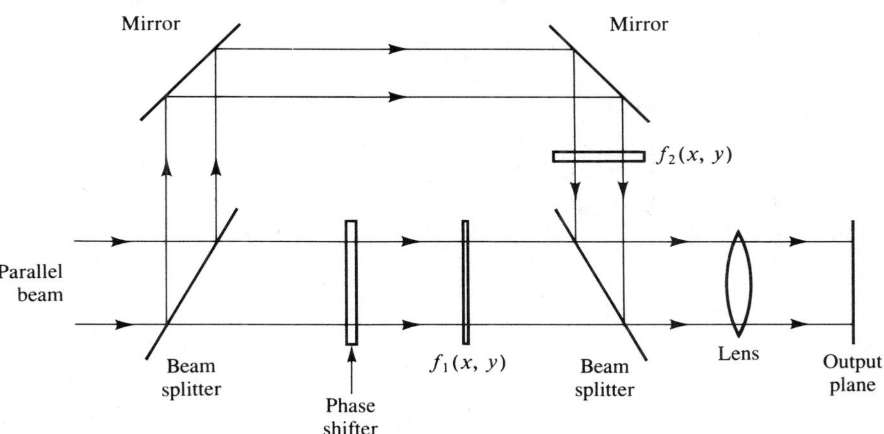

FIGURE 8.26

An optical adder-subtracter system with inputs placed at different locations

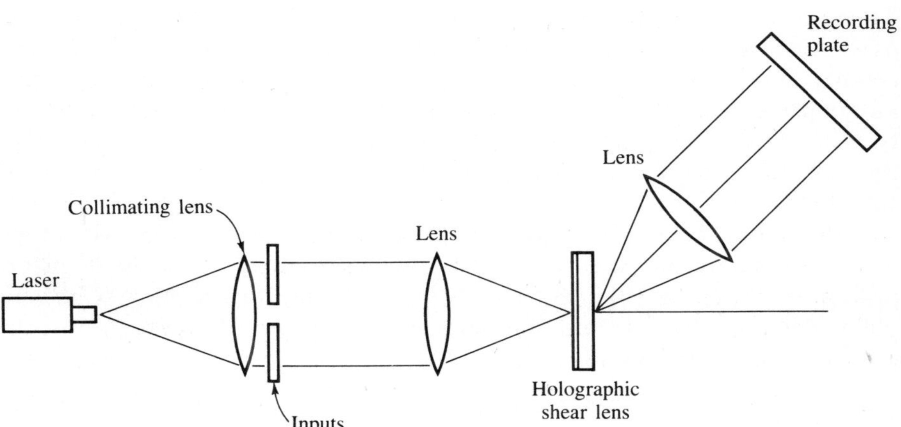

FIGURE 8.27

An optical adder-subtracter using a holographic shear lens

transmittances turn(s) out to be yellow at the output. A red region in the output indicates that the upper image has a higher transmittance at that location than that at the lower image. A green output indicates just the opposite.

Multiplication is relatively simple. The best way to accomplish it is to put the input transparencies in intimate contact. This can be a problem because if the two transparencies are moved with respect to one another, the plates can be

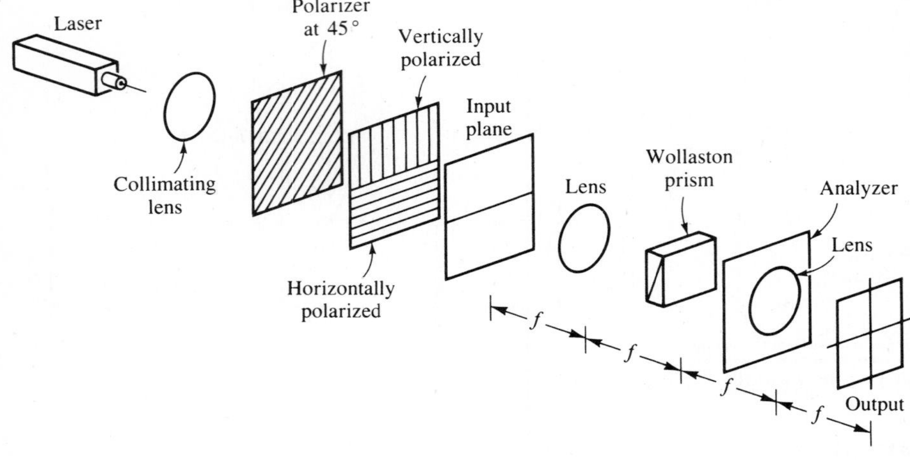

FIGURE 8.28

An optical adder-subtracter using a Wollaston prism

damaged substantially. In such a case, a system like the one in Fig. 8.29 can be used. An inverted real image of $f_1(x, y)$ is projected on $f_2(x, y)$ by the lens combination L_2 and L_3. Finally, lens L_4 projects the product of the two at the output plane. It should be noted that both addition and multiplication can also be realized using holographic means because the contributions proportional to the sum and product are both present in the holographic expression of Eq. (8.112).

The division operation can be expected to use the same set up as that of Fig. 8.27. However, for division, $[f_2(x, y)]^{-1}$, should be introduced at the P_2 plane instead of $f_2(x, y)$. Thus the problem of division reduces to a problem of how to obtain the reciprocal of the transmittance function $f_2(x, y)$. However, the inverse of $f_2(x, y)$ can be written as

$$1/f_2(x, y) = |f_2(x, y)|^{-2} f_2^*(x, y) \tag{8.123}$$

This implies that the inverse is obtainable by multiplying $|f_2(x, y)|^{-2}$ and $f_2^*(x, y)$. The first factor is, however, realizable by the simple photographic process as described by Eq. (4.67). By using a photographic film that has $\gamma = 2$, $|f_2(x, y)|^{-2}$ can be obtained. Again, the $f_2^*(x, y)$ factor can be realized by taking a hologram of $f_2(x, y)$. A multiplication unit such as the one just described can be thus used to realize $1/f_2(x, y)$, while a second multiplication unit can provide the result of division.

A class of optical processors, referred to as discrete, is finding prominence in matrix-vector multiplications. Such processors perform discrete—i.e., sampled—analogs of continuous-time linear operations. While these operations are discrete, they are analog in nature and are subject to the standard limitations

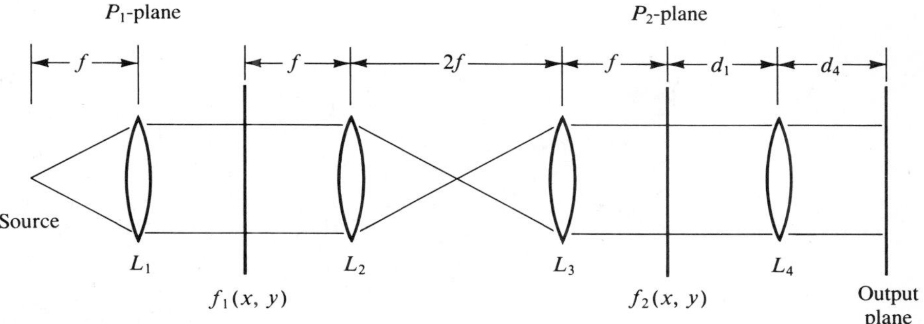

FIGURE 8.29

An optical multiplication unit

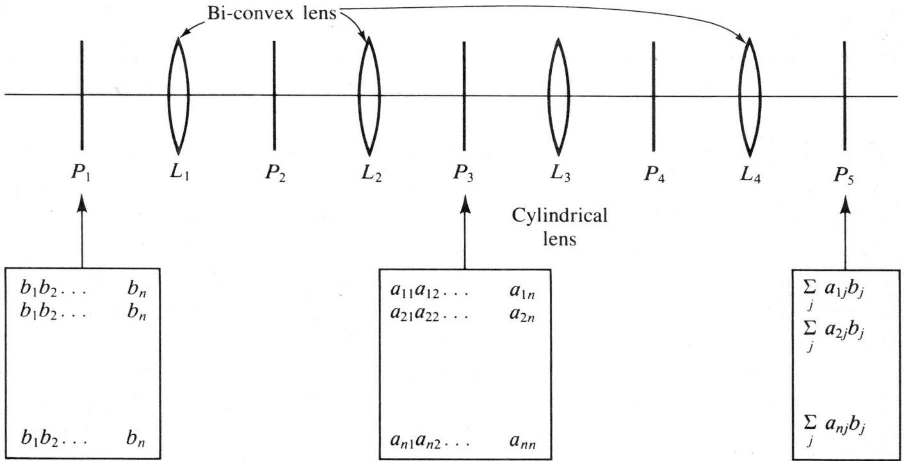

FIGURE 8.30

Parallel matrix-vector multiplication unit

on accuracy and dynamic range constraints. By making use of cylindrical lenses that have finite radii of curvature along one of the axes and infinite radii of curvature along the transverse axis, we can perform multiple one-dimensional operations simultaneously. Consider the matrix-vector multiplier as illustrated in Fig. 8.30. The elements of the incoming vector B are projected in plane P_1. This is accomplished by using an array of n LEDs and an optical system to spread the light from each of the LEDs out into a vertical column. These columns of light from each of the LEDs are then superimposed on a matrix mask representing the $m \times n$ matrix A. The product of the stored matrix and the

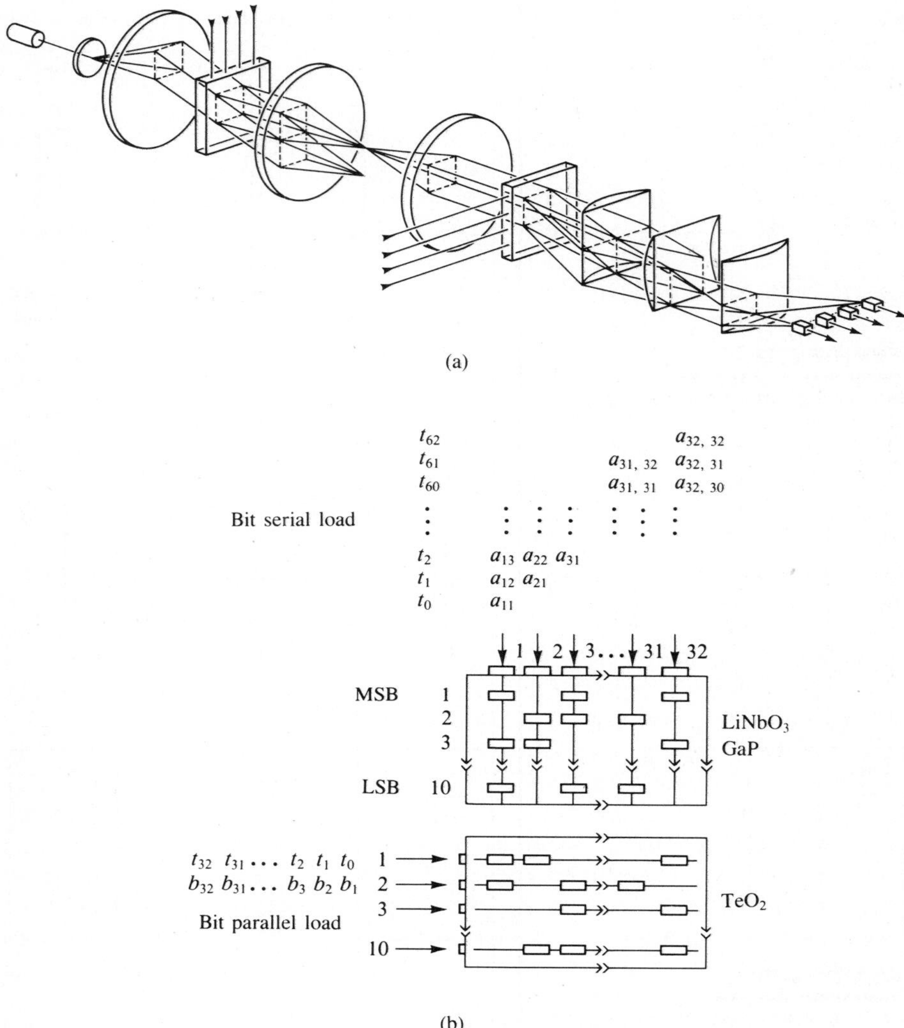

FIGURE 8.31

SAOBIC processor: (a) configuration and (b) its data flow scheme. (P. S. Guilfoyle. "Systolic Acousto-Optic Binary Convolver." *Opt. Engn.*, Vol. 23, p. 20, 1984.)

incoming vector is then Fourier transformed along only one of the directions by the cylindrical lens L_3, which is then Fourier transformed by the bi-convex lens L_4. Consequently, at the output plane P_5, we have along the vertical axis values representing the column vector of the product C such that its mth row consists of the sum of products $\sum a_{mj}b_j$. An array of n detectors may be placed along the vertical axis of P_5 to pick up the elements of the product C. This matrix-vector multiplier system is highly parallel and, consequently, very high data rates can be achieved.

Recently a number of optical-digital computing systems that are based solely on systolic filtering ideas have been proposed. This concept is based on the fact that multiplying two numbers is equivalent to convolving their digital representation, provided the result of the convolution is interpreted in mixed (nonbinary) format. Figure 8.31(a) shows one such matrix-vector multiplication unit, referred to as the **systolic A–O binary convolver** (SAOBIC). The processor consists of two multielectrode A–O devices, detectors, and three cylindrical lenses along with other standard components. The two A–O cells are imaged onto one another, and the cylindrical lenses are arranged in such a way that only light diffracted by the cells reaches the output plane. The number of electrodes in the A–O cells corresponds to the length of the columns of the matrix or vector. The operation of the processor can be described with the help of the data flow diagram of Fig. 8.31(b).

The SAOBIC system is appropriate for handling a 32-component vector-matrix product where each of the elements of the vector and matrix is represented by ten binary bits. Note that the binary bit stream of each matrix element is loaded into the A–O cell serially, whereas the vector is loaded into the other A–O cell in parallel. The instantaneous product from the entire vertical aperture is focused by a single cylindrical lens. All light transmitted by a particular column is collected by one specific detector at the output. Figure 8.31(b) also shows the timing (t_0 through t_{63}) that corresponds to the strobe cycles of the matrix-vector elements as they are continually introduced into the A–O cells. Obviously, the SAOBIC processorlike systems are capable of performing high-accuracy and high-speed matrix vector systolic operations. However, it is important to note that systolic systems result in nonbinary numbers that may have to be converted to the equivalent binary bits before being reused for further processing. The success of such systems thus depends on whether or not a total system (consisting of the systolic processor and a post-processor that converts mixed binary numbers to their binary equivalents) can compete well with the all-electronic systems.

Problems

8.1 Show that for a rectangular aperture of width a and length b, the intensity distribution due to a normally illuminated monochromatic plane wave (under Fraunhofer approximation) takes the form

$$I = I_0\left(\frac{\sin\alpha}{\alpha}\right)^2\left(\frac{\sin\beta}{\beta}\right)^2$$

where $\alpha = ka \sin \phi/2$, $\beta = kb \sin \theta/2$, and the angles ϕ and θ define the direction of the diffracted beam.

8.2 Determine the Fraunhofer intensity distribution function for a double-slit line aperture (of insignificant width) of length b each and separated by a center-to-center distance c. Discuss the consequence of such intensity distribution if the double slit is replaced by a single slit.

8.3 Show that for N-slit apertures, each of which has a length of b and is separated by a center-to-center distance h, the Fraunhofer intensity distribution function is given by

$$I = I_0\left(\frac{\sin\beta}{\beta}\right)^2\left(\frac{\sin N\gamma}{N\sin\gamma}\right)^2$$

where $\beta = kb\sin\theta/2$, $\gamma = kh\sin\theta/2$, and θ defines the direction of the diffracted beam. Plot a graph for the distribution when $N = 20$. Discuss the consequence of having very narrow slits.

8.4 Obtain and discuss the Fraunhofer intensity distribution function of a circular aperture of radius a when an opaque disk of radius b ($b < a$) is placed in its center to block a portion of the normally illuminated monochromatic plane beam.

8.5 Determine the Fraunhofer fractional energy contained in a circle of radius r at the observation plane due to a circular aperture of radius a which is illuminated by a monochromatic plane wave.

8.6 Determine the Fresnel intensity distribution function for a circular aperture in terms of Fresnel's integrals $C(x) = \int_0^x \cos(\pi y^2/2)\, dy$ and $S(x) = \int_0^x \sin(\pi y^2/2)\, dy$.

8.7 Calculate the field distribution in the back focal plane of a thin lens when an object with amplitude transmittance $t(x, y)$ placed a distance d away from the lens is illuminated by monochromatic plane wave.

8.8 Find the CTF and the OTF of a diffraction-limited square aperture of area d^2. Compare the corresponding cut-off spatial frequencies.

8.9 Prove that (a) $\mathcal{Q}(-\nu_x, -\nu_y) = \mathcal{Q}^*(\nu_x, \nu_y)$ and (b) $|\mathcal{Q}(\nu_x, \nu_y)| \leqslant |\mathcal{Q}(0, 0)|$.

8.10 Consider a circular camera shutter with radius a. The distance between the lens and the film plate is d, while the wavelength of light exposing the film is λ. Determine the OTF and sketch it.

8.11 The intensity transmittance of an object is given by $T(x_0, y_0) = \cos^2(2\pi\nu x_0)$, where $\nu_0/2 < \nu < \nu_0$ and ν_0 is the coherent cut-off frequency. Compare the Fourier transforms of the output intensity and comment on the performance of the coherent and the incoherent systems.

8.12 Show that $F\{F\{f(x, y)\}\} = F^{-1}\{F^{-1}\{f(x, y)\}\} = f(-x, -y)$.

8.13 Verify Eq. (8.38).

8.14 Verify Eq. (8.51).

8.15 Verify Eqs. (8.59a) and (8.59b).

8.16 Verify Eq. (8.62).

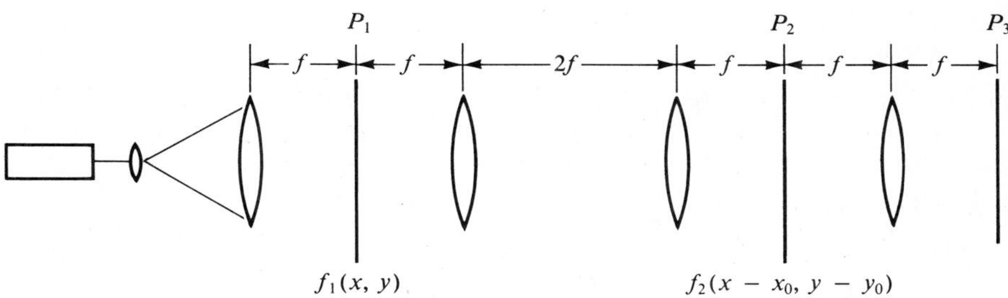

FIGURE 8.P1

8.17 Determine the intensity distribution of the output when the input line diffraction grating (Eq. (8.53)) is subjected to a bandpass binary filter capable of passing only ± 1 diffraction orders. Comment on the periodicity of the output.

8.18 Verify Eq. (8.63).

8.19 Consider the coherent optical set up in Fig. 8.P1, where $f_1(x, y)$ and $f_2(x - x_0, y - y_0)$ are introduced at P_1 and P_2, respectively. Discuss the output obtained in P_3 as a function of x_0 and y_0 and comment on the applicability of this set up.

References and Suggested Readings

Athale, R. A., W. C. Collins, and P. D. Stilwell. "High Accuracy Matrix Multiplication with Outer Product Optical Processor." *Appl. Opt.* Vol. 22 (1983): 368.

Awwal, A. A. S., M. A. Karim, and S. R. Jahan. "Improved Correlation Discrimination Using an Amplitude-Modulated Phase-Only Filter." *Appl. Opt.* Vol. 29 (January 1990).

Baker, A. "Image Detection and Enhancement." *Appl. Opt.* Vol. 20 (1981): 3917.

Berg, N. J., and J. N. Lee, (Ed.). *Acousto-Optic Signal Processing Theory and Implementation.* New York: Marcel Dekker, 1983.

Bocker, R. P., S. R. Clayton, and K. Bromley. "Electrooptical Matrix Multiplication Using the Twos Complement Arithmetic for Improved Accuracy." *Appl. Opt.* Vol. 22 (1983): 2019.

Bracewell, R. N. *The Fourier Transform and Its Applications* (2nd rev. ed.). New York: McGraw-Hill, 1986.

Caulfield, H. J., W. T. Rhodes, M. J. Foster, and S. Horvitz. "Optical Implementation of Systolic Array Processing." *Opt. Commun.* Vol. 40 (1981): 86.

Davis, J. A., T. Day, R. A. Lilly, and H. K. Liu. "Multichannel Optical Correlator/Convolver Utilizing the Magnetooptical Spatial Light Modulator." *Appl. Opt.* Vol. 26 (1987): 969.

Duffiex, P. M. *The Fourier Transform and Its Application to Optics* (2nd ed.). New York: Wiley, 1983.

Flannery, D. L., A. M. Biernacki, J. S. Loomis, and S. L. Cartwright. "Real Time Coherent Correlator Using Binary Magneto-Optic Spatial Light Modulator at Input and Fourier Planes." *Appl. Opt.* Vol. 25 (1986): 466.

Fowles, G. R. *Introduction to Modern Optics.* New York: Holt, Rinehart & Winston, 1975.

Gaskill, J. D. *Linear Systems, Fourier Transforms, and Optics.* New York: Wiley, 1978.

Gerritsen, H. J., and M. L. Jepsen. "Holographic Recordings of Nematic LC Displays." *Appl. Opt.* Vol. 26 (1987): 3717.

Goodman, J. W. *Introduction to Fourier Optics.* New York: McGraw Hill, 1968.

Goutzoulis, A. P. "Systolic Time-Integrating Acoustooptic Binary Processors." *Appl. Opt.* Vol. 23 (1984): 4095.

Guilfoyle, P. S. "Systolic Acousto-Optic Binary Convolver." *Opt. Engn.* Vol. 23 (1984): 20.

Gregory, D. A., and L. L. Huckabee. "Acoustooptically Addressed Fourier Transform Matched Filtering." *Appl. Opt.* Vol. 24 (1985): 859.

Jackson, J., and D. Casasent. "Optical Systolic Array Processor Using Residue Arithmetic." *Appl. Opt.* Vol. 22 (1983): 2817.

Javidi, B., and C. J. Kuo. "Joint Transform Image Correlation Using a Binary Spatial Light Modulator at the Fourier Plane." *Appl. Opt.* Vol. 27 (1988): 663.

Johnson, K. M., M. Armstrong, L. Hesselink, and J. W. Goodman. "Multiple Multiple-Exposure Hologram." *Appl. Opt.* Vol. 24 (1985): 4467.

Karim, M. A., and H. K. Liu. "All-Optical Homomorphic Image-Processing System." *Opt. Lett.* Vol. 7 (1982): 371.

Karim, M. A., and H. K. Liu. "Linear Versus Logarithmic Spatial Filtering in the Removal of Multiplicative Noises." *Opt. Lett.* Vol. 6 (1981): 207.

Karim, M. A., A. M. Hanafi, F. Hussain, S. Mustafa, Z. Samberid, and N. M. Zain. "Realization of a Uniform Circular Source Using a Two-Dimensional Binary Filter." *Opt. Lett.* Vol. 10 (1985): 470.

Kuehls, J. F., A. E. Attard, and V. B. Burke. "Optical Correlator Using a Thin-Film Traveling-Wave Light Modulator." *Appl. Opt.* Vol. 24 (1985): 3842.

Liang, Y. Z., and H. K. Liu. "Optical Matrix-Matrix Multiplication Method Demonstrated by the Use of a Multifocus Hololens." *Opt. Lett.* Vol. 9 (1984): 322.

Lohmann, A. H., and J. Weigelt. "Digital Optical Adder Based on Spatial Filtering." *Appl. Opt.* Vol. 25 (1986): 3047.

Lohmann, A. H., and J. Weigelt. "Spatial Filtering Logic Based on Polarization." *Appl. Opt.* Vol. 26 (1987): 131.

Marom, E. "Real-Time Image Subtraction Using a Liquid Crystal Light Valve." *Opt. Engn.* Vol. 25 (1986): 274.

Mu, G. G., C. K. Chiang, and H. K. Liu. "White-Light Image Addition and Subtraction by Colorimetric Measurement." *Opt. Lett.* Vol. 6 (1981): 389.

Perez, O., and M. A. Karim. "An Efficient Implementation of Joint Fourier Transform Correlation Using a Modified LCTV." *Microwave & Opt. Technol. Lett.* Vol. 2 (1989): 193.

Perlee, C., and D. Casasent. "Negative Base Encoding in Optical Linear Algebra Processors." *Appl. Opt.* Vol. 25 (1986): 168.

Pinson, L. J. *Electro-Optics.* New York: Wiley, 1985.

Psaltis, D. "Optical Image Correlation Using Acoustic and Charge-Coupled Devices." *Appl. Opt.* Vol. 21 (1982): 491.

Psaltis, D. "Incoherent Electrooptic Image Correlator." *Opt. Engn.* Vol. 23 (1984): 12.

Rao, V. V., C. Joenathan, and R. S. Sirohi. "Real-Time Optical Image Subtraction by a Holographic Shear Lens." *Opt. Commun.* Vol. 55 (1985): 83.

Steward, E. G. *Fourier Optics: An Introduction.* United Kingdom: Ellis Horwood Limited, 1983.

Taylor, B. K., and D. P. Casasent. "Twos-Complement Data Processing for Improved Encoded Matrix-Vector Processors." *Appl. Opt.* Vol. 25 (1986): 956.

Verber, C. M., R. P. Kenan, and J. R. Busch. "Correlator Based on an Integrated Optical Spatial Light Modulator." *Appl. Opt.* Vol. 20 (1981): 1626.

Yu, F. T. S. *Optical Information Processing.* New York: Wiley, 1983.

Yu, F. T. S. *White-Light Optical Signal Processing.* New York: Wiley, 1985.

Yu, F. T. S., and M. S. Dymek. "Optical Information Parallel Processing: A Technique." *Appl. Opt.* Vol. 20 (1981): 1450.

Yu, F. T. S., S. Jutamulia, and D. A. Gregory. "Real-Time Liquid Crystal TV XOR- and XNOR-Gate Binary Image Subtraction Technique." *Appl. Opt.* Vol. 26 (1987): 2738.

Zhao, D., and H. K. Liu. "Real-Time White-Light Pseudocolor Image-Difference Detection Technique." *Opt. Lett.* Vol. 8 (1983): 99.

Zhao, D., C. C. Chiang, and H. K. Liu. "Contact Screen Image Subtraction Technique." *Appl. Opt.* Vol. 20 (1981): 4234.

CHAPTER NINE

Fiber-Optics-Based Devices and Systems

9.1 Introduction

The previous chapter dealt primarily with systems based on Fourier transforms. Basically, the Fourier optics-based systems are information processors, but they involve the transmittal of information over short distances. In this chapter, however, we study systems and devices that are essential for signal transmission over longer distances. The availability of reliable laser sources, optical fibers, and detectors provides an attractive alternative to conventional coaxial cables for communication applications. High fiber losses of the earlier days initially seemed to exclude this alternative from competition with the traditional means of electronic communication, but at present optical fiber technology has advanced to a point where relatively low-loss fibers can be drawn without any significant problem. In fact, fibers with an absorption coefficient as low as 0.2 dB/km can now be manufactured for transmission wavelengths in the neighborhood of 1.55 μm.

An optical fiber communication system is similar in concept to any type of a communication system. A block diagram of a typical system is shown in Fig. 9.1. The input data are usually coded by means of a current pulse network that can directly modulate the light source. The optical power that is ultimately received by the photodiode at the other end of the optical fiber is substantially reduced from its initial value because of a variety of loss mechanisms. To maintain reliable operation, therefore, we must use sufficient input power to overcome the losses.

In many situations, the advantages of fiber optics are so compelling that they cannot be ignored. Some of these are: (a) extremely wide bandwidth, (b) small size and weight, (c) immunity to cross-talk, (d) immunity to inductive

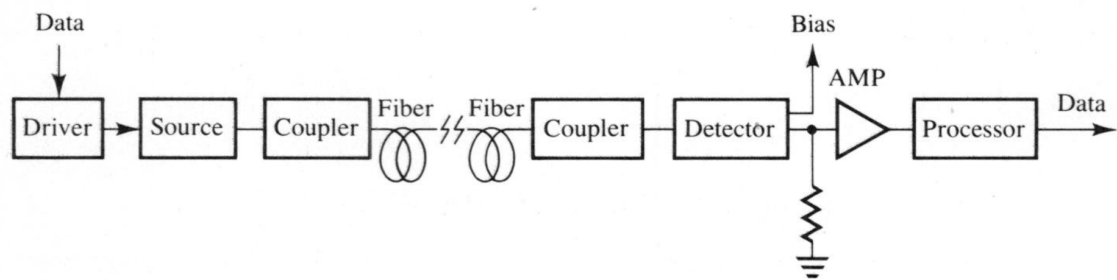

FIGURE 9.1

A typical fiber-optic transmission system

interference, (e) low cost, and (f) low transmission losses. There are also several important secondary advantages that allow: (a) signal security, (b) greater safety, (c) longer life-span, (d) greater reliability and ease of maintenance, (e) ruggedness and flexibility, (f) system expandability, and (g) nonradiative signal transmission.

Theoretically, the bandwidth of a standard coaxial cable can go up to around 500 MHz. By comparison, the optical carrier frequency, in the frequency range of 10^{13} and 10^{16} Hz, yields a far greater transmission bandwidth. With a suitable combination of source and fiber link, it is possible to have a transmission capacity in excess of 100 GHz-km. With high enough bandwidths, it is possible to transmit thousands of voice signals or dozens of video signals together over the same channel. It is anticipated that the fiber system bandwidth can be extended even further to provide an information-carrying capacity far in excess of what is realizable using coaxial cables. Table 9.1 lists and compares the characteristics of major metallic conductors. Note that while a twisted pair of wires can have an attenuation slightly less than that of a fiber, its bandwidth is considerably lower. Further, Fig. 9.2 illustrates the fact that unlike the metallic conductors, the attenuation of fibers is independent of data rate.

One fiber-optic filament, only a few thousandths of an inch in diameter, can effectively replace a copper wire cable that is several inches in diameter. Even when optical fibers are covered with protective sheaths, they are far smaller and lighter than equivalent coaxial cables. Such an improvement in size and weight is especially important for underwater cables, aircraft, satellites, and in systems with over-crowded transmission wires.

Optical transmission is immune to ambient electrical noise, ringing, echoes, electromagnetic interference (EMI), radio-frequency interference (RFI), and switching transients that causes electromagnetic pulses (EMP). In addition, optical transmission produces no electric noise of its own. A fiber cable is not susceptible to lightning when used above ground. A fiber-optic transmission system thus does not present any of the traditional problems of cross-talk. In addition, the number of repeaters used in systems with low attenuation fiber cables is far less than that used in conventional systems. In particular, for short hauls of less than 10 km, no repeaters are necessary. Fiber-optic systems thus

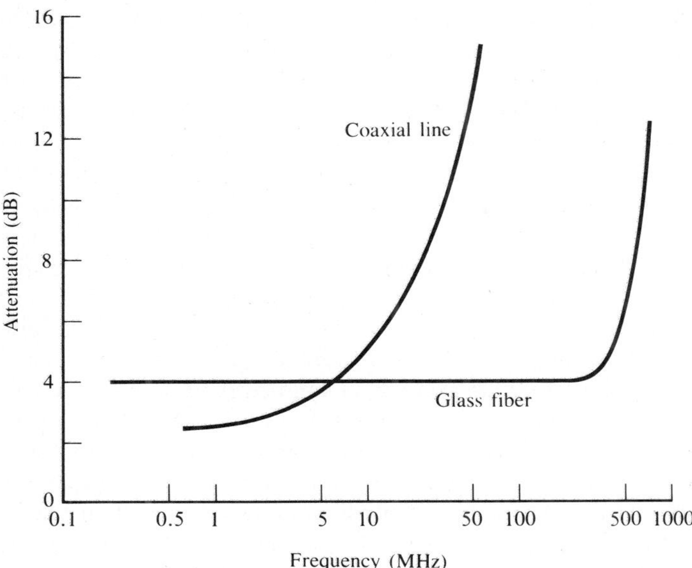

FIGURE 9.2

Typical attenuation of coaxial and silica fibers. (Adapted from J. C. Palais, *Fiber Optic Communications*. Englewood Cliffs, NJ: Prentice-Hall, 1988, p. 21. © 1988 by Prentice-Hall. Credit Alpha Wire Corporation.)

TABLE 9.1

Different Transmission Schemes

Type	*Frequency*	*Channel Capacity*	*Repeater Spacing or Attenuation*
Wire pairs	1–140 kHz	up to 240	0.1–0.3 dB/km
Multipair cable	12–250 kHz	up to 1400	3 dB/km
Circular waveguide	50 GHz	several hundred thousand	0.5–4 dB/km
Coaxial cable	64–3096 kHz	1800	12.6 km
	308–832 kHz	16740	6.4 km
	0.5–17 MHz	32400	3.2 km
	3–51 MHz	90000	1.6 km
Optical fiber	100000 GHz	several hundred thousand	0.2 dB/km

provide a significant cost advantage in long haul links. However, in the case of short haul applications where additional cost is incurred due to electrical-to-optical conversion and vice versa, the cost advantage may not be so attractive.

In general, light signals propagating through optical fibers do not radiate outside the cable, so the only way to eavesdrop into a link is to couple a tap directly into the fiber. If an eavesdropper were to do this, he or she would cause a great power loss that could be detected easily using a device referred to as an optical time-domain reflectometer (OTDR). An OTDR can transmit an optical pulse down the fiber and subsequently measure the reflected signals. Accordingly, we can identify and locate discontinuities that are present in the fiber. Besides the consequential immunity to wiretapping, optical fibers also provide greater safety because there is no short-circuit spark even when the fiber-optic cable is damaged severely. In fact, fibers can be repaired in the field even when the equipment is turned on. In addition, the silica that is used for fabricating optical fibers is not only abundant but also has an extremely high tolerance to temperature fluctuations as well as to liquids and corrosive gases.

The upgradability of a fiber-optic communications system refers to possibility of adapting the system to suit the available data rate and incorporating newer and more reliable system components. Often it may be necessary to improve the attenuation properties of the cable or reduce the losses at the splices, connectors, and couplings by means of new equipment and/or design. Most of the low-loss, fiber-optic cables made today can be upgraded by improved modulation techniques and wavelength multiplexing without replacing the original cable. In particular, this is accomplished by replacing LEDs with injection lasers, and/or by replacing *p–i–n* photodiodes with an APD detector, or simply by operating the system at a longer wavelength.

The use of fibers for optical communication does indeed have some drawbacks. These include (a) connector difficulties that are caused by the physical fiber sizes, (b) the requirement of independent electrical power feed for the repeaters, (c) complicated and expensive test procedures, and (d) the fragility of the bare fibers. Continuing research and development and experience with fiber-based systems are expected to eliminate or reduce many of these problems.

The advantages, however, far outnumber the disadvantages and make optical fiber transmission a very attractive alternative for telecommunication networks. The technology that once was based on multimode step-index fibers operating in shorter wavelength regions (0.8–0.9 μm) has now progressed to produce systems that utilize multimode graded-index fibers operating in both the shorter and longer wavelength region (0.8–1.6 μm). Recently single-mode, fiber-based systems suitable for operation in the longer wavelength region (1.1–1.6 μm) have been adopted for commercial telecommunications networks. Currently, newer systems based both on coherent transmission and integrated optics are being developed and studied. Work involving nonlinear pulse propagation in optical fibers is now being done to provide increased channel capacity.

The source and the detectors were discussed extensively in Chapters 4 and 5.

Several schemes for modulating signals were considered in Chapter 7. To complete the study of fiber-optic communication systems, therefore, we concentrate on the characteristics of optical fibers in this chapter. Following an initial discussion on the various aspects of fibers and the mechanisms of various losses that are encountered during transmission, we return to consider a complete fiber-optic system.

9.2 Optical Waveguides

The most significant of the fiber-optics components is the optical fiber itself, also referred to as an **optical waveguide**. It is generally cylindrical in shape as shown in Fig. 9.3, and it consists of a cylindrical core of refractive index n_1 that is relatively higher than the refractive index n_2 of the annular cladding. Such an optical fiber is referred to as a **step-index fiber** because both n_1 and n_2 are uniform in the core and the cladding regions, respectively. Some of the other variations of refractive index profiles for the fibers are illustrated in Fig. 9.4. For clarity, let us consider the propagation of light rays. Assuming that the refractive index of air is unity, the angle of refraction θ_r of the meridional ray can be related (see Eq. (1.33)) to the angle of incidence θ_i at the air-core interface (entrance) by

$$\sin\theta_i = n_1 \sin\theta_r \tag{9.1}$$

The angle of incidence θ' at which the refracted ray strikes the core-clad interface is thus $\frac{1}{2}\pi - \theta_r$. The ray will continue to propagate through the core provided it is totally reflected at the core-clad interface. But in order to achieve the condition for total internal reflection, the minimum value of θ', represented by θ'_{min}, is given by

$$n_1 \sin(\theta'_{\text{min}}) = n_2 \tag{9.2}$$

Thus Eq. (9.1) can be used to evaluate the maximum permissible value of θ_i, represented by $\theta_{i,\text{max}}$ as

$$\begin{aligned}\sin\theta_{i,\text{max}} &= n_1 \sin\left[\frac{\pi}{2} - \theta'_{\text{min}}\right] \\ &= n_1[1 - \sin^2\theta'_{\text{min}}]^{1/2} \\ &= n_1\left[1 - \left(\frac{n_2}{n_1}\right)^2\right]^{1/2} = [n_1^2 - n_2^2]^{1/2} \simeq n_1(2\Delta)^{1/2}\end{aligned} \tag{9.3}$$

where $\Delta \equiv 1 - (n_2/n_1)$ is often referred to as the core-cladding index difference or, simply, the **index difference**. The maximum angle of acceptance for which light propagates along the core is referred to as the **numerical aperture** (NA).

The electromagnetic wave is required to undergo a constructive interference with itself in order for it to propagate effectively along a step-index fiber. The total phase change encountered by a wave that has been reflected twice thus has

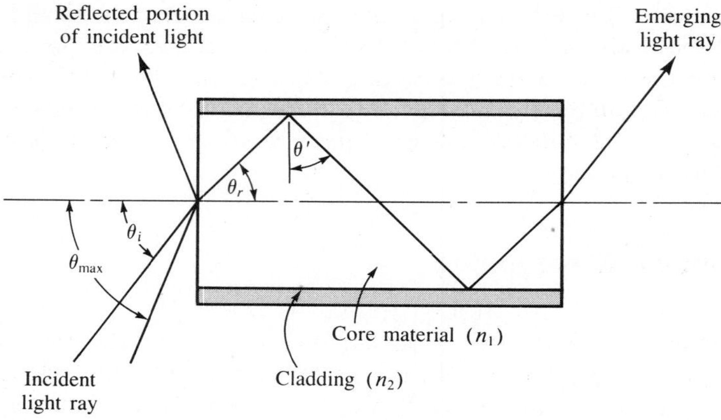

FIGURE 9.3

Light propagation in the fiber

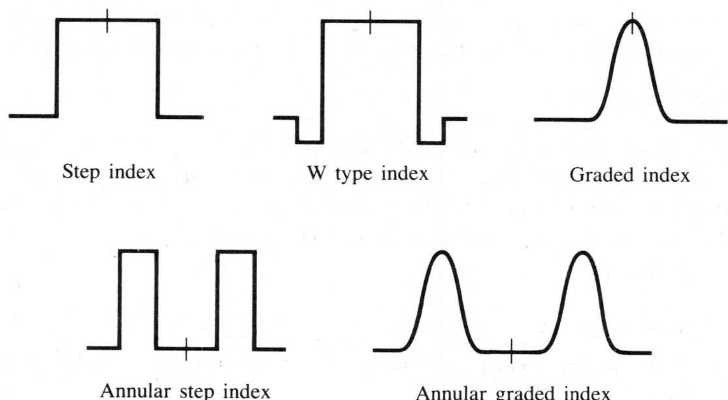

FIGURE 9.4

The refractive index profiles of the fibers

to be an integer multiple of 2π. Note there are two different factors that contribute to the total phase delay: path difference and reflection. The phase change δ, owing to reflection only, is given by

$$\tan\left(\frac{\delta}{2}\right) = \begin{cases} \dfrac{\sqrt{n_1^2 \cos^2\theta' - n_2^2}}{n_1 \sin\theta'} \\[2ex] \dfrac{\sqrt{n_1^2 \cos^2\theta' - n_2^2}}{n_2 \sin\theta'} \end{cases} \tag{9.4}$$

respectively for the wave components normal and parallel to the reflecting surfaces. Thus the condition for the best waveguiding is given by

$$\frac{8\pi n_1 r}{\lambda \sin\theta'} + 2\delta = 2\pi m \tag{9.5}$$

where m is an integer, λ is the wavelength of the propagating light, and r is the radius of the fiber core.

An optical signal becomes increasingly distorted as it travels through the fiber core. The input optical energy is distributed among beams of various quantized angles. The relative difference in propagation durations amounts to a spread in the pulse width. This phenomenon, referred to as **dispersion**, determines the maximum rate at which the pulses can be transmitted through optical waveguides. In order for two successive pulses to be resolvable, the pulse spread Δt should be less than the minimum pulse width. Under normal circumstances, the pulse spread is maintained at a value less than 75% of the pulse width. The reciprocal of the minimum pulse width is often referred to as the **bit rate of transmission.**

Consider the step-index multimode fiber of Fig. 9.3. For an incoming light incident at the air-core interface at an angle of θ_i, the total geometric propagation length per unit length of fiber core is given by $(\cos\theta_r)^{-1}$. The time taken by light to travel a unit length of fiber is often referred to as the **group delay**. The maximum variation $\Delta\tau$ in the group delay is obtained by comparing the light beams incident at 0° with those incident at θ_i and is given by

$$\Delta\tau = \frac{n_1}{c}\left(\frac{1}{\cos\theta_r} - 1\right) \simeq \frac{n_1}{2c}\theta_r^2 \tag{9.6}$$

when θ_i is very small. Using Eqs. (9.1) and (9.3) we can rewrite, however, Eq. (9.6) as

$$\Delta\tau \simeq \frac{(\text{NA})^2}{2cn_1} \tag{9.7}$$

The modal pulse spread $\Delta\tau$ of the multimode step-index fiber, thus, is found to vary as the square of the numerical aperture. The value of the numerical aperture can be decreased, but at the cost of having a very small acceptance angle that limits the system to a low power input. The bit rate of transmission is inversely proportional to $(\text{NA})^2$. This type of mode dispersion, as described by Eq. (9.7), is independent of whether or not the propagating light is monochromatic. Two more types of dispersion are associated with propagating nonmonochromatic light—material dispersion and waveguide dispersion. Whereas the former is caused by the variation of the refractive index with the wavelength, the latter is related to the change of path length with the wavelength. The pulse spreading that is caused by the spread in wavelength is

given by

$$\Delta\tau = \frac{d\tau}{d\lambda}\Delta\lambda = \frac{d}{d\lambda}\left[\frac{n_1}{c}\frac{1}{\cos\theta_r}\right]\Delta\lambda$$

$$= \frac{1}{c\cos\theta_r}\frac{dn_1}{d\lambda}\Delta\lambda + \frac{n_1}{c}\frac{d}{d\lambda}(\cos\theta_r)\,\Delta\lambda \tag{9.8}$$

The first term of Eq. (9.8) refers to material dispersion, while the second term corresponds to waveguide dispersion. Note that these two dispersions are related to one another.

There are three possible ways to reduce fiber dispersions. The first approach involves using a single-mode fiber—i.e., one with a very small fiber core—as this can eliminate mode dispersion. The second approach involves using a source with a narrow spectral width that minimizes both material and waveguide dispersions. The third alternative involves restricting the transmission to a particular wavelength region such that the material dispersion can cancel out the effect of the waveguide dispersion. However, to attain a clearer understanding of optical propagation in a fiber, we investigate the solutions of Maxwell's equations by subjecting them to the pertinent boundary conditions. The following subsections are devoted to these issues. The treatment in the earlier subsections is limited to the most simple fibers. In the last two subsections, however, the analysis is sufficiently general to include graded-index optical fibers.

9.2.1 Cylindrical Waveguide Theory

Refractive index profiles of most of the optical fibers are circularly symmetric, and hence it is easy to develop the solutions for Maxwell's equations in the cylindrical coordinate system $\{r, \phi, z\}$. The electric and magnetic fields in the cylindrical waveguides are expected to be of the form

$$\mathbf{E} = \mathbf{E}(r, \phi)\,e^{j(\omega t - \beta z)} \tag{9.9a}$$

$$\mathbf{H} = \mathbf{H}(r, \phi)\,e^{j(\omega t - \beta z)} \tag{9.9b}$$

where β is the component of the propagation vector along z direction. Substituting the expressions for the fields into Maxwell's equations for a nonconducting medium as given by Eqs. (1.1) and (1.2), we obtain two sets of three equations respectively as

$$\frac{1}{r}\left[\frac{\partial E_z}{\partial\phi} + jr\beta E_\phi\right] = -j\omega\mu H_r \tag{9.10a}$$

$$j\beta E_r + \frac{\partial E_z}{\partial r} = j\omega\mu H_\phi \tag{9.10b}$$

$$\frac{1}{r}\left[\frac{\partial}{\partial r}(rE_\phi) - \frac{\partial E_r}{\partial\phi}\right] = -j\omega\mu H_z \tag{9.10c}$$

and

$$\frac{1}{r}\left[\frac{\partial H_z}{\partial \phi} + j\beta r H_\phi\right] = j\omega\varepsilon E_r \quad \textbf{(9.11a)}$$

$$j\beta H_r + \frac{\partial H_z}{\partial r} = -j\omega\varepsilon E_\phi \quad \textbf{(9.11b)}$$

$$\frac{1}{r}\left[\frac{\partial}{\partial r}(rH_\phi) - \frac{\partial H_r}{\partial \phi}\right] = j\omega\varepsilon E_z \quad \textbf{(9.11c)}$$

where ε and μ represent, respectively, electrical permittivity and magnetic permeability (or susceptivity) of the medium. By eliminating appropriate variables, we can rewrite Eqs. (9.10a) and (9.10b) to provide E_r, E_ϕ, H_r, and H_ϕ in terms of E_z and H_z. Equations describing these relationships are given by

$$E_r = -\frac{j\beta}{k_r^2}\left[\frac{\partial E_z}{\partial r} + \frac{\omega\mu}{\beta r}\frac{\partial H_z}{\partial \phi}\right] \quad \textbf{(9.12a)}$$

$$E_\phi = -\frac{j\beta}{k_r^2}\left[\frac{1}{r}\frac{\partial E_z}{\partial \phi} - \frac{\omega\mu}{\beta}\frac{\partial H_z}{\partial r}\right] \quad \textbf{(9.12b)}$$

$$H_r = -\frac{j\beta}{k_r^2}\left[\frac{\partial H_z}{\partial r} - \frac{\omega\varepsilon}{\beta r}\frac{\partial E_z}{\partial \phi}\right] \quad \textbf{(9.13a)}$$

$$H_\phi = -\frac{j\beta}{k_r^2}\left[\frac{1}{r}\frac{\partial H_z}{\partial \phi} + \frac{\omega\varepsilon}{\beta}\frac{\partial E_z}{\partial r}\right] \quad \textbf{(9.13b)}$$

where k_r represents the radial component of the propagation vector such that

$$k_r = \sqrt{\omega^2\mu\varepsilon - \beta^2} \quad \textbf{(9.14)}$$

Equations (9.12a) and (9.12b) can be substituted in Eq. (9.10c) and, similarly, Eqs. (9.13a) and (9.13b) can be substituted in Eq. (9.11c) to provide the modified wave equations in terms of the cylindrical coordinates:

$$\left[\frac{\partial^2}{\partial r^2} + \frac{1}{r}\frac{\partial}{\partial r} + \frac{1}{r^2}\frac{\partial^2}{\partial \phi^2} + k_r^2\right]\begin{pmatrix} E_z \\ H_z \end{pmatrix} = 0 \quad \textbf{(9.15)}$$

An equation such as this one is often solved by the method of separation of variables. Moreover, we need to solve for only one of the fields because the form of the equations of both E_z and H_z is identical. We assume that

$$E_z(\phi, r) = \Phi(\phi)R(r) \quad \textbf{(9.16)}$$

But, since the fiber has a circular symmetry, it is reasonable to rewrite Eq. (9.16) as

$$E_z(\phi, r) = R(r)e^{jq\phi} \quad \textbf{(9.17)}$$

where q is any integer. Using Eq. (9.17), Eq. (9.15) reduces to

$$\frac{d^2R}{dr^2}+\frac{1}{r}\frac{dR}{dr}+\left(k_r^2-\left[\frac{q}{r}\right]^2\right)R=0 \tag{9.18}$$

It may be noted that the standard Bessel equation is generally expressed in the form of

$$\zeta^2\frac{d^2P}{d\zeta^2}+\zeta\frac{dP}{d\zeta}+(\zeta^2-m^2)P=0 \tag{9.19}$$

Its most general solution, where m is an integer, is given by

$$P(\zeta)=c_1J_m(\zeta)+c_2N_m(\zeta) \tag{9.20}$$

where c_1 and c_2 are two constants, the Bessel function of the first kind of order m is defined as

$$J_m(\zeta)=\sum_{l=0}^{\infty}\frac{(-1)^l\zeta^{2l+m}}{2^{2l+m}l!\,\Gamma(m+l+1)} \tag{9.21}$$

the gamma function for $p>0$ is given by

$$\Gamma(p)=\int_0^{\infty}t^{p-1}e^{-t}\,dt \tag{9.22}$$

and the Bessel function of the second kind, also referred to as the Neumann function, is given by

$$N_m(\zeta)=\frac{J_m(\zeta)\cos m\pi-J_{-m}(\zeta)}{\sin m\pi} \tag{9.23}$$

Note that $\Gamma(n+1)=n\Gamma(n)$ and $\Gamma(1)=0$. It may be also noted that by replacing ζ with $j\eta$ in Eq. (9.19), we can obtain the modified Bessel equation

$$\eta^2\frac{d^2P}{d\eta^2}+\eta\frac{dP}{d\eta}-(\eta^2+m^2)P=0 \tag{9.24}$$

whose general solution is given by

$$P(\zeta)=c_3I_m(\zeta)+c_4K_m(\zeta) \tag{9.25}$$

where the modified Bessel function of the first kind of order m is defined by

$$I_m(\zeta)=(j)^{-m}J_m(j\zeta) \tag{9.26}$$

the modified Bessel function of the second kind of order m is defined by

$$K_m(\zeta)=\tfrac{1}{2}\pi\frac{I_{-m}(\zeta)-I_m(\zeta)}{\sin m\pi} \tag{9.27}$$

and c_3 and c_4 are two arbitrary constants.

Clearly Eq. (9.18) has the form of a Bessel equation. But to obtain valid solutions, we may have to consider the geometry of the fiber once again. The fiber consists of a core and a cladding of refractive index n_1 and n_2, respectively, and of radii r_1 and r_2, respectively. Usually, the cladding radius r_2 is large enough so that the confined mode vanishes at $r = r_2$. Consequently, for the sake of our analysis, r_2 can be set equal to infinity. In addition, the field must be finite within the core, specifically at $r = 0$. However, it is appropriate to note that when $\zeta \ll 1$,

$$N_0(\zeta) = \frac{2}{\pi} \ln(\zeta) \tag{9.28}$$

and for all nonzero m

$$N_m(\zeta) = -\frac{(m-1)!}{\pi}\left(\frac{2}{\zeta}\right)^m \tag{9.29}$$

again, when $\zeta \gg 1$,

$$I_m(\zeta) = \frac{1}{\sqrt{2\pi\zeta}} e^{\zeta} \tag{9.30}$$

Consequently, the solutions of Eq. (9.18) give us

$$E_z = \begin{cases} C_1 J_q(ur) e^{jq\phi} e^{j(\omega t - \beta z)} & r \leqslant r_1 \\ C_2 K_q(vr) e^{jq\phi} e^{j(\omega t - \beta z)} & r > r_1 \end{cases} \tag{9.31}$$

$$H_z = \begin{cases} C_3 J_q(ur) e^{jq\phi} e^{j(\omega t - \beta z)} & r \leqslant r_1 \\ C_4 K_q(vr) e^{jq\phi} e^{j(\omega t - \beta z)} & r > r_1 \end{cases} \tag{9.32}$$

where $u^2 = k_1^2 - \beta^2$ with $k_1 = 2\pi n_1/\lambda$, $v^2 = \beta^2 - k_2^2$ with $k_2 = 2\pi n_2/\lambda$, and C_1, C_2, C_3, and C_4 are arbitrary constants. But in order for Eq. (9.31) to be valid, u must be real such that $R(r)$ is real. Again, since $K_q(vr)$ approaches $\exp(-vr)$ as vr approaches infinity, v must be positive. Thus while $k_1 \geqslant \beta$ is considered to be one of the cut-off conditions, the other cut-off condition is $\beta \geqslant k_2$. The condition necessary for the confined modes to exist is thus provided by

$$n_2 \leqslant \frac{\beta}{k} \leqslant n_1 \tag{9.33}$$

where $k = 2\pi/\lambda$.

9.2.2 Mode Characteristics of Cylindrical Waveguides

Equation (9.32) provides the range of allowable values for β for the electromagnetic wave propagating through the step-index cylindrical waveguide. However, the exact solution of β is determined by the field boundary conditions at the core-clad interface. The boundary conditions require that at $r = r_1$, each of the field components, E_z, E_ϕ, H_z, and H_ϕ, is continuous. The boundary conditions involving E_z and H_z, respectively, yield two simultaneous equations.

$$C_1 J_q(ur_1) - C_2 K_q(vr_1) = 0 \tag{9.34a}$$

$$C_3 J_q(ur_1) - C_4 K_q(vr_1) = 0 \tag{9.34b}$$

Similarly, Eqs. (9.31) and (9.32) can be incorporated into Eqs. (9.12b) and (9.13b) to provide the two remaining boundary condition requirements.

$$\frac{C_1 q\beta}{r_1 u^2} J_q(ur_1) + \frac{C_2 q\beta}{r_1 v^2} K_q(vr_1) + jC_3 \frac{\omega\mu}{u} J_q'(ur_1) + jC_4 \frac{\omega\mu}{v} K_q'(vr_1) = 0 \tag{9.35a}$$

$$-j\frac{C_1 \omega\varepsilon_1}{u} J_q'(ur_1) - j\frac{C_2 \omega\varepsilon_2}{v} K_q'(vr_1) + \frac{C_3 q\beta}{r_1 u^2} J_q(ur_1) + \frac{C_4 q\beta}{r_1 v^2} K_q(vr_1) = 0 \tag{9.35b}$$

where ε_1 and ε_2 are electrical permittivities in the core and clad, respectively. Equations (9.34a), (9.34b), (9.35a), and (9.35b) can next be solved for the four coefficients C_1, C_2, C_3, and C_4. However, for these nontrivial solutions to exist, the value of the determinant corresponding to the four coefficients must be zero. This constraint provides a transcendental eigenvalue equation for the propagating constant β as given by

$$\left[\frac{1}{u}\frac{J_q'(ur_1)}{J_q(ur_1)} + \frac{1}{v}\frac{K_q'(vr_1)}{K_q(vr_1)}\right]\left[\frac{n_1^2}{u}\frac{J_q'(ur_1)}{J_q(ur_1)} + \frac{n_2^2}{v}\frac{K_q'(vr_1)}{K_q(vr_1)}\right] = \left(\frac{q\beta}{r_1 k}\right)^2 \left[\frac{1}{u^2} + \frac{1}{v^2}\right]^2 \tag{9.36}$$

For given values of q and k, β will be found to have discrete values so that both Eqs. (9.33) and (9.36) are satisfied simultaneously. For each of the allowable values of β, the ratios C_2/C_1, C_3/C_1, and C_4/C_1 can then be determined (see Problem 9.3). These values can in turn be used to obtain the six cylindrical field components.

Equation (9.36) can be solved numerically for the propagation constant. But first, let us introduce an important parameter related to the cut-off condition. This parameter, referred to as the **normalized frequency V**, is defined as

$$V = r_1^2(u^2 + v^2)^{1/2} = \frac{2\pi r_1}{\lambda}(n_1^2 - n_2^2)^{1/2} = \frac{2\pi r_1}{\lambda}(\text{NA}) \tag{9.37}$$

where NA is defined in Eq. (9.3). This dimensionless number can be used to estimate the number of modes that can be supported by the waveguide. For all practical purposes, the solid angle of acceptance Ω for the cylindrical waveguide is $\pi\theta_{i,\max}^2 \simeq \pi(\text{NA})^2$ because $\sin\theta_{i,\max} \simeq \theta_{i,\max}$. It may be noted that the number of modes per unit solid angle is given by $2\pi r_1^2/\lambda^2$, where λ is the light wavelength and the factor of 2 accounts for the two polarization conditions of the plane wave. Consequently, the total number of modes that can be accommodated in the cylindrical waveguide is

$$N \simeq \frac{2\pi r_1^2}{\lambda^2}\Omega = \frac{V^2}{2} \tag{9.38}$$

It is important to understand that Eq. (9.38) refers to an approximate result since it has been derived using ray optics. Equation (9.38) is reasonably valid only when N is large.

To understand the eigenvalue condition of Eq. (9.36), consider the characteristics of Bessel functions. For example, Fig. 9.5 shows three of the lower order Bessel functions only. It is noteworthy that the first four zeros of $J_0(\zeta)$ occur at $\zeta = 2.450$, 5.520, 8.654, and 11.790, while the first four zeros of $J_1(\zeta)$ occur at $\zeta = 3.832$, 7.016, 10.173, and 13.323. We make use of these values later. Because of the oscillating nature of Bessel functions, the equation under consideration will have p possible roots for each value of q. The modes corresponding to the roots of the equations are designated as either TE_{qp}, TM_{qp}, EH_{qp}, or HE_{qp}. The TE and TM modes correspond to the meridional rays and are usually defined for $q = 0$. In comparison, the hybrid HE and EH modes are associated with longitudinal electric and magnetic field components and correspond to the propagating skew rays.

First consider the radially symmetric modes when $q = 0$. With $q = 0$, Eq. (9.36) reduces to two families of solutions given by

$$\frac{1}{u}\frac{J'_0(ur_1)}{J_0(ur_1)} + \frac{1}{v}\frac{K'_0(vr_1)}{K_0(vr_1)} = 0 \tag{9.39a}$$

$$\frac{n_1^2}{u}\frac{J'_0(ur_1)}{J_0(ur_1)} + \frac{n_2^2}{v}\frac{K'_0(vr_1)}{K_0(vr_1)} = 0 \tag{9.39b}$$

But

$$J'_q(\zeta) = J_{q-1}(\zeta) - \frac{q}{\zeta}J_q(\zeta) = \frac{q}{\zeta}J_q(\zeta) - J_{q+1}(\zeta) \tag{9.40a}$$

and

$$K'_q(\eta) = -\frac{q}{\eta}K_q(\eta) - K_{q-1}(\eta) = \frac{q}{\eta}K_q(\eta) - K_{q+1}(\eta) \tag{9.40b}$$

Equation (9.39a) can be thus rewritten using Eqs. (9.40a) and (9.40b) to provide the TM_{0_p} mode (corresponding to nonzero H_z) family of solutions as

$$\frac{1}{ur_1}\frac{J_1(ur_1)}{J_0(ur_1)} = -\frac{1}{vr_1}\frac{K_1(vr_1)}{K_0(vr_1)} \tag{9.41}$$

Such a condition, also known as the **transverse magnetic mode condition**, refers to the case when $C_1 = C_2 = 0$, corresponding to a field having nonvanishing H_r, H_z, and E_ϕ components. Similarly, Eq. (9.39b) can be rewritten to provide the

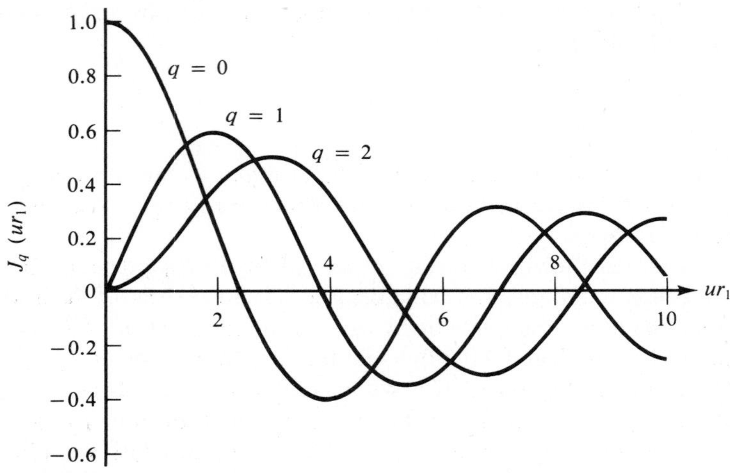

FIGURE 9.5

Bessel function of the first kind of three orders

TE_{0_p} mode (nonzero E_z) family of solutions as

$$\frac{1}{ur_1}\frac{J_1(ur_1)}{J_0(ur_1)} = -\frac{n_2^2}{vr_1 n_1^2}\frac{K_1(vr_1)}{K_0(vr_1)} \tag{9.42}$$

This latter case, known as the **transverse electric mode condition**, refers to the case when $C_3 = C_4 = 0$, corresponding to a field having nonvanishing E_r, E_z, and H_ϕ components.

We explore graphic solutions of Eqs. (9.41) and (9.42) because they are convenient for comprehending mode characteristics. Let us first consider Eq. (9.41). It can be solved from graphic plots corresponding to the expressions on its left- and right-hand sides, respectively. The left-hand side of Eq. (9.41) approaches infinity whenever $J_0(ur_1)$ goes to zero, and that occurs when ur_1 equals 2.405, 5.520, 8.654, 11.790, and so on. Obviously, the left-hand side becomes zero whenever $J_1(ur_1)$ goes to zero. But since

$$\mathrm{Lt}_{\zeta\to\infty} J_q(\zeta) = \sqrt{\frac{2}{\pi\zeta}}\cos\left[\zeta - \left(q + \frac{1}{2}\right)\frac{\pi}{2}\right] \tag{9.43a}$$

$$\mathrm{Lt}_{\zeta\to 0} J_q(\zeta) = \frac{1}{\Gamma(q+1)}\left(\frac{\zeta}{2}\right)^q \tag{9.43b}$$

the left-hand side of Eq. (9.41) is $\frac{1}{2}$ at $ur_1 = 0$, from which it increases monotonically, approaching infinity at $ur_1 = 2.405$. Between the two successive $J_0(ur_1) = 0$ points beyond $ur_1 = 2.405$, the left-hand side of Eq. (9.41) varies between $-\infty$ and $+\infty$ as shown in Fig. 9.6. Note that at large values of ur_1, the

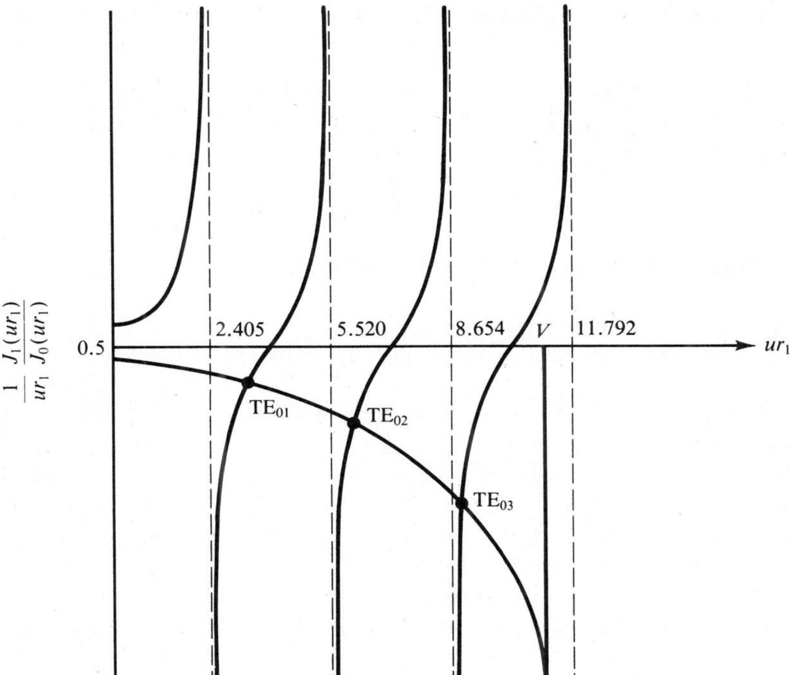

FIGURE 9.6

Graphic evaluation of Eq. (9.41)

left-hand side of Eq. (9.41) is given by

$$\text{Lt}_{ur\to\infty}\frac{1}{ur_1}\frac{J_1(ur_1)}{J_0(ur_1)}=\frac{1}{ur_1}\frac{\cos\left(ur_1-\dfrac{3\pi}{4}\right)}{\cos\left(ur_1-\dfrac{\pi}{4}\right)}=-\frac{1}{ur_1}\tan\left(ur_1-\frac{\pi}{4}\right) \tag{9.44}$$

Equation (9.44) provides for the characteristics of the corresponding asymptotes.

We concentrate next on the plot related to the right side of Eq. (9.41). In this context, it is noteworthy that

$$\text{Lt}_{\eta\to 0}K_q(\eta)=\begin{cases}-\ln(\eta) & q=0\\ \dfrac{\Gamma(q)}{2}\left(\dfrac{2}{\eta}\right)^q & q>0\end{cases} \tag{9.45a}$$

$$\text{Lt}_{\eta\to\infty}K_q(\eta)=\sqrt{\frac{\pi}{2\eta}}\,e^{-\eta} \tag{9.45b}$$

Using Eq. (9.37), the right-hand side of Eq. (9.41), at or near $ur_1 = 0$, thus becomes

$$\mathrm{Lt}_{ur_1 \to 0}\left[-\frac{K_1(vr_1)}{vr_1 K_0(vr_1)}\right] = \mathrm{Lt}_{ur_1 \to 0}\left[-\frac{K_1([V^2 - u^2r_1^2]^{1/2})}{(V^2 - u^2r_1^2)^{1/2} K_0([V^2 - u^2r_1^2]^{1/2})}\right]$$
$$= -\frac{K_1(V)}{VK_0(V)} \tag{9.46}$$

Again, as ur_1 approaches V, the right-hand side is given by

$$\mathrm{Lt}_{ur_1 \to V}\left[-\frac{K_1(vr_1)}{vr_1 K_0(vr_1)}\right] = \mathrm{Lt}_{ur_1 \to V}\frac{2}{(V^2 - u^2r_1^2)\ln(V^2 - u^2r_1^2)} \tag{9.47}$$

which finally becomes $-\infty$ at $ur_1 = V$. Note, however, that the right-hand side of Eq. (9.47) is undefined when $ur_1 > V$. The location of V is therefore crucial in finding out whether or not there is any mode. For example, for $V \leqslant 2.405$, there exists no TE mode, while for $2.405 < V \leqslant 5.520$, there is a single mode, and for $5.520 < V \leqslant 8.654$, there are two modes, and so on. Similar results may be obtained when finding the TM solutions of Eq. (9.42). Note that the right-hand side of Eq. (9.42) has an additional factor of $(n_2/n_1)^2$.

The cut-off conditions for the TE_{0_p} and TM_{0_p} modes are generally obtained from the pth roots of

$$J_0(ur_1) = 0 \tag{9.48}$$

and, accordingly, the corresponding cut-off value of r_1/λ is given by

$$\left.\frac{r_1}{\lambda}\right|_{\text{cut-off}} = \frac{(ur_1)_p}{2\pi\sqrt{n_1^2 - n_2^2}} \tag{9.49}$$

where $(ur_1)_p = 2.405$, 5.520, 8.654, 11.792, and so on, respectively, for $p = 1, 2, 3, 4, \ldots$, and it is $(p\pi - \frac{1}{4}\pi)$ when p is very large.

Consider next the solutions of Eq. (9.36) where q is nonzero. The mode corresponding to nonzero q is no longer of the transverse type, but is referred to as a hybrid type, represented by either EH or HE. Both of the z components of field, E_z and H_z, are present in these modes. When E_z makes a larger contribution to the mode, it is referred to as E-like and is denoted by EH and, similarly, when H_z is more significant, the mode is designated by HE. For simplicity, let us limit ourselves to $q = 1$ for the purpose of illustration, when Eq. (9.36) reduces to two families of solutions such as

$$\frac{1}{ur_1}\frac{J_2(ur_1)}{J_1(ur_1)} = \frac{1}{u^2r_1^2} + \frac{n_1^2 + n_2^2}{2n_1^2}\frac{1}{vr_1}\frac{K_1'(vr_1)}{K_1(vr_1)}$$
$$-\left[\left(\frac{\beta}{n_1r_1k}\right)^2\left(\frac{1}{u^2} + \frac{1}{v^2}\right)^2 + \frac{n_1^2 - n_2^2}{2n_1^2}\left(\frac{1}{vr_1}\frac{K_1'(vr_1)}{K_1(vr_1)}\right)^2\right]^{1/2} \tag{9.50}$$

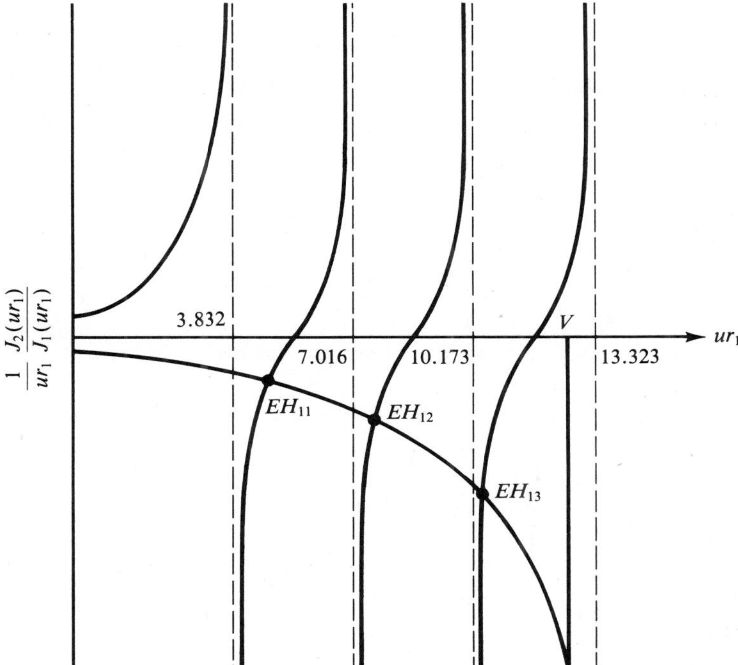

FIGURE 9.7

Graphic evaluation of Eq. (9.50)

for the EH_{1_p} modes and

$$\frac{1}{ur_1}\frac{J_0(ur_1)}{J_1(ur_1)} = \frac{1}{v^2r_1^2} - \frac{n_1^2+n_2^2}{2n_1^2}\frac{1}{vr_1}\frac{K'_1(vr_1)}{K_1(vr_1)} - \left[\left(\frac{\beta}{n_1r_1k}\right)^2\left(\frac{1}{u^2}+\frac{1}{v^2}\right)^2 + \frac{n_1^2-n_2^2}{2n_1^2}\left(\frac{1}{vr_1}\frac{K'_1(vr_1)}{K_1(vr_1)}\right)^2\right]^{1/2} \quad \textbf{(9.51)}$$

for the HE_{1_p} modes. In either case, we can see that the vertical asymptotes corresponding to the plot of the left-hand sides will be given by the cut-off condition

$$J_q(ur_1) = 0 \quad \textbf{(9.52)}$$

This cut-off condition occurs at $ur_1 = 3.832$, 7.016, 10.173, 13.323, ..., respectively, for $p = 1, 2, 3, 4, \ldots$, and at $(p\pi + \frac{1}{4}\pi)$ when p is very large for the EH modes; and at $ur_1 = 0$, 3.832, 7.016, 10.173, ..., respectively, for $p = 1, 2, 3, 4, \ldots$, and at $(p\pi + \frac{1}{4}\pi)$ when p is very large for the HE modes. The HE_{11} mode exists independent of the value of V. Therefore, this mode has no cut-off.

Figures 9.7 and 9.8, respectively, show the graphs corresponding to

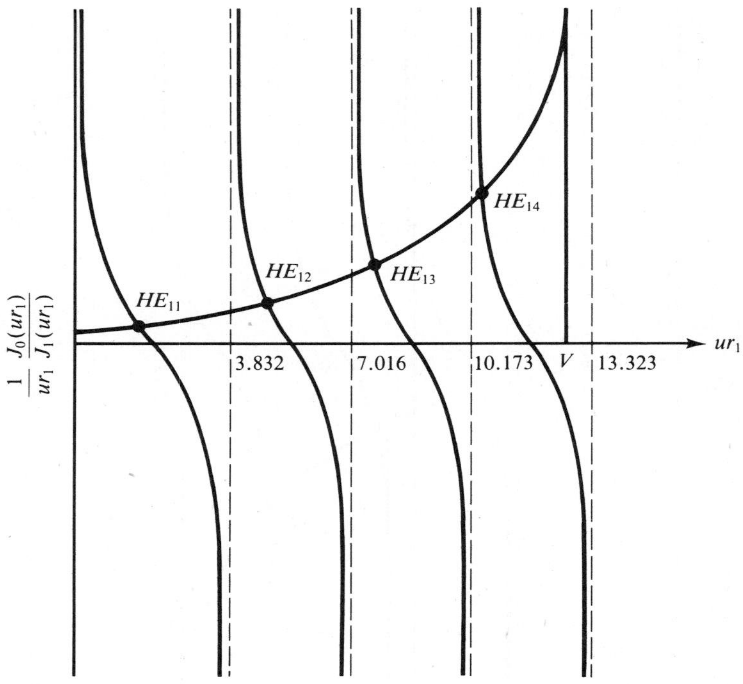

FIGURE 9.8

Graphic evaluation of Eq. (9.51)

Eqs. (9.50) and (9.51). The general characteristics of the left-hand side of Eq. (9.50) are similar to a great extent to those of Eq. (9.41). However, the left-hand side of Eq. (9.51) is distinctly different from that of either Eq. (9.41) or Eq. (9.50). It approaches $+\infty$ as ur_1 approaches 0 and, while decreasing monotonically, it becomes 0 at $ur_1 = 2.405$ and finally $-\infty$ at $ur_1 = 3.832$. Between the successive roots of $J_1(ur_1) = 0$, this trend of oscillation between $+\infty$ and $-\infty$ is repeated. Note that for a V value just less than 13.323, it is possible to have either three EH modes or four HE modes. The cut-off value of r_1/λ is still given by Eq. (9.49), except that $(ur_1)_p = 3.832,\ 7.016,\ 10.173,\ 13.323, \ldots$, respectively, for $p =$ 1, 2, 3, 4, ..., and at $(p\pi + \frac{1}{4}\pi)$ when p is very large for the EH_{1_p} modes; and at $(ur_1)_p = 0,\ 3.832,\ 7.016,\ 10.173, \ldots$, respectively, for $p = 1, 2, 3, 4, \ldots$, and at $(p\pi + \frac{1}{4}\pi)$ when p is very large for the HE_{1_p} modes.

When $q > 1$, the modes are still of the hybrid type. Without going into the details of the corresponding solution, it will suffice to say that the cut-off condition for the EH_{qp} mode is still given by Eq. (9.52) but with the added constraint that ur_1 is now nonzero. The cut-off condition of the HE_{qp} modes is,

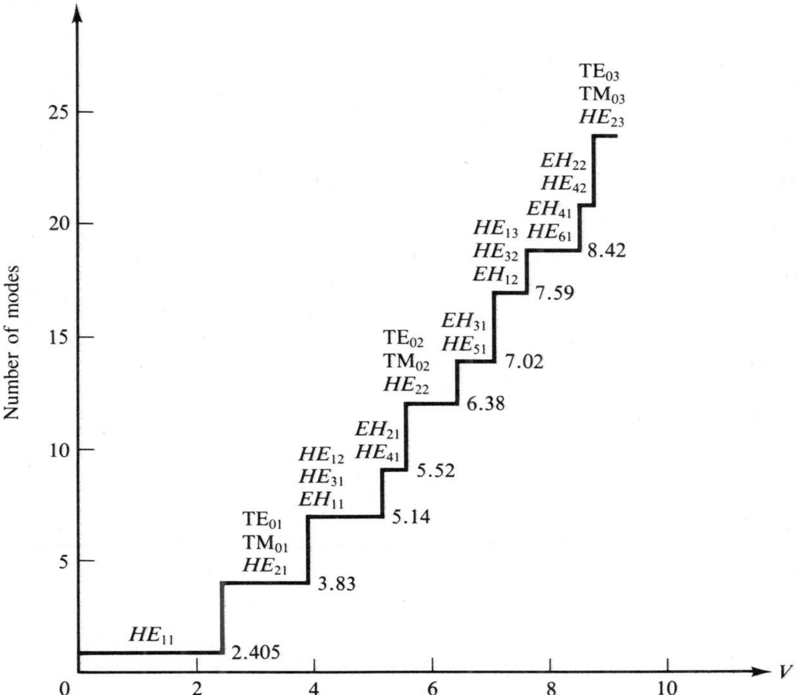

FIGURE 9.9

Number of modes versus normalized frequency

however, quite complex and is given by

$$\left[1+\left(\frac{n_1}{n_2}\right)^2\right]J_{q-1}(ur_1)=\frac{ur_1}{q-1}J_q(ur_1) \tag{9.53}$$

The cut-off values for r_1/λ are given by Eq. (9.49) so that for the EH modes, $(ur_1)_p$ is the pth root of Eq. (9.52) and, for the HE modes, $(ur_1)_p$ is the pth root of Eq. (9.53). Note that in order to calculate the cut-off parameters for the HE_{qp} modes, we need to know the ratio of the refractive indices of the core and the cladding.

To appreciate the significance of the normalized frequency parameter V, we note that the number of modes increases as V increases. And accordingly (as per Eq. (9.37)), the number of propagating modes increases as the core radius increases, as the core refractive index increases, as the cladding refractive index decreases, or as the source wavelength decreases. Figure 9.9 shows the number of propagating modes plotted against the normalized frequency V. This justifies the fact that the scope of ray theory (per Eq. (9.38)) is fairly limited.

The mode characterization of the step-index cylindrical waveguide allows

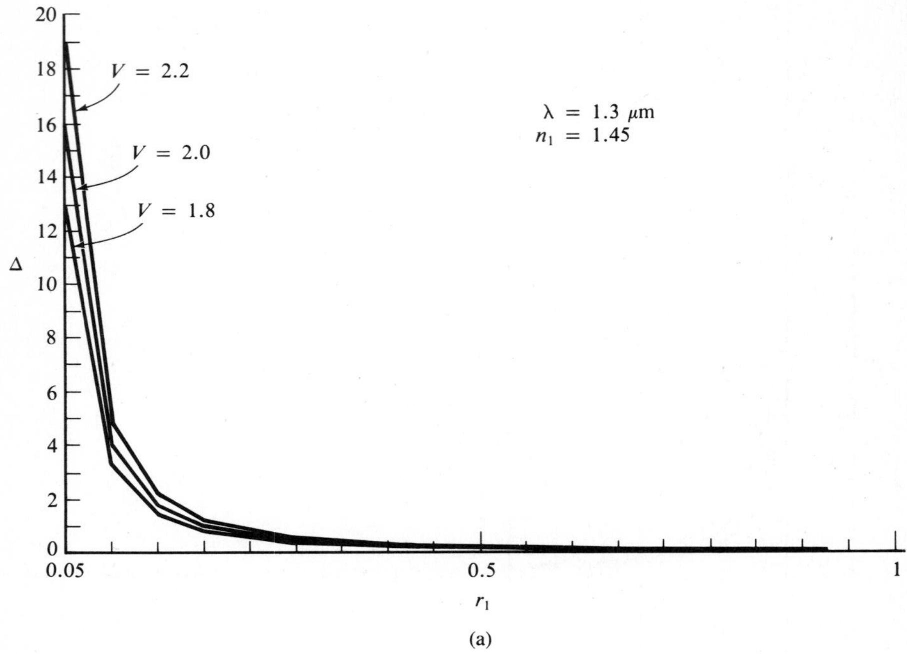
20
18
16
14
12
10
8
6
4
2
0
Δ
V = 2.2
V = 2.0
V = 1.8
λ = 1.3 μm
n_1 = 1.45
0.05
0.5
1
r_1

(a)

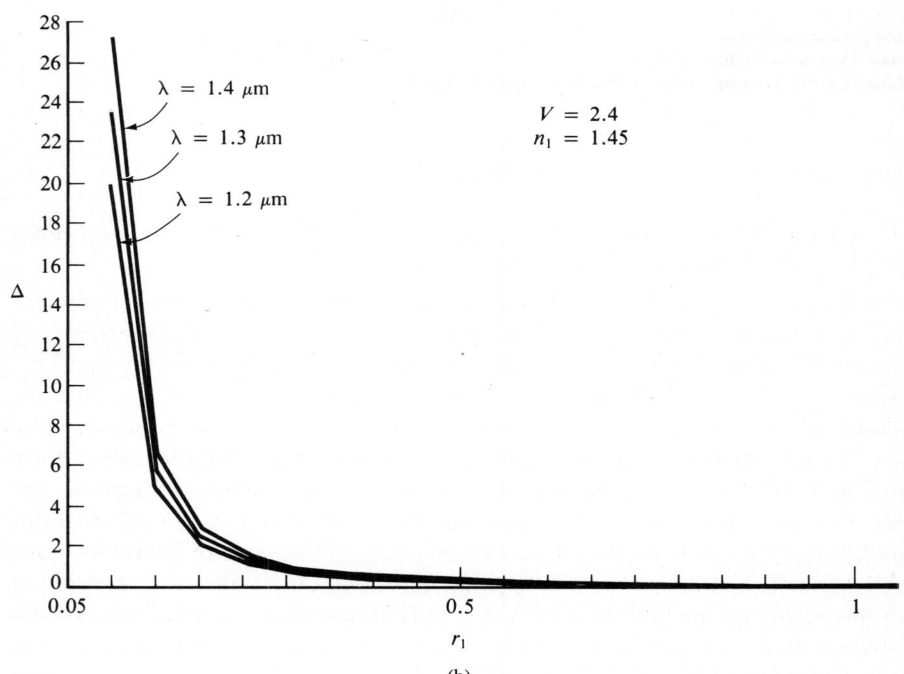
28
26
24
22
20
18
16
14
12
10
8
6
4
2
0
Δ
λ = 1.4 μm
λ = 1.3 μm
λ = 1.2 μm
V = 2.4
n_1 = 1.45
0.05
0.5
1
r_1

(b)

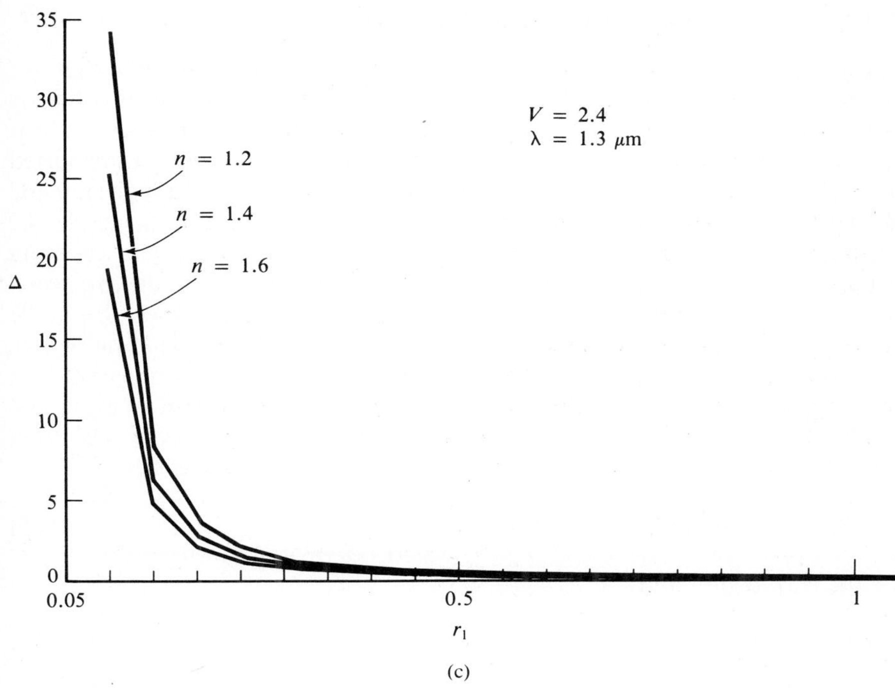

FIGURE 9.10

The characteristics (r_1 versus V) of a single-mode, step-index fiber for different (a) V, (b) wavelength, and (c) core indices

us to design single-mode fibers that support only the hybrid HE_{11} mode as long as $V < 2.405$. Using Eqs. (9.3) and (9.37), we can see that

$$V \simeq \frac{2\pi r_1 n_1 \sqrt{2\Delta}}{\lambda} \tag{9.54}$$

where $\Delta = 1 - (n_2/n_1)$ is the fractional refractive index difference between the core and the cladding. Figure 9.10 shows a family of plots relating Δ and r_1 for a number of (a) V values, (b) wavelengths and (c) core refractive indices. The practical considerations that are taken into account during the design of single-mode fibers involve making both Δ and r_1 large enough to allow for easy splicing and manufacturing. There is no restriction in the selection of cladding radius r_2, except for the fact that r_2 should be large enough so that the confined modes vanish at $r = r_2$. In Section 9.5.1, we list the radii values for a number of different types of fibers.

9.2.3 Power Flow in Cylindrical Waveguides

The design of a complete fiber-optic communication link requires that we account for both a bandwidth budget and a power budget. Power loss occurs not only at the connections but also within the fibers themselves. In this subsection, therefore, we look into the basics of how power is transported through the fibers. It is important to understand that the fields for a given mode do not completely vanish at the core-cladding interface. In general, the oscillatory field present in the core vanishes exponentially in the cladding region. In fact, it is possible to show that the energy is dominant in the core when the corresponding mode is away from its cut-off. Similarly, when the mode approaches its cut-off frequency, the energy is dominant in the cladding region.

From the review of electromagnetic waves in Chapter 1 it is apparent that the net power flow in either the core or the cladding, or both, can be estimated simply by integrating the axial component of the Poynting vector. The total time-averaged power flow present in these fiber regions is given by

$$P_{\text{core}} = \tfrac{1}{2} \int_0^{\infty} \int_0^{2\pi} r(E_x H_y^* - E_y H_x^*)\, d\phi\, dr \tag{9.55}$$

for the core region and

$$P_{\text{clad}} = \tfrac{1}{2} \int_{r_1}^{\infty} \int_0^{2\pi} r(E_x H_y^* - E_y H_x^*)\, d\phi\, dr \tag{9.56}$$

for the cladding region. Before we can proceed further in evaluating the integrals of Eqs. (9.55) and (9.56), it is necessary to solve for the rectangular field components E_x and H_y.

The transcendental eigenvalue relation of Eq. (9.36) is extremely complex. It is difficult, if not impossible, to solve it analytically for a fiber with many modes unless we make some reasonable assumptions. For most practical fibers, the refractive index of the core differs a little from that of the cladding. And, indeed, an assumption of the kind $\Delta \simeq 0$ drastically simplifies Eq. (9.36). A fiber for which this assumption is true is often referred to as a **weakly guiding fiber**. Under such a simplistic, yet good, assumption for communication grade fibers, we get

$$\left[\frac{1}{ur_1}\frac{J_{q+1}(ur_1)}{J_q(ur_1)} - \frac{1}{vr_1}\frac{K_{q+1}(vr_1)}{K_q(vr_1)}\right]\left[\frac{1}{ur_1}\frac{J_{q-1}(ur_1)}{J_q(ur_1)} - \frac{1}{vr_1}\frac{K_{q-1}(vr_1)}{K_q(vr_1)}\right] = 0 \tag{9.57}$$

Since, from the recurrence relations,

$$J_{q-1}(\zeta) + J_{q+1}(\zeta) = \frac{2q}{\zeta} J_q(\zeta) \tag{9.58a}$$

$$K_{q-1}(\zeta) + K_{q+1}(\zeta) = -2K_q'(\zeta) \tag{9.58b}$$

$$J_{q-1}(\zeta) - J_{q+1}(\zeta) = 2J_q'(\zeta) \tag{9.58c}$$

$$K_{q-1}(\zeta) - K_{q+1}(\zeta) = -\frac{2q}{\zeta} K_q(\zeta) \tag{9.58d}$$

Equation (9.57) readily yields two characteristic equations:

$$\frac{1}{ur_1}\frac{J_{q+1}(ur_1)}{J_q(ur_1)} = \frac{1}{vr_1}\frac{K_{q+1}(vr_1)}{K_q(vr_1)} \tag{9.59}$$

for the EH modes and

$$\frac{1}{ur_1}\frac{J_{q-1}(ur_1)}{J_q(ur_1)} = \frac{1}{vr_1}\frac{K_{q-1}(vr_1)}{K_q(vr_1)} \tag{9.60}$$

for the HE modes. An examination of Eqs. (9.59) and (9.60) should convince us that the EH modes of order $m - 1$ are almost degenerate with the HE modes of order $m + 1$. In fact, the data of Fig. 9.9 should support this observation. It may, therefore, sound reasonable to consider a linearly polarized (LP) mode that is formed by superimposing an EH mode on an HE mode. Such simplistic LP modes have only four field components, whereas the hybrid modes have a total of six field components. Note that LP modes are not truly representative of the actual modes of the fiber; however, the concept is used as an effective tool for visualizing the corresponding field characteristics.

The transverse field components corresponding to the LP modes can be expressed as follows:

$$E_y = \begin{cases} \dfrac{1}{n_1}\sqrt{\dfrac{\mu_0}{\varepsilon_0}}H_x = E_{cc}\dfrac{J_q(ur)}{J_q(ur_1)}\cos(q\phi) & r < r_1 \\[2ex] \dfrac{1}{n_2}\sqrt{\dfrac{\mu_0}{\varepsilon_0}}H_x = E_{cc}\dfrac{K_q(vr)}{K_q(vr_1)}\cos(q\phi) & r > r_1 \end{cases} \tag{9.61}$$

where E_{cc} is the value of the electric field strength at the core-cladding interface. The remaining longitudinal components E_z and H_z can be obtained in a similar fashion (Problem 9.5). In fact, by equating the tangential component H_z at the core-cladding interface of the weakly guiding fiber, we obtain

$$u\frac{J_{q\pm1}(ur_1)}{J_q(ur_1)} = \pm v\frac{K_{q\pm1}(vr_1)}{K_q(vr_1)} \tag{9.62}$$

which is identical to Eq. (9.41). Equation (9.61) can be used next to evaluate the net power in either the core or the cladding. For example, Eq. (9.55) can be evaluated as follows:

$$\begin{aligned} P_{\text{core}} &= \frac{1}{2}\sqrt{\frac{\varepsilon_0}{\mu_0}}\frac{n_1 E_{cc}}{[J_q(ur_1)]^2}\int_0^{r_1}\int_0^{2\pi} r[J_q(ur)]^2\cos^2(q\phi)\,dr\,d\phi \\ &= \frac{1}{2}\sqrt{\frac{\varepsilon_0}{\mu_0}}\frac{\pi n_1 E_{cc}}{[J_q(ur_1)]^2}\int_0^{r_1} r[J_q(ur)]^2\,dr \\ &= \frac{1}{4}\sqrt{\frac{\varepsilon_0}{\mu_0}}\pi r_1^2 n_1 E_{cc}\left[1 - \frac{J_{q+1}(ur_1)J_{q-1}(ur_1)}{[J_q(ur_1)]^2}\right] \end{aligned} \tag{9.63}$$

using the Bessel function relationship given by

$$\int_0^{r_1} r[J_q(ur)]^2\,dr = \tfrac{1}{2}r_1^2[\{J_q(ur_1)\}^2 - J_{q+1}(ur_1)J_{q-1}(ur_1)] \tag{9.64}$$

Similarly, Eq. (9.56) can be evaluated to give

$$\begin{aligned} P_{\text{clad}} &= \frac{1}{4}\sqrt{\frac{\varepsilon_0}{\mu_0}}\,\pi r_1^2 n_2 E_{cc}\left[\frac{K_{q+1}(vr_1)K_{q-1}(vr_1)}{[K_q(vr_1)]^2} - 1\right] \\ &= A(\Psi - 1) \end{aligned} \tag{9.65}$$

where

$$A = \frac{1}{4}\sqrt{\frac{\varepsilon_0}{\mu_0}}\,\pi r_1^2 n_2 E_{cc} \tag{9.66a}$$

$$\Psi = \frac{K_{q+1}(vr_1)K_{q-1}(vr_1)}{[K_q(vr_1)]^2} \tag{9.66b}$$

and

$$\int_r^{\infty} r[K_q(vr)]^2\,dr = \tfrac{1}{2}r_1^2[K_{q+1}(vr_1)K_{q-1}(vr_1) - \{K_q(vr_1)\}^2] \tag{9.66c}$$

Consequently, we can rewrite Eq. (9.63) using Eqs. (9.62) and (9.65) as

$$P_{\text{core}} = A\left[1 + \Psi\left(\frac{u}{v}\right)^2\right] \tag{9.67}$$

Now the total power can be found using Eqs. (9.37), (9.64), and (9.66) as

$$P_{\text{tot}} = P_{\text{core}} + P_{\text{clad}} = A\Psi\left(\frac{V}{ur_1}\right)^2 \tag{9.68}$$

It may be worthwhile to find ratios of the core power and the cladding power to the total power as

$$\frac{P_{\text{core}}}{P_{\text{tot}}} = 1 - \left(\frac{ur_1}{V}\right)^2\frac{\Psi - 1}{\Psi} \tag{9.69a}$$

and

$$\frac{P_{\text{clad}}}{P_{\text{tot}}} = \left(\frac{ur_1}{V}\right)^2\frac{\Psi - 1}{\Psi} \tag{9.69b}$$

For modes closer to the cut-off, however, $ur_1 \simeq V$ and $vr_1 \simeq 0$ when

$$\Psi \simeq \frac{\sqrt{q^2+1}}{\sqrt{q^2+1} - 1} \tag{9.70}$$

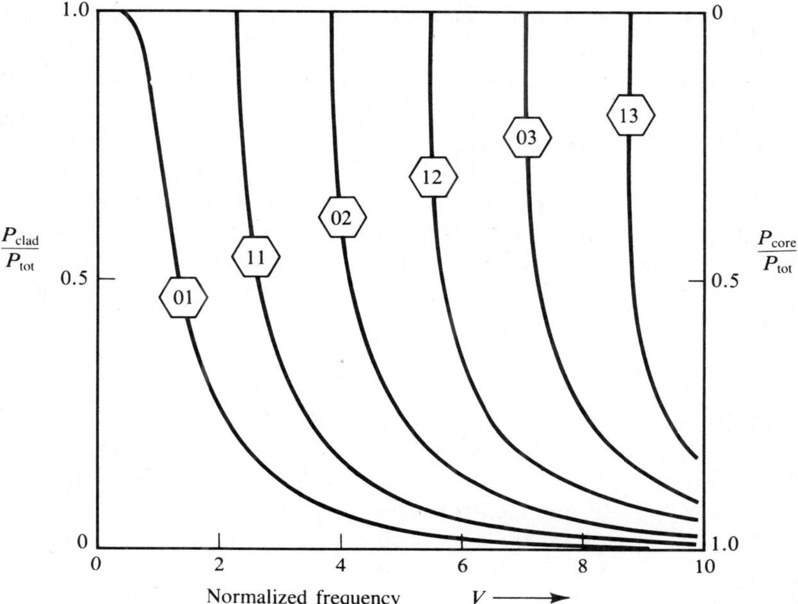

FIGURE 9.11

Fractional power in the core and cladding of a step-index fiber

and thus, at or near the cut-off,

$$\frac{P_{core}}{P_{tot}} \simeq 1 - \frac{1}{\sqrt{q^2+1}} \tag{9.71a}$$

$$\frac{P_{clad}}{P_{tot}} \simeq \frac{1}{\sqrt{q^2+1}} \tag{9.71b}$$

It may be noted that for exceedingly skew rays—i.e., when $q \gg 0$—power in the core remains dominant even at the cut-off. However, for the meridional rays—i.e., when $q = 0$—power in the cladding becomes dominant. From Fig. 9.11 we may conclude that the cladding power normally decreases as V increases. Figures 9.11 and 9.10 can be used concurrently in predicting the performance of the step-index fiber.

Using Eq. (9.62) we can now investigate the characteristics of LP modes. For the cut-off condition $V = 0$, the characteristics equation reduces to

$$J_{q-1}(ur_1) = 0 \tag{9.72}$$

Again, for the lowest order mode, Eq. (9.72) reduces to give us

$$J_{-1}(ur_1) = -J_1(ur_1) = 0 \tag{9.73}$$

implying that $ur_1 = 0$. Thus, for the lowest order LP mode, there is no cut off, implying further that LP_{01} and HE_{11} are identical. Succeeding LP modes are obtained similarly at $ur_1 = 3.832$, 7.016, 10.173, and so on, corresponding to $J_1(ur_1) = 0$. These modes are respectively referred to as LP_{02}, LP_{03}, LP_{04}, and so on. For the next-order mode, $q = 1$ and the cut offs occur when $J_0(ur_1) = 0$. Thus we expect to locate LP_{11}, LP_{12}, LP_{13}, LP_{14}, and so on, respectively, at $ur_1 = 2.405$, 5.520, 8.654, 11.793, and so on. Figure 9.12 shows the regions of these LP modes. A reasonable way to estimate the total number of modes in a step-index circular waveguide is to estimate the cut-off value of V according to Eqs. (9.43) and (9.72). The cut-off value of V for the high order LP modes is given by

$$V \simeq \tfrac{1}{2}(q + 2p)\pi \qquad \textbf{(9.74)}$$

Recall that each combination of q and p represents a particular LP mode of a given polarization and ϕ dependence. However, Eq. (9.74) corresponds to a triangle in the q-p plane bounded by q axis, the p axis, and a line whose intercepts are $(2V/\pi, 0)$ and $(0, V/\pi)$. Thus the total number of modes can be considered to be four times the area of this triangle since each of the modes is associated with two mutually orthogonal polarizations and can have either a cosine or sine dependence of ϕ according to Eq. (9.61). The total number of LP modes that can exist in a step-index fiber is thus given by

$$N \simeq 4\left(\frac{1}{2}\right)\left(\frac{2V}{\pi}\right)\left(\frac{V}{\pi}\right) = \left(\frac{2V}{\pi}\right)^2 \qquad \textbf{(9.75)}$$

This shows that our estimation of Eq. (9.38) based on ray optics is reasonably valid.

The group delay time τ_g needed by the mode energy of a light pulse to propagate through a fiber of length L is given by

$$\tau_g = L\frac{d\beta}{d\omega} \qquad \textbf{(9.76)}$$

in accordance with Eq. (1.144) where the mode index $\beta = n_{qp}\omega/c$ and n_{qp}, the effective refractive index of the mode in question, is a function of n_1, n_2, and ω. At any given frequency, the group delay is different for different modes. In a multimode waveguide, we can thus expect light pulse distortion not only due to group delay dispersion but also due to modal dispersion that is caused by the presence of multiple modes. The modal dispersion is completely absent only in a single-mode fiber.

The group delay in a single-mode, step-index cylindrical waveguide can be expressed as

$$\tau_g = \frac{\omega L}{c}\left(\frac{\partial n_{01}}{\partial n_1}\frac{\partial n_1}{\partial \omega} + \frac{\partial n_{01}}{\partial n_2}\frac{\partial n_2}{\partial \omega} + \frac{\partial n_{01}}{\partial \omega}\right) + \frac{n_{01}L}{c} \qquad \textbf{(9.77)}$$

where the subscript 01 refers to the LP_{01} mode. It is, however, common

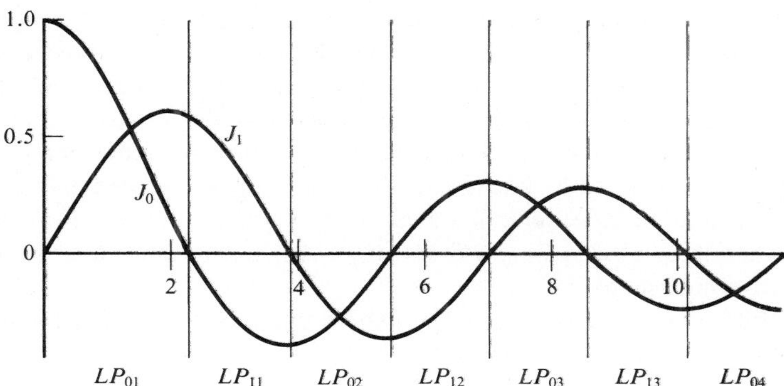

FIGURE 9.12

Solution to the LP cut-off condition

knowledge that changes in n_1^2 and n_2^2 contribute to a change in β^2. Therefore, for a step-index fiber,

$$d\beta^2 = \left(\frac{\omega}{c}\right)^2 \frac{1}{P_{\text{tot}}}[P_{\text{core}}\,dn_1^2 + P_{\text{clad}}\,dn_2^2] \tag{9.78}$$

Accordingly, Eq. (9.77) can be rewritten as

$$\tau_g = \frac{\omega L}{cP_{\text{tot}}}\left[P_{\text{core}}\frac{n_1}{n_{01}}\frac{\partial n_1}{\partial\omega} + P_{\text{clad}}\frac{n_2}{n_{01}}\frac{\partial n_2}{\partial\omega} + \left(\frac{\partial n_{01}}{\partial\omega}\right)_\omega\right] + \frac{n_{01}L}{c} \tag{9.79}$$

The term $(\partial n_{01}/\partial\omega)$ relates to what we have already referred to as waveguide dispersion. For emphasis, we now introduce the subscript ω to this term. For a weakly guiding fiber where $\Delta \simeq 0$, Eq. (9.79) reduces to

$$\begin{aligned}\tau_g &= \frac{\omega L}{c}\left[\left(\frac{\partial n_{01}}{\partial\omega}\right)_m + \left(\frac{\partial n_{01}}{\partial\omega}\right)_\omega\right] + \frac{n_{01}L}{c}\\ &= -\frac{\lambda L}{c}\left[\left(\frac{\partial n_{01}}{\partial\lambda}\right)_m + \left(\frac{\partial n_{01}}{\partial\lambda}\right)_\omega\right] + \frac{n_{01}L}{c}\end{aligned} \tag{9.80}$$

where $(\partial n_{01}/\partial\omega)_m$ relates to material dispersion. For most component materials of the fiber, there are wavelength regions where material dispersion can be disregarded totally. On the other hand, the waveguide dispersion may vanish at a particular wavelength, depending on the values of the fiber parameters Δ and r_1. Further, it is possible to find a range of wavelengths for particular values of Δ and r_1 where the total group delay dispersion is a minimum.

9.2.4 Graded-Index Cylindrical Waveguides

The transmission capacity of most step-index multimode fibers is relatively low because multiple modes are excited simultaneously. Multimodes often contribute to pulse spreading in the order of $\simeq 10^2$ ns/km, which thus limits the maximum data rate. By comparison, the graded-index fiber has a smaller dispersion and hence its information-carrying capacity is appreciably higher. We have already said that while dispersion associated with the single-mode fibers is tolerable, coupling and splicing of single-mode fibers are generally difficult (but not impossible) because of their generally small core diameters. By comparison, graded-index fibers have larger core diameters. For obvious reasons, therefore, graded-index fibers are used in a large majority of telecommunication systems.

Equation (9.18) can be used for expressing the wave equation for the graded-index fibers as

$$\frac{d^2R}{dr^2} + \frac{1}{r}\frac{dR}{dr} + \left[\kappa^2(r) - \beta^2 - \left(\frac{q}{r}\right)^2\right]R = 0 \tag{9.81}$$

where

$$\kappa(r) = \frac{2\pi}{\lambda}n(r) = kn(r) \tag{9.82}$$

and $n(r)$ represents the refractive index profile as a function of the radial coordinate r. An approximate mode solution of Eq. (9.81) is often obtained by employing a technique, referred to as the Wentzel-Kramers-Brillouin (WKB) method, that is encountered frequently in quantum mechanics. The basic assumptions for carrying out the WKB analysis include: (a) the index variation of the fiber is negligible over distances in the order of a wavelength and (b) the total index change within a core is small enough not to consider LP modes. Thus the general solution can be expressed as the superposition of nonuniform plane waves given by

$$R(r) = e^{jk\rho(r)} \tag{9.83}$$

where $\rho(r)$ is a power series expansion given by

$$\rho(r) = \rho_0(r) + \frac{\rho_1(r)}{k} + \frac{\rho_2(r)}{k^2} + \cdots \tag{9.84}$$

Note that k is inversely proportional to the wavelength.

The WKB approximation allows us to neglect all but the first two terms of Eq. (9.84). Consequently, Eq. (9.83) yields

$$jk\frac{d^2\rho_0}{dr^2} + j\frac{d^2\rho_1}{dr^2} + \frac{jk}{r}\frac{d\rho_0}{dr} + \frac{j}{r}\frac{d\rho_1}{dr} - k^2\left(\frac{d\rho_0}{dr}\right)^2 - \left(\frac{d\rho_1}{dr}\right)^2 - 2k\frac{d\rho_0}{dr}\frac{d\rho_1}{dr}$$
$$+ \kappa^2(r) - \beta^2 - \left(\frac{q}{r}\right)^2 = 0 \tag{9.85}$$

which in turn reduces to two equations belonging to the zeroth and the first orders, respectively.

$$\frac{d\rho_0}{dr} = \pm\frac{1}{k}\left[\kappa^2(r) - \beta^2 - \left(\frac{q}{r}\right)^2\right]^{1/2} \tag{9.86a}$$

$$j\left[\frac{d^2\rho_0}{dr^2} + \frac{1}{r}\frac{d\rho_0}{dr}\right] = 2\frac{d\rho_0}{dr}\frac{d\rho_1}{dr} \tag{9.86b}$$

Equation (9.86a) can be solved for $\rho_0(r)$ to estimate the characteristics of the propagation constant of the modes. For a more complete description of the modes, $\rho_1(r)$ can be evaluated from Eq. (9.86b). For simplicity, we limit our treatment to only zero-order WKB approximation, which provides

$$\rho_0(r) = \pm\frac{1}{k}\int\left[\kappa^2(r) - \beta^2 - \left(\frac{q}{r}\right)^2\right]^{1/2} dr \tag{9.87}$$

Note that in order for Eq. (9.83) to have valid propagating modes, $\rho_0(r)$ must be real; i.e., $\kappa^2(r) - \beta^2 - (q/r)^2$ has to be a positive quantity. Consequently, for a fixed value of β, there are two values of r (say, a and b) between $r = 0$ and $r = r_1$ when

$$\kappa^2(r) - \beta^2 - \left(\frac{q}{r}\right)^2 = 0 \tag{9.88}$$

The light ray is thus expected to remain bounded between $r = a$ and $r = b$. For a fixed β, as q increases $|a - b|$ decreases until a q value is reached when both a and b coincide. Beyond this particular value of q the modes are no longer bounded. For a constant q, modes further away from the cut-off have a larger value for β and a smaller value for $|a - b|$. A bound hybrid mode corresponds to a propagating skew ray while it remains constrained between $r = a$ and $r = b$.

In order for the bound modes to exist, the phase difference $k\rho_0(r)$ encountered between any two consecutive reflections should be an integral multiple of 2π. Consequently, we can set

$$\int_a^b\left[\kappa^2(r) - \beta^2 - \left(\frac{q}{r}\right)^2\right]^{1/2} dr = p\pi \tag{9.89}$$

where p is an arbitrary integer. Here we can reemploy the argument involving the q-p plane that initially led to the development of Eq. (9.75) to evaluate the total number of modes in a graded-index waveguide. With $p = 0$ and β approaching its cut-off value kn_2, q takes its largest value. Similarly, with $\beta = kn_2$ and $q = 0$, p becomes a maximum. Thus we can obtain the total number of modes as

$$\begin{aligned} N &= 4\int_0^{q_{\max}} p\,dq = \frac{4}{\pi}\int_0^{q_{\max}}\int_0^b\left[\kappa^2(r) - \beta^2 - \left(\frac{q}{r}\right)^2\right]^{1/2} dr\,dq \\ &= \frac{4}{\pi}\int_0^r [\kappa^2(r) - kn_2]^{1/2}\int_0^{r_1}\left[\kappa^2(r) - \beta^2 - \left(\frac{q}{r}\right)^2\right]^{1/2} dr\,dq \\ &= \int_0^{r_1} [\kappa^2(r) - \{n_2 k\}^2] r\,dr = k^2\int_0^{r_1} [n^2(r) - n_2^2] r\,dr \end{aligned} \tag{9.90}$$

The refractive index profile for the general class of graded-index cylindrical waveguides is described as

$$n(r) = \begin{cases} n_1\left[1 - 2\Delta\left(\dfrac{r}{r_1}\right)^{\alpha}\right]^{1/2} & \text{for } r \leqslant r_1 \\ n_2(1-2\Delta)^{1/2} \simeq n_1(1-\Delta) & \text{for } r > r_1 \end{cases} \tag{9.91}$$

Using Eq. (9.91), we can thus evaluate Eq. (9.90) to provide

$$N = n_1^2 k^2 r_1^2 \Delta \frac{\alpha}{\alpha + 2} \tag{9.92}$$

for the general graded-index fiber. When $\alpha = 2$, the graded-index profile becomes parabolic, and when $\alpha = \infty$ the profile approaches the characteristics of a step-index fiber. Therefore,

$$\frac{N_{\text{parabolic}}}{N_{\text{step-index}}} = \frac{1}{2} \tag{9.93}$$

because, using Eq. (9.54),

$$N_{\text{step-index}} = n_1^2 k^2 r_1^2 \Delta = \frac{V^2}{2} \tag{9.94}$$

Note that Eq. (9.94) is identical to Eq. (9.38), which justifies the fact that the WKB approximation provides an acceptable description of the actual fiber. More importantly, the number of modes in a parabolic index fiber is about one-half of that in the step-index fiber.

It is interesting that the number of modes having a propagation constant greater than β is given by

$$\mathcal{N} = \int_0^b [\kappa^2(r) - \beta^2] r\,dr = N\left[\frac{n_1^2 k^2 - \beta^2}{2k^2 n_1^2 \Delta}\right]^{(\alpha+2)/\alpha} \tag{9.95}$$

using Eqs. (9.91) and (9.92). Equation (9.95) can also be used to provide information about modal spacing. This is made possible by first evaluating β in terms of the total number of $\mathcal{N}$-fold degeneracies and then differentiating it with respect to $\mathcal{N}$ as follows:

$$\beta = kn_1\left\{1 - 2\Delta\left(\frac{\mathcal{N}}{N}\right)^{2\alpha/(\alpha+2)}\right\}^{1/2} \tag{9.96}$$

which can be then differentiated to yield

$$\frac{d\beta}{d\mathcal{N}} = \frac{2}{r_1}\left(\frac{\alpha\Delta}{\alpha+2}\right)^{1/2}\left(\frac{\mathcal{N}}{N}\right)^{(\alpha-2)/(\alpha+2)} \tag{9.97}$$

where all but the most dominant term has been neglected. We see by examining Eq. (9.97) that while the modal spacing for the step-index fibers varies directly as the mode order, that for the parabolic-index fiber is independent of the mode number. Note that Eq. (9.95) can be used to evaluate the mode confinement by relating β to the axial component of the propagation vector whose magnitude is kn_1. The component of the propagation vector normal to the fiber axis is thus given by

$$\gamma = kn_1 \sin\left[\cos^{-1}\left(\frac{\beta}{kn_1}\right)\right] = kn_1(2\Delta)^{1/2}\left(\frac{\mathcal{N}}{N}\right)^{\alpha/(\alpha+2)} \tag{9.98}$$

The γ parameter is a measure of mode confinement. A smaller value of γ generally corresponds to a higher degree of confinement. It is evident that the modes in a graded-index waveguide are relatively more confined than those in a step-index waveguide. Note also that at cut-off—i.e., when $\mathcal{N} = N$—the value γ/k provides the numerical aperture of a step-index fiber as in Eq. (9.3).

It may be proper now to obtain the numerical aperture of a graded-index fiber. The maximum angle of acceptance corresponding to the numerical aperture is estimated to be $\cos^{-1}[\beta/\kappa(r)]$ when β approaches its cut-off value kn_2. The local numerical aperture in the receiving end of the fiber is given as a function of the radial location by

$$\mathrm{NA}(r) = n(r)\sin\left[\cos^{-1}\left(\frac{\beta}{\kappa(r)}\right)\right] = [n^2(r) - n_2^2]^{1/2} \tag{9.99}$$

which can be then rewritten using Eq. (9.91) to give us

$$\mathrm{NA}(r) = \begin{cases} \mathrm{NA}(0)\left[1 - \left(\dfrac{r}{r_1}\right)^{\alpha}\right]^{1/2} & \text{for } r \leqslant r_1 \\ 0 & \text{for } r > r_1 \end{cases} \tag{9.100}$$

where $\mathrm{NA}(0) = n_1(2\Delta)^{1/2}$, which turns out to be n_1 times the numerical aperture of the step-index waveguide. The normalized received power is given as a function of the radial location by

$$P(r) = P(0)\left[\frac{\mathrm{NA}(r)}{\mathrm{NA}(0)}\right]^2 = P(0)\left[1 - \left(\frac{r}{r_1}\right)^{\alpha}\right] \tag{9.101}$$

as long as $r \leqslant r_1$.

9.3 Signal Degradation in Fibers

We have already established that both intramodal dispersion and intermodal delay cause distortion of optical signals. **Intramodal dispersion**, also referred to as **chromatic dispersion**, results from the fact that group velocity is wavelength-dependent. The two main types of intramodal dispersion are **material dispersion** and **waveguide dispersion**. The former relates to the variation of the refractive index in the core material with changing wavelength, while the latter is a consequence of having a variation in the modal propagation constant with changing wavelength. Intermodal delay, on the other hand, is a result of the fact that at a single frequency, each mode has a different group velocity.

In multimode fibers, it is generally possible to ignore waveguide dispersion; however, in single-mode fibers, this effect may not be negligible at lower wavelengths. But at longer wavelengths, waveguide dispersion plays a significant role in signal distortion. Again, there are spectral regions between 1.3 μm and 1.7 μm for which the total single-mode dispersion becomes a minimum due to the mutual cancellation of the effects of material and waveguide dispersions. Note that intermodal delay distortion can be minimized by carefully tailoring the refractive-index profile of the core. Thus, while mode delay causes the most pulse distortion in the step-index fiber, material dispersion plays the most significant role in the graded-index waveguide. In any event, dispersion-induced signal distortion causes light pulses to broaden as they propagate, ultimately leading to a situation when the neighboring pulses begin to overlap each other. Consequently, dispersion affects the information capacity of the waveguide.

An important consideration in fiber design is the attenuation of the light signal. Attenuation, on the other hand, determines the maximum allowable transmission distance between a transmitter and either a receiver or a repeater. Basic waveguide attenuation can be classified into two categories: (a) radiative losses and (b) intrinsic losses. Radiative losses include those losses that are caused by physical defects or the arrangement of the waveguide whereby the waveguide is either twisted or bent. Intrinsic losses, on the other hand, include scattering and absorption losses that are inherent in the waveguide itself. Attenuation α of a fiber of length L is measured in terms of the ratio of the optical output power P_{out} to the optical input power P_{in} as

$$\alpha = \frac{10}{L} \log\left(\frac{P_{in}}{P_{out}}\right) \quad \text{(in dB/m)} \tag{9.102}$$

Attenuation is generally a function of wavelength. A reasonable way to model the spectral response of a fiber is to represent the attenuation by

$$\alpha = A + B(\lambda) + \frac{C}{\lambda^4} \tag{9.103}$$

where A corresponds to the launch conditions, waveguide imperfections, and bending; $B(\lambda)$ includes absorption losses due to impurities; and C accounts for scattering losses from density fluctuations.

Radiative losses are brought about whenever an optical fiber is bent. When a fiber is bent as shown in Fig. 9.13, the off-axis field tails move nonuniformly. The field tail on the far side of the center of the curvature must move faster to keep up with the corresponding field tail on the near side of the center of curvature. Consequently, a portion of the modes in the cladding are required to travel at a speed greater than that of light in that medium. This causes the energy associated with this particular portion of the mode to be radiated. The most

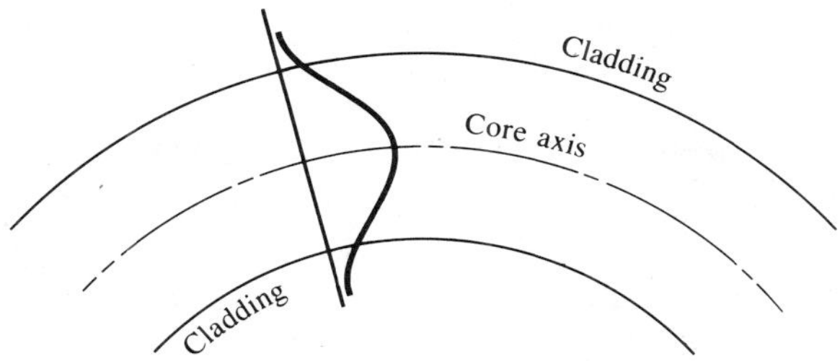

FIGURE 9.13

Field distribution in bent fibers

common radiation loss occurs either when a fiber cable turns a corner with bends having radii larger than the fiber's diameter or when microscopic bends of the fiber's axis occur randomly inside the fiber cables. Generally, the loss is greater in fibers with smaller radii of curvature and for those modes that extend further into the cladding.

Scattering losses are caused by irregularity in the fiber index, which in turn is caused by the inhomogeneity of the glass material used to make the fibers. The presence of oxides, such as GeO_2 and P_2O_5 in glass, is responsible for its inhomogeneity. Scattering losses, also known as Rayleigh scattering, are identical to the scattering of sunlight in the atmosphere and are therefore proportional to λ^{-4}. Consequently, scattering loss decreases dramatically with increasing wavelength. In fact, for a wavelength below 1 μm, it is the most dominant loss mechanism present in a fiber.

Absorption losses are of three types: (a) OH^- ion absorption loss, (b) inherent absorption loss, and (c) impurity absorption loss. The OH^- ion absorption loss is caused by the presence of water molecules in the composition of the fiber's material. Water concentrations of less than a few parts per billion are enough to introduce an attenuation loss on the order of 20 dB/km. The OH^- ion has its fundamental absorption peak at 2.75 μm with the first, second, and third overtones at 1.4, 0.95, and 0.73 μm, respectively. A glass fiber, referred to as dry glass, with lower absorption peaks is preferable for low-loss communication. An inherent loss, on the other hand, refers to that which is caused by the dopants used in fiber preparation. Generally, dopants of the type GeO_2 or P_2O_5 are added to silicon to increase the refractive index, whereas dopants such as fluorine or B_2O_3 are used to decrease the refractive index. These dopants are associated with different molecular absorption peaks. Finally, impurity absorption loss is caused by the impurities, such as iron and chromium, introduced unintentionally during the fabrication of fibers. Figure 9.14 shows a typical attenuation versus a wavelength plot for a glass fiber. Losses related to fiber geometry, such as boundary irregularities, core irregularities, and imperfect splicing, are often added to these losses.

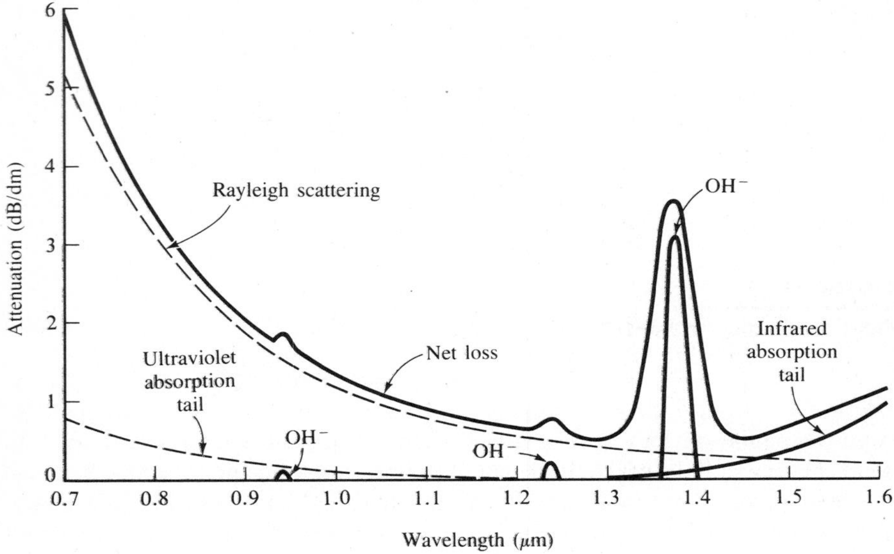

FIGURE 9.14

Attenuation characteristics as a function of wavelength. [Adapted from A. H. Cherin, *An Introduction to Optical Fibers* (New York: McGraw-Hill, 1983). Courtesy of McGraw-Hill.]

9.4 Power Launching and Coupling

We have already considered losses that are encountered in optical waveguides. An integral part of the optical fiber link is a transmitting source from which light pulses are introduced at the input end of the waveguide. At times, fibers are connected by means of splices to make a longer fiber. In either of these cases, not all of the power being transmitted by the transmitting side reaches the receiving side of the receiver.

The coupling of the power from an optical source into a fiber is usually characterized by coupling efficiency defined as

$$\eta_{sf} = \frac{P_f}{P_s} \tag{9.104}$$

where P_s is the source output power and P_f is the power coupled into the fiber. The coupling efficiency is thus a measure of the illumination loss. It accounts for the unintercepted light that misses the receiving end of the fiber. Such a loss is also referred to as the **numerical aperture loss**. When either the source and the fibers are misaligned or when the emitting area of the source is larger than the area of the fiber core, a portion of the emitted illumination is unintercepted.

Numerical aperture loss results from the fact that the radiation profile of the source may lie in part outside the acceptance cone of the cylindrical waveguide.

The output of a source is conveniently characterized by its radiances that measure the optical power radiated into a solid angle per unit surface area. Two types of light sources, LEDs and laser diodes (LDs), are typically used in fiber optics communication systems. In particular, LEDs can be either surface-emitting or edge-emitting. Surface-emitting LEDs are nearly Lambertian in nature, and their emission profile is given by

$$L_e(\theta, \phi) = L_{eo} \cos\theta \tag{9.105}$$

where L_{eo} is the radiance along the normal to the radiating surface, θ is the angle between the radiation direction and the source axis, and ϕ is the angle between the projection of the radiation direction along the emitting surface and a reference axis in the plane of the emitting surface. However, sources such as LDs and edge-emitting LEDs have a complex radiation profile given by

$$\frac{1}{L_e(\theta, \phi)} = \frac{\sin^2\phi}{L_{eo}\cos^m\theta} + \frac{\cos^2\phi}{L_{eo}\cos^n\theta} \tag{9.106}$$

where the values of m and n characterize the particular type of source involved. In practice, a higher value of m and n corresponds to sources that are typically narrower. Narrower sources are generally preferable in fiber optics systems because only then can more light be coupled into waveguides.

For simplicity, consider a Lambertian source that is coupling power into a fiber as shown in Fig. 9.15. Assume further that the emitting surface area is circular with a radius r_s and is placed in close contact with the receiving end of the fiber. The total power emitted by the source of area πr_s^2 is given by

$$\begin{aligned} P_s &= \pi r_s^2 \int L_e(\theta, \phi)\, d\Omega \\ &= \pi r_s^2 \int_0^{2\pi}\int_0^{\pi/2} L_{eo}\cos\theta \sin\theta\, d\theta\, d\phi = \pi^2 r_s^2 L_{eo} \end{aligned} \tag{9.107}$$

assuming that the source is emitting into a total of a 2π solid angle. We can next calculate the coupled power by integrating the contribution from each area element dA_s of the source by

$$\begin{aligned} P_f &= \iint L_e(\theta, \phi)\, dA\, d\Omega \\ &= \int_0^{2\pi}\int_0^{r_c}\left[\int_0^{2\pi}\int_0^{\theta_a} L_{eo}\cos\theta\,(\sin\theta\, \mathrm{d}\theta\, \mathrm{d}\phi)\right] r\, dr\, d\phi \end{aligned} \tag{9.108}$$

In this integral the angle θ_a corresponds to the maximum acceptance angle of the waveguide. The upper limit for the integral involving r depends on whether or not $r_s > r_1$ where r_1 is the radius of the core. When $r_s > r_1$, the upper limit $r_c = r_1$; but if $r_s \leq r_1$, then $r_c = r_s$. For an emitting area larger than the area of the core, light not intercepted by the core is not coupled into the fiber, whereas if the reverse is true, emitted light is coupled into the fiber, depending on the exact

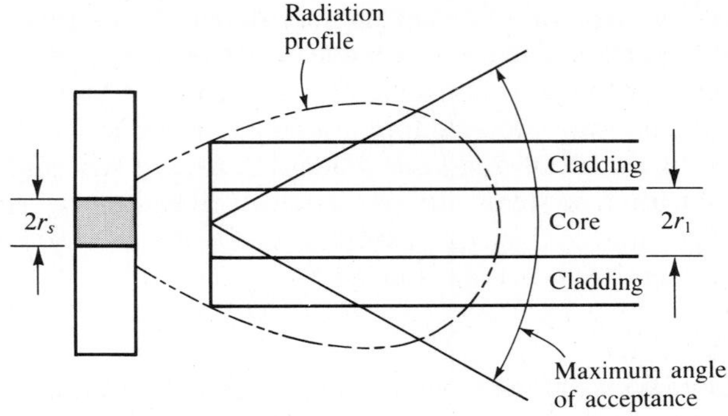

FIGURE 9.15

Coupling of a Lambertian source to an optical waveguide

value of the numerical aperture. Using the definition of Eq. (9.3), Eq. (9.108) is expressible in terms of the numerical aperture of the waveguide as follows:

$$P_f = \int_0^{2\pi} \int_0^{r_c} [\pi L_{eo} \sin^2 \theta_c] r \, dr \, d\phi$$

$$= \pi L_{eo} \int_0^{2\pi} \int_0^{r_c} (\text{NA})^2 r \, dr \, d\phi \tag{9.109}$$

Let us now consider the case of the step-index fiber whose numerical aperture $n_1(2\Delta)^{1/2}$ is independent of both r and ϕ. Equation (9.109) can be evaluated using Eq. (9.107) as follows:

$$P_f = \begin{cases} P_s(\text{NA})^2 & r_s \leqslant r_1 \\ \left[\dfrac{r_1(\text{NA})}{r_s}\right]^2 P_s & r_s > r_1 \end{cases} \tag{9.110}$$

Accordingly, the coupling efficiency is given by

$$\eta_{S\text{-}F} = \begin{cases} (\text{NA})^2 & r_s \leqslant r_1 \\ \left[\dfrac{r_1(\text{NA})}{r_s}\right]^2 & r_s > r_1 \end{cases} \tag{9.111}$$

Similarly, using Eq. (9.100), we can evaluate Eq. (9.109) for the case of the graded-index fiber. Consequently, for $r_s \leqslant r_1$, we obtain the coupled power into a graded-index fiber as follows:

$$P_f = 2\pi^2 L_{eo} \int_0^{r_s} 2\Delta n_1^2 \left[1 - \left(\frac{r}{r_1}\right)^{\alpha}\right]^{1/2} r \, dr$$

$$= 2P_s n_1^2 \Delta \left[1 - \frac{2}{\alpha + 2}\left(\frac{r_s}{r_1}\right)^{\alpha}\right] \tag{9.112}$$

For simplicity, we can consider a coupling system with $r_s \simeq r_1$. For such a coupling, the coupling efficiency of a parabolic-index (i.e., $\alpha = 2$) waveguide is given by

$$\eta_{S\text{-}F} = n_1^2 \Delta \tag{9.113}$$

whereas that for the step-index fiber is found to be

$$\eta_{S\text{-}F} = n_1^2 2\Delta \tag{9.114}$$

Note that as expected, Eqs. (9.111) and (9.114) are identical. It can be seen further that the coupling efficiency for the step-index fiber is twice that of the parabolic-graded fiber. In either case, however, the coupling can be increased only when Δ is increased.

The treatment carried out in this section is simple enough since we tacitly assumed that the medium separating the source and the fiber has a refractive index of n_1. If this were not so and the medium had a refractive index of n_m, then the coupled power would be reduced by a factor known as Fresnel's reflection coefficient, $[(n_1 - n_m)/(n_1 + n_m)]^2$. You should also note that coupling efficiency is generally independent of light wavelength. Equation (9.92) convinces us that the number of propagating modes in a fiber varies inversely as the square of the wavelength. Consequently, more power needs to be launched into a fiber when the wavelength is lower.

Not all of the sources are flat. In fact, we have discussed LEDs that are intended to have a hemispheric emitting surface. With such LEDs, it may be appropriate to introduce a bi-convex or spherical lens between the source and the fiber so that it becomes possible to collect light from all regions of the emitter. In case the emitting surface area is smaller than the area of the core, such imaging optics can improve the overall coupling efficiency. The effective solid angle of collection of light can be improved by a factor equal to the ratio of the two areas.

Up until now, we have dealt with losses encountered during the transmittal of light from the source to the fiber. A major loss occurs *between* fibers where they are coupled by means of mechanical connectors. Fiber-to-fiber losses are classified broadly into three subclasses: (a) transverse offset loss, (b) longitudinal offset loss, and (c) axial tilt loss. Transverse offset occurs whenever the core axes of the transmitting and the receiving fibers are not collinear, even though the axes may be parallel to each other. Longitudinal offset loss occurs whenever the two axes remain collinear but the facing ends remain physically separated by a finite distance. Finally, the last of the fiber-to-fiber losses is caused by a connection whereby the angle between the two axes is nonzero. In practice, however, these three losses can occur in combination. Of these three, the axial offset loss is the most disastrous. An accurate theoretical analysis of these losses is quite complex and we shall, therefore, confine ourselves to only a few simpler cases.

Consider two identical step-index fibers (each of core radius r_1) facing each other in such a way that the two axes are offset by a lateral distance d as shown in Fig. 9.16. For simplicity, let us assume further that the modal power distribution is reasonably uniform at the transmitting fiber end. The calculation of the coupling efficiency for such an ideal system is quite simple because the coupled power is proportional to the area that is common between the two end faces. Thus the resultant fiber-to-fiber coupling efficiency is given by

$$\eta_{F\text{-}F,T,\text{step}} = \frac{\text{common area}}{\pi r_1^2} = \frac{2}{\pi}\cos^{-1}\left(\frac{\delta}{2}\right) - \left(\frac{\delta}{\pi}\right)\left[1 - \left(\frac{\delta}{2}\right)^2\right]^{1/2} \tag{9.115}$$

where $\delta \equiv d/r$. The simplicity of the calculation of this coupling efficiency springs from the fact that the numerical aperture in step-index fiber core is invariant over all the end area. Thus, if we replace step-index fibers with graded-index fibers, our calculations must account for the variation of the numerical aperture. In fact, Eq. (9.101) relates the local power at a radial distance r away from the axis to the local numerical aperture. The local power is a maximum at $r = 0$ and zero at $r = r_1$ and beyond. The total power in the transmitting fiber is given by

$$P_{\text{transmitting}} = \int_0^{2\pi}\int_0^{r_1} P(r) r\, dr\, d\theta = P(0)\int_0^{2\pi}\int_0^{r_1}\left[1 - \left(\frac{r}{r_1}\right)^{\alpha}\right] r\, dr\, d\theta \tag{9.116}$$

In the particular case of the parabolic-index fiber, Eq. (9.116) reduces to

$$P_{\text{transmitting,parabolic}} = \tfrac{1}{2}\pi r_1^2 P(0) \tag{9.117}$$

The calculation of the coupled power is, however, much more involved. Along the dotted vertical line, the numerical apertures due to the two graded-index fibers are equivalent. This line splits the common area into two regions. The left of this line corresponds to a region where the numerical aperture of the transmitting fiber is larger than that of the receiving fiber. Correspondingly, on the right side of this line the reverse is true. Consequently, all light transmitted by the right half of the common area is coupled. But only that portion of the light transmitted by the left half which falls within the numerical aperture of the fiber is accepted. Taking symmetry into consideration, we evaluate the coupling efficiency for a parabolic-index waveguide using Eq. (9.117) to give us

$$\eta_{F\text{-}F,T,\text{parabolic}} = 4\frac{P(0)}{P_{\text{transmitting,parabolic}}}\int_0^{\cos^{-1}[\delta/2]}\int_{d/(2\cos\theta)}^{r_1}\left(1 - \frac{r^2}{r_1^2}\right) r\, dr\, d\theta = \frac{2}{\pi}\left[\cos^{-1}(\tfrac{1}{2}\delta) - (\tfrac{1}{6}\delta)\{1 - (\tfrac{1}{2}\delta)^2\}^{1/2}\left\{5 - \left(\frac{1}{\sqrt{2}}\delta\right)^2\right\}\right] \tag{9.118}$$

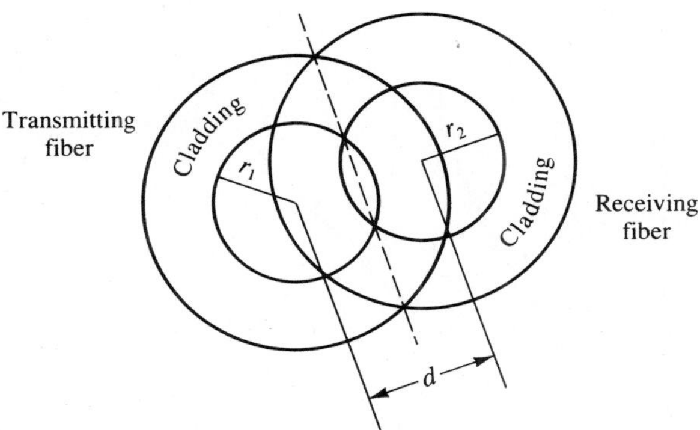

FIGURE 9.16

Two fibers offset laterally by distance *d*

We consider next the effect of longitudinal offset on coupling. The more a receiving fiber is physically separated from the transmitting fiber, the larger the portion of the transmitted power that is lost. We can show easily that the coupling efficiency due to longitudinal offset for the step-index fiber is given by

$$\eta_{F\text{-}F,L,\text{step}} = \frac{1}{1 + \delta' \tan \theta_c} \tag{9.119}$$

where $\delta' = d'/r$, d' is the longitudinal offset, and θ_c is the critical acceptance angle of the fiber.

Finally, the axial tilt angle θ will also cause a power loss, since a portion of the transmitted power will fall beyond the acceptance angle of the receiving fiber. After a rigorous calculation, the axial tilt coupling efficiency is found to be given by

$$\eta_{F\text{-}F,A,\text{step}} = 10^{-0.0217[\theta \omega n V/(\text{NA})]^2} \tag{9.120}$$

where

$$\omega = 2^{-1/2}[0.65 + 1.62V^{-1.5} + 2.88V^{-6}] \tag{9.121}$$

In practice, axial tilt angle is limited to a maximum of about 1° in the currently available splices and connectors. Figure 9.17 shows the coupling losses for all three of these misalignments. Note that the most dominant loss is caused by the transverse offset. Again, it may not be possible to completely eliminate the longitudinal offset because otherwise the fiber ends would rub against one another. You should keep in mind the fact that there can be intrinsic splice losses too. The intrinsic losses are caused by the mismatch of core radius, α, and Δ. Although less significant, other mismatch parameters—such as ellipticity and concentricity—may also contribute to the total loss of a splice.

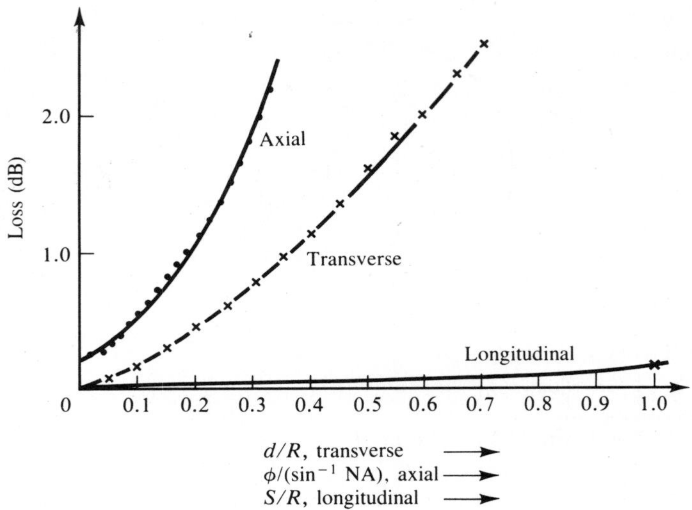

FIGURE 9.17

Losses due to transverse offset, longitudinal offset, and axial tilt

9.5 Fiber-Optic Communication Systems

Fiber-optic communication systems contain a number of individual components, fibers, light sources, repeaters, receivers, connectors, etc., that must be efficiently linked together to make the system competitive with existing electrical communication systems. In this section we review the significance of the various fiber-optics components and the mechanisms associated with them. The major parameters that are often measured for the evaluation of fiber-optic transmission system performance are as follows:

1. *Reflection coefficient*—the ratio of the forward-to-reverse power along the fiber at a given location.
2. *Signal level*—the peak and average carrier power within the waveguide at a given location.
3. *Loss*—the difference between the transmitted and the received power.
4. *Pulse spread*—the nonsimultaneous arrival of various signal components at a given location.
5. *SNR*—the ratio of a transmitted signal to all stray power in the channel.
6. *Temperature immunity*—the ability of the transmission system to withstand temperature fluctuations.
7. *Stress immunity*—the ability of the transmission line to withstand mechanical stresses under dynamic and static conditions.

8. *Radiation immunity*—the ability to withstand radiation that usually affects the attenuation and dispersion characteristics of fibers.
9. *Corrosion immunity*—the ability to withstand various chemical interactions, including that of moisture.

Other types of measurement are also important. For example, variation in the refractive index profile of a fiber will alter its transmission capabilities. Thus it may be appropriate to determine the actual refractive index profile of the fiber.

9.5.1 Fibers and Cables

The major system characteristics that we need to consider in the design of optical fiber transmission lines are:

1. *Attenuation* and its variance with input transmission wavelength, modal distribution, and temperature.
2. *Distortion* and its variance with bandwidth, modal distribution, the amplitudes and wavelength of the input light, length of the fiber, and temperature.
3. *Radiation* and its variance with bend radius and temperature.
4. *Physical parameters*, including size, weight, total volume, ease of installation, splicing, and coupling.
5. *Environmental parameters*, including resistance to water, mechanical stress, chemical corrosion, and temperature.

The various factors that contribute to the preceding fiber characteristics are as follows:

1. *Numerical aperture:* It determines the coupling efficiency between the LED or LD source and the optical fiber. Small values of NA tend to provide low pulse dispersions. Fibers with large NA values have large losses but are easier and cheaper to fabricate. For long-haul transmission, an NA of about 0.2 is usually acceptable.
2. *Refractive index:* Increasing the difference in the refractive index of the core and the cladding permits better source coupling; but, at the same time, it increases the degree of intermodal dispersion. Loss due to fiber bending decreases as the refractive index differential between the core and the cladding increases and as the core radius decreases. However, as the index differential increases, the waveguide bandwidth decreases, and as the core radius decreases, the coupling efficiency decreases as well.
3. *Fiber bending:* The macrobending of an optical waveguide causes a radiation loss that becomes increasingly severe with decreasing bend radius. The smallest permissible curvature of bend radius is often limited by the actual fiber strength.
4. *Attenuation and pulse dispersion:* Attenuation in fibers is caused by material

absorption and scattering, waveguide scattering, and radiation losses, all of which limit the data rate capability. These losses can be partially avoided by increasing the numerical aperture. Light entering a fiber at an oblique angle takes a longer time to traverse the length of the fiber because it must undergo numerous internal reflections. The difference in the total distance traveled by meridional and oblique rays contributes to broadening the input pulse. Both attenuation and pulse spreading limit the maximum allowable length of transmission lines between repeaters.

Many different types of communications-grade fibers, as shown in Fig. 9.18, are available. Designers of fiber links should become familiar with the characteristics of each type of fiber.

1. Multimode Step-Index Fiber:

Core diameter:	50–400 μm
Cladding diameter:	125–500 μm
Buffer jacket diameter:	250–1000 μm
Numerical aperture:	0.16–0.5
Bandwidth:	6–25 MHz-km
Attenuation:	4–50 dB/km. The lower values correspond to silica-clad silica fiber, and the higher values correspond to glass-clad glass fiber.
Applications:	Short-haul, limited bandwidth, and low-cost systems

2. Multimode Graded-Index Fiber:

Core diameter:	30–60 μm
Cladding diameter:	100–150 μm
Buffer jacket diameter:	250–1000 μm
Numerical aperture:	0.2–0.3
Bandwidth:	150 MHz-km to 2 GHz-km
Attenuation:	2–10 dB/km
Applications:	Medium-haul, medium-to-high bandwidth systems using LEDs and injection lasers

3. Step-Index, Single-Mode Fiber:

Core diameter:	3–10 μm
Cladding diameter:	50–125 μm
Buffer jacket diameter:	250–1000 μm
Numerical aperture:	0.08–0.15
Bandwidth:	500 MHz-km to 40 GHz-km
Attenuation:	2–5 dB/km
Applications:	Long-haul, high bandwidth systems using single-mode injection lasers

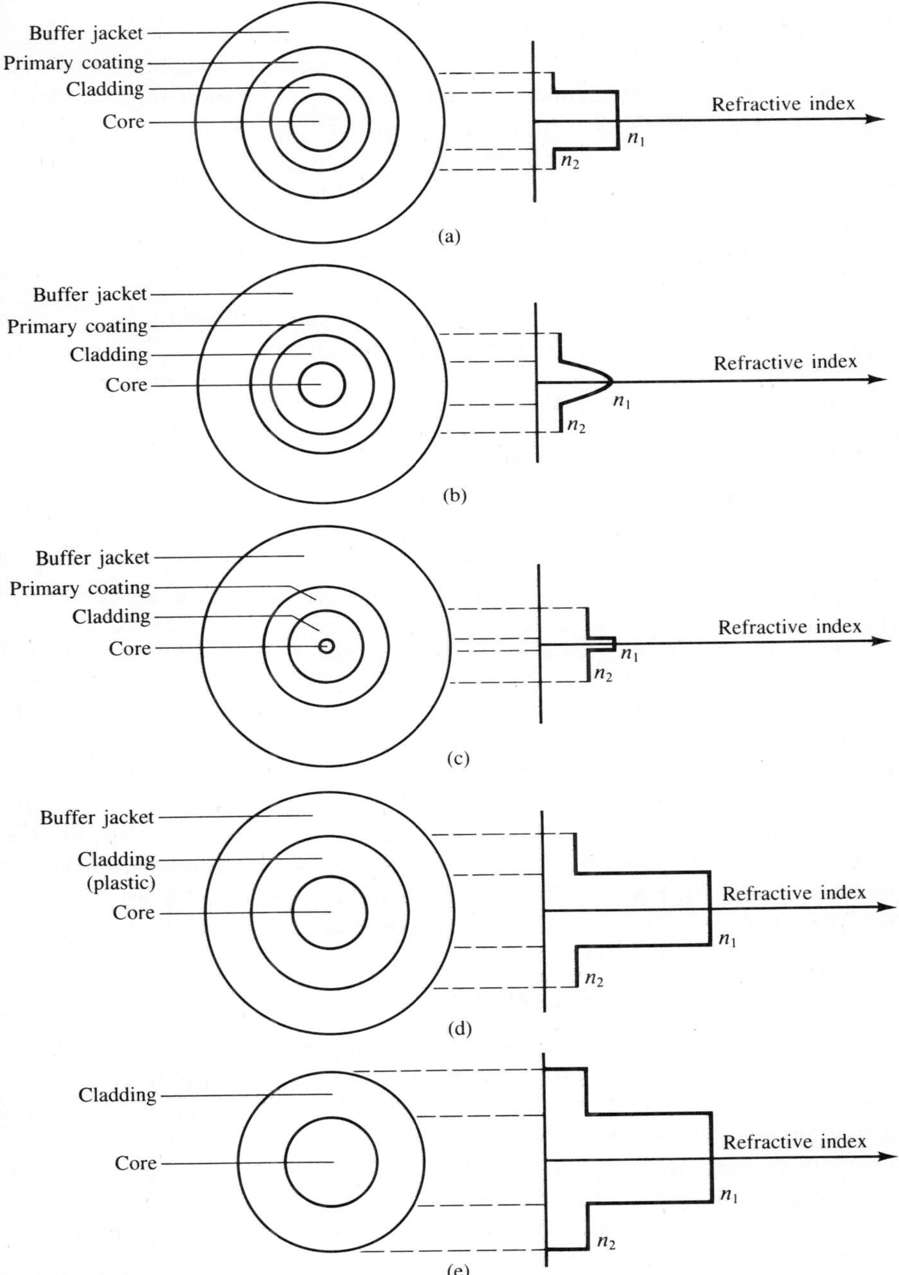

FIGURE 9.18

Typical optical fibers: (a) multimode step-index, (b) multimode graded-index, (c) step-index single mode, (d) multimode plastic-clad, and (e) multimode all-plastic

4. Multimode Plastic-Clad Fiber:
   For each of the following categories, the first value corresponds to the step-index fiber, and the second value corresponds to the graded-index fiber.

Core diameter:	100–500 μm/50–100 μm
Cladding diameter:	300–500 μm/125–150 μm
Buffer jacket diameter:	500–1000 μm/250–1000 μm
Numerical aperture:	0.2–0.5/0.2–0.3
Bandwidth:	5–25 MHz-km/200–400 MHz-km
Attenuation:	5–50 dB/km/4–15 dB/km
Applications:	Short-haul, lower bandwidth, and less-expensive systems

5. Multimode Step-Index, All-Plastic Fiber:

Core diameter:	200–600 μm
Cladding diameter:	450–1000 μm
Numerical aperture:	0.5–0.6
Bandwidth:	Transmission is limited to $\simeq$100 m
Attenuation:	350–1000 dB/km at 0.65 μm
Applications:	Very low cost and very short-haul links.

The unprotected fiber has several disadvantages with regard to strength and durability. Bare glass fibers are brittle and have relatively small cross-sectional areas that make them extremely susceptible to damage. The tensile strength of freshly made fibers is high, but it deteriorates with surface damage and exposure. Fibers are thus coated with protective sheaths, referred to as coating and cabling. When fibers are assembled in the form of a cable, they are coated and then wrapped helically around a central strength member. Numerous designs and configurations have been worked out to meet these requirements, some of which are shown in Fig. 9.19. These include coatings to protect fibers, strength-bearing materials, filler or buffer materials, and external protective jackets.

9.5.2 Transmitters and Receivers

Sources (transmitters) such as LEDs, LDs, and lasers and receivers such as *p-i-n* diodes and APDs were covered in detail in earlier chapters. Of the transmitters, LDs and lasers are narrow-band whereas LEDs have broader radiation wavelength spectra. Again, the peak power of LDs and lasers is generally several milliwatts, and the peak power of LEDs is only a milliwatt or much less. LED output is usually linear with bias current, whereas LD output is nonlinear in behavior, particularly near the threshold current. Both *p-i-n* photodiodes and APDs are prime candidates for receivers. A *p-i-n* photodiode is more stable in the presence of fluctuations of temperature and applied voltage, but the excessive sensitivity of an APD can be exploited advantageously to control APD output by means of an automatic gain control signal. APDs have very high gain, but they require comparatively higher voltage.

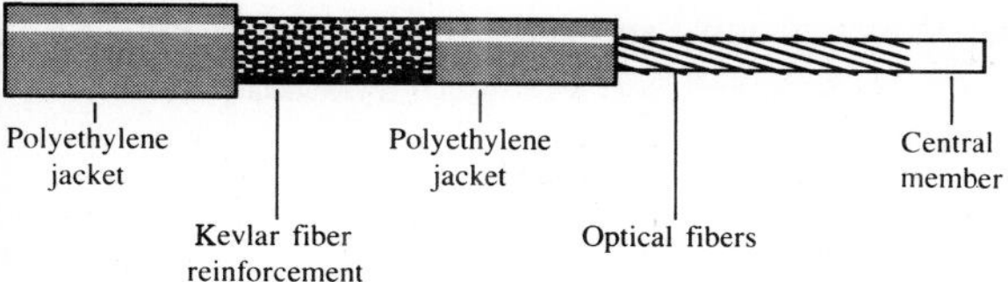

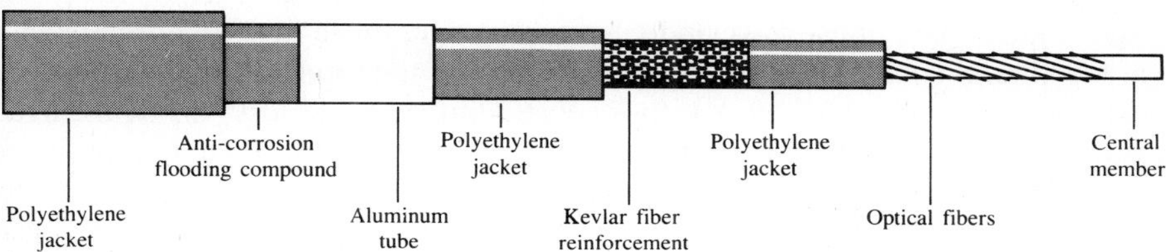

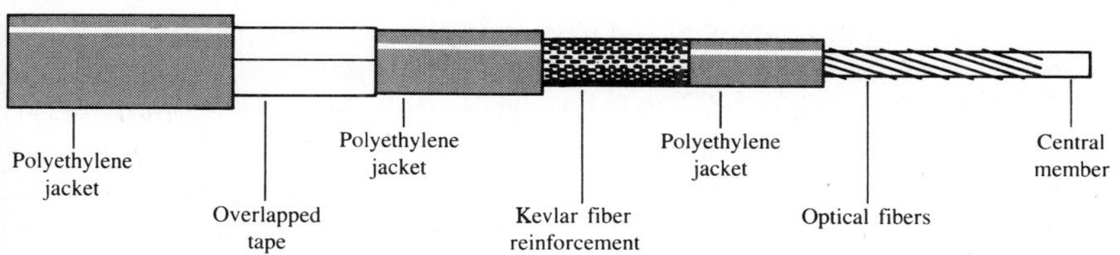

FIGURE 9.19

Typical fiber cable configurations

In general, LEDs are available in various geometric configurations and can be divided into two types: the small-area, high-brightness type and the large-area, low-brightness type. Typical low-brightness LEDs have emission areas ranging from about 0.2 to 5.6 mm^2. The core diameters of typical low-loss fibers range from 50 to 100 μm, which thus correspond to cross-sectional areas in the range of 0.008 to 0.03 mm^2. This mismatch in area of several orders of magnitude between the fiber and the light source results in large coupling losses. Consequently, the low-brightness, large-area LEDs are unacceptable for use with single fibers. However, in cables containing up to a thousand fibers, these emitters can be used with reasonable efficiency. LEDs having a relatively small modulation bandwidth (50 MHz or so) are generally prone to harmonic distortions. However, LEDs have several advantages when compared to LDs:

1. Simpler fabrication.
2. Low cost.

3. Immune to catastrophic degradation, self-pulsation, and modal noise problem.
4. Less sensitive to temperature.
5. Simpler drive circuitry.
6. Device linearity that is suitable for analog modulation.

These advantages have made LEDs more popular than LDs as optical sources in communications applications.

The amount of energy that can be coupled into a fiber depends on its numerical aperture as well as on the angular directional characteristics of the light source. An optical waveguide can accept only those rays that are incident at less than the maximum angle as determined by the total internal reflection at the core-cladding interface. A rather substantial coupling loss can be incurred when the angular distribution of the light source exceeds the angular acceptance angle as described by the NA.

The efficiency of coupling between either source and fiber, or between detector and fiber, can be dramatically improved by using some form of transfer optics, such as a lens between either the transmitter and the fiber or between the fiber and the receiver. The purpose of this optical device in front of the fiber end is to make the effective NA of the source and the fiber approximately similar. Assuming that no losses occur in the lens or in the spaces in between, the lens can project all of the emitted beam to the receiving end.

The coupling between the photodetector and the fiber is more easily accomplished than that between the fiber and the light source. This is because photodetectors have a relatively large active surface area and a large acceptance angle. By using index-matching fluids or thin films, we can reduce the insertion losses of the detectors to less than 1 dB. In many fiber system applications, we can make a tradeoff among coupling efficiency between the source and fiber, attenuation, and delay distortion. In order to have a sufficient optical signal at the photodetector, it is necessary that the source be able to unload as large a signal as possible into the fiber. We thus attempt to achieve a high coupling efficiency between the transmitter and the fiber, and at the same time we must also attempt to have as low a loss as possible in the fiber link.

9.5.3 Couplers, Splices, and Connectors

Couplers, splices, and connectors are functionally similar in that they link two or more fibers together to provide two or more paths for the transmission of optical signals. In particular, a coupler is referred to as active if it provides a switching mechanism for route selection, and it is referred to as passive if routing is predetermined by the geometry of the coupler. A splice is a form of coupler that permanently joins two fibers or two fiber bundles. On the other hand, a connector serves as a nonpermanent link between fibers or between a fiber and a repeater, or between a fiber and a receiver such that the original signal is transmitted with the least loss into the second fiber, fiber bundle, receiver, or end device.

When we join two fibers—a light source to a fiber or a detector to a fiber—the alignment of the two components becomes critical. Longitudinal and angular misalignments have little effect on coupling efficiency as long as reflection losses are a minimum. However, because of the small surface area of the fiber core, lateral or transverse misalignments can have a significant impact by increasing coupling loss. The expected loss caused by coupling two fibers also depends on whether the refractive index profile is a step-index or a graded-index type. Couplers, in addition to connecting one fiber to another, can be designed to connect a fiber to a light source, a detector, a repeater, an end device, or a signal-splitting device. It is desirable that these couplings should have a minimum amount of loss associated with them so that a minimum of energy is reflected back into the optical waveguide.

Splicing an optical fiber is a very complicated task. Special training, practice, and equipment—together with patience and good coordination—are necessary to produce an acceptable splice. In the fusion technique, two fiber ends are placed between two electrodes that cause the tips of the fibers to melt, and then a transparent epoxy is used to glue the ends together. Fusion splicing is more expensive than mechanical splicing, but many believe that it results in a stronger splice and lower loss.

The development of ideal connectors has lagged behind that of most other fiberoptic components; however, numerous practical connectors are available at reasonable prices. Connectors are now available in several varieties: (a) tube, (b) straight sleeve, (c) double eccentric, (d) tapered sleeve, (e) three-rod, (f) four-pin, (g) ferrules, and (h) jewel bushing. Figures 9.20(I) through 9.20(III) show some of the various types of available connectors.

In fiber-optics systems, major losses can occur at three optical junctions: (a) from source to fiber, (b) from fiber to fiber, and (c) from fiber to photo-detector. Whether these junctions are permanent splices or demountable connectors, considerable care must be taken to keep the loss to a minimum.

9.5.4 Distribution Systems

In the systems we have considered so far, only two terminals are used—a transmitting source and a receiver sink. In multiterminal systems, a single common data bus can be used instead of running a separate line from each terminal to each of the other terminals. Multiple signal transmission on a common bus is made possible through the multiplexing of signals. Access to the bus is accomplished using directional couplers that divert part of the signal power from the bus to the terminals and couple signals from the terminals onto the bus.

A directional coupler forms the basis of distribution networks. The parameters used to characterize a directional coupler are as follows:

1. *Throughput loss:* The amount of loss between the input port and the favored port.

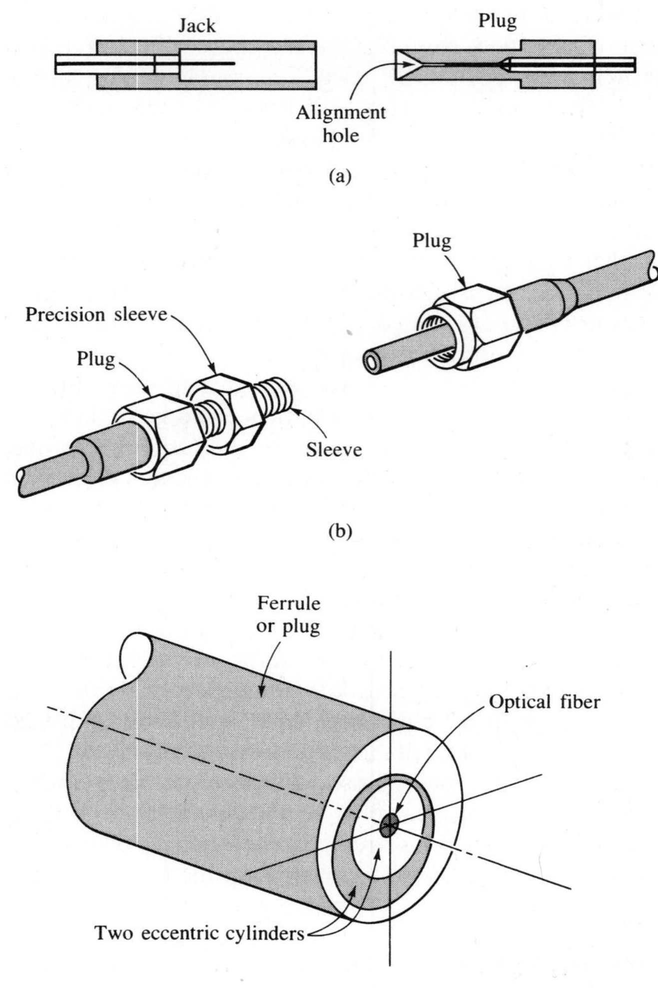

FIGURE 9.20(I)

Connectors: (a) tube, (b) straight sleeve, and (c) eccentric

2. *Tap loss:* The amount of loss between the input port and the tap.
3. *Directionality:* The amount of loss between the input port and the port that is to be isolated.
4. *Excess loss:* The amount of power loss within the coupler.
5. *Splitting ratio:* The ratio of powers at the output ports.

A network with many terminals requires a large coupling ratio (such that throughput power is greater than tapped power) for the coupler so that signals

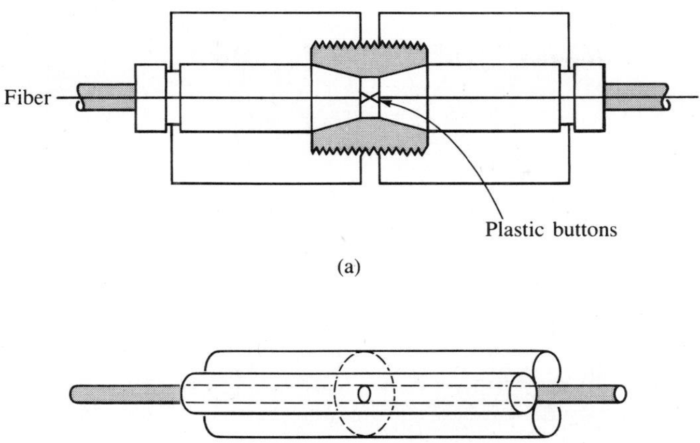

(b)

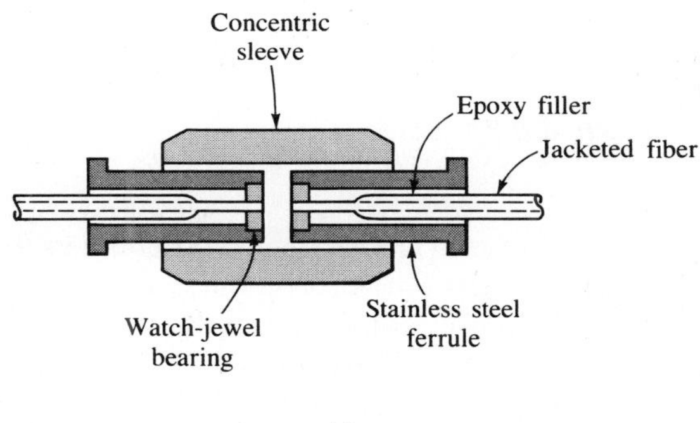

(c)

FIGURE 9.20(II)

Connectors: (a) tapered sleeve, (b) three-rod, and (c) jewel bushing

reaching receivers at a distance far away from the transmitter will have sufficient strength. Some of the possible network configurations are: *ring*, *star*, *tee*, and *tree*. These networks are shown in Fig. 9.21.

In a **ring network**, any information fed into the network by a user passes through all of the network nodes. Hence, in a ring network, a number of channels must be available between the nodes. This is often realized by a time-division multiplex system.

The **star network** provides a separate link for every user. Star couplers, as shown in Fig. 9.22, distribute the power received from any one input port to multiple output ports. Depending on the design, a coupler can be either a

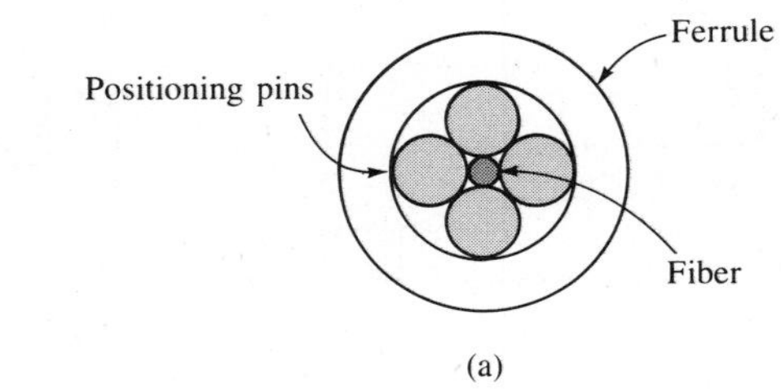

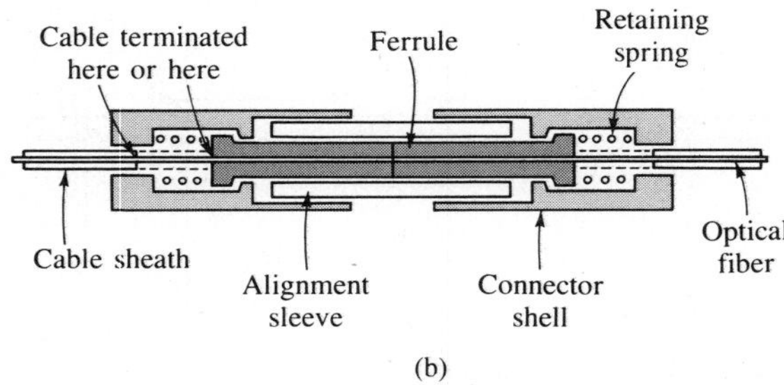

FIGURE 9.20(III)

Connectors: (a) four-pin and (b) ferrule

transmission or a reflection type. Transmission star couplers are directional in nature and provide isolation between ports on the same side of the coupler. In a transverse star coupler, the power of the transmitter port is distributed equally among each of the receiver ports. Reflection couplers distribute input power among all of the ports, including the port that was the input.

The **tee network**, as shown in Fig. 9.23, interconnects multiple terminals by means of taps provided by tee couplers. The tee couplers permit a bidirectional flow of signals in the bus fiber. A terminal in a tee network will receive more power from an adjacent terminal than from a distant one. Consequently, the receiver should be able to process signals over a wide range of power levels. Localized damage in a tee network does not necessarily shut off the total system. A break in the bus splits the system into two parts. Again, new terminals can be added to the network simply by cutting the bus fiber and inserting a tee coupler. Whereas in the star network power loss varies as the logarithm of the number of ports, in the tee network power loss varies linearly as the number of ports. Thus, with an increasing number of ports, fluctuation in the power loss is less critical in the case of the star network. By comparison, the star network provides better efficiency when only a few terminals are interconnected. For every new terminal added to a tee system, the signal must pass through two more connectors. In a star system, however, an added terminal does not increase the number of connectors through which the signal must pass on its path from the transmitter to the receiver.

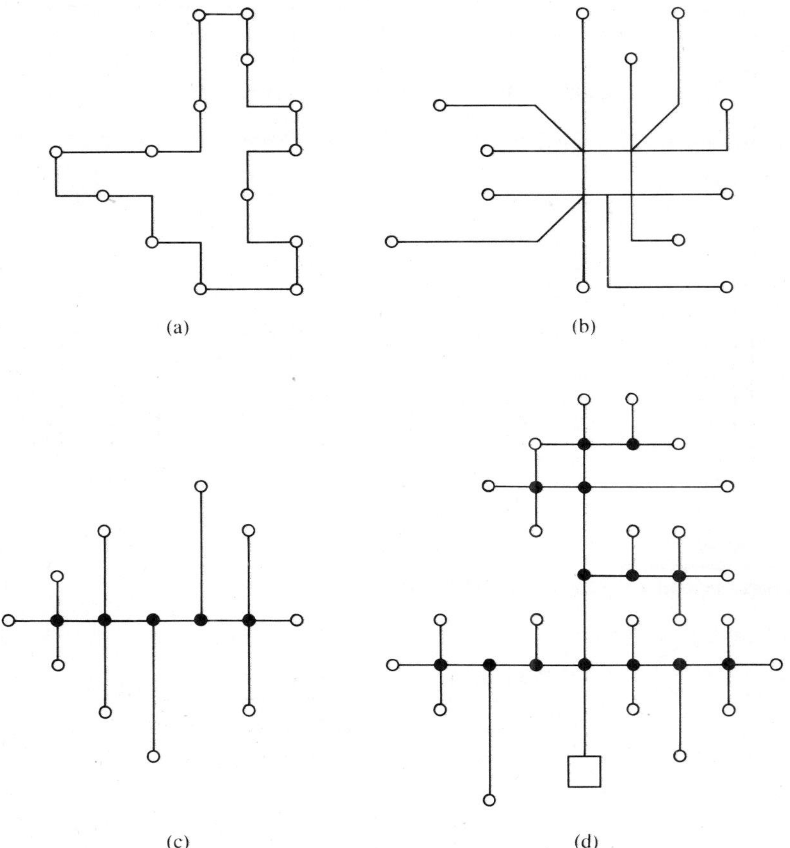

FIGURE 9.21

Network scheme: (a) ring, (b) star, (c) tee, and (d) tree

The **tree network**, on the other hand, consists of several branches. It is a hybrid of both star and tee configurations. As expected, it provides a number of transmission channels on its common links. The star portion links the closely spaced units, whereas the tee portion connects the more distant terminals. Often an active repeater is placed between the star and the tee to boost signal levels. However, by comparison with other networks the tree network has the advantage of greater flexibility in terms of overall topological enlargements.

Fiber-optic couplers are devices that add multiple inputs and generate more than one output in some fixed format. Couplers can be classified as monitoring-type, directional, or distribution-type. The sole aim of monitoring couplers is to monitor the optical power in an optical link. Directional couplers may have either three or four ports, although in many cases the fourth port is not used for

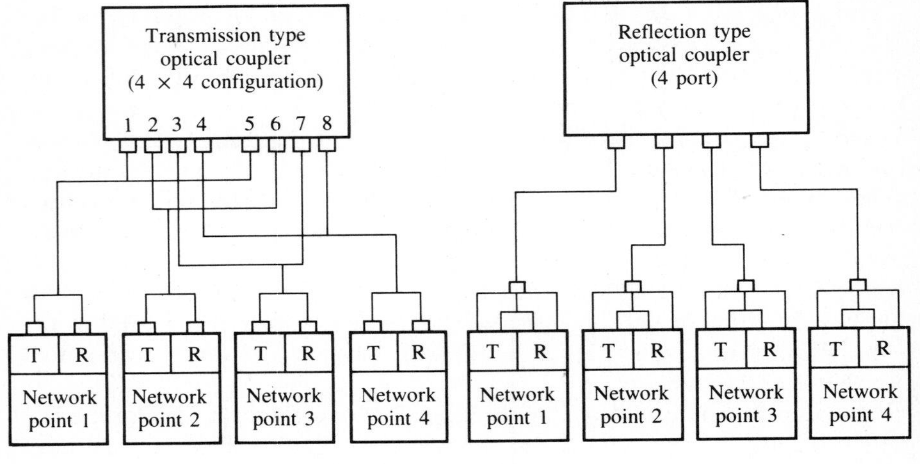

FIGURE 9.22

Transmission and reflection-type star couplers

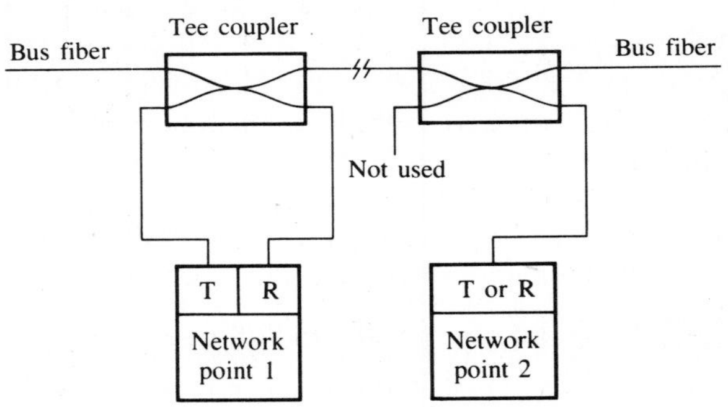

FIGURE 9.23

Bus system using tee couplers

any meaningful purpose. These couplers are used for simple bidirectional duplexing, wavelength bidirectional duplexing, feed-forward source linearization, fiber attenuation measurement, etc. Central distribution couplers are designed primarily for star-type topology and are either fused or micro-optic. The fused-type coupler is fabricated from fused waveguides, whereas the micro-optic-type coupler uses reflective techniques to distribute optical power. Often a mirror is used as a beam-splitter to partially reflect and redistribute optical input.

In general, an $m \times n$ star coupler has m input ports and n output ports. The input and output ports may be physically separated, as in transmissive couplers, or they may be identical, as in reflective couplers. However, in accordance with the principles employed, couplers can be classified into three categories: (a) diffusion-type, (b) area-splitting type, and (c) beam-splitting type. **Diffusion-type monitoring couplers**, as shown in Fig. 9.24, function by picking up part of the primary energy by means of diffusion. Diffusion-based coupling is caused by the evanescent field that spreads deep into the cladding. When two single-mode fibers run parallel to each other, separated only by a critical distance over a finite length, the evanescent field of one can penetrate into the other, thus coupling the primary fiber into the secondary one. A very sharp bend in the primary fiber can cause a severe radiation loss, which in turn can be transmitted into the secondary fiber.

Evanescent field coupling and **radiation field coupling** can also be used in multimode fibers. However, the coupling efficiency turns out to be very weak unless the thickness of the cladding is significantly thinned in the coupling region. In addition, this may result in differential mode coupling—i.e., most of the coupled power arises from the higher order modes of the primary fiber and is subsequently transformed into higher order modes in the secondary fiber.

A rather simple type of coupler for multimode fibers is realized using a different diffusion scheme, as shown in Fig. 9.25. It is basically a fused biconical taper coupler just suited for a duplex link. It is realized by twisting a pair of all-glass fibers together under tension and by softening the silica during fusion. The incoming light is led out of the core at the location of the coupler. The power lost from the core goes into radiation modes and is trapped by the higher refractive index of the cladding. This trapped light is then mixed at the junction and distributed between the claddings of the two fibers. The fusion technique has already been used to couple more than just two fibers to provide components suitable for star networking.

An area-splitting-type coupler is able to redistribute the optical power of a cylindrical waveguide into several channels when its cross section is smoothly transformed into multiple, physically separate light paths. Provided the circular geometry of the primary fiber has been smoothly transformed into several separate circular cross-sectional geometries, the coupling loss is minimal. The two classes of optical data buses that have evolved from such area-splitting-type couplers are the star bus and the tee coupler bus. A tee coupler is the simplest example of an area ratio splitter, formed as shown in Fig. 9.26. The star couplers shown in Fig. 9.27 are perhaps more conventional examples of a signal distribution scheme.

The reflection star coupler is more versatile because the ratio between the number of input and output ports may be selected or changed even after the device has been constructed. However, the scheme is slightly inefficient because a fraction of the injected light is always rejected. A transmission star is twice as efficient as a reflection star that has the identical number of input and output

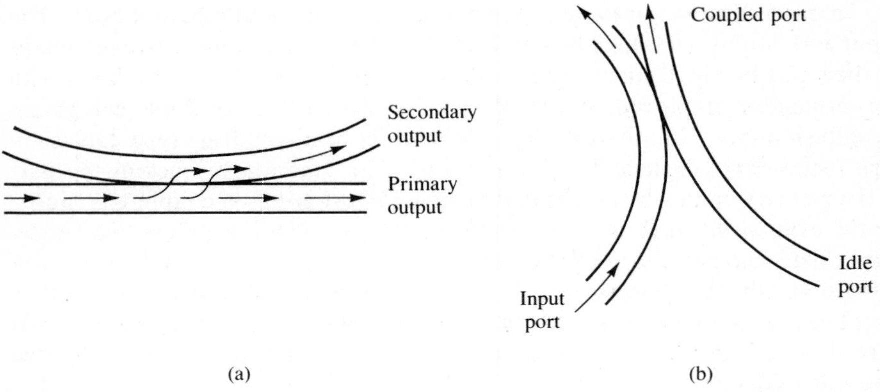

FIGURE 9.24

Diffusion-type couplers using (a) an evanescent field and (b) a radiation field

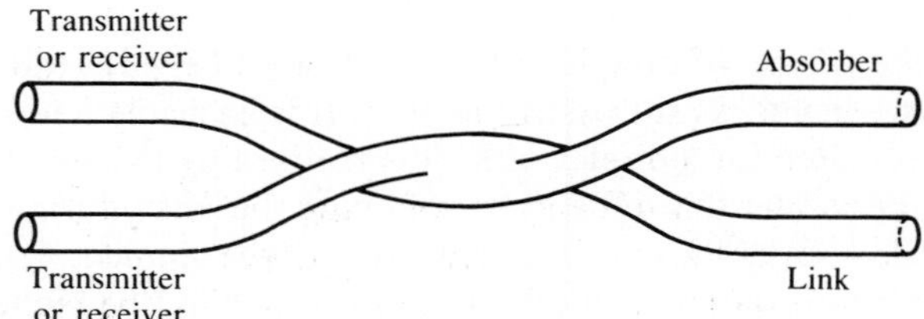

FIGURE 9.25

A fused biconical taper coupler

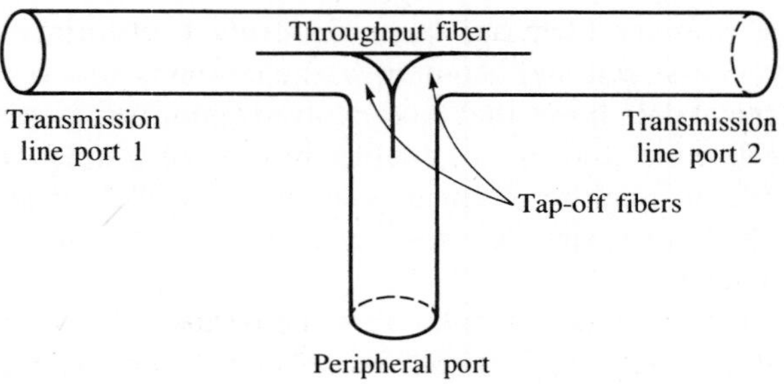

FIGURE 9.26

Tee coupler

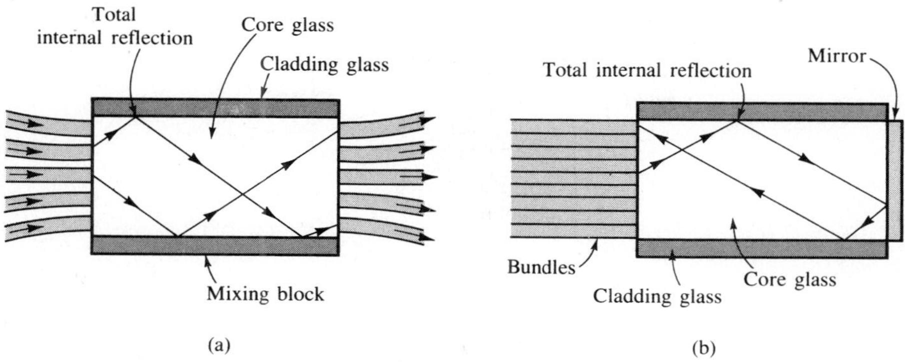

FIGURE 9.27

Star coupler: (a) transmissive and (b) reflective

ports. In the case of a star coupler having N terminals, the difference between the received and transmitted powers varies as $\log(N)$, whereas for a tee coupler the power difference varies linearly with N. This consequently limits the maximum number of terminals in a tee-coupled network.

The third mechanism for realizing couplers is partial beam splitting. In this scheme, a partially reflecting surface is placed at an angle to the direction of propagation. Oftentimes, partial reflection is also accomplished by using solid dielectrics or metals or liquid. The possibility of using a variety of materials allows us to achieve coupling over a broad range. It is possible to use a dichroic beam splitter between a pair of graded refractive index (selfoc) lenses, as shown in Fig. 9.28(a), to realize wavelength duplexing. Figure 9.28(b) shows a monitoring prism coupler that allows transmission between ports A and B and isolation between ports C and D. Figure 9.28(c) shows a beam-splitter-type directional coupler. When light is transmitted from port A to port B, some of the light originating from port A is coupled into port C. But when the direction is reversed, port C is isolated completely.

Wavelength division multiplexing is employed often in optical communication links for transmitting more than one wavelength of an optical signal through a fiber. Such schemes therefore increase the transmission capability of a fiber. Figure 9.29(a) shows a dispersive, prism-type multiplexing, while Fig. 9.29(b) shows a dispersive, grating-type multiplexing that combines more than one wavelength of light in an optical fiber. Figures 9.29(c) through 9.29(e) show interference filter-type multiplexings. While Fig. 9.29(c) shows an interference filter that splits the input signal in terms of its wavelength characteristics, Fig. 9.29(d) shows how several wavelengths of light can be combined, transmitted, and separated. Finally, Fig. 9.29(e) shows a scheme for the bidirectional transmission of light.

Until recently, the high cost of fiber-optic couplers has prevented the

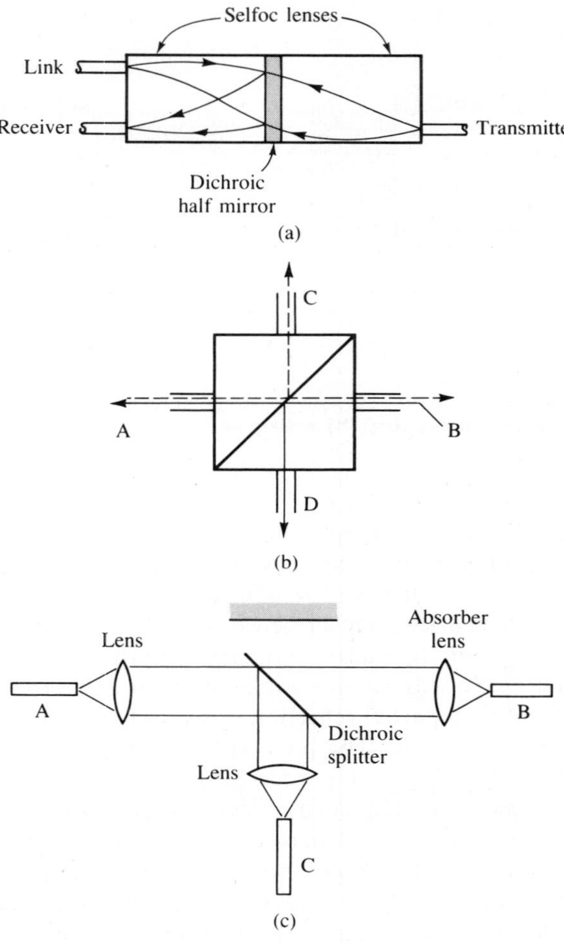

FIGURE 9.28

(a) A dichroic beam-splitter wavelength duplexing coupler, (b) a prism coupler, and (c) a beam-splitter coupler

widespread application of fiber-optic multiplexing in short-haul systems. They were reserved exclusively for the more expensive long-haul links. However, newer low-cost multiplexers are now being used to redefine system configurations. The advantages of fiber-optic multiplexing include increased effective bandwidth, the possibility of two-way communication, simpler system upgrading, cost reductions, and passive multiplexing/demultiplexing for high reliability at low cost. However, there are several disadvantages associated with it. They include increased insertion loss, cross-talk, and the fact that the low-cost multiwavelength sources and detectors are rare.

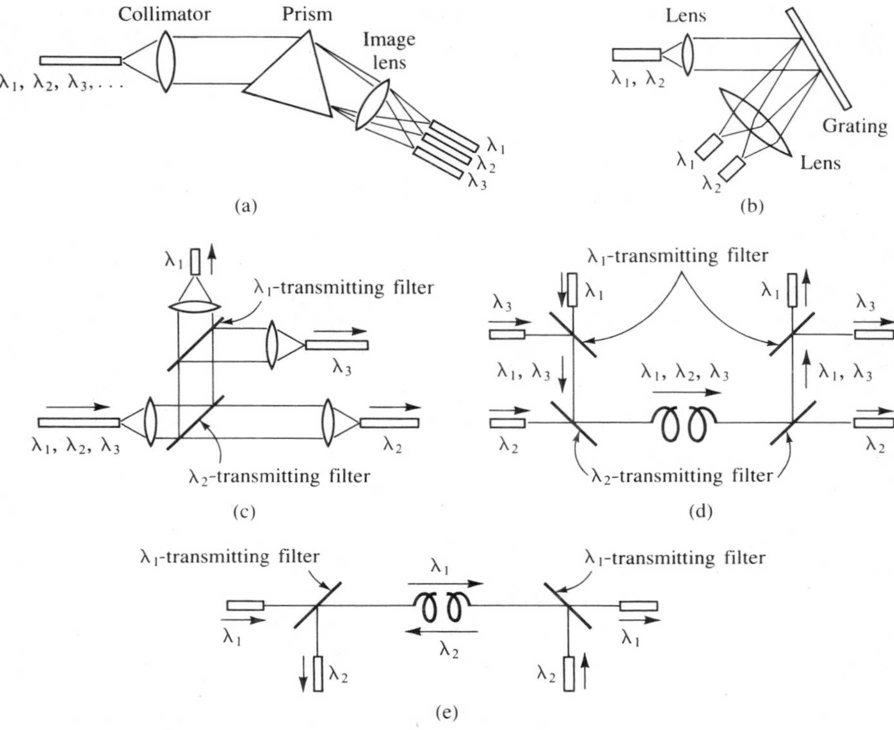

FIGURE 9.29

Wavelength division multiplexing using (a) a dispersive prism, (b) a dispersive grating, and (c) through (e) interference filters

Of all the available coupling mechanisms, couplers using filter technology have the most advantages. In general, they use an index-matching medium on either side of the filter. This solves two problems: it eliminates the Fresnel loss, and it protects the filter from being subjected to performance degrading moisture. The two types of filters that are in common use are bandpass and dichroic filters. It should be noted, however, that filter performance for an incidence angle greater than 30° is quite limited. In addition, it is often difficult to construct dichroic and bandpass filters having sharp cut-on and cut-off characteristics because of polarization constraints. Problems such as these are thus solved by designing couplers to operate at low incidence angles.

9.6 Integrated Optics

The benefit derived from the tight tolerances of electronic integrated circuits has stimulated the optics community to fabricate optical systems on an integrated scale. Such a system offers many possibilities. It may very well eliminate the alignment problem that is so crucial in optical set ups, and at the same time it may reduce the power requirement. Several of the electro-optic devices—such as sources, waveguides, modulators, detectors, bistable elements, and many others—have already been fabricated using integrated optics (IO). Optical integration is now being geared toward bringing together various electro-optic devices on a single chip. Most device elements are, however, expected to be based on single-mode optical waveguides organized in either thin-slab or stripe form. To explore the characteristics of a slab waveguide, we need to make use of the rectangular solutions of Maxwell's equation. In the next subsection, therefore, we attempt to characterize the field solutions in a slab waveguide in terms of the familiar TE and TM modes. And then, finally, in the subsection following that, we explore various IO device possibilities.

9.6.1 The Characteristics of a Slab Waveguide

Figure 9.30 shows a number of basic geometries for making slab waveguides. They include the raised or embedded strip, ridge, and strip-loaded waveguides. For the purpose of analysis, however, we consider the slab waveguide of Fig. 9.31 having three parallel dielectric layers—cover, film, and substrate—of refractive indexes n_c, n_f, and n_s, respectively. In practice, the cover may consist of air, while the substrate may be considered infinite in thickness such that $n_f > n_s > n_c$. The field solutions can be derived for such a geometry in either the TE or TM mode. The solutions are expected to be exponentially decaying as x approaches $\pm\infty$. Thus the wavenumber is imaginary in character in both cover and substrate regions, whereas it is real in character in the film, guaranteeing the existence of a standing wave field therein.

Let us first concentrate on finding the TE mode solutions such that the electric field is assumed polarized along the y axis while the propagation direction is taken to be along the z axis. We can express the y component of the electric field to be of the form

$$E_y = \begin{cases} A_c e^{-(k_c x + jk_z z)} & x > \tfrac{1}{2}d \\ A_f \cos(k_f x + \phi) e^{-jk_z z} & \tfrac{1}{2}d \geqslant x \geqslant -\tfrac{1}{2}d \\ A_s e^{k_s x - jk_z z} & x < -\tfrac{1}{2}d \end{cases} \qquad \textbf{(9.122)}$$

where

$$k_c = \left[k_z^2 - \left(\frac{\omega n_c}{c} \right)^2 \right]^{1/2} \qquad \textbf{(9.123a)}$$

$$k_f = \left[\left(\frac{\omega n_f}{c} \right)^2 - k_z^2 \right]^{1/2} \qquad \textbf{(9.123b)}$$

$$k_s = \left[k_z^2 - \left(\frac{\omega n_s}{c} \right)^2 \right]^{1/2} \qquad \textbf{(9.123c)}$$

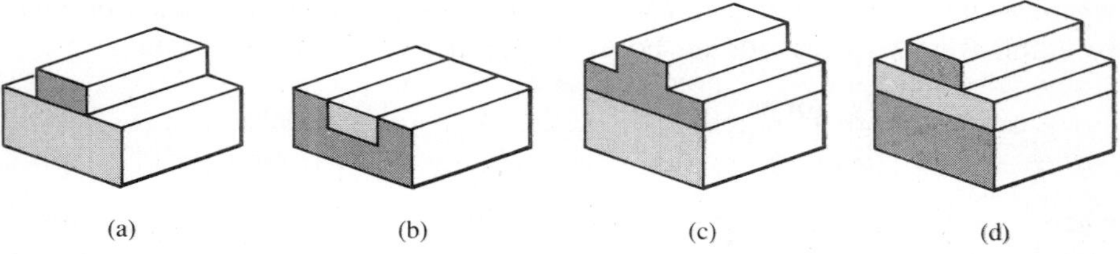

FIGURE 9.30

Geometry for integrated optic waveguides: (a) raised strip, (b) embedded strip, (c) ridge, and (d) strip-loaded

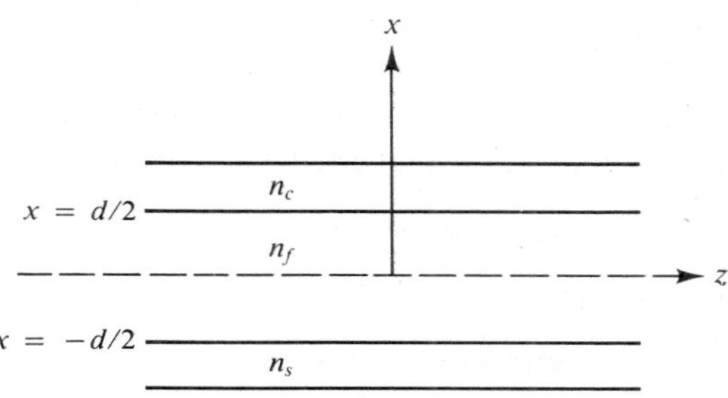

FIGURE 9.31

A slab waveguide

and k_z is the propagation constant along z axis. The corresponding magnetic field components can be obtained easily using Maxwell's equation. By equating field components (E_y and H_z) at $x = \pm\frac{1}{2}d$, we find that

$$\tan\left(\tfrac{1}{2}k_f d + \phi\right) = \frac{k_c}{k_f} \tag{9.124a}$$

$$\tan\left(\tfrac{1}{2}k_f d - \phi\right) = \frac{k_s}{k_f} \tag{9.124b}$$

But Eqs. (9.124a) and (9.124b) can be reexpressed, respectively, as

$$\tfrac{1}{2}k_f d + \phi = \tan^{-1}\left(\frac{k_c}{k_f}\right) \pm m\pi \tag{9.125a}$$

$$\tfrac{1}{2}k_f d - \phi = \tan^{-1}\left(\frac{k_s}{k_f}\right) \pm n\pi \tag{9.125b}$$

where m and n are any integers. The characteristic equation can be next obtained by eliminating ϕ. Thus, adding Eqs. (9.125a) and (9.125b), we obtain

$$k_f d - \tan^{-1}\left(\frac{k_c}{k_f}\right) - \tan^{-1}\left(\frac{k_s}{k_f}\right) = p\pi \tag{9.126}$$

where $p = 0, 1, 2, 3, \ldots$. Therefore, we can find a distinct TE solution corresponding to each value of p.

Using the boundary conditions that lead to Eqs. (9.124a) and (9.124b), we can rewrite Eq. (9.122) as

$$E_y = \begin{cases} A_f \cos\left(\frac{1}{2}k_f d + \phi\right) e^{-[k_c(x - \{d/2\}) + jk_z z]} & x > \frac{1}{2}d \\ A_f \cos(k_f x + \phi)\, e^{-jk_z z} & \frac{1}{2}d \geqslant x \geqslant -\frac{1}{2}d \\ A_f \cos\left(\frac{1}{2}k_f d - \phi\right) e^{k_s(x + \{d/2\}) - jk_z z} & x < -\frac{1}{2}d \end{cases} \tag{9.127}$$

We can carry out the derivation of TM solutions in exactly the same manner. The y component of the magnetic field is given by

$$H_y = \begin{cases} A_f \cos\left(\frac{1}{2}k_f d + \phi'\right) e^{-[k_c(x - \{d/2\}) + jk_z z]} & x > \frac{1}{2}d \\ A_f \cos(k_f x + \phi') e^{-jk_z z} & \frac{1}{2}d \geqslant x \geqslant -\frac{1}{2}d \\ A_f \cos\left(\frac{1}{2}k_f d - \phi'\right) e^{k_s(x + \{d/2\}) - jk_z z} & x < -\frac{1}{2}d \end{cases} \tag{9.128}$$

where ϕ' is given by the dual of either Eq. (1.125a) or Eq. (1.125b). The corresponding characteristic equation is found to be

$$k_f d - \tan^{-1}\left[\left(\frac{n_f}{n_c}\right)^2\left(\frac{k_c}{k_f}\right)\right] - \tan^{-1}\left[\left(\frac{n_f}{n_s}\right)^2\left(\frac{k_s}{k_f}\right)\right] = p\pi \tag{9.129}$$

A close examination of the characteristic relationships (Eqs. (9.126) and (9.129)) indicates that they relate to the fact that the phase shift incurred by an oblique ray is equal to an integral multiple of 2π. Note that the net phase shift is equivalent to the sum of the phase shifts corresponding to the path difference and that incurred due to total internal reflection at the cover-film and film-substrate interfaces. And indeed, while the first term of the characteristic equations corresponds to half of the phase difference owing to the path difference, the second and the third terms correspond to half of the phase difference due to the total internal reflections at the cover-film and film-substrate interfaces, respectively. Whenever the frequency is decreased, the total phase shift decreases. Thus, to maintain an equivalent total phase shift, we need to increase the path length. This is easily accomplished by allowing the rays to strike the $x = \pm d/2$ interface at a steeper angle. With decreasing frequency, the angle of incidence decreases continually until it reaches the cut-off, which in turn limits the number of higher order propagating modes.

Using normalized frequency between the film and the substrate, as defined in Eq. (9.37), we can write Eq. (9.126) as

$$V\sqrt{1-\mathrm{b}}-\tan^{-1}\frac{\sqrt{\dfrac{b}{1-b}}+\sqrt{\dfrac{a+b}{1-b}}}{1-\dfrac{\sqrt{b(a+b)}}{1-b}}=p\pi \tag{9.130}$$

where the normalized propagation parameter b is given by

$$b=\frac{k_z-\left(\dfrac{\omega n_s}{c}\right)^2}{\left(\dfrac{\omega n_f}{c}\right)^2-\left(\dfrac{\omega n_s}{c}\right)^2}=\frac{\left(\dfrac{k_z}{k}\right)^2-n_s^2}{n_f^2-n_s^2} \tag{9.131}$$

and the asymmetry parameter a is defined as

$$a=\frac{n_s^2-n_c^2}{n_f^2-n_s^2} \tag{9.132}$$

A plot of b versus V for different values of a is shown in Fig. 9.32. It shows that the dominant mode, corresponding to $p=0$, is effective at all frequencies only when $a=0$—i.e., when the cover and the substrate have identical refractive indices. The plot corresponding to the TM mode of Eq. (9.129) would be exactly similar to the one shown in Fig. 9.32, provided the asymmetry parameter is multiplied by the factor $(n_f/n_c)^4$. The slab waveguide can support a mode of order p such that $V \geqslant (m+\frac{1}{2})\pi$.

Next we consider a set up of two parallel optical waveguides placed side by side. Such set ups are extremely useful in designing devices such as couplers, switches, and modulators. Bringing a slab waveguide into the vicinity of another slab waveguide causes the propagation constant to modify itself and allows an exchange of energy to take place between the waveguides. The coupled differential equations (corresponding to the two waveguides designated with respective identifying subscripts) for such an arrangement become

$$\frac{dE_1}{dz}=jk_1'E_1+k_{12}E_2 \tag{9.133a}$$

$$\frac{dE_2}{dz}=jk_2'E_2+k_{21}E_1 \tag{9.133b}$$

where $k_1'=k_1-k_{11}$, $k_2'=k_2-k_{22}$, k_1 is the propagation constant of waveguide 1 when it is all by itself, k_2 is the propagation constant of waveguide 2 when it is all by itself, k_{12} and k_{21} are the coupling coefficients related to the respective energy transfers; and k_{11} and k_{22} are the changes in the propagation constants k_1 and k_2, respectively, when the other waveguide is introduced into the region of its evanescent wave.

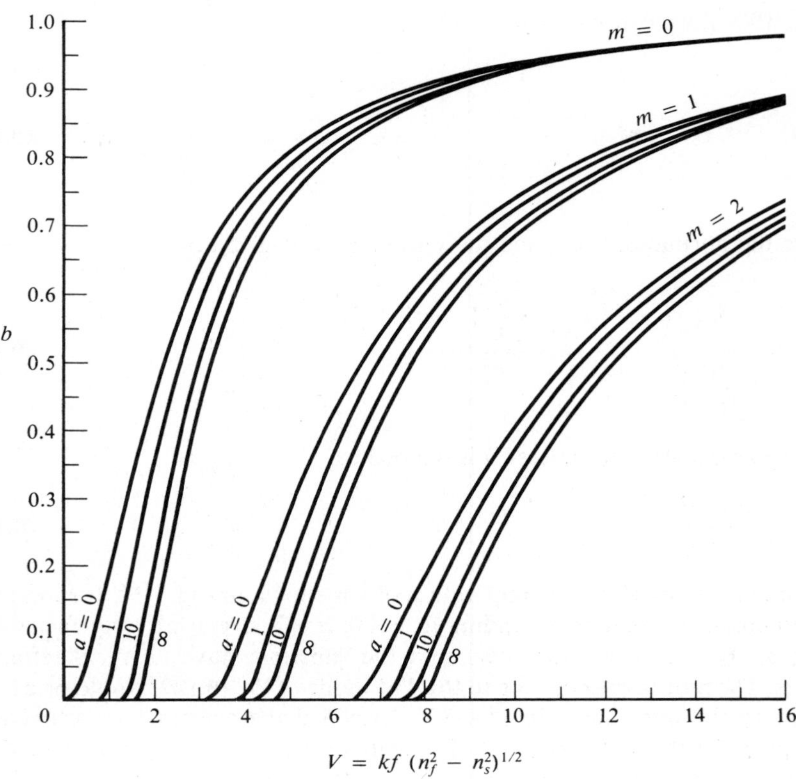

FIGURE 9.32

Normalized propagation parameter versus normalized frequency for the slab waveguide. (H. Kogelnik and V. Ramaswamy. "Scaling Rules for Thin Film Optical Waveguides." *Appl. Opt.*, Vol. 13, p. 1857, 1974.)

By solving Eqs. (9.133a) and (9.133b), we can find the coupled fields to be

$$E_1 = A_1 e^{k_{z1}z} + A_2 e^{k_{z2}z} \tag{9.134a}$$

$$E_2 = \frac{1}{k_{12}}[A_1(k_{z1} - jk_1')e^{k_{z1}z} + A_2(k_{z2} - jk_1')\,e^{k_{z2}z}] \tag{9.134b}$$

where

$$k_{z1} = [k_{12}k_{21} - \{\tfrac{1}{2}(k_1' - k_2')\}^2]^{1/2} + \tfrac{1}{2}j(k_1' + k_2') \tag{9.135a}$$

$$k_{z2} = -[k_{12}k_{21} - \{\tfrac{1}{2}(k_1' - k_2')\}^2]^{1/2} + \tfrac{1}{2}j(k_1' + k_2') \tag{9.135b}$$

For simplicity, we may assume that the energy lost by any one of the waveguides is picked up completely by the other, so that the total energy is conserved between the two waveguides. Consequently, it is appropriate to say

$$\frac{d}{dz}[E_1 E_1^* + E_2 E_2^*] = 0 \tag{9.136}$$

Using Eqs. (9.133a) and (9.133b), we can see that in order for Eq. (9.136) to be true for all values of E_1 and E_2,

$$k_{12} = -k_{21}^* \tag{9.137}$$

Consequently, Eqs. (9.135a) and (9.135b) can be rewritten in accordance with the coupling condition of Eq. (9.137) as

$$k_{z1} = j\left\{\left[|k_{12}|^2 + \left(\frac{k_1' - k_2'}{2}\right)^2\right]^{1/2} + \frac{1}{2}(k_1' + k_2')\right\} \tag{9.138a}$$

$$k_{z2} = j\left\{-\left[|k_{12}|^2 + \left(\frac{k_1' - k_2'}{2}\right)^2\right]^{1/2} + \frac{1}{2}(k_1' + k_2')\right\} \tag{9.138b}$$

The coupled field solutions of Eqs. (9.134a) and (9.134b) are applicable to the cases of various IO devices. The constants A_1 and A_2 are determined by conditions of device excitation, such as the presence or absence of input excitation, in coupled waveguides. For example, in the case of identical waveguides, all energy is transferred to the second waveguide within a transmission distance corresponding to $|k_{12}|z = \pi/2$. The energy is again retransferred back to the first waveguide when $|k_{12}|z = \pi$. Such energy transfer continues back and forth between the two waveguides. But if both of the waveguides are excited by the same optical field, the coupled field continues to have the same amplitude but different phase velocities. In the next subsection we see a manifestation of these principles.

9.6.2 IO Devices

Consider the modulator of Fig. 9.33, also referred to as the Mach-Zehnder modulator, as a basic example of a slab waveguide application. A slab waveguide suitable for fundamental mode operation is laid out on electro-optic substrate. The input arm is split into two branches while one of its arms is subjected to a potential V by means of two electrodes placed on its two sides. The two guides are subsequently recombined to form the output arm. Care is taken to make sure that the two branches are exactly the same length. Recall from Eqs. (7.37) and (7.38) that a voltage applied to one of the arms introduces a phase shift, the exact amount of which depends on the length and separation of the electrodes. Whenever the phase shift introduced is an odd multiple of π radians, the interferometer output vanishes due to destructive interference. And if the phase shift is just equal to an even multiple of π radians, a nonzero output is produced. For an arbitrary phase difference of ϕ, the modulator output is given by

$$I_{\text{out}} = \tfrac{1}{2}I_{\text{in}}(1 + \cos\phi) \tag{9.139}$$

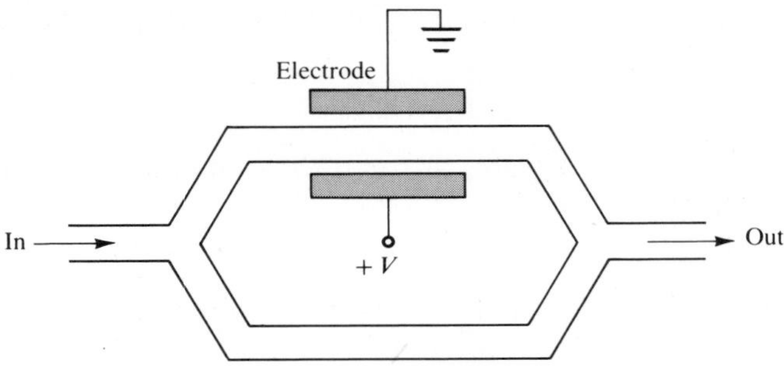

FIGURE 9.33

Mach-Zehnder interferometric modulator

Consequently, by changing voltage, it is possible to modulate the output optical power. Devices of this type can respond with switching times in the order of several hundred picoseconds.

The modulator just described can be modified to function as exclusive-OR (X-OR) or exclusive-NOR (X-NOR) logic. Such an organization is shown in Fig. 9.34, where one of the branches has two control variables while the other has only one control electrode, and each branch is able to introduce a phase shift of π. Consider also that V is fixed to bias the lower branch with a phase shift of π. Consequently, whenever either only V_1 or only V_2 is turned on, the output power is present. But if *both* V_1 and V_2 are turned on or off, the output is completely diminished. These functional characteristics are identical to the behavior of an X-OR logic operation. However, if V is turned off, the set up behaves in the exact opposite manner—with reversed outputs for all of the input conditions. Thus when V is turned off, the device works as an X-NOR logic gate.

Figure 9.35 shows a particular arrangement of a modulator suitable for digital-to-analog optical conversion. Using this set up, it is possible to convert a four-bit digital input to its equivalent analog optical output. Considering the fact that the positional weight of the most significant bit is 2^3 and that corresponding to the least significant bit is 2^0, the corresponding lengths of the electrodes are progressively decreased by a factor of 2. Accordingly, the resulting analog output power corresponds to the digital input.

Figure 9.36 shows a possible system for realizing an analog-to-digital (A-D) converter. This four-bit A-D converter consists of four interferometric parallel modulators that are fed from a common input through three layers of branching. The length of the modulator electrodes is decreased in length, successively, by a factor of 2 as it moves from the least significant bit to the most significant bit. The to-be-digitized analog signal is fed to the modulators, one for each bit of resolution. The sampling of the analog signal is realized by feeding pulsed laser light (at the desired sampling rate) to the modulators through the branching circuit.

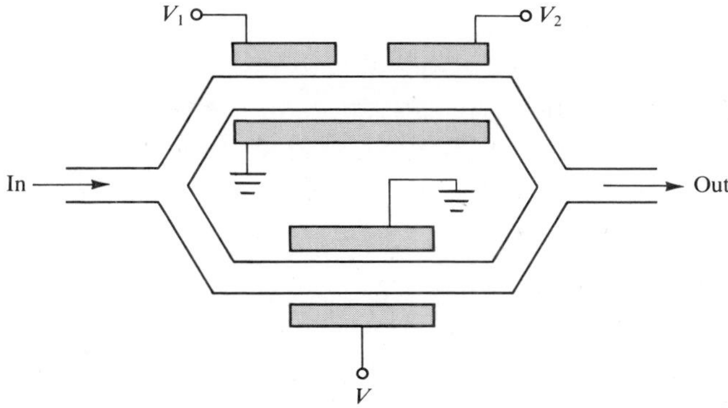

FIGURE 9.34

Set up for optical X-OR/X-NOR logic operations

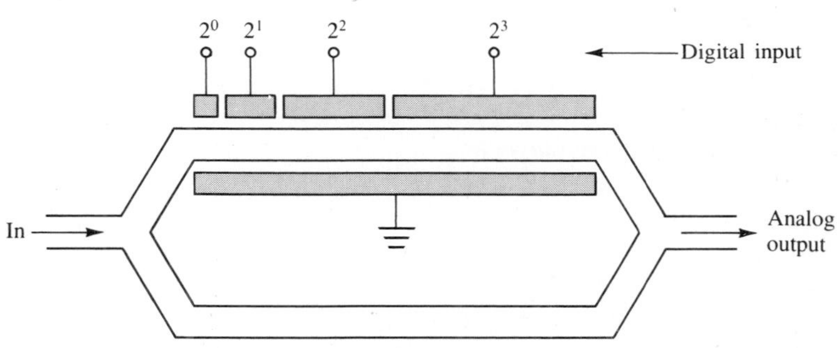

FIGURE 9.35

A digital-to-analog converter using a Mach-Zehnder modulator

Figure 9.37 shows a directional coupler switch that incorporates the principles of coupled mode theory. The two input arms are brought close but parallel to each other for a length of d in an electro-optic substrate. The two arms are equipped with electrodes so that the external electric field can be employed across the two waveguides. Depending on the magnitude of the applied voltage, the incoming voltage propagating through arm 1 can flow either through arm 3 or arm 4. It can be shown that whenever $d[|k_{12}|^2 + \{(k'_1 - k'_2)/2\}^2]^{1/2}$ is an odd multiple of $\pi/2$, all incoming energy is transferred to arm 4, and when it is equivalent to an even multiple of $\pi/2$, no transfer of power takes place. Note that to have a complete transfer of power, k'_1 and k'_2 must be equal. The main problem with this set up is that it may not be possible to have a complete transfer of energy because of the difficulty of ensuring an exact

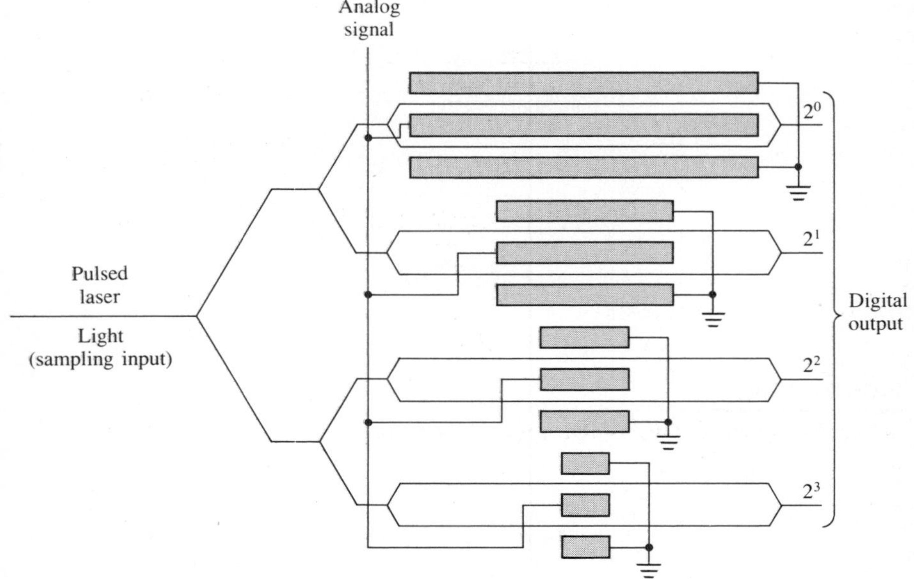

FIGURE 9.36

A four-bit analog-to-digital converter

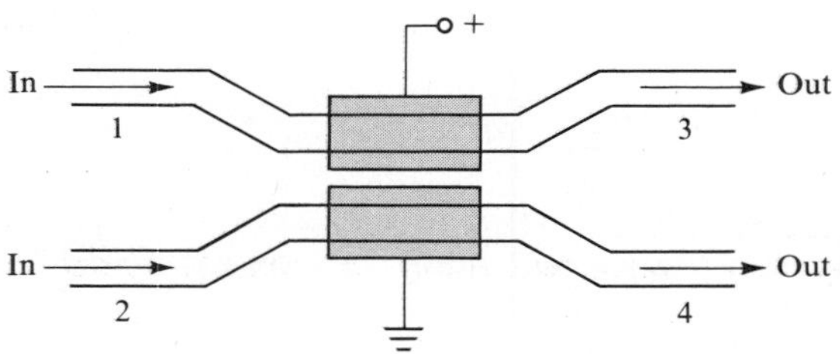

FIGURE 9.37

A directional coupler using IO principles

coupling length. This difficulty can be partially avoided by using a reversed directional coupler as shown in Fig. 9.38. The midsection of the system is split into two portions such that each half is subjected to a reversed voltage. Thus while $(k_1' - k_2')/2$ is positive in one of the halves, it is negative in the other half. Conditions of complete transmission are thereby achieved for any length of the waveguide. In fact, by increasing the number of such subsections, we can decrease the voltage requirement substantially.

Coupled mode theory can also be applied to make a *Y* junction, as shown in Fig. 9.39. The incoming single guide is symmetrically divided into two output

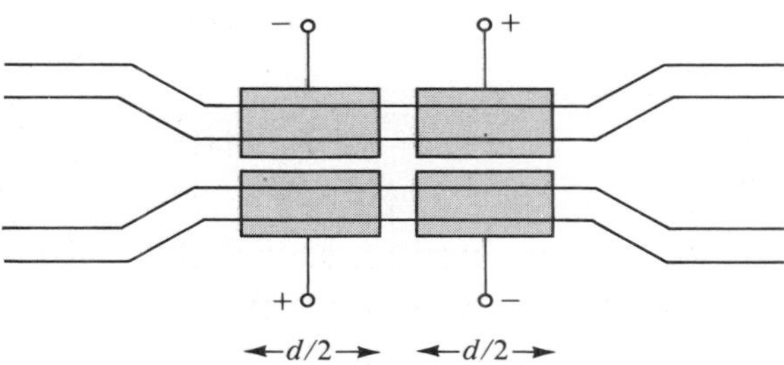

FIGURE 9.38

A reversed directional coupler

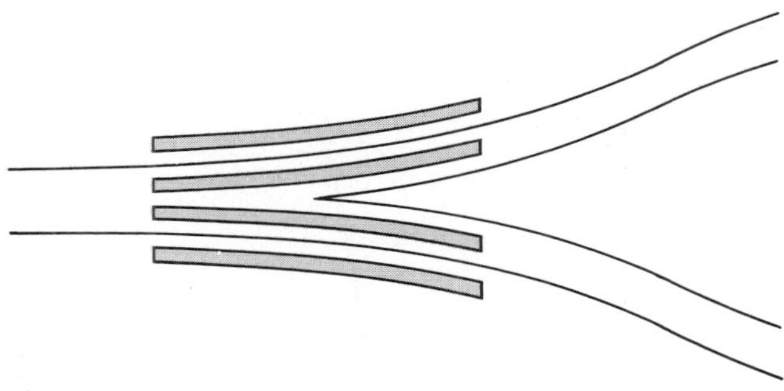

FIGURE 9.39

A *Y* junction

guides. Two pairs of external electrodes are deposited on the top of the guides as shown. When the external field is absent, the incoming light is divided into equal portions. Consider now that an external field is activated so that the difference between the indices of refraction of the two branches becomes significant. Under such an arrangement, the incoming light bends toward the branch that has the higher refractive index. Thus, by making use of a suitably controlled electric field, it is possible to turn on one guide and turn off the other.

An important class of IO devices having the property of optical bistability can be designed by allowing a portion of the output to control the reversed directional coupler of Fig. 9.39. Figure 9.40(a) shows a configuration where a portion of the output is withdrawn by means of a tee coupler and is then detected by a photodiode and later amplified before being applied to the electrodes. As a consequence of optical bistability, there can be two levels of device output depending on the history of the device. Figure 9.40(b) shows the characteristics for a bistable device. Such devices are considered extremely valuable for fabricating optical memory elements, logic gates, and amplifiers.

In particular, the device shown in Fig. 9.40(a) can be organized to function

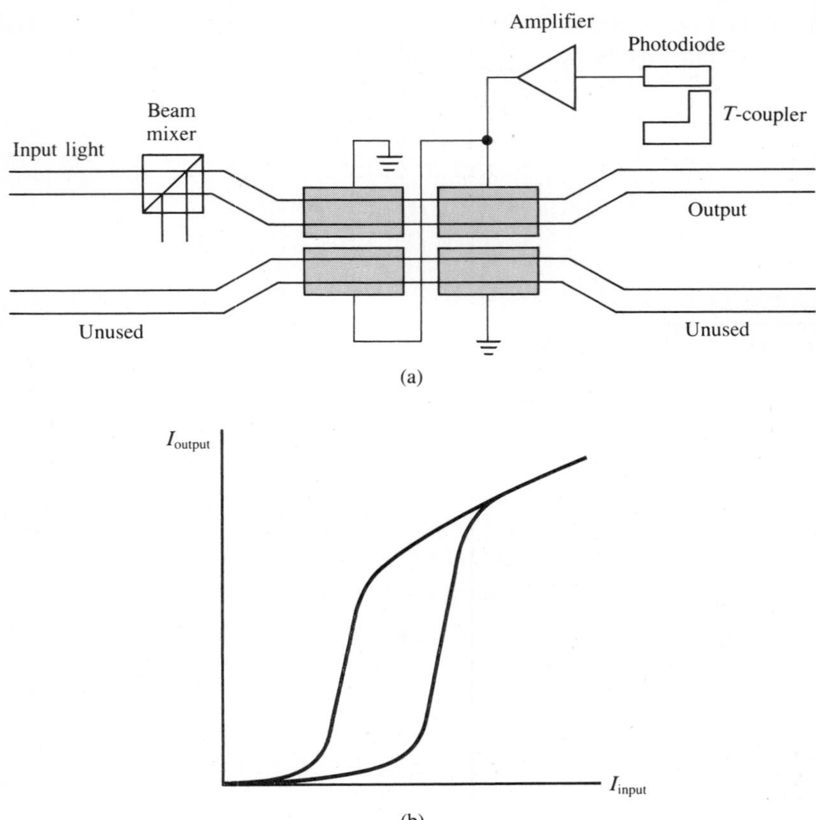

FIGURE 9.40

(a) An IO bistable device and (b) its bistable characteristics

as either an AND or an OR logic gate. Provided that the optical input power is maintained below the lower threshold point of the hysteresis curve, it is possible that the output will not be turned on unless both of the inputs are turned on. The device thus acts as an AND gate. Similarly, if the input power is maintained just above the lower threshold point, the output will be turned on whenever at least one of the inputs is on. This latter arrangement thus functions as an OR logic gate.

Problems

9.1 Verify Eqs. (9.12) and (9.13).

9.2 Obtain Eq. (9.36) from Eqs. (9.34) and (9.35).

9.3 Show that for a fixed value of q and ω, the coefficients C_1, C_2, C_3, and C_4 satisfying Eqs. (9.34) and (9.35) are given by

$$\frac{C_2}{C_1} = \frac{J_q(ur_1)}{K_q(vr_1)}$$

$$\frac{C_3}{C_1} = \frac{j\beta q}{\omega\mu r_1^2}\left(\frac{1}{u^2} + \frac{1}{v^2}\right)\left[\frac{1}{ur_1}\frac{J_q'(ur_1)}{J_q(ur_1)} + \frac{1}{vr_1}\frac{K_q'(vr_1)}{K_q(vr_1)}\right]$$

$$\frac{C_4}{C_1} = \frac{C_3}{C_1}\frac{J_q(ur_1)}{K_q(vr_1)}$$

9.4 Justify that Eqs. (9.52) and (9.53) are the cut-off conditions necessary for the hybrid modes in a step-index cylindrical fiber when $q > 1$.

9.5 Considering that E_x and H_y are both negligible corresponding to E_y and H_x in LP modes, obtain E_z and H_z for weakly guiding fibers.

9.6 (a) From the LP longitudinal modes (Problem 9.5), obtain the cylindrical transverse components E_ϕ and H_ϕ. (b) Obtain the eigenvalue equations using the tangential coefficients H_z at the core-cladding interface.

9.7 Solve Eq. (9.86b) for $\rho_1(r)$. At what values of r does the first order WKB approximation fail? Discuss its physical significance.

9.8 Verify Eqs. (9.92) and (9.95).

9.9 Using Eq. (9.76), evaluate the group delay time of the modes for the graded-index waveguide and show that when

$$\alpha = 2 - \frac{2n_1\lambda}{\left(n_1 - \lambda\dfrac{dn_1}{d\lambda}\right)\Delta}\frac{d\Delta}{d\lambda}$$

the delay differences among the modes vanish (to first order) in Δ.

9.10 Consider a non-Lambertian source whose emission profile is given by $L_e = L_{eo}\cos^n\phi$, where $n > 1$. Determine the coupling efficiency when such a source is coupled with a cylindrical waveguide. Consider the cases for step-index and graded-index fibers separately.

9.11 Verify Eqs. (9.118) and (9.119).

9.12 Determine the longitudinal fiber-to-fiber coupling efficiency for graded index fibers.

9.13 Verify Eqs. (9.124a) and (9.124b).

9.14 Starting with Eq. (9.126), discuss the consequence of TE modes in a symmetric slab waveguide where both the cover and substrate have equal refractive indices.

9.15 Obtain Eqs. (9.128) and (9.129) for the slab waveguide of Fig. 9.31.

9.16 For a slab waveguide of width W, show that the total power is given by

$$P = \begin{cases} \dfrac{Wk_z}{4\omega\mu_0}|A_z|^2\left\{d + \dfrac{1}{k_c} + \dfrac{1}{k_s}\right\} & \text{TE modes} \\ \dfrac{Wk_z}{4\omega\varepsilon_f}|A_z|^2\left\{d + \dfrac{q_c}{k_c} + \dfrac{q_s}{k_s}\right\} & \text{TM modes} \end{cases}$$

where

$$q_i = \frac{k_f^2 + k_i^2}{k_f^2 + \left(\dfrac{n_f}{n_i}\right)^4 k_i^2}\left(\frac{n_f}{n_i}\right)^2$$

and ε_f is the dielectric constant of the film.

9.17 Obtain the characteristics of the coupled mode solutions for the following cases: (a) waveguides are identical but only one of them is excited, (b) waveguides are identical and both are excited with the same light signal, (c) waveguides are identical but each of them is excited with different amplitudes of light signal, and (d) waveguides are nonidentical and only one of them is excited.

References and Suggested Readings

Baker, D. G. *Fiber Optic Design and Applications.* Reston, VA: Reston Pub. Co., 1985.

Basch, E. E., (Ed). *Optical-Fiber Transmission.* Indianapolis, IN: Howard W. Sams & Co., 1986.

Becker, R. A., C. E. Woodward, F. J. Leonberger, and R. C. Williamson. "Wideband Electro-Optic Guided-Wave Analog-to-Digital Converters." *Proc. IEEE.* Vol. 72 (1984): 802.

Bjerken, L., and H. Nordby. "Pulse Distortion in Optical Fibers with Transverse Offset Splices." *Appl. Opt.* Vol. 20 (1981): 435.

Cheo, P. K. *Fiber Optics Devices and Systems.* Englewood Cliffs, NJ: Prentice-Hall, 1985.

Cherin, A. H. *An Introduction to Optical Fibers.* New York: McGraw-Hill, 1983.

Coppa, S., B. Costa, P. Di Vita, and U. Rossi. "Single-Mode Optical Fiber Characterization." *Opt. Engn.* Vol. 24 (1985): 676.

Daido, Y., E. Miyauchi, and T. Iwama. "Measuring Fiber Connection Loss Using Steady-State Power Distribution: A Method." *Appl. Opt.* Vol. 20 (1981): 451.

Danko, J. L., and D. P. Ryan-Howard. "Index Profile of Graded-Index Channel Waveguides." *Appl. Opt.* Vol. 25 (1986): 1505.

Davis, C. M. "Fiber Optic Sensors: An Overview." *Opt. Engn.* Vol. 24 (1985): 347.

Frost, B. S., P. M. Gourlay, N. R. Heckenberg, and S. T. Shanahan. "Geometrical Optics Treatment of Circular Lightguides." *Appl. Opt.* Vol. 24 (1985): 4395.

Gabardi, D. R., and D. L. Shealy. "Coupling of Domed Light-Emitting Diodes with a Multimode Step-Index Optical Fiber." *Appl. Opt.* Vol. 25 (1986): 3435.

Gidon, P., S. Valette, and P. Mottier. "Integrated Lenses on Silicon Nitride Waveguides." *Opt. Engn.* Vol. 24 (1985): 235.

Golge, D. "Weakly Guiding Fibers." *Appl. Opt.* Vol. 10 (1971): 2252.

Dziedzic, J. M., R. H. Stolen, and A. Ashkin. "Optical Kerr Effect in Long Fibers." *Appl. Opt.* Vol. 20 (1981): 1403.

Goodman, J. W., F. J. Leonberger, S. Y. Kung, and R. A. Athale. "Optical Interconnections for VLSI Systems." *Proc. IEEE.* Vol. 72 (1984): 850.

Gravel, R. L., and A. R. Nelson. "Asymmetrical Star Coupler for Optical Fiber Systems." *Appl. Opt.* Vol. 20 (1981): 2735.

Haus, H. A. *Waves and Fields in Optoelectronics.* Englewood Cliffs, NJ: Prentice-Hall, 1984.

Johnson, E. L., and M. A. Karim. *Digital Design: A Pragmatic Approach.* Boston: PWS Publishers, 1987.

Keiser, G. *Optical Fiber Communications.* New York: McGraw-Hill, 1983.

Klocek, P., and M. Sparks. "Theoretical Overview of Limitations of Light Propagation in Infrared Optical Fibers." *Opt. Engn.* Vol. 24 (1985): 1098.

Kogelnik, H., and V. Ramaswamy. "Scaling Rates for Thin-Film Optical Waveguides." *Appl. Opt.* Vol. 13 (1974): 1857.

Kuehls, J. F., A. E. Attard, and V. B. Burke. "Optical Correlator Using a Thin-Film Traveling-Wave Light Modulator." *Appl. Opt.* Vol. 24 (1985): 3842.

Lam, D. K. W., and B. K. Garside. "Characterization of Single-Mode Optical Fiber Filters." *Appl. Opt.* Vol. 20 (1981): 440.

Lee, D. L. *Electromagnetic Principles of Integrated Optics.* New York: Wiley, 1986.

Lee, H. "Optical Isolator Using Acoustooptic and Faraday Effects." *Appl. Opt.* Vol. 26 (1987): 969.

Leonberger, F. J., C. E. Woodward, and R. A. Becker. "4-Bit 828-Megasample/s Electro-Optic Guided-Wave Analog-to-Digital Converter." *Appl. Phys. Lett.* Vol. 40 (1982): 565.

Liao, F. J., and J. T. Boyd. "Single-Mode Fiber Coupler." *Appl. Opt.* Vol. 20 (1981): 2731.

Lizuka, K. *Engineering Optics.* Berlin: Springer-Verlag, 1985.

MacDonald, R. I. "Frequency Domain Optical Reflectometer." *Appl. Opt.* Vol. 20 (1981): 1840.

MacDonald, R. I., and D. K. W. Lam. "Optoelectronic Switch Matrices: Recent Developments." *Opt. Engn.* Vol. 24 (1985): 220.

Matthijsse, P., and C. M. De Blok. "Optical Fiber Power-Splitter with Selectable Splitting Ratio." *Appl. Opt.* Vol. 21 (1982): 4290.

Miyagi, M. "Bending Losses in Hollow and Dielectric Tube Leaky Waveguides." *Appl. Opt.* Vol. 20 (1981): 1221.

Mueller, C. T., and E. Garmire. "Coupling Characteristics of $LiNbO_3$ Directional Couplers." *Appl. Opt.* Vol. 24 (1985): 2846.

Nicia, A. "Lens Coupling in Fiber-Optic Devices: Efficiency Limits." *Appl. Opt.* Vol. 20 (1981): 3136.

Palais, J. C. *Fiber Optic Communications* (2nd ed). Englewood Cliffs, NJ: Prentice-Hall, 1988.

Polky, J. N., and D. D. Miller. "Optical Waveguide Circuit Design of an Adaptive Filter in the Residue Number System." *Appl. Opt.* Vol. 21 (1982): 3539.

Revelli, Jr., J. F. "Mode Analysis and Prism Coupling for Multilayered Optical Waveguides." *Appl. Opt.* Vol. 20 (1981): 3158.

Sauter, G. F., R. W. Honebrink, and J. A. Krawczak. "Alterable Grating Fiber-Optic Switch." *Appl. Opt.* Vol. 20 (1981): 3566.

Shadaram, M., and W. L. Kuriger. "Using the Optical Frequency Domain Technique for the Analysis of Discrete and Distributed Reflections in an Optical Fiber." *Appl. Opt.* Vol. 23 (1984): 1092.

Shadaram, M., and R. Hippenstiel. "Fourier Analysis of the Complex Envelope of the Echoes in an OFDR." *Appl. Opt.* Vol. 25 (1986): 1083.

Shirasaki, M., H. Takamatsu, and T. Obokata. "Bistable Magnetooptic Switch for Multimode Optical Fibers." *Appl. Opt.* Vol. 21 (1982): 1943.

Soref, R. A. "Electrooptic 4×4 Matrix Switch for Multimode Fiber-Optic Systems." *Appl. Opt.* Vol. 21 (1982): 1386.

Southwell, W. H. "Planar Optical Waveguide Lens Design." *Appl. Opt.* Vol. 21 (1982): 1985.

Syms, R. R. A. "Perturbation Analysis of Nearly Uniform Waveguide Arrays." *Appl. Opt.* Vol. 25 (1986): 2988.

Takato, N., and T. Kurokawa. "Polymer Waveguide Star Coupler." *Appl. Opt.* Vol. 21 (1982): 1940.

Tamir, T., (Ed). *Integrated Optics.* New York: Springer-Verlag, 1975.

Verber, C. M. "Integrated-Optical Approaches to Numerical Optical Processing." *Proc. IEEE.* Vol. 72 (1984): 942.

Wilson, J., and J. F. B. Hawkes, *Optoelectronics: An Introduction.* Englewood Cliffs, NJ: Prentice-Hall Int., 1983.

Wilson, J., and J. F. B. Hawkes. *Lasers: Principles and Applications.* Hertfordshire, United Kingdom: Prentice-Hall International (UK) Ltd, 1987.

Witte, H., and V. Kulich. "Branching Elements for Optical Data Buses." *Appl. Opt.* Vol. 20 (1981): 715.

Zizzo, C. "Optical Circulator for Fiber-Optic Transceivers." *Appl. Opt.* Vol. 26 (1987): 3470.

APPENDIX A

Periodic Table

The number above the symbol of each element is its atomic weight, and that below is its atomic number. The atomic weight of those elements that do not exist in nature are given in parentheses.

Period	Group I	Group II											Group III	Group IV	Group V	Group VI	Group VII	Group VIII
1	1.00 H 1																	4.00 He 2
2	6.94 Li 3	9.01 Be 4											10.82 B 5	12.01 C 6	14.01 N 7	16.00 O 8	19.00 F 9	20.18 Ne 10
3	22.99 Na 11	24.32 Mg 12											26.98 Al 13	28.09 Si 14	30.98 P 15	32.07 S 16	35.46 Cl 17	39.94 Ar 18
4	39.10 K 19	40.08 Ca 20	44.96 Sc 21	47.90 Ti 22	50.95 V 23	52.01 Cr 24	54.94 Mn 25	55.85 Fe 26	58.94 Co 27	58.71 Ni 28	63.54 Cu 29	65.38 Zn 30	69.72 Ga 31	72.60 Ge 32	74.91 As 33	78.96 Se 34	79.92 Br 35	83.8 Kr 36
5	85.48 Rb 37	87.63 Sr 38	88.92 Y 39	91.22 Zr 40	92.91 Nb 41	95.95 Mo 42	(99) Tc 43	101.1 Ru 44	102.91 Rh 45	106.7 Pd 46	107.88 Ag 47	112.41 Cd 48	114.82 In 49	118.70 Sn 50	121.76 Sb 51	127.61 Te 52	126.01 I 53	131.30 Xe 54
6	132.91 Cs 55	137.36 Ba 56	57–71 *	178.58 Hf 72	180.95 Ta 73	183.86 W 74	186.22 Re 75	190.2 Os 76	192.2 Ir 77	195.09 Pt 78	197.0 Au 79	200.61 Hg 80	204.39 Tl 81	207.21 Pb 82	209.00 Bi 83	210 Po 84	(211) At 85	222 Rn 86
7	(223) Fr 87	226.05 Ra 88	89–102 **															
*Rare earths			138.92 La 57	140.13 Ce 58	140.91 Pr 59	144.27 Nd 60	(147) Pm 61	150.35 Sm 62	152.0 Eu 63	157.26 Gd 64	158.93 Tb 65	162.51 Dy 66	164.94 Ho 67	167.27 Er 68	168.94 Tm 69	173.04 Yb 70	174.99 Lu 71	
**Actinides			227 Ac 89	232.05 Th 90	231 Pa 91	238.07 U 92	(237) Np 93	(242) Pu 94	(243) Am 95	(245) Cm 96	(249) Bk 97	(249) Cf 98	(254) Es 99	(253) Fm 100	(256) Md 101	(253) No 102		

APPENDIX B

Important Integrals

The function $F(\alpha)$ referred to as the integral transform of $f(x)$ by the kernel $K(\alpha, x)$ is given by

$$F(\alpha) = \int_a^b f(x) K(\alpha, x)\, dx$$

The kernels associated with Fourier, Laplace, Hankel, and Mellin transforms are, respectively,

$$F(\alpha) = \int_{-\infty}^{\infty} f(x) e^{j\alpha x}\, dx$$

$$F(\alpha) = \int_0^{\infty} f(x) e^{-\alpha x}\, dx$$

$$F(\alpha) = \int_0^{\infty} f(x) x J_n(\alpha x)\, dx$$

$$F(\alpha) = \int_0^{\infty} f(x) x^{\alpha - 1}\, dx$$

where $J_n(\cdot)$ is the nth order Bessel function of the first kind.

In particular, the Fourier transform of a function m of two cartesian variables x and y is represented by $F\{m\}$ and is expressed as

$$F\{m\} = \iint_{-\infty}^{\infty} m(x, y)^{-j2\pi(f_x x + f_y y)}\, dx\, dy$$

On the other hand, the inverse Fourier transform of a function $M(f_x, f_y)$ is represented by $F^{-1}\{M\}$ and is given by

$$F^{-1}\{M\} = \iint_{-\infty}^{\infty} M(f_x, f_y) e^{j2\pi(f_x x + f_y y)}\, df_x\, df_y$$

A few of the important properties of the Fourier transforms are listed as follows:

1. $F\{\alpha m + \beta n\} = \alpha F\{m\} + \beta F\{n\}$ where α and β are two constants.
2. $F\{F^{-1}\{m(x, y)\}\} = F^{-1}\{F\{m(x, y)\}\} = m(x, y)$.
3. $FF\{m(x, y)\} = F^{-1}F^{-1}\{m(x, y)\} = m(-x, -y)$.
4. Defining $F\{m(x, y)\} = M(f_x, f_y)$, $F\{m(\alpha x, \beta y)\} = M(f_x/\alpha, f_y/\beta)/|\alpha\beta|$.
5. Defining $F\{m(x, y)\} = M(f_x, f_y)$,

   $$F\{m(x - \alpha, y - \beta)\} = M(f_x, f_y) \exp\,[-j2\pi(f_x\alpha + f_y\beta)]$$

6. Defining $F\{m(x, y)\} = M(f_x, f_y)$ and $F\{n(x, y)\} = N(f_x, f_y)$,

   $$F\left\{\iint_{-\infty}^{\infty} m(\zeta, \eta) n(x - \zeta, y - \eta)\, d\zeta\, d\eta\right\} \equiv F\{m(x, y) * n(x, y)\}$$

   $$= M(f_x, f_y) N(f_x, f_y)$$

   where $*$ is the convolution operation and, therefore,

   $$F\{m(x, y) n(x, y)\} = M(f_x, f_y) * N(f_x, f_y)$$

7. Defining $F\{m(x, y)\} = M(f_x, f_y)$,

   $$F\{\iint_{-\infty}^{\infty} m(\zeta, \eta) m*(\zeta - x, \eta - y)\, d\zeta\, d\eta\} = |M(f_x, f_y)|^2$$

   and

   $$\iint_{-\infty}^{\infty} M(\zeta, \eta) M*(\zeta - f_x, \eta - f_y)\, d\zeta\, d\eta = F\{|m(\zeta, \eta)|^2\}.$$

APPENDIX C

Important Functions

Rectangle function:

$$\text{rect}(x) = \begin{cases} 1 & |x| \leqslant \frac{1}{2} \\ 0 & \text{otherwise} \end{cases}$$

Sinc (pronounced "sink") function:

$$\text{sinc}(x) = (\pi x)^{-1} \sin(\pi x)$$

Sign (pronounced "signum") function:

$$\text{sign}(x) = \begin{cases} -1 & x < 0 \\ 0 & x = 0 \\ 1 & x > 0 \end{cases}$$

Ramp function:

$$\text{ramp}(x) = \begin{cases} 0 & x \leqslant 0 \\ |x| & x > 0 \end{cases}$$

Dirac Delta function:

$$\delta(x) = \lim_{N \to \infty} N^2 e^{-N^2 \pi x^2}$$

or

$$\delta(x) = \begin{cases} \infty & x = 0 \\ 0 & \text{otherwise} \end{cases}$$

Circle function:

$$\operatorname{circ}((x^2+y^2)^{1/2}) = \begin{cases} 1 & (x^2+y^2)^{1/2} \leqslant 1 \\ 0 & \text{otherwise} \end{cases}$$

Comb function:

$$\operatorname{comb}(x) = \sum_{n=-\infty}^{\infty} \delta(x-n)$$

Cylindrical function:

$$\operatorname{cyl}(x^2+y^2)^{1/2}/d = \begin{cases} 1 & 0 \leqslant (x^2+y^2)^{1/2} \leqslant \frac{1}{2}d \\ \frac{1}{2} & (x^2+y^2)^{1/2} = \frac{1}{2}d \\ 0 & (x^2+y^2)^{1/2} > \frac{1}{2}d \end{cases}$$

Gaussian function:

$$\operatorname{Gaus}(r/d) = e^{-\pi(r/d)^2}$$

Step function:

$$\operatorname{step}(x) = \begin{cases} 0 & x < 0 \\ \frac{1}{2} & x = 0 \\ 1 & x > 0 \end{cases}$$

Triangle function:

$$\operatorname{tri}(x) = \begin{cases} 1-|x| & |x| \leqslant 1 \\ 0 & \text{otherwise} \end{cases}$$

APPENDIX D

Fourier Transform Pairs

$m(x)$	$M(f_x)$
1	$\delta(f_x)$
$\delta(x)$	1
$\delta(x \pm x_0)$	$e^{\pm j2\pi x_0 f_x}$
$\text{rect}(x)$	$\text{sinc}(f_x)$
$\text{sinc}(x)$	$\text{rect}(f_x)$
$\text{tri}(x)$	$\text{sinc}^2(f_x)$
$\text{sinc}^2(x)$	$\text{tri}(f_x)$
$\text{sgn}(x)$	$(j\pi f_x)^{-1}$
$\text{comb}(x)$	$\text{comb}(f_x)$
$\text{Gaus}(x)$	$\text{Gaus}(f_x)$

INDEX